Problem Solving in Electrochemical Impedance Spectroscopy

THE ELECTROCHEMICAL SOCIETY SERIES

The Electrochemical Society
Advancing solid state & electrochemical science & technology

ECS - The Electrochemical Society
65 South Main Street
Pennington, NJ 08534-2839
http://www.electrochem.org

A complete list of the titles in this series appears at the end of this volume.

Problem Solving in Electrochemical Impedance Spectroscopy: Methods, Data Analysis and Case Studies

Mark E. Orazem
University of Florida

Bernard Tribollet
UMR8235 CNRS, Sorbonne Université

WILEY

Published by John Wiley & Sons, Inc., Hoboken, New Jersey.

For general information on our other products and services or for technical support, please contact our Customer Care Department within the United States at (800) 762-2974, outside the United States at (317) 572-3993 or fax (317) 572-4002.

Wiley also publishes its books in a variety of electronic formats. Some content that appears in print may not be available in electronic formats. For more information about Wiley products, visit our web site at www.wiley.com.

Library of Congress Cataloging-in-Publication Data applied for:
Hardback ISBN: 9781119779216

Cover Design: Wiley
Cover Image: Courtesy Ada Lang

Dedicated to our families:
Jennifer and Athéna,
Françoise, Julie, and Benjamin

Contents

Preface

This book is intended for use both as a professional reference and as a resource suitable for training new scientists and engineers. As a textbook, this work is suitable for graduate students in a variety of disciplines, including electrochemistry, materials science, physics, electrical engineering, and chemical engineering.

Organization

The book has been prepared in three parts:

Part I Problems and Solutions

Solutions are provided for homework problems taken from the second edition of our book[1] and for other problems developed during the teaching of courses based on the book. Sufficient explanation is provided to allow the solutions to facilitate learning of the material. While derived from the second edition of our book,[1] problems have been restated to make them independent. To facilitate the use of this material as a companion to the second edition of our book,[1] problems are arranged according to the presentation of chapters in the book. Supplementary problems are presented in an appropriate chapter and are separated from the homework problems taken from the second edition of our book[1] by a trail of elephants.

As in our previous books, the elephant is used as a symbol for impedance spectroscopy. Application of impedance spectroscopy is very much like feeling an elephant that we cannot see. Measurement of current and potential under a steady state yields some information concerning a given system. By adding frequency dependence to the macroscopic measurements, impedance spectroscopy expands the information that can be extracted from the measurements. Impedance measurements, however, are not sufficient. Additional observations are needed to gain confidence in the model identification.

Part II Application to Experimental Data

Impedance spectroscopy represents a rich and inter-related area of science that is applied to a large number of important areas of research, including corrosion and corrosion control by coatings, electrochemical kinetics and mechanisms, energy storage systems such as batteries and capacitors, fuel cells, and biological, biocellular, and biomedical systems. While electrical circuit analogs are commonly employed, interpretation of impedance spectra is best achieved by developing a model that accounts for the chemistry and physics of the system under study. The objective of this presentation is to describe a systematic approach to model development that includes the use of graphical methods to guide model development, the use of electrical circuits to account for the geometry of a given system, and the use of a phasor notation to facilitate calculation of a faradaic impedance based on proposed reaction mechanisms and mass transfer.

1. A preliminary graphical inspection of the data includes use of the traditional Nyquist and Bode formats. Estimates for parameters such as capacitance, ohmic resistance and polarization resistances may be extracted from the plots. Knowledge of the ohmic resistance allows preparation of ohmic-resistance-corrected Bode plots. Values of the extracted capacitance, the shape of the ohmic-resistance-corrected Bode plots, and the shape of the Nyquist plots guide development of a preliminary model. If a set of spectra are obtained at different elapsed times or different conditions, superposition of scaled plots may be used to determine whether the differences may be attributed to changes in system chemistry or to simply changes in parameters such as active area.

2. Evaluation of consistency with the Kramers–Kronig relations allows the determination of the frequency range suitable for analysis. Replicated measurements allow determination of error structure, useful for the regression of experimental data. The measurement model, developed by our group, provides a useful tool for evaluating the error structure of impedance data.[2–4]

3. Graphical methods may be used to extract parameters. This step is an extension of the preliminary graphical approach but has emphasis on the self-consistent part of the spectrum. Plots of complex capacitance may be used to extract meaningful values for capacitance.[5,6] An estimate for the characteristic frequency associated with geometrical effects may be established for well-defined cell geometries.[7]

4. The interpretation or process model development takes place in two stages: development of circuit analogues and derivation of the faradaic impedance. Electrical circuits are prepared to account for the geometry of the electrochemical system under study. These circuits provide a framework for the faradaic impedance, which is developed based on the proposed reaction chemistry. Models for the faradaic impedance are prepared for proposed reaction kinetics. The faradaic impedance includes the influence of mass transfer, coupled reactions, and even homogeneous reactions.[1]

5. The process model is regressed to the Kramers–Kronig-consistent portion of the spectrum. Simplex regression is insensitive to initial guessed values for parameters but does not provide confidence intervals for the regressed parameters. Thus, regression by simplex algorithms may yield values for parameters that are statistically insignificant. Levenberq–Marquardt regression algorithms provide confidence intervals that can be used to assess the statistical significance of parameter values. The limitation of Levenberq–Marquardt regression is that it is very sensitive to initial parameter values. Some regression programs provide simplex algorithms to obtain initial parameter values and Levenberq–Marquardt algorithms to yield final values and their respective confidence intervals.

Impedance spectroscopy is not a stand-alone technique, and models for impedance are not unique. The model identified by the procedure described in this presentation represents a process model intended to account for the hypothesized physical and chemical character of the system under study. The objective of the model is not to provide a good fit with the smallest number of parameters. The objective is rather to use the model to gain a physical understanding of the system. The model should be able to account for, or at least be consistent with, all experimental observations. The proposed model can suggest experiments needed to validate model hypotheses.

These ideas will be illustrated by examples, including the impedance of aluminum showing constant-phase-element behavior, lithium-ion batteries subject to anomalous diffusion, iron in presence of H_2S yielding apparently large double-layer capacitance, corrosion of copper nickel alloys showing an apparent rotation-speed dependence of CPE parameters, and delamination of coated silver mirrors, which can be explained in terms of a porous electrode model.

Part III Reference Material

The reference material includes helpful tables taken from the second edition of our book,[1] listings of computer programs, a list of symbols, and a list of references.

Pedagogical Approach

As can be found in any field, the notation used in the impedance spectroscopy literature is inconsistent. In treatments of diffusion impedance, for example, the symbol θ is used to denote the dimensionless oscillating concentration variable; whereas, the symbol θ used in kinetic studies denotes the fractional surface coverage by a reaction intermediate. Compromises were necessary to create a consistent notation for this book. For example, the dimensionless oscillating concentration variable was given the symbol θ, and γ was used to denote the fractional surface coverage by a reaction intermediate.

Following the second edition of our book,[1] the present work deviates from the IUPAC convention for the notation used to denote the imaginary number and the real and imaginary parts of impedance. The IUPAC convention, as described in the overview by Sluyters-Rehbach,[8] is that $\sqrt{-1}$ should be denoted by the symbol i. To avoid confusion with current density, given the symbol i, and the index i used to indicate specific chemical species, we have chosen here to follow the electrical engineering convention in which $\sqrt{-1}$ is given the symbol j.

We also depart from the IUPAC convention in the notation used to denote real and imaginary parts of the impedance. The IUPAC convention is that the real part of the impedance is given by Z' and the imaginary part is given by Z''. We consider that the IUPAC notation can be confused with the use of primes and double primes to denote first and second derivatives, respectively. Thus, we choose to identify the real part of the impedance by Z_{r} and the imaginary part of the impedance by Z_{j}.

This book is intended to provide a background and training suitable for application of impedance spectroscopy to a broad range of applications, such as corrosion, biomedical devices, semiconductors and solid-state devices, sensors, batteries, fuel cells, electrochemical capacitors, dielectric measurements, coatings, electrochromic materials, analytical chemistry, and imaging. The emphasis is on generally applicable fundamentals rather than on detailed treatment of applications. The reader is referred to other sources for discussion of specific applications of impedance.[9–12]

Impedance spectroscopy represents the confluence of a significant number of disciplines, and successful training in the use and interpretation of impedance requires a coherent education in the application of each of these disciplines to the subject. In addition to learning about impedance spectroscopy, the student will gain a better understanding of a general philosophy of scientific inquiry.

Mark E. Orazem
Gainesville, Florida

Bernard Tribollet
Paris, France

August 4, 2026

Acknowledgments

The authors met for the first time in 1981 in the research group of John Newman at the University of California, Berkeley. **Mark Orazem** was a graduate student, and **Bernard Tribollet** was a visiting scientist on sabbatical leave from the *Centre National de la Recherche Scientifique* (CNRS) in Paris. We have maintained a fruitful collaboration ever since, and our careers, as well as the content of this book, build on the foundation we received from John. We owe an additional debt of gratitude to many people, including

- Our families, to whom this book is dedicated and who have embraced our collaboration over the years. We appreciate their immense patience and support.
- *The Electrochemical Society* (ECS), which encouraged Mark Orazem to teach the ECS short course on impedance spectroscopy on an biannual basis, thus allowing us to test the pedagogical approach developed in this book.
- Graduate students, who provided suggestions and content. In particular, Ming Gao, Chen You, Cynthia Eluagu, and Yunhan Chuai provided solutions to many of the problems presented in Part I.
- Bob Esposito, our former editor from John Wiley & Sons, who helped us through the initial steps associated with publishing this book. Bob reassured us that it was good to be considered difficult authors.
- Ada Lang, University of Florida, who provided the cover art for this book.
- Michael Leventhal, Editor, Elizabeth Rose Amaladoss and Harishwar Nesamani, Managing Editors, and the rest of the editorial team at John Wiley & Sons.
- Our many colleagues and friends who assured us that there would be a demand for this book.

While we have received help and support from many people, the remaining errors and omissions in this text are ours. We will gratefully receive corrections and suggestions from our readers to be implemented in possible future editions of this book.

About the Companion Website

This book is accompanied by a companion website.

www.wiley.com/go/orazem/problemsolving

To access the website materials, click on the link under Related Resources on the Book's page on wiley.com. You can also find this by searching www.wiley.com for the ISBN 9781119779216.

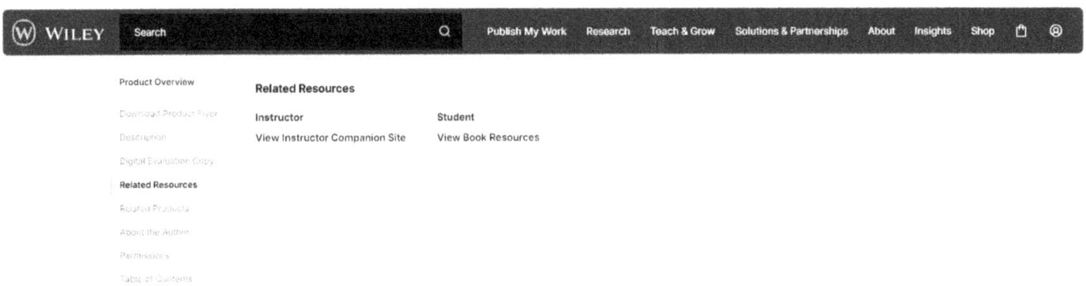

This website includes:

- Computer programs
- Experimental data

Why an Elephant?

The distorted impedance diagram, shown on the cover of this book, was intended to resemble an elephant and, in doing so, evoke the lessons of the Buddhist tradition of the blind men and the elephant. The multiple loops resemble the Nyquist plots obtained in some cases for the impedance of corroding systems influenced by formation of surface films. The low-frequency inductive loop was deformed to evoke the image of the elephant's trunk, and the capacitive loops resemble the head and body of the elephant.

The elephant was adopted by members of the Orazem group to be, as shown in Figure 1, a symbol for the group. The trail of elephants shown in Figure 1 is the inspiration

Figure 1: Elephants drawn by doctoral student Ming Gao to welcome prospective students to the Orazem lab.

for the line of elephants used to separate solutions to problems found in Orazem and Tribollet[1] from solutions to problems created after the book's publication. An example can be seen on page 13.

The logo for the measurement model program, described in Chapter 25, is an elephant. Members of the Orazem group have defined a constant-phase elephant, which is referenced in Problem 14.1.

As discussed in Orazem and Tribollet,[1] "impedance spectroscopy is an application of a frequency-domain measurement to a complex system that cannot be easily visualized. The quantities measured, for example current and potential for electrochemical or electronic systems and stress and strain for mechanical systems, are macroscopic values that represent the spatial average of individual events. These quantities are influenced by the desired physical properties, such as diffusivity, rate constants, and viscosity, but do not provide a direct measure."

The point is that every experimental measurement we use gives us one view of a system. From the results of this measurement, we can develop a model of the system under investigation. Our confidence in the validity of the model can be improved by performing new observations of the system, often guided by the model we developed. Impedance spectroscopy is a powerful tool, but the elephant reminds us that impedance spectroscopy is not a stand-alone technique.

Part I

Problems and Solutions

Chapter 1

Complex Variables

A working understanding of complex variables is essential for the analysis of experiments conducted in the frequency domain, such as impedance spectroscopy. The objective of problems presented in this chapter is to introduce the subject of complex variables at a level sufficient to understand the development of interpretation models for impedance spectroscopy. Problems 1.1–1.6 are based on problems taken from Orazem and Tribollet.[1]

Problems

1.1 Calculate the phase angle and modulus for the following:

(a) $Z = 1/(\mathrm{j}\omega C)$

(b) $Z = R$

Solution 1.1.

(a) The equation $Z = 1/(\mathrm{j}\omega C)$ corresponds to the electrical behavior of a capacitor. As $1/\mathrm{j} = -\mathrm{j}$, the impedance is an imaginary number $Z = -\mathrm{j}/\omega C$ with phase $\varphi = -90°$ and magnitude $|Z| = 1/\omega C$. This result may also be obtained from the definitions of phase angle and magnitude. The phase angle is given by

$$\varphi = \tan^{-1}\frac{Z_\mathrm{j}}{Z_\mathrm{r}} = \tan^{-1}(-\infty) = -\pi/2 \tag{1.1}$$

or $\varphi = -90°$. The magnitude is given by

$$|Z| = \sqrt{Z_\mathrm{r}^2 + Z_\mathrm{j}^2} = 1/\omega C \tag{1.2}$$

(b) The equation $Z = R$ corresponds to the electrical behavior of a resistor and is a real number. For a real number, the phase angle is equal to zero, and the magnitude is the value of the real number, thus $|Z| = R$. These results may also be

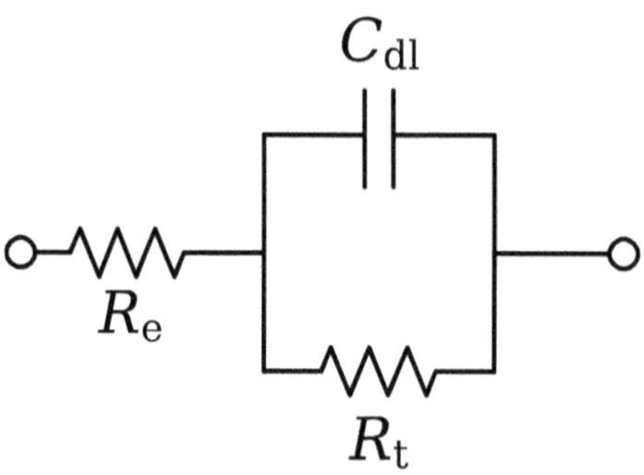

Figure 1.1: Circuit representation of equation (1.5). Taken from Orazem and Tribollet[1] with permission of John Wiley & Sons.

obtained from the definitions of phase angle and magnitude. The phase angle is given by

$$\varphi = \tan^{-1}\frac{Z_{\text{j}}}{Z_{\text{r}}} = \tan^{-1} 0 = 0 \tag{1.3}$$

and the magnitude is given by

$$|Z| = \sqrt{Z_{\text{r}}^2 + Z_{\text{j}}^2} = R \tag{1.4}$$

1.2 The impedance associated with a single electrochemical reaction on a uniform surface can be expressed as

$$Z(\omega) = R_{\text{e}} + \frac{R_{\text{t}}}{1 + \text{j}\omega R_{\text{t}} C_{\text{dl}}} \tag{1.5}$$

where R_{e} is the electrolyte resistance, R_{t} is the charge-transfer resistance, and C_{dl} is the capacity of the double layer.

(a) Find expressions for the real and imaginary parts of the impedance as a function of frequency.

(b) Find expressions for the magnitude and phase angle of the impedance.

(c) Find expressions for the real and imaginary parts of the admittance as a function of frequency.

Solution 1.2. Equation (1.5) is the response of the electrical circuit presented in Figure 1.1, which can also account for the impedance response of a simple potential-dependent electrochemical reaction on an electrode surface.

(a) Expressions for the real and imaginary parts of Z are found by multiplying numerator and denominator by the complex conjugate, i.e.,

$$Z = R_e + \frac{R_t}{1 + j\omega R_t C_{dl}} \frac{1 - j\omega R_t C_{dl}}{1 - j\omega R_t C_{dl}} \tag{1.6}$$

or

$$Z = R_e + \frac{R_t}{1 + \omega^2 R_t^2 C_{dl}^2} - j\frac{R_t^2 \omega C_{dl}}{1 + \omega^2 R_t^2 C_{dl}^2} \tag{1.7}$$

The real part of the impedance is given by

$$Z_r = R_e + \frac{R_t}{1 + \omega^2 R_t^2 C_{dl}^2} \tag{1.8}$$

and the imaginary part is given by

$$Z_j = -\frac{\omega R_t^2 C_{dl}}{1 + \omega^2 R_t^2 C_{dl}^2} \tag{1.9}$$

As the frequency becomes small, i.e., $\omega \rightarrow 0$, the real part of the impedance approaches $R_e + R_t$ and the imaginary part of the impedance approaches zero. As the frequency becomes large, i.e., $\omega \rightarrow \infty$, the real part of the impedance approaches R_e and the imaginary part of the impedance again approaches zero.

(b) The magnitude is given by

$$|Z| = \sqrt{Z_r^2 + Z_j^2} \tag{1.10}$$

A compact notation may be used in which

$$x = \left(1 + \omega^2 R_t^2 C_{dl}^2\right) \tag{1.11}$$

Application of equation (1.10) yields

$$|Z| = \frac{1}{x}\sqrt{R_t^2 + \omega^2 R_t^4 C_{dl}^2 + 2R_e R_t x + R_e^2 x^2} \tag{1.12}$$

which can be simplified to

$$|Z| = \frac{1}{x}\sqrt{R_t^2 x + 2R_e R_t x + R_e^2 x^2} \tag{1.13}$$

The phase angle is given by

$$\varphi = \tan^{-1}\left(\frac{Z_j}{Z_r}\right) \tag{1.14}$$

which can be expressed as

$$\varphi = \tan^{-1}\left(\frac{-R_t^2 \omega C_{dl}}{R_e\left(1 + \omega^2 R_t^2 C_{dl}^2\right) + R_t}\right) \tag{1.15}$$

As the frequency becomes small, i.e., $\omega \rightarrow 0$, the magnitude of the impedance approaches $R_e + R_t$ and the phase angle approaches zero. As the frequency becomes large, i.e., $\omega \rightarrow \infty$, the magnitude of the impedance approaches R_e and the phase angle again approaches zero.

(c) Expressions for the real and imaginary parts of the admittance can be obtained in terms of the impedance by using the complex conjugate, i.e.,

$$Y = \frac{1}{Z} = \frac{1}{Z_r + jZ_j}\left(\frac{Z_r - jZ_j}{Z_r - jZ_j}\right) \tag{1.16}$$

yielding

$$Y = \frac{1}{|Z|^2}\left(Z_r - jZ_j\right) \tag{1.17}$$

The real part of the resulting admittance is

$$Y_r = \frac{R_e + R_t + \omega^2 R_e R_t^2 C_{dl}^2}{R_e^2 + 2R_e R_t + R_t^2 + \omega^2 R_e^2 R_t^2 C_{dl}^2} \tag{1.18}$$

and the imaginary part is given by

$$Y_j = \frac{\omega R_t^2 C_{dl}}{R_e^2 + 2R_e R_t + R_t^2 + \omega^2 R_e^2 R_t^2 C_{dl}^2} \tag{1.19}$$

As the frequency becomes small, i.e., $\omega \to 0$, the real part of the admittance approaches $1/(R_e + R_t)$ and the imaginary part of the admittance approaches zero. As the frequency becomes large, i.e., $\omega \to \infty$, the real part of the admittance approaches $1/R_e$ and the imaginary part of the admittance again approaches zero.

1.3 The impedance associated with an ideally polarized (blocking) electrode can be expressed as

$$Z(\omega) = R_e + \frac{1}{j\omega C_{dl}} \tag{1.20}$$

where R_e is the electrolyte resistance and C_{dl} is the capacity of the double layer.

(a) Find expressions for the real and imaginary parts of the impedance as a function of frequency.

(b) Find expressions for the magnitude and phase angle of the impedance.

(c) Find expressions for the real and imaginary parts of the admittance as a function of frequency.

Solution 1.3. This problem is very much like Problem 1.1.a with the exception that an ohmic resistance has been added to the capacitance.

(a) As $1/j = -j$, the impedance is given by

$$Z = R_e - j\frac{1}{\omega C_{dl}} \tag{1.21}$$

where

$$Z_r = R_e \tag{1.22}$$

is the real part of the impedance and

$$Z_j = -\frac{1}{\omega C_{dl}} \tag{1.23}$$

is the imaginary part.

(b) The phase angle, expressed by equation (1.14), can be written

$$\varphi = \tan^{-1}\left(\frac{-1}{\omega R_e C_{dl}}\right) \tag{1.24}$$

At a frequency $\omega = 1/R_e C_{dl}$, $\varphi = -\pi/4 = -45°$. As the frequency becomes small, i.e., $\omega \to 0$, the phase angle approaches $-90°$. As the frequency becomes large, i.e., $\omega \to \infty$, the phase angle approaches zero.

The magnitude, expressed by equation (1.10), can be written

$$|Z| = \sqrt{\frac{1+\omega^2 R_e^2 C_{dl}^2}{\omega^2 C_{dl}^2}} \tag{1.25}$$

When $\omega = 1/R_e C_{dl}$, $|Z| = \sqrt{2}R_e$. As the frequency becomes small, i.e., $\omega \to 0$, the magnitude approaches $1/\omega C_{dl}$. As the frequency becomes large, i.e., $\omega \to \infty$, the magnitude approaches R_e.

(c) Following equation (1.17),

$$Y_r = \frac{\omega^2 R_e C_{dl}^2}{1+\omega^2 R_e^2 C_{dl}^2} \tag{1.26}$$

and

$$Y_j = \frac{\omega C_{dl}}{1+\omega^2 R_e^2 C_{dl}^2} \tag{1.27}$$

As the frequency becomes small, i.e., $\omega \to 0$, the real and imaginary parts of the admittance approach zero. As the frequency becomes large, i.e., $\omega \to \infty$, the real part of the admittance approaches $1/R_e$ and the imaginary part of the admittance again approaches zero.

1.4 The impedance associated with a constant-phase element can be expressed as

$$Z(\omega) = R_e + \frac{1}{(j\omega)^{\alpha} Q} \tag{1.28}$$

where α and Q are parameters associated with a constant-phase element (CPE). When $\alpha = 1$, Q has units of a capacitance, i.e., $\mu F/cm^2$, and represents the capacity of the interface. When $\alpha \neq 1$, the system shows behavior that has been attributed to surface heterogeneity, oxide films, or to continuously distributed time constants for charge-transfer reactions.

(a) Find expressions for the real and imaginary parts of the impedance as a function of frequency.

(b) Find expressions for the magnitude and phase angle of the impedance.

(c) Find expressions for the real and imaginary parts of the admittance as a function of frequency.

Solution 1.4.

(a) Equation (1.28) may be expressed as

$$Z(\omega) = R_e + \frac{j^{-\alpha}}{\omega^{\alpha} Q} \tag{1.29}$$

Introduction of

$$j^{-\alpha} = \cos\left(\frac{\alpha\pi}{2}\right) - j\sin\left(\frac{\alpha\pi}{2}\right) \tag{1.30}$$

yields

$$Z = R_e + \frac{\cos\left(\frac{\alpha\pi}{2}\right)}{\omega^{\alpha} Q} - j\frac{\sin\left(\frac{\alpha\pi}{2}\right)}{\omega^{\alpha} Q} \tag{1.31}$$

Thus, the real part of the impedance is given by

$$Z_r = R_e + \frac{\cos(\alpha\pi/2)}{\omega^{\alpha} Q} \tag{1.32}$$

and imaginary part of the impedance is given by

$$Z_j = -\frac{\sin(\alpha\pi/2)}{\omega^{\alpha} Q} \tag{1.33}$$

When $\alpha = 1$, the parameter Q can be expressed as a capacitance C, and

$$\lim_{\alpha \to 1} Z_{CPE} = -j\frac{1}{\omega C} \tag{1.34}$$

which is the impedance for a capacitor.

(b) Following equation (1.14),

$$\varphi = \tan^{-1}\left(\frac{-\sin(\alpha\pi/2)}{\omega^{\alpha}R_{e}Q + \cos(\alpha\pi/2)}\right) \quad (1.35)$$

If $R_e = 0$, $\varphi = -\alpha\pi/2 = -\alpha 90°$. Following equation (1.10),

$$|Z| = \frac{1}{\omega^{\alpha}Q}\sqrt{1 + R_e^2\omega^{2\alpha}Q^2 + 2R_e\omega^{\alpha}Q\cos(\alpha\pi/2)} \quad (1.36)$$

If $R_e = 0$, $|Z| = 1/\omega^{\alpha}Q$.

(c) Following equation (1.17),

$$Y_r = \frac{\omega^{2\alpha}R_eQ^2 + \omega^{\alpha}Q\cos(\alpha\pi/2)}{1 + \omega^{2\alpha}R_e^2Q^2} \quad (1.37)$$

and

$$Y_j = \frac{\omega^{\alpha}Q\sin(\alpha\pi/2)}{1 + \omega^{2\alpha}R_e^2Q^2} \quad (1.38)$$

1.5 Consider a situation where the impedance of one layer is given by

$$Z_1(\omega) = \frac{R_1}{1 + j\omega R_1C_1} \quad (1.39)$$

and the impedance of a second layer is given by

$$Z_2(\omega) = \frac{R_2}{1 + j\omega R_2C_2} \quad (1.40)$$

Add the two impedances to find the overall impedance of the two layers.

Solution 1.5. Sum impedances by adding the real parts and adding the imaginary parts. The sum of complex numbers z and w may be expressed as

$$z + w = (z_r + w_r) + j(z_j + w_j) \quad (1.41)$$

As

$$Z_1 = \frac{R_1}{1 + j\omega R_1C_1} = \frac{R_1}{1 + \omega^2R_1^2C_1^2} - j\frac{\omega R_1^2C_1}{1 + \omega^2R_1^2C_1^2} \quad (1.42)$$

and

$$Z_2 = \frac{R_2}{1 + j\omega R_2C_2} = \frac{R_2}{1 + \omega^2R_2^2C_2^2} - j\frac{\omega R_2^2C_2}{1 + \omega^2R_2^2C_2^2} \quad (1.43)$$

the real part of the sum of Z_1 and Z_2 is given by

$$\mathrm{Re}\{Z_1 + Z_2\} = \frac{R_1}{1 + \omega^2R_1^2C_1^2} + \frac{R_2}{1 + \omega^2R_2^2C_2^2} \quad (1.44)$$

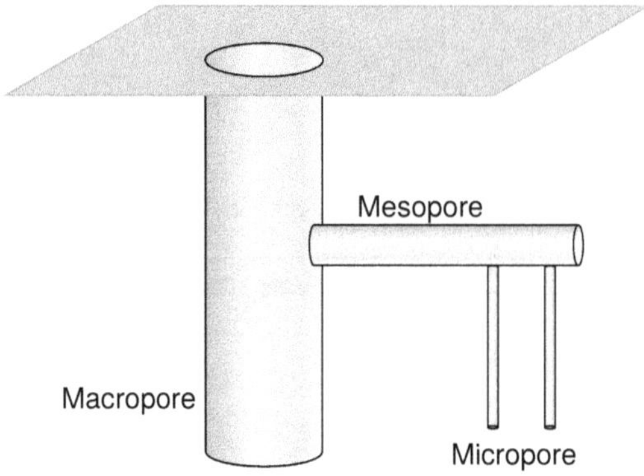

Figure 1.2: Schematic representation of a pore-within-a-pore hierarchy as suggested by Itagaki et al.[13] Taken from Orazem and Tribollet[1] with permission of John Wiley & Sons.

and the imaginary part is given by

$$\operatorname{Im}\{Z_1 + Z_2\} = -\left(\frac{\omega R_1^2 C_1}{1+\omega^2 R_1^2 C_1^2} + \frac{\omega R_2^2 C_2}{1+\omega^2 R_2^2 C_2^2}\right) \tag{1.45}$$

1.6 Gourbeyre et al.[14] and Itagaki et al.[13] have proposed that pores can appear inside a macropore structure as illustrated in Figure 1.2. The pore-in-pore model, involving macropores and mesopores, yields an impedance

$$Z = \sqrt{R_0\sqrt{R_1 Z_1}} \tag{1.46}$$

Develop an expression for the real and imaginary parts of the associated impedance and the associated phase angle:

(a) The impedance Z_1 is given by

$$Z_1 = \frac{1}{j\omega C_{\mathrm{dl}}} \tag{1.47}$$

(b) The impedance Z_1 is given by

$$Z_1 = \frac{R_{\mathrm{t}}}{1 + j\omega R_{\mathrm{t}} C_{\mathrm{dl}}} \tag{1.48}$$

Solution 1.6. Two approaches were taken to solve this problem. The solution for part (a) makes use of the relationships between rectangular and polar coordinates for complex variables to derive real and imaginary parts of the impedance from which the modulus and phase angle may be derived. The solution for part (b) derives directly the modulus and phase angle from which the real and imaginary parts of the impedance may be derived.

(a) The imaginary number may be expressed in polar coordinates as

$$\mathrm{j} = \cos(\frac{\pi}{2}) + \mathrm{j}\sin(\frac{\pi}{2}) \tag{1.49}$$

Therefore, Z_1 can be expressed as

$$Z_1 = \frac{1}{\mathrm{j}\omega C_{\mathrm{dl}}} = -\mathrm{j}\frac{1}{\omega C_{\mathrm{dl}}} = \frac{\cos(-\frac{\pi}{2}) + \mathrm{j}\sin(-\frac{\pi}{2})}{\omega C_{\mathrm{dl}}} \tag{1.50}$$

Equation (1.46) may be expressed as

$$\begin{aligned} Z &= \sqrt{R_0\sqrt{R_1 Z_1}} \\ &= R_0^{1/2}(R_1 Z_1)^{1/4} \\ &= R_0^{1/2}R_1^{1/4}\left(\frac{\cos(-\frac{\pi}{2}) + \mathrm{j}\sin(-\frac{\pi}{2})}{\omega C_{\mathrm{dl}}}\right)^{1/4} \\ &= R_0^{1/2}R_1^{1/4}\frac{\cos(\frac{-\pi}{8}) + \mathrm{j}\sin(\frac{-\pi}{8})}{(\omega C_{\mathrm{dl}})^{1/4}} \\ &= \left(\frac{R_0^2 R_1}{\omega C_{\mathrm{dl}}}\right)^{1/4}\left(\cos(\frac{\pi}{8}) - \mathrm{j}\sin(\frac{\pi}{8})\right) \end{aligned} \tag{1.51}$$

The expressions of real and imaginary parts of Z are

$$Z_{\mathrm{r}} = (\frac{R_0^2 R_1}{\omega C_{\mathrm{dl}}})^{1/4}\cos(\frac{\pi}{8}) \tag{1.52}$$

$$Z_{\mathrm{j}} = -(\frac{R_0^2 R_1}{\omega C_{\mathrm{dl}}})^{1/4}\sin(\frac{\pi}{8}) \tag{1.53}$$

The phase angle is expressed as

$$\begin{aligned} \varphi &= \tan^{-1}\left(\frac{Z_{\mathrm{j}}}{Z_{\mathrm{r}}}\right) \\ &= -\frac{\pi}{8} \end{aligned} \tag{1.54}$$

The phase angle is $-22.5°$ (or $-\pi/8$), which is half of the $-45°$ (or $-\pi/4$) phase angle associated with a single pore.

(b) The impedance Z_1 given by equation (1.48) may be expressed in terms of real and imaginary parts as

$$Z_1 = \frac{R_{\mathrm{t}}}{1 + \mathrm{j}\omega R_{\mathrm{t}} C_{\mathrm{dl}}}\frac{1 - \mathrm{j}\omega R_{\mathrm{t}} C_{\mathrm{dl}}}{1 - \mathrm{j}\omega R_{\mathrm{t}} C_{\mathrm{dl}}} \tag{1.55}$$

where the real part of the impedance is given by

$$Z_{1,r} = \frac{R_t}{1+\omega^2 R_t^2 C_{dl}^2} \tag{1.56}$$

and the imaginary part is given by

$$Z_{1,j} = -\frac{\omega R_t^2 C_{dl}}{1+\omega^2 R_t^2 C_{dl}^2} \tag{1.57}$$

The modulus of Z_1 may be expressed as

$$|Z_1| = \sqrt{\frac{R_t^2}{\left(1+\omega^2 R_t^2 C_{dl}^2\right)^2} + \frac{\omega^2 R_t^4 C_{dl}^2}{\left(1+\omega^2 R_t^2 C_{dl}^2\right)^2}} = \sqrt{\frac{R_t^2 + \omega^2 R_t^4 C_{dl}^2}{\left(1+\omega^2 R_t^2 C_{dl}^2\right)^2}} \tag{1.58}$$

and the modulus of Z may be obtained as

$$|Z| = R_0^{1/2} R_1^{1/4} \sqrt{\sqrt{|Z_1|}} \tag{1.59}$$

The phase of Z_1 is given by

$$\varphi_1 = \tan^{-1}\left(-\omega R_t C_{dl}\right) \tag{1.60}$$

and the phase of Z may be obtained as

$$\varphi = \frac{\varphi_1}{4} = \frac{\tan^{-1}\left(-\omega R_t C_{dl}\right)}{4} \tag{1.61}$$

In the high-frequency limit, the impedance given as equation (1.48) approaches equation (1.47), and the phase angle approaches a value of $-\pi/8$ or $-22.5°$, just as was seen for part (a).

The real part of Z is obtained from the modulus and phase angle as

$$Z_r = R_0^{1/2} R_1^{1/4} \left(\frac{R_t^2 + \omega^2 R_t^4 C_{dl}^2}{\left(1+\omega^2 R_t^2 C_{dl}^2\right)^2}\right)^{1/8} \cos\left(\frac{\varphi_1}{4}\right) \tag{1.62}$$

and the imaginary part is given as

$$Z_j = R_0^{1/2} R_1^{1/4} \left(\frac{R_t^2 + \omega^2 R_t^4 C_{dl}^2}{\left(1+\omega^2 R_t^2 C_{dl}^2\right)^2}\right)^{1/8} \sin\left(\frac{\varphi_1}{4}\right) \tag{1.63}$$

A similar treatment can be applied for a pore-in-pore-in-pore model in which macro-, meso-, and micropores are considered.

1.7 A phasor notation is commonly used in the development of mathematical models for the transfer function, or impedance, response of electrochemical systems.

(a) Show that a time-dependent variable $i(t)$, expressed in terms of the steady-state value $\bar{i}$ and a sinusoidal time-dependent contribution as

$$i(t) = \bar{i} + |\Delta i| \cos(\omega t + \varphi) \tag{1.64}$$

can be expressed as

$$i(t) = \bar{i} + \mathrm{Re}\left\{\tilde{i} \exp(\mathrm{j}\omega t)\right\} \tag{1.65}$$

where the phasor

$$\tilde{i} = |\Delta i| \exp(\mathrm{j}\varphi) \tag{1.66}$$

is a complex number for $\varphi \neq 0$.

(b) Under galvanostatic modulation, $\tilde{i}$ may be considered a real number. Explain.

(c) Under potentiostatic modulation, the difference between the working and the reference electrode potentials $\widetilde{\Phi}_{\mathrm{m}} - \widetilde{\Phi}_{\mathrm{ref}}$ may be considered a real number, but the interfacial potential $\widetilde{\Phi}_{\mathrm{m}} - \widetilde{\Phi}_{0}$, where $\widetilde{\Phi}_{0}$ corresponds to a position just outside the diffuse part of the double layer, is a complex number. Explain.

Solution 1.7.

(a) Following

$$\exp(\mathrm{j}x) = \cos(x) + \mathrm{j}\sin(x) \tag{1.67}$$

the oscillating component of equation (1.64) can be expressed in terms of an exponential as

$$|\Delta i| \cos(\omega t + \varphi) = |\Delta i| \exp(\mathrm{j}(\omega t + \varphi)) - \mathrm{j}|\Delta i| \sin(\omega t + \varphi) \tag{1.68}$$

or

$$|\Delta i| \cos(\omega t + \varphi) = \tilde{i} \exp(\mathrm{j}\omega t) - \mathrm{j}|\Delta i| \sin(\omega t + \varphi) \tag{1.69}$$

where $\tilde{i}$ is given by equation (1.66). The quantity on the left-hand side of equation (1.66) must be a real number. Equation (1.69) is formally equivalent to

$$\begin{aligned} |\Delta i| \cos(\omega t + \varphi) = \mathrm{Re}\left\{|\Delta i| \cos(\omega t + \varphi)\right\} = \\ \mathrm{Re}\left\{\tilde{i} \exp(\mathrm{j}\omega t) - \mathrm{j}\sin(\omega t + \varphi)\right\} \end{aligned} \tag{1.70}$$

The imaginary term $\mathrm{j}\sin(\omega t + \varphi)$ does not contribute to the real part of the complex number inside the brackets; thus,

$$|\Delta i| \cos(\omega t + \varphi) = \mathrm{Re}\left\{\widetilde{i} \exp(\mathrm{j}\omega t)\right\} \tag{1.71}$$

(b) The phase angle is referenced to the input signal. Under galvanostatic modulation, current is the input signal and the phase angle of the current signal is equal to zero, as it is referenced to itself. The phasor for current $\widetilde{I}$ (or current density $\widetilde{i}$) is given by equation (1.66). The real and imaginary parts are given by

$$\widetilde{i}_{\mathrm{r}} = |\Delta i| \cos(\varphi) \tag{1.72}$$

and

$$\widetilde{i}_{\mathrm{j}} = |\Delta i| \sin(\varphi) \tag{1.73}$$

For $\varphi = 0$, $\widetilde{i}_{\mathrm{r}} = |\Delta i|$ and $\widetilde{i}_{\mathrm{j}} = 0$.

(c) Under potentiostatic modulation, the potential difference between the working and reference electrodes is controlled. Thus, using the arguments presented in part (b), $\widetilde{\Phi}_{\mathrm{m}} - \widetilde{\Phi}_{\mathrm{ref}}$ may be considered a real number. The difference between the potential of the working electrode and the potential outside the diffuse part of the double layer cannot be controlled. Thus, the phase of $\widetilde{\Phi}_{\mathrm{m}} - \widetilde{\Phi}_0$ is referenced to the input signal $\widetilde{\Phi}_{\mathrm{m}} - \widetilde{\Phi}_{\mathrm{ref}}$ and cannot be assumed to be equal to zero. The difference $\widetilde{\Phi}_{\mathrm{m}} - \widetilde{\Phi}_0$ is therefore considered a complex number.

1.8 The impedance may be expressed in the form of a complex capacitance following

$$\mathcal{C} = \frac{1}{\mathrm{j}\omega\,(Z - R_{\mathrm{e}})} \tag{1.74}$$

Find an expression for the complex capacitance for:

(a) $Z = R_{\mathrm{e}} + 1/\mathrm{j}\omega C$.

(b) $Z = R_{\mathrm{e}} + R_{\mathrm{t}}/(1 + \mathrm{j}\omega R_{\mathrm{t}} C)$.

(c) $Z = R_{\mathrm{e}} + 1/(\mathrm{j}\omega)^{\alpha} Q$.

Solution 1.8.

(a) For $Z = R_{\mathrm{e}} + 1/\mathrm{j}\omega C$, equation (1.74) yields

$$\mathcal{C} = \frac{1}{\mathrm{j}\omega\left(\frac{1}{\mathrm{j}\omega C}\right)} = C \tag{1.75}$$

The complex capacitance for an impedance associated with an ohmic resistance in series with a capacitor is a real number with the value of the capacitance. The imaginary part is equal to zero.

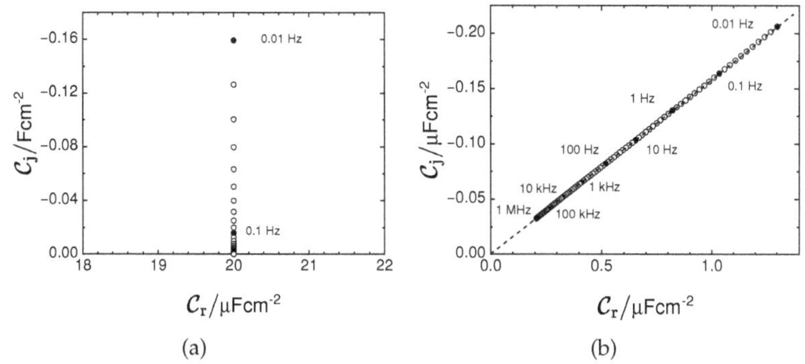

Figure 1.3: Complex capacitance plots for Problem 1.8: (a) solution to part b for parameters $R_t = 100\ \Omega\text{cm}^2$ and $C = 20\ \mu\text{F/cm}^2$ and (b) solution to part c for parameters $\alpha = 0.9$ and $Q = 10^{-6}\ \text{F/cm}^2\text{s}^{1-\alpha}$.

(b) For $Z = R_e + R_t/(1 + j\omega R_tC)$,

$$\mathcal{C} = \frac{1}{j\omega\left(\frac{R_t}{1+j\omega R_t C}\right)} = C + \frac{1}{j\omega R_t} = C - j\frac{1}{\omega R_t} \tag{1.76}$$

The real part of the complex capacitance is C and the imaginary part is $-1/\omega R_t$. The calculated complex capacitance presented in Figure 1.3(a) was obtained for parameters $R_t = 100\ \Omega\text{cm}^2$ and $C = 20\ \mu\text{F/cm}^2$. These calculations confirm that the real part of the complex capacitance yields the input value for capacitance.

(c) For an ohmic resistance in series with a constant-phase elements (CPE), e.g., $Z = R_e + 1/(j\omega)^\alpha Q$,

$$\mathcal{C} = \frac{1}{j\omega\left(\frac{1}{(j\omega)^\alpha Q}\right)} = \frac{Q}{(j\omega)^{1-\alpha}} \tag{1.77}$$

Introduction of equation (1.30) yields

$$\mathcal{C} = \frac{Q}{\omega^{1-\alpha}}\cos\left(\frac{\pi(\alpha-1)}{2}\right) + \frac{Q}{\omega^{1-\alpha}}j\sin\left(\frac{\pi(\alpha-1)}{2}\right) \tag{1.78}$$

or

$$\mathcal{C} = \frac{Q}{\omega^{1-\alpha}}\cos\left(\frac{\pi(1-\alpha)}{2}\right) - j\frac{Q}{\omega^{1-\alpha}}\sin\left(\frac{\pi(1-\alpha)}{2}\right) \tag{1.79}$$

Equation (1.79) shows that, for a CPE, where $0 < \alpha < 1$, $\mathcal{C}_r \to 0$ as frequency tends toward infinity. Thus, the pure CPE with $0 < \alpha < 1$ cannot be represented by a capacitance. A value for capacitance is seen only if $\alpha = 1$, for which $\mathcal{C}_r = Q$ and $\mathcal{C}_j = 0$.

The calculated complex capacitance presented in Figure 1.3(b) was obtained for $\alpha = 0.9$ and $Q = 10^{-6}\ \text{F/cm}^2\text{s}^{1-\alpha}$. In contrast to the results shown in Figure 1.3(a), the real part of the complex capacitance does not approach a finite value for capacitance.

Chapter 2

Differential Equations

Development of models for impedance requires solution of differential equations. The method of solution requires two steps. In the first, a steady-state solution is obtained, which generally requires solution of ordinary differential equations. In the second, a solution is obtained for the sinusoidal steady state. These two steps require solutions of ordinary differential equations. While in some cases numerical solution is required, analytic solutions are possible for a large number of problems. The problems presented in this chapter explore analytic solutions associated with impedance spectroscopy. Problems 2.1–2.4 are based on problems taken from Orazem and Tribollet.[1]

Problems

2.1 Find the general solution for c as a function of y for:

(a) $\frac{\mathrm{d}c}{\mathrm{d}y} + \mathrm{k}c = 0$

(b) $\frac{\mathrm{d}^2c}{\mathrm{d}y^2} + \mathrm{k}c = 0$

(c) $\frac{\mathrm{d}^2c}{\mathrm{d}y^2} + \mathrm{a}\frac{\mathrm{d}c}{\mathrm{d}y} + \mathrm{k}c = 0$

(d) $\frac{\mathrm{d}^2c}{\mathrm{d}y^2} + \mathrm{a}\frac{\mathrm{d}c}{\mathrm{d}y} + \mathrm{k}c = \mathrm{b}$

Solution 2.1.

(a) Equation

$$\frac{\mathrm{d}c}{\mathrm{d}y} + \mathrm{k}c = 0 \tag{2.1}$$

is a homogeneous first-order differential equation. In the first step, the term kc is moved to the right-hand side of the equation, yielding

$$\frac{\mathrm{d}c}{\mathrm{d}y} = -\mathrm{k}c \tag{2.2}$$

Division by c and multiplication of both sides by $\mathrm{d}y$ yields

$$\frac{1}{c}\mathrm{d}c = (-1)\mathrm{d}y \tag{2.3}$$

Integration of both sides yields

$$\int \frac{1}{c}\mathrm{d}c = \int (-1)\mathrm{d}y \tag{2.4}$$

and

$$\ln c = -y + \alpha_1 \tag{2.5}$$

where α_1 is an arbitrary constant. The solution

$$c = \mathrm{e}^{\alpha_1}\mathrm{e}^{-y} \tag{2.6}$$

is obtained by taking the exponential of both sides.

(b) Equation

$$\frac{\mathrm{d}^2 c}{\mathrm{d}y^2} + \mathrm{k}c = 0 \tag{2.7}$$

is a homogeneous second-order differential equation. Because the equation is linear, the general solution is in the form of

$$c(y) = \mathrm{A}c_1(y) + \mathrm{B}c_2(y) \tag{2.8}$$

where A and B are arbitrary constants. The quantities $c_1(y)$ and $c_2(y)$ are the two solutions to equation (2.7), and they are both in the form:

$$c(y) = \mathrm{e}^{\lambda}y \tag{2.9}$$

The second derivative of equation (2.9) may be inserted into the differential equation (2.7) to yield

$$\lambda^2\mathrm{e}^{\lambda}y + k\mathrm{e}^{\lambda}y = 0 \tag{2.10}$$

The quantity $\mathrm{e}^{\lambda}y$ can be eliminated, which yields the quadratic characteristic equation

$$\lambda^2 + k = 0 \tag{2.11}$$

The solution is

$$\lambda = \pm \mathrm{j}\sqrt{k} \tag{2.12}$$

Therefore, the general solution is

$$c = \mathrm{A}\mathrm{e}^{\mathrm{j}\sqrt{k}y} + \mathrm{B}\mathrm{e}^{-\mathrm{j}\sqrt{k}y} \tag{2.13}$$

where A and B are arbitrary constants to be determined by use of boundary conditions.

(c) The equation to solve is

$$\frac{d^2c}{dy^2} + a\frac{dc}{dy} + kc = 0 \tag{2.14}$$

This is also a homogeneous linear second-order differential equation. According to the method used above, substitution of equation (2.9) yields the quadratic characteristic equation

$$\lambda^2 + a\lambda + k = 0 \tag{2.15}$$

The two roots of this quadratic equation are:

$$\lambda_1 = \frac{-a + \sqrt{a^2 - 4k}}{2} \tag{2.16}$$

and

$$\lambda_2 = \frac{-a - \sqrt{a^2 - 4k}}{2} \tag{2.17}$$

Therefore, the general solution to the differential equation (2.14) is

$$c = Ae^{-a+\sqrt{a^2-4k}y} + Be^{-a-\sqrt{a^2-4k}y} \tag{2.18}$$

where A and B are arbitrary constants determined through application of the boundary conditions.

(d) The equation to solve is

$$\frac{d^2c}{dy^2} + a\frac{dc}{dy} + kc = b \tag{2.19}$$

This is a heterogeneous linear second order differential equation. Solution re quires the corresponding homogeneous equation to obtain the solution c_h. The second step is to find a specific solution c_s to the heterogeneous equation. As the problem is linear, the eventual general solution is the addition of c_h and c_s. The corresponding homogeneous equation is obtained following the solution to Problem 2.1.c. Thus,

$$\frac{d^2c}{dy^2} + a\frac{dc}{dy} + kc = 0 \tag{2.20}$$

The characteristic equation for equation (2.20) is

$$\lambda^2 + a\lambda + k = 0 \tag{2.21}$$

The solution is

$$\lambda = \frac{-a \pm \sqrt{a^2 - 4k}}{2} \tag{2.22}$$

The general solution of the homogeneous equation is

$$c_h = Ae^{-a+\sqrt{a^2-4k}y} + Be^{-a-\sqrt{a^2-4k}y} \tag{2.23}$$

where A and B are arbitrary constants determined through application of the boundary conditions.

To find the specific solution, the specific solution is assumed to take the form

$$c_s = my + n \tag{2.24}$$

The first and second derivatives are

$$\frac{dc}{dy} = m \tag{2.25}$$

and

$$\frac{d^2c}{dy^2} = 0 \tag{2.26}$$

respectively. Insertion of equations (2.25) and (2.26) into the original equation (2.19) yields

$$0 + am + k(my + n) = b \tag{2.27}$$

A solution is obtained for $m = 0$ and $n = b/k$. Thus, the specific solution is

$$c_s = \frac{b}{k} \tag{2.28}$$

and the general solution of heterogeneous equation (2.19) is

$$c = Ae^{-a+\sqrt{a^2-4k}y} + Be^{-a-\sqrt{a^2-4k}y} + \frac{b}{k} \tag{2.29}$$

Again, the constants A and B are obtained by application of suitable boundary conditions.

2.2 Show that

$$\theta_i = A_i e^{\xi\sqrt{jK_i}} + B_i e^{-\xi\sqrt{jK_i}} \tag{2.30}$$

satisfies both real and imaginary parts of

$$\frac{d^2\theta_i}{d\xi^2} - jK_i\theta_i = 0 \tag{2.31}$$

Solution 2.2. Verification that equation (2.30) represents the solution for equation (2.31) can be obtained by calculating the first and second derivatives of equation (2.30), i.e.,

$$\frac{d\theta_i}{d\xi} = A_i\sqrt{jK_i}e^{\xi\sqrt{jK_i}} - B_i\sqrt{jK_i}e^{-\xi\sqrt{jK_i}} \tag{2.32}$$

and

$$\frac{d^2\theta_i}{d\xi^2} = A_i jK_i e^{\xi\sqrt{jK_i}} + B_i jK_i e^{-\xi\sqrt{jK_i}} \tag{2.33}$$

respectively. Multiplication of both sides of equation (2.30) by $\mathrm{j}K_\mathrm{i}$, yields

$$\mathrm{j}K_\mathrm{i}\theta_\mathrm{i} = \mathrm{j}K_\mathrm{i}A_\mathrm{i}\mathrm{e}^{\xi\sqrt{\mathrm{j}K_\mathrm{i}}} + \mathrm{j}K_\mathrm{i}B_\mathrm{i}\mathrm{e}^{-\xi\sqrt{\mathrm{j}K_\mathrm{i}}} \tag{2.34}$$

The difference of equations (2.33) and (2.34), yields equation (2.31). Thus, equation (2.30) is the solution to equation (2.31).

2.3 Verify that

$$\theta_\mathrm{i} = \frac{\sinh\left((\xi - 1)\sqrt{\mathrm{j}K_\mathrm{i}}\right)}{\sinh\left(-\sqrt{\mathrm{j}K_\mathrm{i}}\right)} \quad \text{for} \quad \xi < 1$$

$$\theta_\mathrm{i} = 0 \quad \text{for} \quad \xi \geq 1 \tag{2.35}$$

represents the solution for equation (2.31) for diffusion in a finite domain.

Solution 2.3. For the present problem, the boundary conditions are

$$\begin{aligned} \theta_\mathrm{i}(1) &= 0 \\ \theta_\mathrm{i}(0) &= 1 \end{aligned} \tag{2.36}$$

As shown in the solution to Problem 2.2, equation (2.30) is the general solution to equation (2.31), which may be written as

$$\theta_\mathrm{i} = \mathrm{A}_\mathrm{i}\left(\cosh\xi\sqrt{\mathrm{j}K_\mathrm{i}} + \sinh\xi\sqrt{\mathrm{j}K_\mathrm{i}}\right) + \mathrm{B}_\mathrm{i}\left(\cosh\xi\sqrt{\mathrm{j}K_\mathrm{i}} - \sinh\xi\sqrt{\mathrm{j}K_\mathrm{i}}\right) \tag{2.37}$$

The boundary conditions (2.36) for diffusion in a finite medium may be applied to equation (2.37) to yield, for $\xi = 0$,

$$\theta_\mathrm{i}(0) = \mathrm{A}_\mathrm{i} + \mathrm{B}_\mathrm{i} = 1 \tag{2.38}$$

or

$$\mathrm{B}_\mathrm{i} = 1 - \mathrm{A}_\mathrm{i} \tag{2.39}$$

For $\xi = 1$,

$$\theta_\mathrm{i}(1) = \mathrm{A}_\mathrm{i}(\cosh\sqrt{\mathrm{j}K_\mathrm{i}} + \sinh\sqrt{\mathrm{j}K_\mathrm{i}}) + \mathrm{B}_\mathrm{i}(\cosh\sqrt{\mathrm{j}K_\mathrm{i}} - \sinh\sqrt{\mathrm{j}K_\mathrm{i}}) = 0 \tag{2.40}$$

or

$$\mathrm{A}_\mathrm{i}(\cosh\sqrt{\mathrm{j}K_\mathrm{i}} + \sinh\sqrt{\mathrm{j}K_\mathrm{i}}) + \mathrm{B}_\mathrm{i}(\cosh\sqrt{\mathrm{j}K_\mathrm{i}} - \sinh\sqrt{\mathrm{j}K_\mathrm{i}}) = 0 \tag{2.41}$$

The solution of equations (2.39) and (2.41) yields

$$\mathrm{A}_\mathrm{i} = \frac{\cosh\sqrt{\mathrm{j}K_\mathrm{i}} - \sinh\sqrt{\mathrm{j}K_\mathrm{i}}}{2\sinh\sqrt{\mathrm{j}K_\mathrm{i}}} \tag{2.42}$$

and

$$\mathrm{B_i} = \frac{\cosh\sqrt{\mathrm{j}K_\mathrm{i}} + \sinh\sqrt{\mathrm{j}K_\mathrm{i}}}{2\sinh\sqrt{\mathrm{j}K_\mathrm{i}}} \tag{2.43}$$

Insertion of equations (2.42) and (2.43) into equation (2.37) yields

$$\theta_\mathrm{i} = \frac{\cosh\sqrt{\mathrm{j}K_\mathrm{i}} - \sinh\sqrt{\mathrm{j}K_\mathrm{i}}}{2\sinh\sqrt{\mathrm{j}K_\mathrm{i}}}\left(\cosh\xi\sqrt{\mathrm{j}K_\mathrm{i}} + \sinh\xi\sqrt{\mathrm{j}K_\mathrm{i}}\right) \tag{2.44}$$

which simplifies to equation (2.35).

2.4 Verify that

$$\theta_\mathrm{i} = \mathrm{e}^{-\xi\sqrt{\mathrm{j}K_\mathrm{i}}} \tag{2.45}$$

represents the solution for equation (2.31) for diffusion in an infinite domain.

Solution 2.4. The boundary conditions for diffusion in an infinite domain are

$$\theta_\mathrm{i}(\infty) = 0 \tag{2.46}$$

and

$$\theta_\mathrm{i}(0) = 1 \tag{2.47}$$

As shown in the solution to Problem 2.2, equation (2.30) is the general solution to equation (2.31). Application of the boundary conditions (2.46) and (2.47) for diffusion in an infinite medium to equation (2.31) yields

$$\theta_i(0) = \mathrm{A_i} + \mathrm{B_i} = 1 \tag{2.48}$$

for $\xi = 0$, and

$$\theta_i(\infty) = \mathrm{A_i}\mathrm{e}^{\xi\sqrt{\mathrm{j}K_\mathrm{i}}} = 0 \tag{2.49}$$

for $\xi = \infty$. Thus, $\mathrm{A_i} = 0$, $\mathrm{B_i} = 1$, and θ_i is given by equation (2.45).

2.5 The line voltage $\widetilde{u}(x)$ and the current $\widetilde{i}(x)$ for the transmission line can be expressed in the frequency domain as

$$\mathrm{d}\widetilde{u}(x) = -Z_1\widetilde{i}(x)\mathrm{d}x \tag{2.50}$$

and

$$\mathrm{d}\widetilde{i}(x) = -\frac{\widetilde{u}(x)}{Z_2}\mathrm{d}x \tag{2.51}$$

respectively, where parameters Z_1 and Z_2 are independent of the position variable x. At $x = 0$, the current is equal to zero. We seek an expression for the impedance of the transmission line.

Solution 2.5. Under the assumption that Z_1 and Z_2 are independent of x, equations (2.50) and (2.51) may be expressed as

$$\frac{\mathrm{d}^2\widetilde{u}(x)}{\mathrm{d}x^2} = \frac{Z_1}{Z_2}\widetilde{u}(x) \tag{2.52}$$

The homogeneous solution to equation (2.52) is

$$\widetilde{u}(x) = A\exp\left(x\sqrt{Z_1/Z_2}\right) + B\exp\left(-x\sqrt{Z_1/Z_2}\right) \tag{2.53}$$

and the derivative of $\widetilde{u}(x)$ is expressed as

$$\mathrm{d}\widetilde{u}(x) = \sqrt{Z_1/Z_2}\left(A\exp\left(x\sqrt{Z_1/Z_2}\right) - B\exp\left(-x\sqrt{Z_1/Z_2}\right)\right)\mathrm{d}x \tag{2.54}$$

The condition that at $x = 0$, $\widetilde{i}(0) = 0$ means that, from equation (2.50), $\mathrm{d}\widetilde{u}(0) = 0$ and, from equation (2.54), $A = B$.

The overall current produced on a length ℓ is given by

$$\begin{aligned}\widetilde{i}(\ell) &= \int_0^\ell \frac{\widetilde{u}(x)}{Z_2}\mathrm{d}x \\ &= \frac{A}{Z_2}\sqrt{\frac{Z_2}{Z_1}}\left(\exp\left(\ell\sqrt{Z_1/Z_2}\right) - \exp\left(-\ell\sqrt{Z_1/Z_2}\right)\right)\end{aligned} \tag{2.55}$$

The impedance is given by

$$Z = \frac{\widetilde{u}(\ell)}{\widetilde{i}(\ell)} = \frac{A\left(\exp\left(\ell\sqrt{Z_1/Z_2}\right) + \exp\left(-\ell\sqrt{Z_1/Z_2}\right)\right)}{\left(A/\sqrt{Z_1 Z_2}\right)\left(\exp\left(\ell\sqrt{Z_1/Z_2}\right) - \exp\left(-\ell\sqrt{Z_1/Z_2}\right)\right)} \tag{2.56}$$

From the hyperbolic relationships for complex variables given in Table A.6 on page 418,

$$Z = \sqrt{Z_1 Z_2}\coth\left(\ell\sqrt{Z_1/Z_2}\right) \tag{2.57}$$

The development presented here supports the treatment of porous electrodes presented in Problems 13.16–13.18.

Chapter 3

Statistics

A working understanding of statistics is essential for the analysis of experiments conducted in the frequency domain, such as impedance spectroscopy. The problems presented in this chapter explore statistical concepts with application to impedance spectroscopy. Problems 3.1–3.8 are based on problems taken from Orazem and Tribollet.[1]

Problems

3.1 The steady-state mass-transfer-limited current density for a rotating disk can be expressed as

$$i_{\text{lim}} = 0.62 n \text{F} c_{\text{i}}(\infty) D_{\text{i}}^{2/3} \nu^{-1/6} \Omega^{1/2} \tag{3.1}$$

where $c_{\text{i}}(\infty)$ is the concentration of the limiting reacting species i far from the disk, D_{i} is the diffusivity of the reacting species i, ν is the kinematic viscosity, and Ω is the rotation speed for the disk. Consider that $i_{\text{lim}} = 60 \pm 0.5$ mA/cm^2, $c_{\text{i}}(\infty) = 0.1 \pm 0.005$ M, $\nu = 10^{-2} \pm 10^{-4}$ cm^2/s, and $\Omega = 1{,}000 \pm 1$ rpm. Find the value and standard deviation for the diffusivity of species i in units of cm^2/s.

Solution 3.1. Under the assumption that the standard deviations are small, the standard deviation of a quantity derived from other parameters, i.e.,

$$D_{\text{i}} = \left(\frac{i_{\text{lim}} \nu^{1/6}}{0.62 n \text{F} c_{\text{i}}(\infty) \Omega^{1/2}} \right)^{3/2} \tag{3.2}$$

may be obtained from

$$\sigma_{D_{\text{i}}}^2 \cong \sigma_{x_1}^2 \left(\frac{\partial D_{\text{i}}}{\partial x_1} \right)^2 + \sigma_{x_2}^2 \left(\frac{\partial D_{\text{i}}}{\partial x_2} \right)^2 + \ldots \tag{3.3}$$

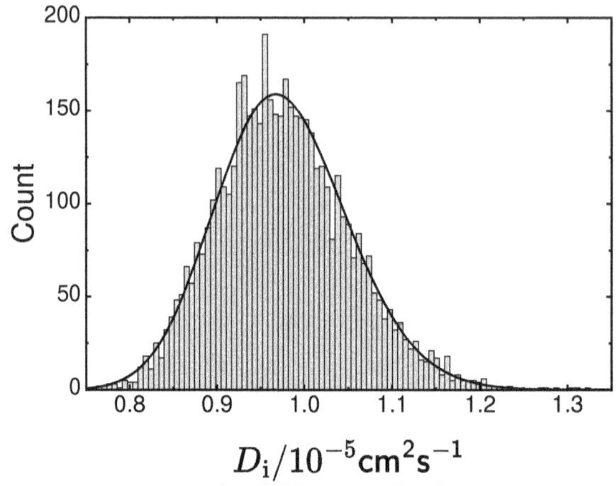

Figure 3.1: Histogram showing the calculated distribution of diffusivity with $\mu_{D_i} = 9.76 \times 10^{-6}$ cm^2/s and $\sigma_{D_i} = 7.6 \times 10^{-7}$ cm^2/s. The line represents the lognormal distribution.

where x_k represents variables in equation (3.2).

A preferred approach is to use numerical evaluation. A normally distributed random number $N_{x_k}(\mu_{x_k}\sigma x_k)$ is assigned to each parameter with μ_{x_k} is the value and σx_k is the standard deviation, respectively, for parameter x_k obtained from the problem statement. Equation (3.2) is used to calculate the instance of D_i. The process is repeated until a satisfactory statistical distribution is obtained.

The diffusivity was calculated under the assumption that $n = 1$. The result of 5,000 calculations is presented in Figure 3.1. The distribution is not perfectly normal, the line in Figure 3.1 is the corresponding lognormal distribution. The mean and standard deviation under the assumption of a normal distribution are $\mu_{D_i} = 9.76 \times 10^{-6}$ cm^2/s and $\sigma_{D_i} = 7.6 \times 10^{-7}$ cm^2/s, respectively.

3.2 The Schmidt number is expressed as $\mathrm{Sc_i} = \nu / D_i$. Find the value and standard deviation for the Schmidt number if the kinematic viscosity is $\nu = 10^{-2} \pm 10^{-4}$ cm^2/s and the diffusivity of species i is $D_i = 10^{-5} \pm 10^{-7}$ cm^2/s.

Solution 3.2. The Schmidt number resulting from 5,000 calculations is presented in Figure 3.2. Unlike Figure 3.1, the distribution is normal, and the line in Figure 3.2 is the corresponding normal distribution. The mean and standard deviation under the assumption of a normal distribution are $\mu_{\mathrm{Sc}} = 1{,}000$ and $\sigma_{\mathrm{Sc}} = 14$, respectively.

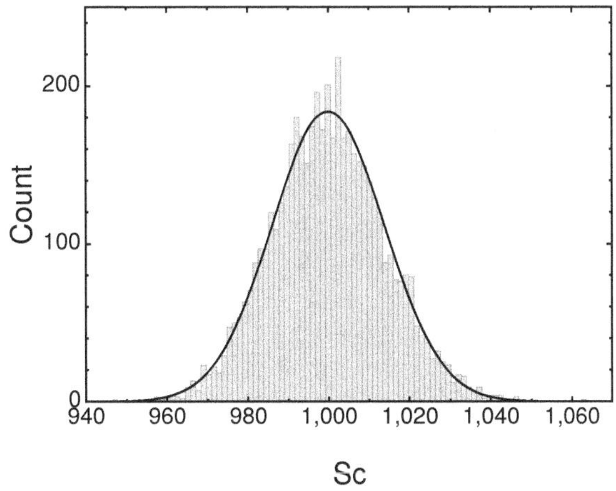

Figure 3.2: Histogram showing the calculated distribution of diffusivity with $\mu_{Sc} = 1,000$ and $\sigma_{Sc} = 14$. The line represents the normal distribution.

3.3 Estimate the 95.45% confidence interval for the equivalent circuit model given in Figure 1.1 over the frequency range of 10 mHz to 10 kHz if the regressed parameter values are expressed as $R_e = 10 \pm 1\ \Omega\text{cm}^2$, $R_t = 100 \pm 15\ \Omega\text{cm}^2$, and $\tau = R_t C_{dl} = 0.01 \pm 0.001$ s.

Solution 3.3. A Monte Carlo calculation was performed using 5,001 evaluations for each parameter. For example, the ohmic resistance was calculated using

$$R_e = R_{e,input} + \sigma_{R_e} N_1 \tag{3.4}$$

the charge-transfer resistance was calculated using

$$R_t = R_{t,input} + \sigma_{R_t} N_2 \tag{3.5}$$

and the $\tau = RC$ time constant was calculated using

$$\tau = \tau_{input} + \sigma_{\o} N_3 \tag{3.6}$$

where independent normal distributions N_1, N_2, and N_3 had mean values equal to zero and standard deviations equal to unity. The impedance was calculated 5,001 times for each frequency using

$$Z_k = R_{e,k} + \frac{R_{t,k}}{1 + 2\pi f \tau_k} \tag{3.7}$$

The corresponding values for the mean and standard deviation of real and imaginary parts of the impedance are presented in Table 3.1.

Table 3.1: Calculated confidence intervals for equation (3.7) with $R_e = 10 \pm 1$ Ωcm^2, $R_t = 100 \pm 15$ Ωcm^2, and $\tau = R_t C_{dl} = 0.01 \pm 0.001$ s. Values presented as mean plus or minus one standard deviation.

f, Hz	Z_r, Ωcm^2	Z_j, Ωcm^2
0.01	110 ± 15	-0.063 ± 0.011
0.1	110 ± 15	-0.63 ± 0.11
1	110 ± 15	-6.3 ± 1.1
10	82 ± 12	-44.8 ± 7.0
100	12.6 ± 1.2	-15.7 ± 2.8
1,000	10.03 ± 0.99	-1.61 ± 0.29
10,000	10.00 ± 0.99	-0.161 ± 0.029

3.4 Estimate the 95.45% confidence interval for the film thickness extracted by use of the power-law model given as

$$Q = \frac{(\varepsilon\varepsilon_0)^\alpha}{g\delta\rho_\delta^{1-\alpha}} \tag{3.8}$$

where

$$g = 1 + 2.88(1-\alpha)^{2.375} \tag{3.9}$$

and $\alpha = 0.91 \pm 0.01$, $Q = 11 \pm 1$ $\mu\text{F/s}^{(1-\alpha)}\text{cm}^2$, and $\rho_\delta = 450$ Ωcm. You may assume that $\varepsilon = 12$.

Solution 3.4. This problem was modified from that in Orazem and Tribollet[1] by changing $\alpha = 0.91 \pm 0.1$ to $\alpha = 0.91 \pm 0.01$, which is closer to the typical regression result. The film thickness resulting from 5,000 calculations is presented in Figure 3.3. This distribution is lognormal. Use of $\alpha = 0.91 \pm 0.1$ yields a nonphysical distribution for α that is occasionally greater than unity.

3.5 Erol and Orazem[15] reported regression results that yielded the parameter estimates given in Table 3.2. Calculate the corresponding 95.45% confidence intervals and determine whether zero is included within the confidence interval.

Solution 3.5. The confidence interval for a parameter X is given by $X - 2\sigma_X$ to $X + 2\sigma_X$. The results reported in Table 3.3 show that the parameter estimates do not include zero.

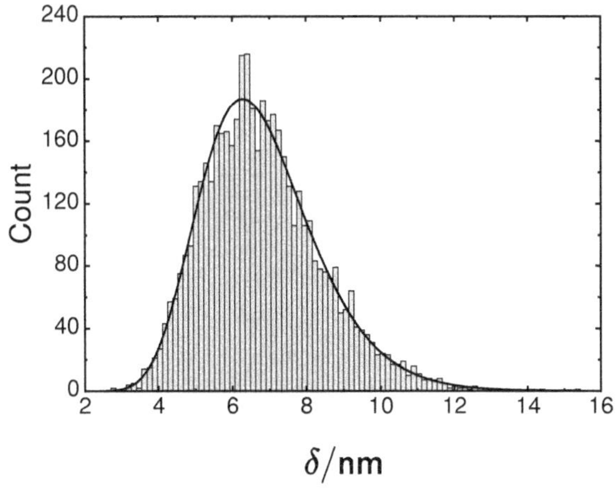

Figure 3.3: Histogram showing the calculated distribution of diffusivity with $\mu_\delta = 6.8$ nm and $\sigma_\delta = 1.6$ nm. The line represents the lognormal distribution.

Table 3.2: Estimated parameter values and standard deviations for a $LiCoO_2|C$ battery at temperatures of 10 and 20 °C at a potential of 4 V.

Parameter	10 °C	20 °C
$C_{a\|\|S}$, μF/cm^2	28 ± 1.8	28.8 ± 0.84
$A_{W,a}$, Ωcm^2 s$^{-0.5}$	58.8 ± 0.87	52.5 ± 0.51
$R_{t,a}$, Ωcm^2	0.256 ± 0.0066	0.195 ± 0.0045
$R_{t,S}$, Ωcm^2	1.93 ± 0.013	1.383 ± 0.0073
R_e. Ωcm^2	1.357 ± 0.0074	0.845 ± 0.0022
$R_{t,c}$, Ωcm^2	1.19 ± 0.018	0.954 ± 0.0094
α_c	0.912 ± 0.0074	0.938 ± 0.0052
Q_c, F/cm^{2}s$^{1-\alpha_c}$	0.0662 ± 0.00088	0.0608 ± 0.00065
$\gamma_{d,c}$	0.72 ± 0.015	0.785 ± 0.0087
$A_{d,c}$, Ω cm^2 s$^{-\gamma_{d,c}/2}$	0.254 ± 0.0042	0.203 ± 0.0023

Table 3.3: Estimated $\pm 2\sigma$ (95.4%) confidence intervals parameter values for the results reported in Table 3.2 .

Parameter	10 °C	20 °C
$C_{a\|\|S}$, μF/cm^2	24.4–31.6	27.12–30.48
$A_{W,a}$, Ωcm^2 s$^{-0.5}$	57.06–60.54	51.48–53.52
$R_{t,a}$, Ωcm^2	0.2428–0.2692	0.186–0.204
$R_{t,S}$, Ωcm^2	1.904–1.956	1.3684–1.3976
R_e, Ωcm^2	1.3422–1.3718	0.8406–0.8494
$R_{t,c}$, Ωcm^2	1.154–1.226	0.9352–0.9728
α_c	0.8972–0.9268	0.9276–0.9484
Q_c, F/cm^2 s$^{1-\alpha_c}$	0.06444–0.06796	0.6067–0.6093
$\gamma_{d,c}$	0.69–0.75	0.7676–0.8024
$A_{d,c}$, Ωcm^{2}s$^{-\gamma_{d,c}/2}$	0.2456–0.2624	0.1984–0.2076

Table 3.4: Replicated impedance collected at a frequency of 1 Hz on the same polymer electrolyte membrane (PEM) fuel cell using two different sets of instrumentation, the 850C provided by Scribner Associates and the FC350 provided by Gamry Instruments.

	850C		**FC350**	
Replicate	Z_r, Ω	Z_j, Ω	Z_r, Ω	Z_j, Ω
1	0.17117	−0.041791	0.167993	−0.038176
2	0.18236	−0.043494	0.173985	−0.040603
3	0.18606	−0.044666	0.176629	−0.041786
4	0.18941	−0.045946	0.1827	−0.044806

3.6 Provide the equation for a 95% confidence interval that corresponds to

$$\mu_x = \overline{x} \pm 2\frac{s_x}{\sqrt{n_x}} \tag{3.10}$$

for a 95.45% confidence interval.

Solution 3.6. The entry in Table A.10 for the 95% confidence interval at $\nu/p = 100,000$ is 1.96. Thus, the equation for a 95% confidence interval is

$$\mu_x = \overline{x} \pm 1.9600\frac{s_x}{\sqrt{n_x}} \tag{3.11}$$

In other words, the 95.4% confidence interval corresponds to $\pm 2\sigma$, whereas the 95% confidence interval corresponds to $\pm 1.96\sigma$.

3.7 Consider the replicated impedance measurements presented in Table 3.4 for a 5 cm^2 polymer electrolyte membrane (PEM) fuel cell operating at a current of 1 A. The measurements were collected at a frequency of 1 Hz on the same cell using two different sets of instrumentation, the 850C provided by Scribner Associates and the FC350 provided by Gamry Instruments.

(a) Test the hypothesis that the impedance values obtained by the two instruments are not distinguishable.

(b) Test the hypothesis that the variances obtained by the two instruments are not distinguishable.

Solution 3.7.

(a) A test of the null hypothesis that the population mean with $\nu = n_x - 1$ degrees of freedom is equal to a specified value μ_0 is obtained in terms of

$$t = \frac{\overline{x} - \mu_0}{s_x / \sqrt{n_x}} \tag{3.12}$$

where n_x is the number of experimental data points and $s_x / \sqrt{n_x}$ represents the standard error for the data set. The corresponding expression for comparison of unpaired variates is given as

$$t = \frac{\overline{x}_1 - \overline{x}_2}{\sqrt{\dfrac{s_{x_1}^2}{n_{x_1}} + \dfrac{s_{x_2}^2}{n_{x_2}}}} \tag{3.13}$$

with $\nu = n_{x_1} + n_{x_2} - 2$ degrees of freedom. The value of the t-statistic for a comparison of the real parts of the impedance is t $= 1.384$, and the value for a comparison of the imaginary part is t $= 1.384$. The null hypothesis is that the sample means are equal. The p-value is the probability of observing the given sample result under the assumption that the null hypothesis is true. In other words, p is the value of α that will yield the observed statistical test value. For the present problem, $p = 0.219$ for the real part of the impedance, and $p = 0.167$ for the imaginary part of the impedance. At $\alpha = 0.05$, the null hypothesis cannot be rejected, and the sample means are equal.

(b) The value for F is obtained from the experimentally determined variances as

$$\mathrm{F} = s_1^2 / s_2^2 \tag{3.14}$$

such that F > 1, i.e., $s_1^2 > s_2^2$. The null hypothesis is that the two variances are from the same population, i.e., they are not statistically different. The value of the F-statistic for a comparison of the real parts of the impedance is $F = 1.384$, and the value for a comparison of the imaginary part is $F = 1.384$. The corresponding probability of observing the given sample result under the assumption that the null hypothesis is true is $p = 0.677$ for the real part of the impedance and $p = 0.484$ for the imaginary part.

The present results suggest that the data obtained using the 850C provided by Scribner Associates and the FC350 provided by Gamry Instruments cannot be distinguished by statistical tests.

Table 3.5: Replicated impedance collected at a frequency of 10 Hz on the same PEM fuel cell but using two flow channels.

	Serpentine		Interdigitated	
Replicate	Z_r, Ω	Z_j, Ω	Z_r, Ω	Z_j, Ω
1	0.055822	−0.048797	0.073113	−0.039505
2	0.0583180	−0.053851	0.074073	−0.04012
3	0.05861	−0.05494	0.074244	−0.040369
4	0.058842	−0.055843	0.074362	−0.040548

3.8 Consider the replicated impedance measurements presented in Table 3.5 for a 5 cm^2 PEM fuel cell operating at a current of 1 A. The measurements were collected at a frequency of 10 Hz on the same cell using a serpentine channel and an interdigitated channel that is believed to be more efficient but also more susceptible to flooding.

(a) Test the hypothesis that the impedance values obtained by the two instruments are not distinguishable.

(b) Test the hypothesis that the variances obtained by the two instruments are not distinguishable.

Solution 3.8.

(a) Application of equation (3.13) yields a value for a comparison of the real parts of the impedance of t = 21.23, and the value for a comparison of the imaginary part is t = 8.315. The corresponding probabilities that the result could be obtained under the null hypothesis are $p = 3.12 \times 10^{-5}$ for the real part of the impedance and $p = 0.00312$ for the imaginary part of the impedance. Thus, null hypothesis that the data come from the same distribution can be rejected.

(b) Application of equation (3.14) yields $F = 6.054$ for the real part of the impedance and $F = 47.75$ for the imaginary part. The corresponding probability of observing the given sample result under the assumption that the null hypothesis is true is $p = 0.173$ for the real part of the impedance and $p = 0.00991$ for the imaginary part.

The present results suggest that the impedance measured for serpentine and interdigitated flow channels can be distinguished by statistical tests. The variances of the measurements are more equivocal. The null hypothesis can be rejected for the imaginary part of the impedance, but cannot be rejected for the real part.

Chapter 4

Electrical Circuits

Transfer function approaches are general and can be applied to a large variety of electrical, mechanical, and optical systems. Electrochemists take advantage of this similarity by comparing the behavior of electrochemical systems to that of known electrical circuits. The problems presented in this chapter explore properties of electrical circuits. Problems 5.1–5.9 are based on problems taken from Orazem and Tribollet.[1]

Problems

4.1 Show that

$$\mathrm{Re}\left\{\widetilde{V}\exp(\mathrm{j}\omega t)\right\} = L\mathrm{Re}\left\{\mathrm{j}\omega\widetilde{I}\exp(\mathrm{j}\omega t)\right\} \tag{4.1}$$

for an inductor can lead to

$$\mathrm{Re}\left\{\widetilde{V}\exp(\mathrm{j}\omega t)\right\} = L\mathrm{Re}\left\{\omega\widetilde{I}\exp\left(\mathrm{j}(\omega t + \pi/2)\right)\right\} \tag{4.2}$$

Show that this result demonstrates that current and potential signals are out-of-phase for an inductor.

Solution 4.1. For an inductor:

$$\mathrm{Re}\{\widetilde{V}\exp\left(\mathrm{j}\omega t\right)\} = L\mathrm{Re}\{\mathrm{j}\omega\widetilde{I}\exp\left(\mathrm{j}\omega t\right)\} \tag{4.3}$$

As

$$\exp \mathrm{j}\pi/2 = \cos\frac{\pi}{2} + \mathrm{j}\sin\frac{\pi}{2} = \mathrm{j} \tag{4.4}$$

$$\begin{aligned}\mathrm{Re}\{\widetilde{V}\exp\left(\mathrm{j}\omega t\right)\} &= L\mathrm{Re}\{\omega\widetilde{I}\exp\left(\mathrm{j}\pi/2\right)\exp\left(\mathrm{j}\omega t\right)\} &(4.5)\\ &= L\mathrm{Re}\{\omega\widetilde{I}\exp\left(\mathrm{j}\left(\omega t + \pi/2\right)\right)\} &(4.6)\end{aligned}$$

The expression above shows that the potential is out of phase with the current.

4.2 Derive

$$\mathrm{Re}\left\{\widetilde{I}\exp(\mathrm{j}\omega t)\right\} = \mathrm{CRe}\left\{\omega\widetilde{V}\exp\left(\mathrm{j}(\omega t + \pi/2)\right)\right\} \tag{4.7}$$

and verify that equation (4.7) demonstrates that current and potential signals are out of phase in a capacitor.

Solution 4.2. For a capacitor,

$$\mathrm{Re}\{\widetilde{I}\exp(\mathrm{j}\omega t)\} = \mathrm{CRe}\{\mathrm{j}\omega\widetilde{V}\exp(\mathrm{j}\omega t)\} \tag{4.8}$$

Substitution for j into equation (4.8) yields

$$\begin{aligned} \mathrm{Re}\{\widetilde{I}\exp(\mathrm{j}\omega t)\} &= \mathrm{CRe}\{\omega\widetilde{V}\exp(\mathrm{j}\pi/2)\exp(\mathrm{j}\omega t)\} && (4.9) \\ &= \mathrm{CRe}\{\omega\widetilde{V}\exp(\mathrm{j}(\omega t + \pi/2))\} && (4.10) \end{aligned}$$

This expression shows that the potential is out of phase with the current.

4.3 With the aid of Table A.17, show that the units of impedance given in

$$Z_{\mathrm{C}} = \frac{1}{\mathrm{j}\omega C} \tag{4.11}$$

and

$$Z_{\mathrm{L}} = \mathrm{j}\omega L \tag{4.12}$$

are ohms.

Solution 4.3. The units of angular frequency are

$$\omega\ [=]\ \frac{1}{\mathrm{s}} \tag{4.13}$$

and the units of capacitance are

$$C\ [=]\ \mathrm{F}\ [=]\ \frac{\mathrm{s}}{\Omega} \tag{4.14}$$

Thus, the units of the impedance given in equation (4.11) are

$$Z_{\mathrm{C}}\ [=]\ [\mathrm{s}]\left[\frac{\Omega}{\mathrm{s}}\right]\ [=]\ \Omega \tag{4.15}$$

In this presentation, the symbol [=] is used to represent the dimensions associated with a given quantity.

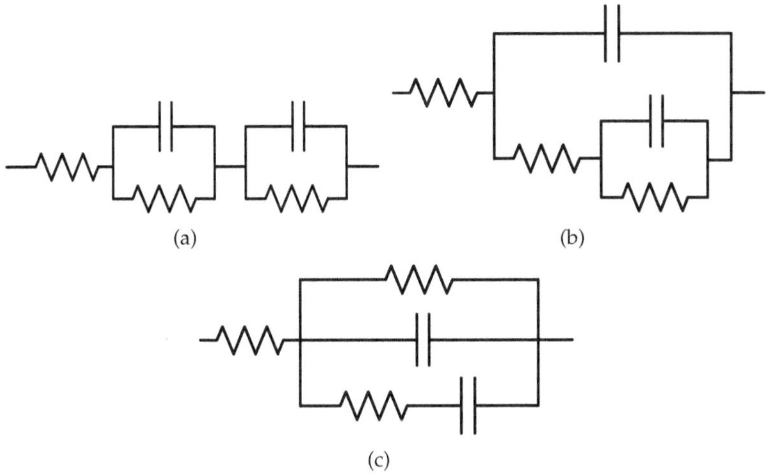

Figure 4.1: Three mathematically equivalent electrical circuits with two time constants: (a) serially connected Voigt elements typically used to model energy storage or conversion devices; (b) a nested geometry commonly used for modeling a porous corrosion protective coating; and (c) a Maxwell geometry commonly used for modeling multiple reaction pathways. Taken from Orazem and Tribollet[1] with permission of John Wiley & Sons.

The units of inductance are

$$L \ [=] \ \mathrm{H} \ [=] \ \mathrm{s}\Omega \tag{4.16}$$

Thus, the units of the impedance given in equation (4.12) are

$$Z_{\mathrm{L}} \ [=] \ \left[\frac{1}{\mathrm{s}}\right] \ [\mathrm{s}\Omega] \ [=] \ \Omega \tag{4.17}$$

Note that the imaginary number $\mathrm{j} = \sqrt{-1}$ is dimensionless.

4.4 Develop an expression for the impedance response of the circuit presented in Figure 4.1(a).

Solution 4.4. The circuit presented in Figure 4.1(a) in the text may be represented as shown in Figure 4.2, where dashed lines are used to represent combined units. The impedance may be expressed as

$$Z = R_1 + Z_2 + Z_3 \tag{4.18}$$

The impedance of Z_2 and Z_3 may be expressed in the form

$$Z_2 = \frac{1}{1/R_2 + \mathrm{j}\omega C_2} \tag{4.19}$$

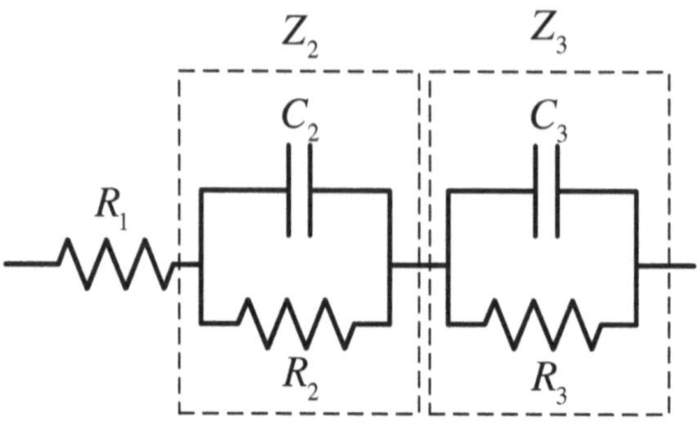

Figure 4.2: The circuit presented in Figure 4.1(a) in the text.

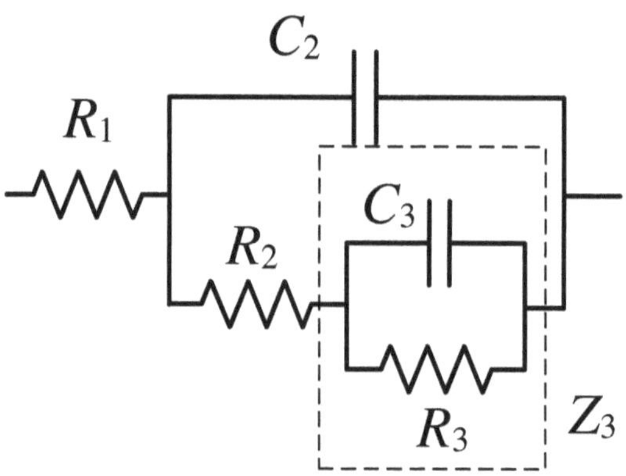

Figure 4.3: The circuit presented as Figure 4.1(b) in the text.

or

$$Z_2 = \frac{R_2}{1 + j\omega R_2 C_2} \tag{4.20}$$

Thus

$$Z = R_1 + \frac{R_2}{1 + j\omega R_2 C_2} + \frac{R_3}{1 + j\omega R_3 C_3} \tag{4.21}$$

4.5 Develop an expression for the impedance response of the circuit presented in Figure 4.1(b).

Solution 4.5. The circuit presented as Figure 4.1(b) in the text may be represented as shown in Figure 4.3, where dashed lines are used to represent a combined unit. The impedance may be expressed as

$$Z = R_1 + \frac{1}{j\omega C_2 + 1/\left(R_2 + Z_2\right)} \tag{4.22}$$

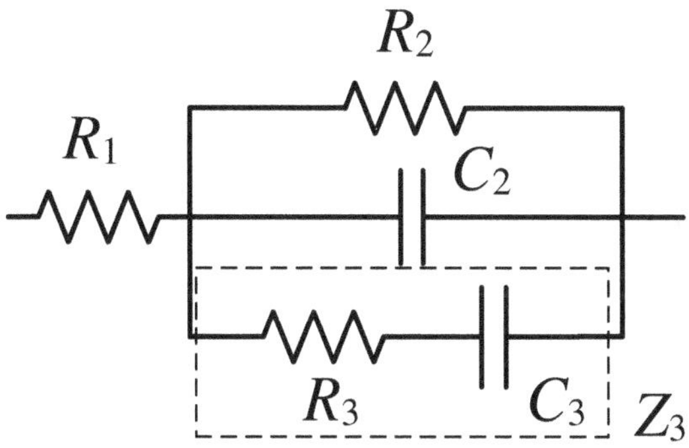

Figure 4.4: The circuit presented in Figure 4.1(c) in the text.

The impedance of Z_3 may be expressed in the form

$$Z_3 = \frac{1}{1/R_3 + j\omega C_3} \tag{4.23}$$

or

$$Z_3 = \frac{R_3}{1 + j\omega R_3 C_3} \tag{4.24}$$

Thus

$$Z = R_1 + \frac{1}{j\omega C_2 + \dfrac{1 + j\omega R_3 C_3}{R_2 + R_3 + j\omega R_3 C_3}} \tag{4.25}$$

Equation (4.25) may be rearranged as

$$Z = R_1 + \frac{R_2 + R_3 + j\omega R_3 C_3}{1 + j\omega R_3 C_3 + j\omega C_2 \left(R_2 + R_3 + j\omega R_3 C_3\right)} \tag{4.26}$$

4.6 Develop an expression for the impedance response of the circuit presented in Figure 4.1(c).

Solution 4.6. The circuit presented in Figure 4.1(c) in the text may be represented as shown in Figure 4.4, where dashed lines are used to represent a combined unit. The impedance may be expressed as

$$Z = R_1 + \frac{1}{1/R_2 + j\omega C_2 + 1/Z_3} \tag{4.27}$$

where

$$Z_3 = R_3 + \frac{1}{j\omega C_3} \tag{4.28}$$

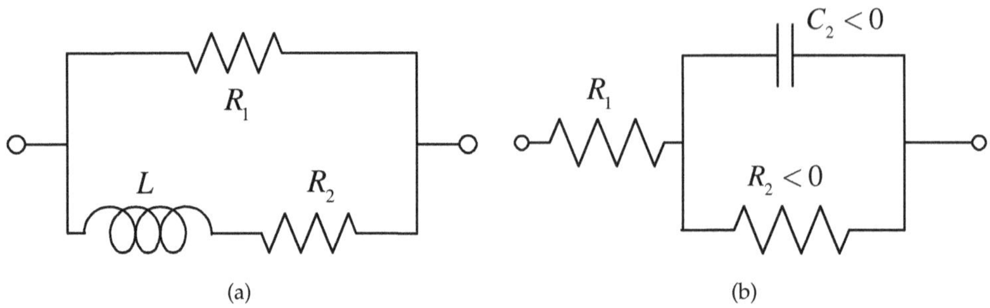

Figure 4.5: Electrical circuits resulting in an inductive loop: (a) circuit with an inductor and (b) circuit with a negative resistance in parallel with a negative capacitance. Taken from Orazem and Tribollet[1] with permission of John Wiley & Sons.

Thus

$$Z = R_1 + \frac{1}{1/R_2 + \mathrm{j}\omega C_2 + \dfrac{\mathrm{j}\omega C_3}{1 + \mathrm{j}\omega R_3 C_3}} \tag{4.29}$$

Equation (4.29) may be written as

$$Z = R_1 + \frac{R_2\left(1 + \mathrm{j}\omega R_3 C_3\right)}{\left(1 + \mathrm{j}\omega R_2 C_2\right)\left(1 + \mathrm{j}\omega R_3 C_3\right) + \mathrm{j}\omega R_2 C_3} \tag{4.30}$$

4.7 The use of inductors to model low-frequency inductive loops in an impedance response is somewhat controversial. Such a circuit is presented in Figure 4.5(a). Demonstrate that the circuit presented in Figure 4.5(b) with negative R and C in the nested element can be mathematically equivalent to a circuit containing an inductor.

Solution 4.7. The impedance of the circuit in Figure 4.5(a) can be expressed as

$$Z = \frac{1}{1/R_1 + 1/\left(R_2 + \mathrm{j}\omega L\right)} \tag{4.31}$$

or

$$Z = \frac{R_1\left(R_2 + \mathrm{j}\omega L\right)}{R_1 + R_2 + \mathrm{j}\omega L} \tag{4.32}$$

multiplication and division by the complex conjugate of the denominator yields

$$Z_{\mathrm{r}} = \frac{R_1 R_2\left(R_1 + R_2\right) + \omega^2 R_1 L^2}{\left(R_1 + R_2\right)^2 + \omega^2 L^2} \tag{4.33}$$

and

$$Z_{\mathrm{j}} = \frac{\omega R_1^2 L}{\left(R_1 + R_2\right)^2 + \omega^2 L^2} \tag{4.34}$$

where the imaginary impedance is a positive number. In the high-frequency limit, the real part of the impedance tends toward

$$\lim_{f \to \infty} Z_r = R_1 \tag{4.35}$$

and the low-frequency limit is

$$\lim_{f \to 0} Z_r = \frac{R_1 R_2}{R_1 + R_2} \tag{4.36}$$

The imaginary part has a positive value at all frequencies and tends toward zero at high and low frequencies.

The impedance of the circuit in Figure 4.5(b) can be expressed as

$$Z = R_1 + \frac{R_2}{1 + j\omega R_2 C_2} \tag{4.37}$$

with real and imaginary parts given as

$$Z_r = R_1 + \frac{R_2}{1 + \omega^2 R_2^2 C_2^2} \tag{4.38}$$

and

$$Z_j = -\frac{\omega R_2^2 C_2}{1 + \omega^2 R_2^2 C_2^2} \tag{4.39}$$

respectively. For $R_2 < 0$ and $C_2 < 0$, the product $\tau_{RC} = R_2 C_2 > 0$ and the imaginary impedance is a positive number for all frequencies, similar to the impedance for the circuit given in Figure 4.5(a). In the low-frequency limit, the real part of the impedance tends toward

$$\lim_{f \to 0} Z_r = R_1 + R_2 \tag{4.40}$$

where $R_2 < 0$. Thus, to be consistent with equation (4.36), $R_1 + R_2 > 0$. In the high-frequency limit, the real impedance tends toward

$$\lim_{f \to \infty} Z = R_1 \tag{4.41}$$

The imaginary part has a positive value at all frequencies and tends toward zero at high and low frequencies. Thus, the circuit presented in Figure 4.5(b) with negative R and C can be mathematically equivalent to a circuit containing an inductor.

A simpler way to demonstrate the equivalence of the two circuits is to fit one circuit to the other. The calculated impedance response of the circuit in Figure 4.5(a) is presented in Figure 4.6 with $R_1 = 10\ \Omega$, $R_2 = 10\ \Omega$, and $L_2 = 1$ H. This spectrum was fit by the circuit in Figure 4.5(b) to yield $R_1 = 10\ \Omega$, $R_2 = -5\ \Omega$, and $C = -0.01$ F. The measurement model software[16] was used to perform the regression. The high-frequency limit given by equation (4.35) is 10 Ω, in agreement with equation (4.41). The low-frequency limit given by equation (4.36) is 5 Ω, in agreement with equation (4.40). The results shown in Figure 4.6 provide a demonstration of the mathematical equivalence between circuits shown in Figures 4.5(a) and 4.5(b).

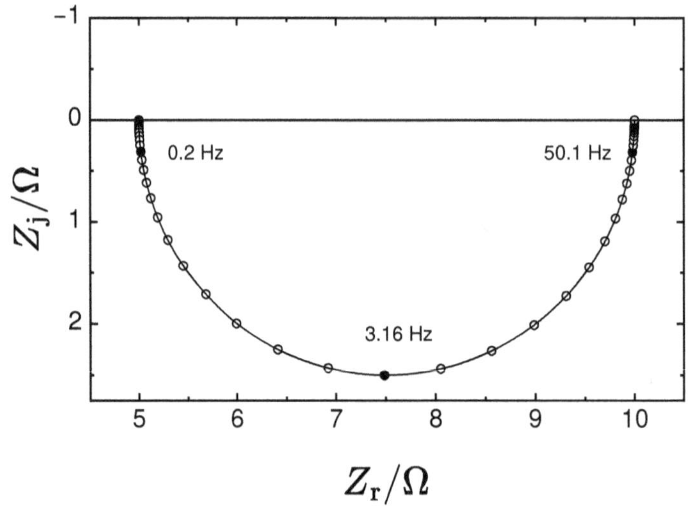

Figure 4.6: Calculated impedance response of the circuit in Figure 4.5(a) given by equation (4.32) with $R_1 = 10\ \Omega$, $R_2 = 10\ \Omega$, and $L_2 = 1$ H. The line represents the fit of equation (4.37) yielding $R_1 = 10\ \Omega$, $R_2 = -5\ \Omega$, and $C = -0.01$ F.

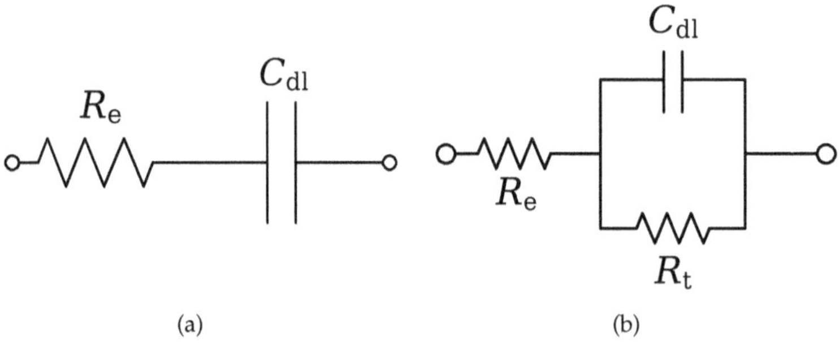

Figure 4.7: Electrical circuit consisting of a) a solution resistance in series with a capacitor and b) a solution resistance in series with a Voigt element. Taken from Orazem and Tribollet[1] with permission of John Wiley & Sons.

Table 4.1: Sample results from the spreadsheet calculations performed to prepare the solution to Problem 4.8.

f, Hz	Z_r, Ω	Z_j, Ω	$\|Z\|$, Ω	φ, degree
0.01	10	−795774.7155	795774.7155	−89.99928
0.1	10	−79577.47155	79577.47217	−89.9928
1	10	−7957.74715	7957.75344	−89.928
10	10	−795.77472	795.83754	−89.28004
100	10	−79.57747	80.20333	−82.83754
10^3	10	−7.95775	12.7799	−38.51189
10^4	10	−0.79577	10.03161	−4.54987
10^5	10	−0.07958	10.00032	−0.45594

4.8 Use a spreadsheet program to plot the impedance response of the circuit presented in Figure 4.7(a) using $R_e = 10\ \Omega$ and $C_{dl} = 20\ \mu F$.

(a) Plot the real and imaginary parts of the impedance as functions of frequency.

(b) Plot the results in Bode format (magnitude and phase angle as functions of frequency).

Solution 4.8. The impedance, given by

$$Z = R_e - j\frac{1}{\omega C_{dl}} \tag{4.42}$$

was input into a spreadsheet. Sample results are shown in Table 4.1. The results for the impedance are presented in Figure 4.8(a). The real part is independent of frequency, and the imaginary part tends toward $-\infty$ according to $1/\omega$ as frequency tends toward zero. The scale for the imaginary impedance was truncated in Figure 4.8(a) to allow comparison to the results from Problem 4.9. The corresponding Bode representation is shown in Figure 4.8(b). The phase angle tends toward $-90°$ at low frequency, where the magnitude tends toward ∞. At high frequency, the phase angle tends toward zero for both Problems 4.8 and 4.9.

4.9 Use a spreadsheet program to plot the impedance response of the circuit presented as Figure 4.7(b) using $R_e = 10\ \Omega$, $C_{dl} = 20\ \mu F$, and $R_t = 100\ \Omega$.

(a) Plot the real and imaginary parts of the impedance as functions of frequency.

(b) Plot the results in Bode format (magnitude and phase angle as functions of frequency).

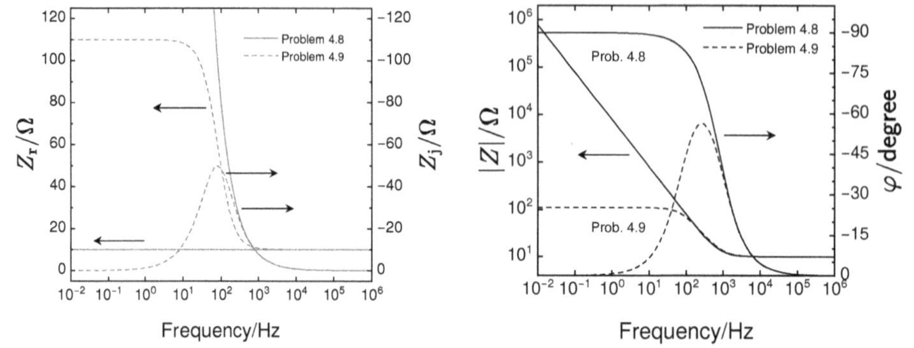

(a) −0.2 V (b) −0.3 V

Figure 4.8: Impedance response for the circuits presented in Problems 4.8 and 4.9: (a) real and imaginary impedance as functions of frequency and (b) Bode representation, i.e., magnitude and phase as functions of frequency.

Table 4.2: Sample results from the spreadsheet calculations performed to prepare the solution to Problem 4.9.

f, Hz	Z_r, Ω	Z_j, Ω	$\|Z\|$, Ω	φ, degree
0.01	110	−0.01257	110	−0.00655
0.1	109.99984	−0.12566	109.99991	−0.06545
1	109.98421	−1.25644	109.99139	−0.65451
10	108.44541	−12.37102	109.14875	−6.50794
100	48.77266	−48.72317	68.93997	−44.97091
10^3	10.62927	−7.90767	13.24812	−36.64749
10^4	10.00633	−0.79572	10.03792	−4.54671
10^5	10.00006	−0.07958	10.00038	−0.45593

Solution 4.9. The impedance given by

$$Z = R_e + \frac{R_t}{1 + j\omega R_t C_{dl}} \tag{4.43}$$

was input into a spreadsheet. Sample results in terms of $f = \omega/2\pi$ are shown in Table 4.2. The results are presented in Figure 4.8. The real part tends toward 110 Ω at low frequency and 10 Ω at high frequency. The imaginary part tends toward zero at high frequency and at low frequency and has a maximum magnitude at $f_c = 1/(2\pi R_t C_{dl}) =$ 79.6 Hz. The magnitude tends toward 110 Ω at low frequency and 10 Ω at high frequency. The phase angle tends toward zero at high frequency and at low frequency. The peak value is not seen at $f_c = 79.6$ Hz due to the influence of the ohmic resistance in the calculation of the phase angle.

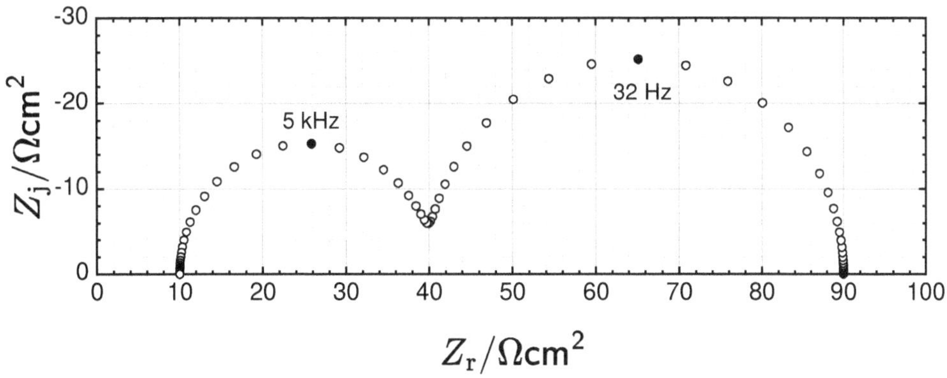

Figure 4.9: Impedance data to be used for Problem 4.10.

4.10 Nonlinear regression using the Levenberg–Marquart algorithm is very sensitive to the initial guesses for parameter values. Use the plot in Figure 4.9 to suggest values for a model of the form

$$Z = R_0 + \frac{R_1}{1 + j2\pi f R_1 C_1} + \frac{R_2}{1 + j2\pi f R_2 C_2} \tag{4.44}$$

and find values for R_0, R_1, C_1, R_2, and C_2. Please show your work as appropriate.

Solution 4.10. The value of R_0 is obtained from the high-frequency asymptote for the real part of the impedance, i.e.,

$$R_0 = \lim_{f \to \infty} Z_r = 10\ \Omega\text{cm}^2 \tag{4.45}$$

The value of $R_0 + R_1 + R_2$ is obtained from the zero-frequency asymptote for the real part of the impedance, i.e.,

$$R_0 + R_1 + R_2 = \lim_{f \to 0} Z_r = 90\ \Omega\text{cm}^2 \tag{4.46}$$

The value of R_1 can be obtained by estimating the width of the high-frequency loop. If the extrapolation of the semicircle yields 40 Ωcm^2, $R_1 = 40 - R_0 = 30\ \Omega$cm^2. The corresponding capacitance C_1 can be obtained from the characteristic frequency of the high frequency loop, i.e.,

$$C_1 = \frac{1}{2\pi R_1 f_{c,1}} \tag{4.47}$$

Table 4.3: Paramater values extracted from Figure 4.9.

Parameter	Value	Units
R_0	10	Ωcm^2
R_1	30	Ωcm^2
C_1	1	$\mu F/cm^2$
R_2	50	Ωcm^2
C_2	100	$\mu F/cm^2$

where $f_{c,1} = 5$ kHz. The value of R_2 is obtained from

$$R_2 = \lim_{f \to 0} Z_r - R_0 - R_1 = 90 - 10 - 30 = 40\ \Omega cm^2 \tag{4.48}$$

The value of C_2 is obtained by application of equation (4.47), where $f_{c,2} = 32$ Hz. The results are presented in Table 4.3.

Chapter 5

Electrochemistry

The problems presented in this chapter cover aspects of electrochemistry associated with impedance spectroscopy. Problems 5.1–5.9 are based on problems taken from Orazem and Tribollet.[1]

Problems

5.1 Estimate the electrolyte resistance for a 0.25 cm radius disk electrode in 0.1 M NaCl solution at 25 °C. Typical values of diffusion coefficients for ions at infinite dilution in water at 25 °C are presented in Table A.13.

Solution 5.1. The conductivity of a 0.1 M NaCl solution at 25 °C is given by

$$\kappa = \frac{\mathrm{F}^2}{\mathrm{R}T}\sum_{\mathrm{i}} z_{\mathrm{i}}^2 D_{\mathrm{i}} c_{\mathrm{i}} \tag{5.1}$$

where, from Table A.13, $D_{\mathrm{Na}^+} = 1.334 \times 10^{-5}$ cm^2/s and $D_{\mathrm{Cl}^-} = 2.032 \times 10^{-5}$ cm^2/s. The conductivity is $\kappa = 1.26 \times 10^{-2}$ S/cm. The ohmic resistance for a disk electrode is given by

$$R_{\mathrm{e}} = \frac{1}{4\kappa r_0} \tag{5.2}$$

which yields $R_{\mathrm{e}} = 79\ \Omega$. Multiplication by the disk area $A = \pi r_0^2$ yields $R_{\mathrm{e}} = 15.5\ \Omega\mathrm{cm}^2$.

5.2 Obtain a relationship for the charge-transfer resistance in the Tafel regime.

Solution 5.2. An anodic current is expressed for the Tafel regime as

$$\bar{i} = K \exp\left(b\overline{V}\right) \tag{5.3}$$

where $\overline{V}$ is the steady-state interfacial potential. The phasor $\widetilde{i}_{\mathrm{F}}$ is obtained from

$$\widetilde{i}_{\mathrm{F}} = \left.\frac{\partial i}{\partial V}\right|_{\overline{V}} \widetilde{V} = bK\exp\left(b\overline{V}\right)\widetilde{V} \tag{5.4}$$

such that $Z_{\mathrm{F}} = \widetilde{V}/\widetilde{i}_{\mathrm{F}}$. Following Ohm's law, $V = IR$ (or $R = V/I$), the expression for charge-transfer resistance is given by

$$R_{\mathrm{t}} = \frac{1}{bK\exp\left(b\overline{V}\right)} \tag{5.5}$$

From equation (5.3),

$$\overline{V} = \frac{1}{b}\ln\left(\frac{\overline{i}}{K}\right) \tag{5.6}$$

Introduction of equation (5.6) to equation (5.5) yields

$$R_{\mathrm{t}} = \frac{1}{b\overline{i}} \tag{5.7}$$

This important result provides the relationship between charge-transfer resistance obtained from impedance, the steady-state current density, and the Tafel slope expressed as $\beta = 2.303 \times 10^3/b$.

5.3 Verify the Koutecky–Levich equation, given as

$$i^{-1} = i_{\mathrm{lim}}^{-1} + i_{\mathrm{k}}^{-1} \tag{5.8}$$

Solution 5.3. Verification of the general expression given as equation (5.8) is demonstrated here for a cathodic reaction. The faradaic current can be expressed as

$$i = -Kc_{\mathrm{i}}(0)\exp\left(-bV\right) \tag{5.9}$$

where the concentration of the reacting species at the electrode surface $c_{\mathrm{i}}(0)$ is related to the faradaic current by Fick's first law, i.e.,

$$i = -n\mathrm{F}D_{\mathrm{i}}\left.\frac{\mathrm{d}c_{\mathrm{i}}}{\mathrm{d}y}\right|_{y=0} \tag{5.10}$$

Under assumption of a Nernst stagnant diffusion layer, which can approximate more sophisticated configurations such as those associated with a rotating disk electrode,

$$i = -n\mathrm{F}D_{\mathrm{i}}\frac{c_{\mathrm{i}}(\infty) - c_{\mathrm{i}}(0)}{\delta_{\mathrm{i}}} \tag{5.11}$$

where δ_i is the diffusion-layer thickness and $c_i(\infty)$ is the uniform concentration in the bulk region outside the diffusion layer.

Equations (5.9) and (5.11) can be manipulated to yield the concentration at the electrode surface, i.e.,

$$c_i(0) = \frac{nFD_ic_i(\infty)/\delta_i}{K\exp\left(-b\overline{V}\right) + nFD_i/\delta_i} \tag{5.12}$$

and equation (5.9) becomes

$$i = -\frac{KnFD_ic_i(\infty)\exp\left(-bV\right)/\delta_i}{K\exp\left(-b\overline{V}\right) + nFD_i/\delta_i} \tag{5.13}$$

Multiplication and division by $c_i(\infty)$ yields

$$i = \frac{i_k i_{lim}}{i_k + i_{lim}} \tag{5.14}$$

where

$$i_k = -Kc_i(\infty)\exp\left(-bV\right) \tag{5.15}$$

and

$$i_{lim} = -\frac{nFD_ic_i(\infty)}{\delta_i} \tag{5.16}$$

Equation (5.14) is mathematically equivalent to equation (5.8).

5.4 Show that the concentration of a reactant at the surface of an electrode can be expressed as a function of current density as

$$\frac{c_i(0)}{c_i(\infty)} = 1 - \frac{i}{i_{lim}} \tag{5.17}$$

Solution 5.4. The starting point is

$$i = -nFD_i\frac{c_i(\infty) - c_i(0)}{\delta_i} \tag{5.18}$$

At the mass-transfer-limited condition, $c_i(0)$ tends toward zero and

$$i_{lim} = -nFD_i\frac{c_i(\infty)}{\delta_i} \tag{5.19}$$

Division of equation (5.18) by equation (5.19) yields

$$\frac{i}{i_{lim}} = \frac{c_i(\infty) - c_i(0)}{c_i(\infty)} \tag{5.20}$$

or

$$\frac{c_i(0)}{c_i(\infty)} = 1 - \frac{i}{i_{lim}} \tag{5.21}$$

5.5 Following equation (5.17), use a spreadsheet program to plot the current density as a function of surface overpotential for $K_c c_i(\infty) = 1$ mA/cm^2, $b_c = 20$ V^{-1}, and i_{lim} equal to -0.1, -1.0, and -10 mA/cm^2.

Solution 5.5. The current density is given by

$$i = -K_c c_i(0)\exp(-b_c V) \tag{5.22}$$

where

$$\frac{c_i(0)}{c_i(\infty)} = 1 - \frac{i}{i_{lim}} \tag{5.23}$$

Thus,

$$i = -K_c c_i(\infty)\left(1 - \frac{i}{i_{lim}}\right)\exp(-b_c V) \tag{5.24}$$

Equation (5.24) may be expressed as

$$i = i_k - i_k\frac{i}{i_{lim}} \tag{5.25}$$

where

$$i_k = -K_c c_i(\infty)\exp(-b_c V) \tag{5.26}$$

An expression for current density is obtained as

$$i = \frac{i_k}{1 + i_k/i_{lim}} \tag{5.27}$$

The resulting spreadsheet is shown in Table 5.1, and the plot is shown in Figure 5.1.

5.6 Use a Taylor's series expansion for the faradaic current about the equilibrium potential to obtain a relationship for the charge-transfer resistance in terms of the parameter $J = (\alpha_a + \alpha_c) i_0 F r_0 / \kappa RT$, where α_a and α_c are apparent transfer coefficients.

Solution 5.6. The Butler–Volmer equation can be expressed as

$$i = i_0\left\{\exp(b_a\eta_s) - \exp(-b_c\eta_s)\right\} \tag{5.28}$$

where η_s represents the departure from the equilibrium potential. A Taylor series expansion, neglecting terms of order of η_s^2 and higher, provides

$$i = i_0\left\{1 + b_a\eta_s - (1 - b_c\eta_s)\right\} \tag{5.29}$$

Table 5.1: Results from spreadsheet showing the calculations performed to plot the impedance response shown in Figure 5.1.

	i_{lim} =	-0.1 mA/cm^2	-1 mA/cm^2	-10 mA/cm^2
V, V	i_k, mA/cm^2	i_1, mA/cm^2	i_2, mA/cm^2	i_3, mA/cm^2
0	−1.00000	−0.09091	−0.50000	−0.90909
−0.01	−1.22140	−0.09243	−0.54983	−1.08846
−0.02	−1.49182	−0.09372	−0.59869	−1.29816
−0.03	−1.82212	−0.09480	−0.64566	−1.54128
−0.04	−2.22554	−0.09570	−0.68997	−1.82040
−0.05	−2.71828	−0.09645	−0.73106	−2.13730
−0.06	−3.32012	−0.09708	−0.76852	−2.49256
−0.07	−4.05520	−0.09759	−0.80218	−2.88520
−0.08	−4.95303	−0.09802	−0.83202	−3.31239
−0.09	−6.04965	−0.09837	−0.85815	−3.76933
−0.1	−7.38906	−0.09866	−0.88080	−4.24926
−0.11	−9.02501	−0.09890	−0.90025	−4.74376
−0.12	−11.02318	−0.09910	−0.91683	−5.24334
−0.13	−13.46374	−0.09926	−0.93086	−5.73810
−0.14	−16.44465	−0.09940	−0.94268	−6.21852
−0.15	−20.08554	−0.09950	−0.95257	−6.67614

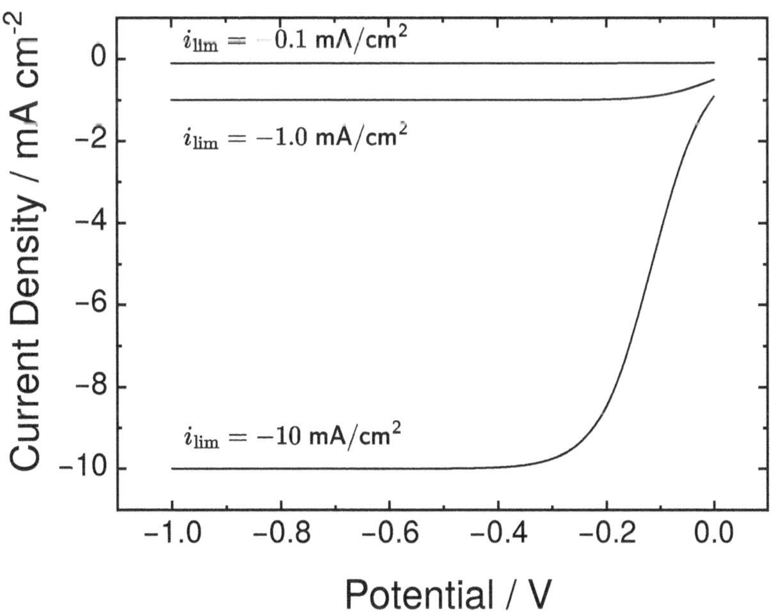

Figure 5.1: Current density calculated from equation (5.27) as a function of surface overpotential with mass-transfer-limited current density as a parameter.

or

$$i = i_0(b_a + b_c)\eta_s \tag{5.30}$$

Following the approach shown in Problem 5.2,

$$R_t = \frac{1}{i_0(b_a + b_c)} \tag{5.31}$$

The term $J = (\alpha_a + \alpha_c)i_0 F/\kappa RT$ can be expressed in terms of the coefficients b_a and b_c as $J = (b_a + b_c)i_0/\kappa$. Thus,

$$R_t = \frac{r_0}{J\kappa} \tag{5.32}$$

For a disk electrode, $R_e = \pi r_0/4\kappa$; thus, by replacing κ in equation (5.32),

$$R_t = \frac{4R_e}{J\pi} \tag{5.33}$$

or

$$J = \frac{4R_e}{\pi R_t} \tag{5.34}$$

The parameter J is seen to be proportional to the ratio of ohmic and charge transfer resistances.

5.7 Estimate the capacitance for an oxide layer on a steel surface that is 50 Å thick.

Solution 5.7. The capacitance of a film is expressed by

$$C = \frac{\varepsilon\varepsilon_0}{\delta} \tag{5.35}$$

The dielectric constant for oxides varies with composition. For hematite, the dielectric constant is $\varepsilon = 12$. For an oxide layer thickness of $\delta = 50$ Å, the capacitance is $C = 2.1\ \mu F/cm^2$.

5.8 Estimate the surface area for an electrode in a 0.1 M NaCl solution if the capacitance is measured to be 120 μF. What might be a reasonable confidence interval for this estimate?

Solution 5.8. The capacitance of a bare electrode C_0 is typically between 10-50 $\mu F/cm^2$. The area is estimated from

$$A = \frac{C_{meas}}{C_0} \tag{5.36}$$

If the measured capacitance is 120 $\mu F/cm^2$, the corresponding area could be from 2.4 to 12 cm^2.

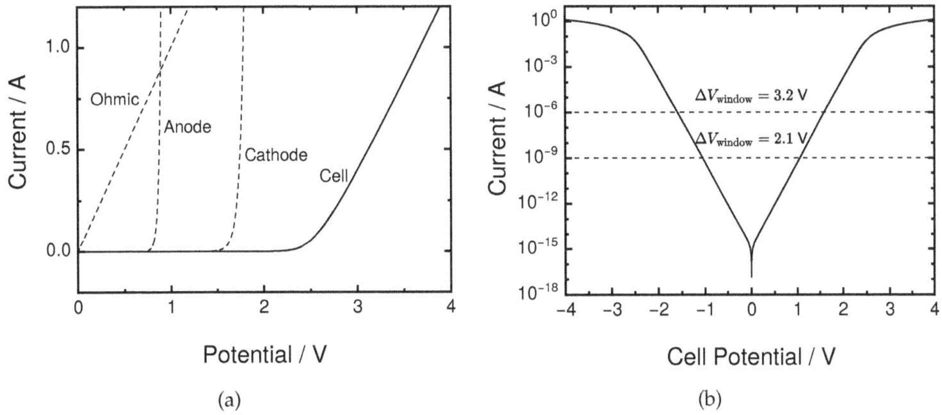

Figure 5.2: Polarization curve for oxygen and hydrogen evolution reactions on inert electrodes in an electrolyte that contains no dissolved oxygen or hydrogen gasses: (a) linear scale for current density and (b) logarithmic scale for the absolute value of current density. Taken from Orazem and Tribollet[1] with permission of John Wiley & Sons. Kinetic parameters were taken from McCafferty[17] and Conway.[18]

5.9 Use a spreadsheet program to reproduce Figure 5.2.

Solution 5.9. The anodic reaction is oxygen evolution

$$H_2O \rightarrow \frac{1}{2}O_2 + 2H^+ + 2e^- \tag{5.37}$$

the cathodic reaction is hydrogen evolution

$$2H_2O + 2e^- \rightarrow H_2 + 2OH^- \tag{5.38}$$

which applies for pH values greater than 7. The current associated with the corrosion and hydrogen evolution reactions may be expressed as

$$i = K^*_{O_2} \exp\left(b_{O_2}(V - V_{0,O_2})\right) - K^*_{H_2} \exp\left(-b_{H_2}(V - V_{0,H_2})\right) \tag{5.39}$$

The potential for which the net current is equal to zero is obtained by setting $i = 0$, yielding

$$K^*_{O_2} \exp\left(b_{O_2}(V_{\text{mixed}} - V_{0,O_2})\right) = K^*_{H_2} \exp\left(-b_{H_2}(V_{\text{mixed}} - V_{0,H_2})\right) \tag{5.40}$$

The potential at zero current may be obtained as

$$V_{\text{mixed}} = \frac{1}{b_{O_2} + b_{H_2}} \left(\ln\left(\frac{K^*_{H_2}}{K^*_{O_2}}\right) + b_{O_2} V_{0,O_2} + b_{H_2} V_{0,H_2} \right) \tag{5.41}$$

Table 5.2: Kinetic parameters for electrochemical reactions considered in Problem 5.9.

Anodic			Cathodic		
$K^*_{O_2}$ or i_{0,O_2}	1×10^{-7}	A/cm^2	$K^*_{H_2}$ or i_{0,H_2}	1×10^{-10}	A/cm^2
V_{O_2}	1.223	V	V_{H_2}	0	V
β_{O_2}	60	mV/decade	β_{H_2}	120	mV/decade
b_{O_2}	38.38	V^{-1}	b_{H_2}	19.19	V^{-1}

The apparent exchange current density i_{mixed} may be obtained from either the anodic or cathodic reaction at the potential V_{mixed}, i.e.,

$$i_{\text{mixed}} = K^{*\,\frac{b_{H_2}}{b_{O_2}+b_{H_2}}}_{O_2} K^{*\,\frac{b_{O_2}}{b_{O_2}+b_{H_2}}}_{H_2} \exp\left(\frac{b_{O_2}b_{H_2}}{b_{O_2}+b_{H_2}}\left(V_{0,H_2} - V_{0,O_2}\right)\right) \tag{5.42}$$

The same expression is obtained by starting from the cathodic part of equation (5.39).

The surface overpotential is defined to be

$$\eta_s = V - V_{\text{mixed}} \tag{5.43}$$

Thus, after algebraic manipulation, an expression for the current density is found as

$$i = i_{\text{mixed}}\left\{\exp(b_{O_2}\eta_s) - \exp(-b_{H_2}\eta_s)\right\} \tag{5.44}$$

Equation (5.44) takes the form of the Butler–Volmer equation, but the exchange current density i_0 is replaced by an apparent exchange current density i_{mixed}, the coefficients b_a and b_c that correspond to the anodic and cathodic parts of a given reaction, respectively, are replaced by coefficients b_{O_2} and b_{H_2} that correspond to different reactions, and the surface overpotential η_s represents a departure from the corrosion potential V_{mixed} rather than the equilibrium potential for a given reaction. Under the assumption that the kinetic parameters are those given in Table 5.2, equation (5.41) yields a value of $V_{\text{mixed}} = 0.706$ V(NHE), and equation (5.42) yields a value of $i_{\text{mixed}} = 1.4 \times 10^{-15}$ A/cm^2.

The cell potential can be expressed for a given current to be

$$V_{\text{cell}} = \eta_{\text{s,anode}} + i_{\text{avg}}R_e - \eta_{\text{s,cathode}} \tag{5.45}$$

where i_{avg} is the current density averaged over the cross-section of the cell. For the anode, the surface overpotential may be obtained from equation (5.44). At large overpotentials, the Tafel equation applies, and the surface overpotential may be expressed as

$$\eta_{\text{s,anode}} = \frac{1}{b_{O_2}} \ln\left(\frac{i_{\text{avg}}}{i_{\text{mixed}}}\right) \tag{5.46}$$

at smaller overpotentials, the influence of the cathodic reaction may not be neglected, and

$$\eta_{\text{s,anode},n+1} = \frac{1}{b_{O_2}} \ln\left(\frac{i_{\text{avg}}}{i_{\text{mixed}}} + \exp\left(-b_{H_2}\eta_{\text{s,anode},n}\right)\right) \tag{5.47}$$

which may be solved iteratively. The overpotential at the cathode is obtained, keeping in mind that the current at the anode must also flow through the cathode, from

$$\eta_{\text{s,cathode}} = -\frac{1}{b_{\text{H}_2}} \ln\left(\frac{i_{\text{avg}}}{i_{\text{mixed}}}\right) \tag{5.48}$$

Again, an iterative approach is required to obtain accurate solutions when the overpotential is small.

Figure 5.2(a) applies for positive overpotentials, for which the current density is positive. The ohmic contribution is displayed as a plot of total current density as a function of the ohmic potential drop. The anode contribution is the total current as a function of anode overpotential. The cathode contribution is a plot of the absolute value of the total current as a function of the absolute value of cathode overpotential. Figure 5.2(b) is obtained by plotting the absolute value of total current as a function of cell potential. The potential encompasses both positive and negative values.

Chapter 6

Electrochemical Instrumentation

Operational amplifiers provide the foundation for electrochemical instrumentation. The problems presented in this chapter explore the properties of operational amplifiers in the context of potentiostats, galvanostats, and impedance measurements. Problems 6.1–6.4 are based on problems presented by Orazem and Tribollet.[1]

Problems

6.1 Devise a circuit for adding a steady baseline potential, a ramped potential, and a sinusoidal perturbation.

Solution 6.1. A voltage adder may be represented as shown in Figure 6.1. The currents for the inputs are

$$I_1 = \frac{V_1}{R_1} \tag{6.1}$$

$$I_2 = \frac{V_2}{R_2} \tag{6.2}$$

and

$$I_3 = \frac{V_3}{R_3} \tag{6.3}$$

The current through R_0 can be expressed as

$$I_0 = -\frac{V_0}{R_0} = I_1 + I_2 + I_3 \tag{6.4}$$

Thus,

$$V_0 = -R_0\left(\frac{V_1}{R_1} + \frac{V_2}{R_2} + \frac{V_3}{R_3}\right) \tag{6.5}$$

If all resistor values are equal,

$$V_0 = -\left(V_1 + V_2 + V_3\right) \tag{6.6}$$

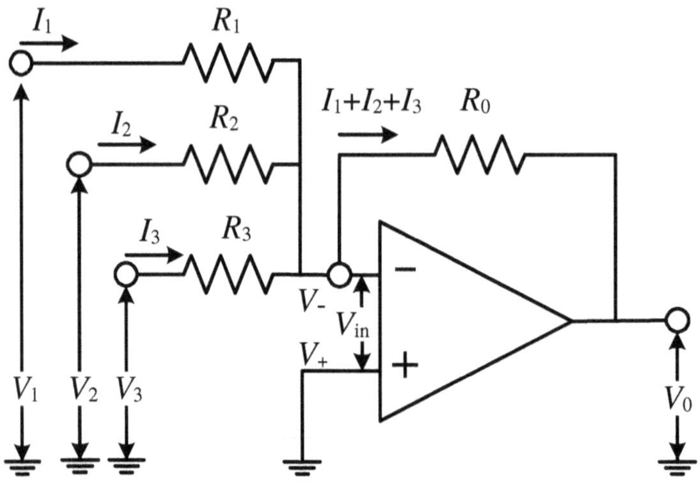

Figure 6.1: Schematic representation of a voltage adder for adding a steady baseline potential, a ramped potential, and a sinusoidal perturbation.

This result is a straightforward method for adding three potentials that could consist of a steady baseline potential, a ramped potential, and a sinusoidal perturbation.

6.2 Estimate the maximum error in the term $V - \Phi_{\text{ref}}$ if the operational amplifiers used in Figure 6.2 had an open-loop gain of 10^5 and power source of ± 10 V.

Solution 6.2. Operation of an operational amplifier within the linear regime requires that

$$|V_+ - V_-| < \frac{V_s}{A_{\text{op}}} \tag{6.7}$$

For the values given, the maximum error is 0.1 mV.

6.3 Show that the output potential for the noninverting amplifier shown in Figure 6.3 can be expressed as

$$V_0 = V_A \left(\frac{R_1 + R_2}{R_2} \right) \tag{6.8}$$

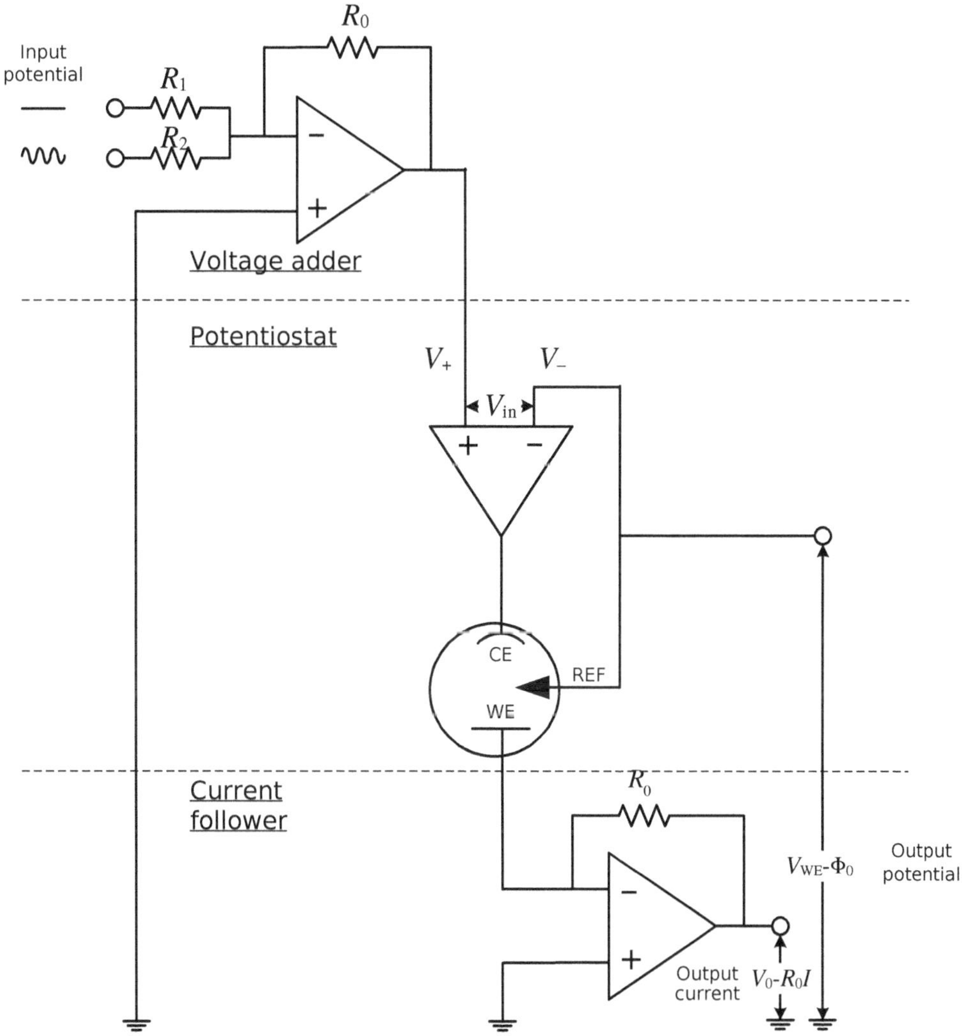

Figure 6.2: Schematic representation of a potentiostat for EIS measurement. Taken from Orazem and Tribollet[1] with permission of John Wiley & Sons.

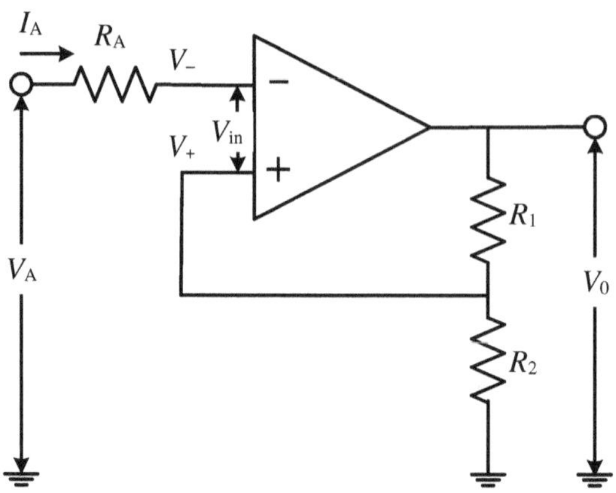

Figure 6.3: Schematic representation of a noninverting amplifier. Taken from Orazem and Tribollet[1] with permission of John Wiley & Sons.

Solution 6.3. The current through R_1 and R_2 can be expressed as

$$I_0 = \frac{V_0}{R_1 + R_2} \tag{6.9}$$

The current through R_2 can be expressed as

$$I_2 = \frac{V_+}{R_2} \tag{6.10}$$

The characteristic of the operational amplifier is that $V_+ = V_-$ and, as the input impedance is very large, $I_A = 0$ and $V_- = V_A$. Thus,

$$V_0 = V_A \frac{R_1 + R_2}{R_2} \tag{6.11}$$

This result may be compared to that of an ideal operational amplifier with negative feedback shown schematically in Figure 6.4.

6.4 Devise a system similar to that shown in Figure 6.2 in which impedance measurements are performed under galvanostatic regulation.

Solution 6.4. A simple circuit for a galvanostat is presented in Figure 6.5. The current passing through the working electrode is given by

$$I_{WE} = I_A = \frac{V_A - V_-}{R_A} \tag{6.12}$$

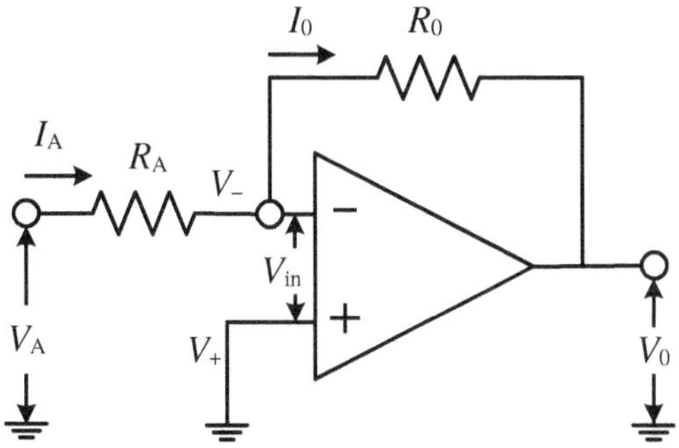

Figure 6.4: Schematic representation of a negative feedback inverting voltage amplifier. Taken from Orazem and Tribollet[1] with permission of John Wiley & Sons.

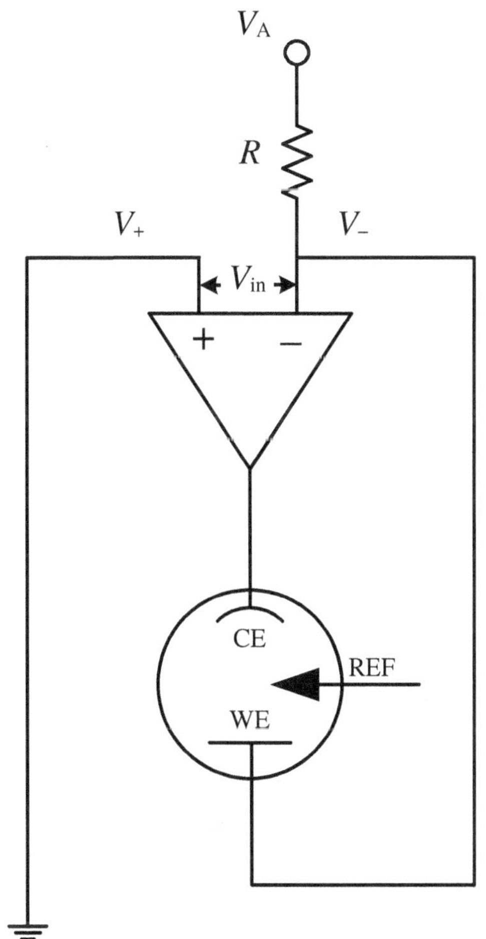

Figure 6.5: Schematic representation of a galvanostat. Taken from Orazem and Tribollet[1] with permission of John Wiley & Sons.

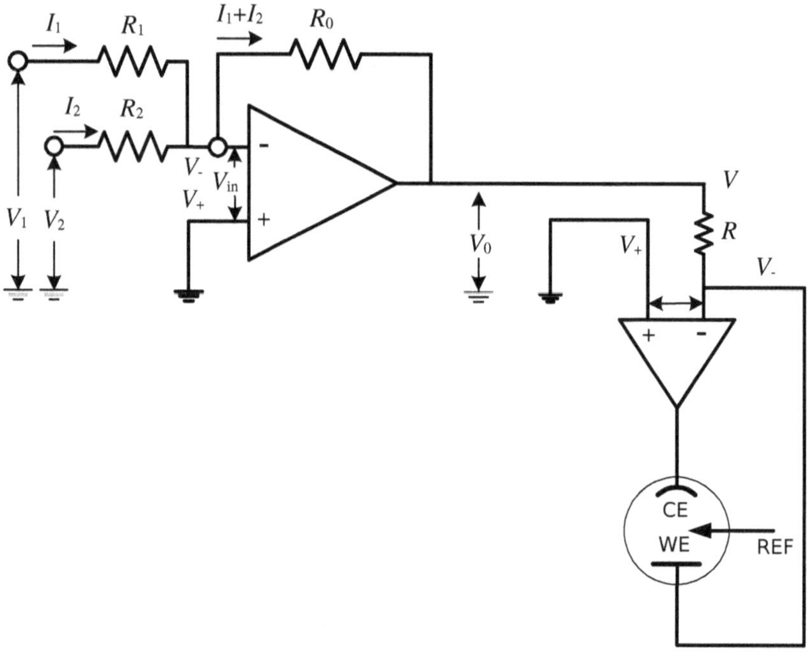

Figure 6.6: Schematic representation of a system capable of impedance analysis under galvanostatic modulation.

As $V_+ = V_- = 0$, referenced to ground,

$$I_{\mathrm{WE}} = \frac{V_{\mathrm{A}}}{R_{\mathrm{A}}} \tag{6.13}$$

In addition, the working electrode potential is held at ground, and the potential at the counterelectrode is given as

$$V_{\mathrm{CE}} = V_0 = |Z| I_{\mathrm{A}} \tag{6.14}$$

where $|Z|$ is the magnitude of the cell impedance.

For impedance analysis, a voltage adder may be placed to add the oscillating component to the steady-state potential. The associated system is shown in Figure 6.6.

Chapter 7

Experimental Methods

Impedance experiments involve the conversion of time-domain input and output signals into a complex quantity that is a function of frequency. The first step in measuring impedance is typically measurement of the steady-state polarization curve. The impedance is then obtained for a specified point on the polarization curve. The problems presented in this chapter explore methods to obtain impedance from time-domain measurements.

Problems 7.1–7.7 are based on problems presented by Orazem and Tribollet.[1] Some of the following problems require use of a spreadsheet program such as Microsoft Excel® or a computational programming environment such as MATLAB®.

Problems

7.1 Beginning with the system shown in Figure 7.1, in which the potential is expressed as

$$V = \Delta V \cos(\omega t) \tag{7.1}$$

faradaic current density can be expressed as

$$i_{\mathrm{F}} = K_{\mathrm{a}} \exp(b_{\mathrm{a}} V) - K_{\mathrm{c}} \exp(-b_{\mathrm{c}} V) \tag{7.2}$$

and the current density associated with the capacitive charging of the electrode can be expressed as

$$i_{\mathrm{C}} = C_{\mathrm{dl}} \frac{\mathrm{d}V}{\mathrm{d}t} = -\omega \Delta V C_{\mathrm{dl}} \sin(\omega t) \tag{7.3}$$

Reproduce the results presented in Figures 7.2 and 7.3.

Solution 7.1. In the Fourier analysis, the potential and total current $i = i_{\mathrm{F}} + i_{\mathrm{C}}$

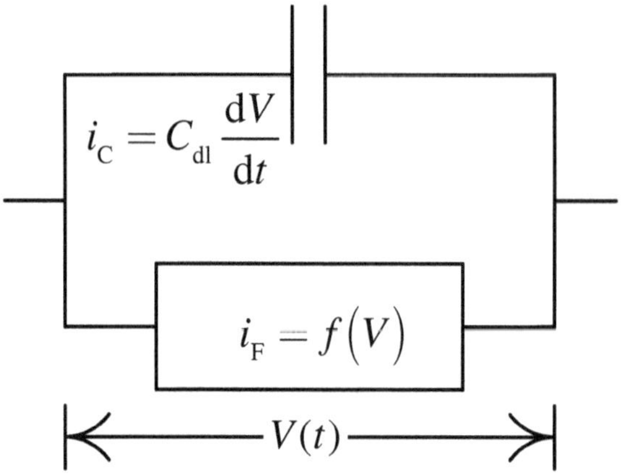

Figure 7.1: Schematic representation of an electrode interface demonstrating the contribution of charging and faradaic current densities. Taken from Orazem and Tribollet[1] with permission of John Wiley & Sons.

are expressed in terms of in-phase and out-of-phase parts. Thus,

$$V_r = \frac{1}{T}\int_0^T V(t)\cos(\omega t)dt \tag{7.4}$$

and

$$V_j = -\frac{1}{T}\int_0^T V(t)\sin(\omega t)dt \tag{7.5}$$

As potential is expressed by equation (7.1), equation (7.5) yields a value of zero for V_j. The real and imaginary parts of current density may be expressed as

$$i_r = \frac{1}{T}\int_0^T i(t)\cos(\omega t)dt \tag{7.6}$$

and

$$i_j = -\frac{1}{T}\int_0^T i(t)\sin(\omega t)dt \tag{7.7}$$

respectively. The impedance is obtained from

$$Z = -\frac{\widetilde{V}}{\widetilde{i}} \tag{7.8}$$

where $\widetilde{V} = V_r + jV_j$ and $\widetilde{i} = i_r + ji_j$.

This problem could be solved using a spreadsheet or by writing a script in a programming language such as MATLAB®. The MATLAB® code used to generate Figures 8.1 and 8.2 is presented in Appendix B. The input for the code is presented in Table 7.1. A series of plots are generated from which Figures 7.2 and 7.3 can be produced.

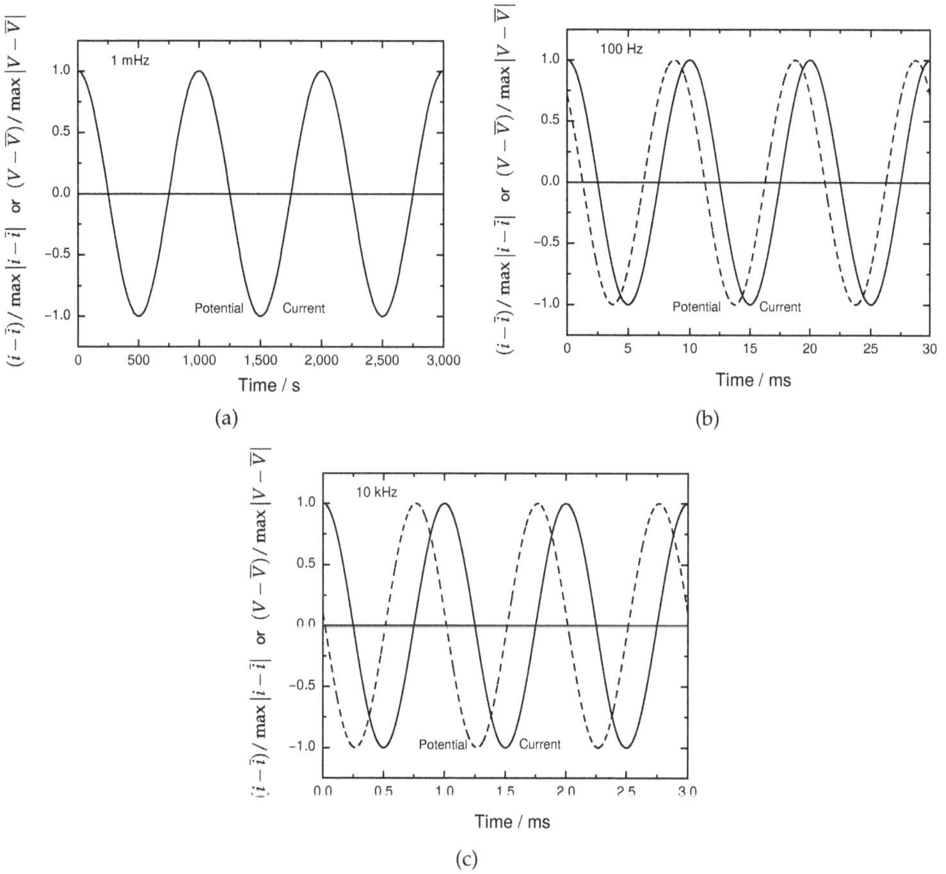

Figure 7.2: The current density response to a sinusoidal potential input for a system with parameters $C_{\rm dl} = 31$ μF/cm^2, $n_{\rm a}{\rm F}k_{\rm a} = n_{\rm c}{\rm F}k_{\rm c} = 0.5$ mA/cm^2, $b_{\rm a} = 19.5$ V^{-1}, $b_{\rm c} = 19.5$ V^{-1}, $\overline{V} = 0$ V, and $\Delta V = 1$ mV: (a) 1 mHz; (b) 100 Hz; and (c) 10 kHz. Taken from Orazem and Tribollet[1] with permission of John Wiley & Sons.

Table 7.1: Queries and input values for the MATLAB® code presented in Section B.1.

Query	Response
Voltage perturbation in mV	1
Steady-state potential in mV	0
Capacity in μF/cm^2	31
Anodic rate constant in mA/cm^2	0.5
Cathodic rate constant in mA/cm^2	0.5
Anodic transfer coefficient $b_{\rm a}$ in V^{-1}	19.5
Cathodic transfer coefficient $b_{\rm c}$ in V^{-1}	19.5

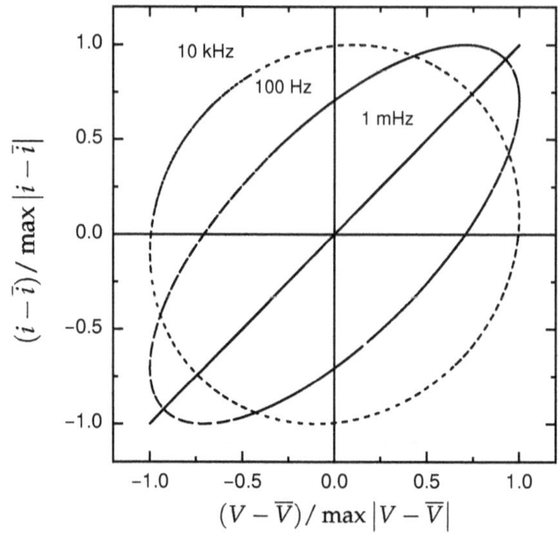

Figure 7.3: Lissajous representation of the signals presented in Figure 7.2 with frequency as a parameter. The signals were normalized by the perturbation amplitude such that the values for current and potential ranged between ± 1. Taken from Orazem and Tribollet[1] with permission of John Wiley & Sons.

7.2 Use a Lissajous plot to calculate the phase angle and magnitude of the impedance for a linear system with capacity $C_{\mathrm{dl}} = 31\ \mu\mathrm{F/cm^2}$, charge-transfer resistance $R_{\mathrm{t}} = 51.34\ \Omega\mathrm{cm^2}$, and a potential perturbation $\Delta V = 0.01$ V at frequencies of 1 Hz, 10 Hz, and 10 kHz. This approach requires calculating the potential perturbation, the charging current, and the faradaic current as functions of time.

Solution 7.2. For the potential expressed in

$$V = \Delta V \cos(\omega t) \tag{7.9}$$

the charging current is given by

$$i_{\mathrm{C}} = C\frac{\mathrm{d}V}{\mathrm{d}t} = -\omega \Delta V C \sin(\omega t) \tag{7.10}$$

and the faradaic current is given by

$$i_{\mathrm{F}} = \frac{V}{R_{\mathrm{t}}} \tag{7.11}$$

The total current is given by $i = i_{\mathrm{C}} + i_{\mathrm{F}}$. A spreadsheet program was used to generate Lissajous figures. The results are presented in Figure 7.4. For 1 Hz, the magnitude can

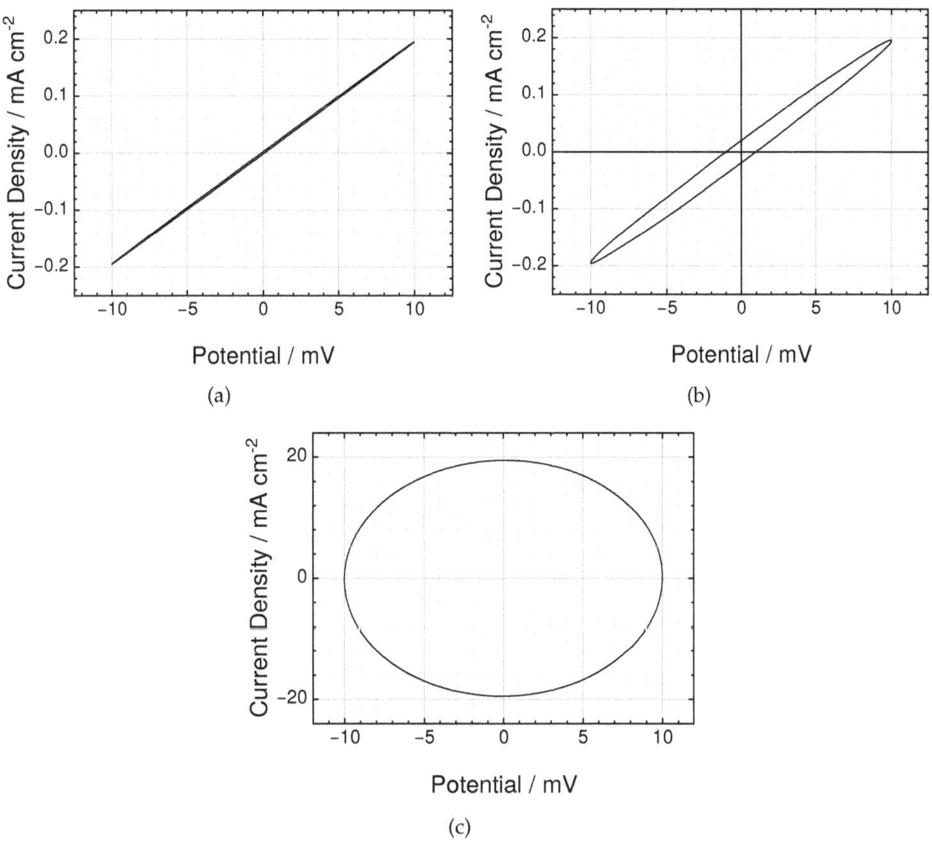

Figure 7.4: The current density response to a sinusoidal potential input for a linear system with capacity $C_{dl} = 31$ μΓ/cm^2, charge-transfer resistance $R_t = 51.34$ Ωcm^2, and a potential perturbation $\Delta V = 0.01$ V: (a) 1 Hz; (b) 10 Hz; and (c) 10 kHz.

Table 7.2: Selected impedance response calculated for a linear system with capacity $C_{\mathrm{dl}} = 31\ \mu\mathrm{F/cm^2}$ and charge-transfer resistance $R_{\mathrm{t}} = 51.34\ \Omega\mathrm{cm}^2$.

f, Hz	Z_{r}, $\Omega\mathrm{cm}^2$	Z_{j}, $\Omega\mathrm{cm}^2$	$\lvert Z\rvert$, $\Omega\mathrm{cm}^2$	φ, degree
1	51.33487	−0.51335	51.33743	−0.57294
10	50.83169	−5.08314	51.08521	−5.71056
10,000	0.00513	−0.51335	0.51338	−89.42706

be obtained as

$$|Z| = \frac{10\ \mathrm{mV}}{0.1948\ \mathrm{mA/cm^2}} = 51.33\ \Omega\mathrm{cm}^2 \tag{7.12}$$

and the phase can be obtained as

$$\varphi = -\sin^{-1}\left(\frac{0.1\ \mathrm{mV}}{10\ \mathrm{mV}}\right) = -0.573\ \text{degree} \tag{7.13}$$

For 10 Hz, the magnitude can be obtained as

$$|Z| = \frac{10\ \mathrm{mV}}{0.1957\ \mathrm{mA/cm^2}} = 51.1\ \Omega\mathrm{cm}^2 \tag{7.14}$$

and the phase can be obtained as

$$\varphi = -\sin^{-1}\left(\frac{0.1\ \mathrm{mV}}{10\ \mathrm{mV}}\right) = -5.74\ \text{degree} \tag{7.15}$$

For 10 kHz, the magnitude can be obtained as

$$|Z| = \frac{10\ \mathrm{mV}}{19.48\ \mathrm{mA/cm^2}} = 0.513\ \Omega\mathrm{cm}^2 \tag{7.16}$$

and the phase can be obtained as

$$\varphi = -\sin^{-1}\left(\frac{10\ \mathrm{mV}}{10\ \mathrm{mV}}\right) = -90\ \text{degree} \tag{7.17}$$

These values can be compared to the calculated impedance response presented in Table 7.2.

7.3 Calculate the ratio of the amplitude of the charging current to the amplitude of the faradaic current for a linear system with capacity $C_{\mathrm{dl}} = 31\ \mu\mathrm{F/cm^2}$, charge-transfer resistance $R_{\mathrm{t}} = 51.34\ \Omega\mathrm{cm}^2$, and a potential perturbation $\Delta V = 0.01$ V at frequencies of 1 Hz, 100 Hz, and 10 kHz.

Solution 7.3. For a potential expressed as equation (7.9), the charging current is given by equation (7.10), and the faradaic current is given by equation (7.11). The ratio of charging current to faradaic current may be expressed as

$$\frac{i_{\mathrm{C}}}{i_{\mathrm{F}}} = -\omega R_{\mathrm{t}} C \sin(\omega t) \tag{7.18}$$

Table 7.3: Ratio of the amplitudes of charging and faradaic current densities for a linear system with capacity $C_{dl} = 31$ μF/cm^2 and charge-transfer resistance $R_t = 51.34$ Ωcm^2.

f, Hz	$2\pi f$, s^{-1}	i_C/i_F
1	6.2832	0.001
100	628.32	1.000
10,000	62,832	100

Table 7.4: Queries and input values for the MATLAB® code presented in Section B.2.

Query	Response
Voltage perturbation in mV	1
Steady-state potential in mV	0
Capacity in μF/cm^2	31
Charge-transfer resistance in Ωcm^2	51.34

and the amplitude is

$$\left|\frac{i_C}{i_F}\right| = \omega R_t C = 2\pi f R_t C \tag{7.19}$$

The results are presented in Table 7.3. The characteristic frequency for this system is $f_c = 1/(2\pi R_t C_{dl})$. The characteristic frequency for parameters used in this problem is $f = 100$ Hz. At this frequency, the charging and faradaic currents have the same amplitude.

7.4 Use a Fourier analysis to calculate the impedance as a function of frequency for a linear system, with capacity $C_{dl} = 31$ μF/cm^2, charge-transfer resistance $R_t = 51.34$ Ωcm^2, and potential perturbation $\Delta V = 0.01$ V. Compare your results to the theoretical value obtained using the methods described in Chapter 4.

Solution 7.4. The Fourier analysis requires calculation of complex potential and current following equations (7.4)–(7.7). The faradaic current used in Problem 7.1 was given by equation (7.2), in which current is an exponential function of potential. In the present problem, the faradaic current is expressed by

$$i_F = \frac{V}{R_t} \tag{7.20}$$

for which current is a linear function of potential.

This problem could be solved by writing a script in a programming language such as MATLAB.® The MATLAB® code is presented in Appendix B. The input for the code is presented in Table 7.4. A series of plots are generated. The results are identical to those generated by $Z = R_t/(1 + j\omega R_t C_{dl})$.

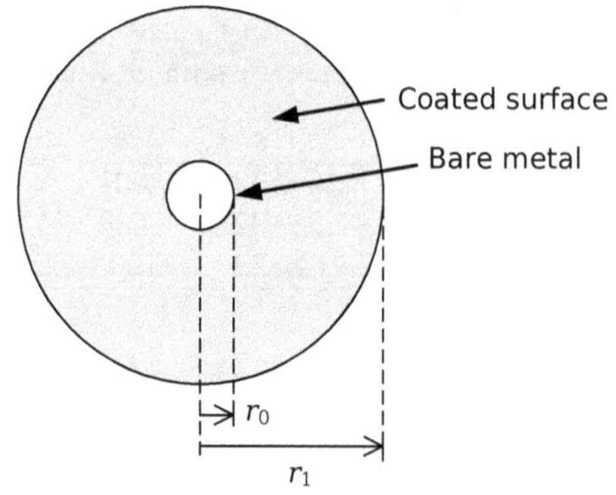

Figure 7.5: Schematic representation of a disk electrode corresponding to Problem 7.5, with bare metal exposed from the origin to radius r_0 and a coated metal surface from radius r_0 to r_1. Taken from Orazem and Tribollet[1] with permission of John Wiley & Sons.

7.5 Consider the disk electrode shown in Figure 7.5, with bare metal exposed from the origin to radius r_0 and a coated metal surface from radius r_0 to r_1. Estimate the global interfacial impedance response of this electrode for the following geometries if the coating consists of an organic material of 100 μm thickness and the charge-transfer resistance on the bare metal is 100 Ωcm^2. Hint: Guidelines for estimation of capacitance are given in Table A.14.

(a) $r_0 = 0.25$ cm and $r_1 = 1.0$ cm.

(b) $r_0 = 0.5$ cm and $r_1 = 1.0$ cm.

(c) $r_0 = 0.75$ cm and $r_1 = 1.0$ cm.

Solution 7.5. The system presented in Figure 7.5 may be represented as the circuit presented in Figure 7.6, where γ is the fractional coverage of the surface by the dielectric film. The corresponding impedance is given by

$$Z = \frac{1}{\gamma j\omega C_\ell + \dfrac{(1-\gamma)(1+j\omega R_t C_{dl})}{R_t}} \tag{7.21}$$

or

$$Z = \frac{R_t}{1-\gamma+j\omega R_t(\gamma C_\ell + (1-\gamma)C_{dl}} \tag{7.22}$$

Following Table A.14, the capacitance on the bare metal may be assumed to have a value on the order of $C_{dl} = 3 \times 10^{-5}$ F/cm^2. The capacitance of the film may be calculated from

$$C_\ell = \frac{\varepsilon\varepsilon_0}{\delta} \tag{7.23}$$

where, following Table A.14, the dielectric constant for the coating may be assigned a value of $\varepsilon = 5$, yielding $C_\ell = 4.43 \times 10^{-11}$ F/cm^2. The values for the fractional

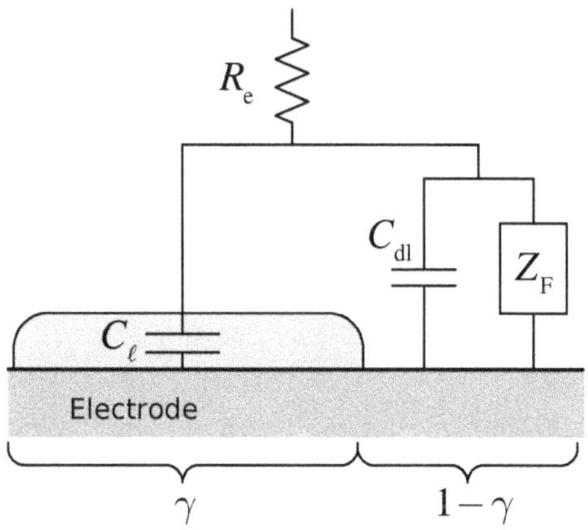

Figure 7.6: Schematic electrical circuit representation for the disk electrode presented in Figure 7.5, where γ is the fractional coverage of the surface by the dielectric film. Taken from Orazem and Tribollet[1] with permission of John Wiley & Sons.

Table 7.5: Dimensions and values of γ for Problem 7.5.

Problem	r_0, cm	r_1, cm	γ	Equation (7.29)
(a)	0.25	1	0.9375	0.0022%
(b)	0.5	1	0.75	0.00044%
(c)	0.75	1	0.4375	0.00012%

coverage of the electrode by the coating are given in Table 7.5. The impedance response corresponding to equation (7.22) is presented in Figure 7.7(a), showing a significant influence of the coated portion of the metal. The impedance increases as the effective area of the electrode $1-\gamma$ decreases. A coating may be applied to a large metal coupon to expose a smaller area of bare metal to the electrolyte. In this case, the impedance should be scaled to the active area. The corresponding impedance is given in Figure 7.7(b). The data overlap, showing that the coating provides adequate insulation and that the differences in Figure 7.7(a) may be attributed to changes in the active area of the electrode.

7.6 Develop an expression for the error in the global interfacial impedance response of the bare metal referenced in Problem 7.5 as a function of the relative area of the coated metal r_1^2/r_0^2 and the coating property ε/δ.

Solution 7.6. If the coating is applied to reduce the active area of the electrode, the appropriate comparison, as suggested by Figure 7.7(b), is between the nominal

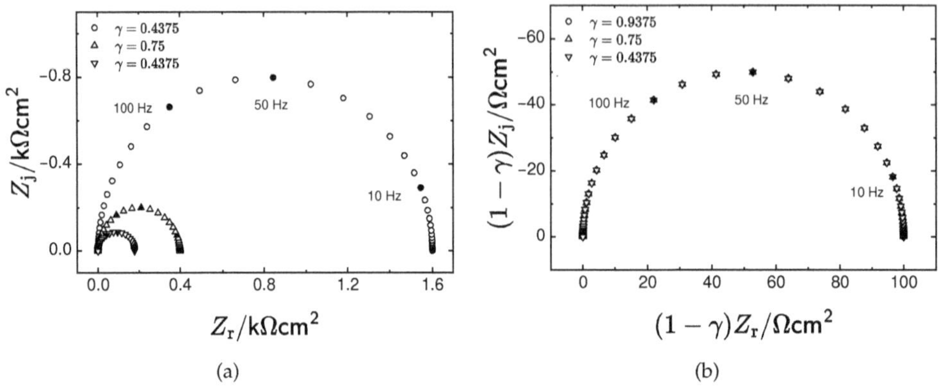

Figure 7.7: The impedance response corresponding to equation (7.22) for $R_t = 100\ \Omega\text{cm}^2$, $C_\ell = 4.43 \times 10^{-11}$ F/cm^2, and $C_{dl} = 3 \times 10^{-5}$ F/cm^2: (a) impedance scaled to the total area and (b) impedance scaled to the area of the bare portion of the electrode. Taken from Orazem and Tribollet[1] with permission of John Wiley & Sons.

impedance of the metal

$$Z_{\text{bare}} = \frac{R_t}{1 + j\omega R_t C_{dl}} \tag{7.24}$$

and the impedance of the bare metal surrounded by the coating, i.e.,

$$(1-\gamma)Z = \frac{R_t}{1 + j\omega R_t \left(\dfrac{\gamma}{1-\gamma} C_\ell + C_{dl}\right)} \tag{7.25}$$

At frequency $\omega \to 0$, both equations (7.24) and (7.25) yield values of R_t, showing that the low-frequency behavior is independent of the coating properties so long as the coating does not allow faradaic reactions.

At frequency $\omega \to \infty$, equation (7.24) yields

$$\lim_{\omega\to\infty} Z_{\text{bare}} = \frac{1}{j\omega C_{dl}} \tag{7.26}$$

and equation (7.25) yields

$$\lim_{\omega\to\infty} (1-\gamma)Z = \frac{1}{j\omega \left(\dfrac{\gamma}{1-\gamma} C_\ell + C_{dl}\right)} \tag{7.27}$$

An error term may be defined by

$$\epsilon = 100\frac{(1-\gamma)Z - Z_{\text{bare}}}{Z_{\text{bare}}} = 100\frac{(1-\gamma)Z}{Z_{\text{bare}}} - 1 \tag{7.28}$$

As only the imaginary parts have value in equations (7.26) and (7.27), cancellation of $j\omega$ yields a real number. Thus,

$$\epsilon = 100\frac{\gamma}{1-\gamma}\frac{C_\ell}{C_{dl}} \tag{7.29}$$

where C_ℓ is given by equation (7.23) and $\gamma = 1 - r_0^2/r_1^2$. The errors corresponding to Problem 7.5, presented in Table 7.5, are small. A thinner coating ($\delta = 1$ µm) with the same dielectric constant as employed in the solution to Problem 7.5 increases the error to 0.22% for Problem 7.5.a, 0.044% for Problem 7.5.b, and 0.011% for Problem 7.5.c. The practice of masking a portion of the electrode is shown to introduce only small errors in the impedance result. It should be noted that the error increases as γ approaches unity.

7.7 Use a Lissajous analysis to calculate the magnitude and phase angle for the plots given in Figures 7.8.

Solution 7.7. The magnitude for Figure 7.8(a) can be obtained as

$$|Z| = \frac{10\ \mathrm{mV}}{0.1948\ \mathrm{mA/cm^2}} = 51.33\ \Omega\mathrm{cm}^2 \tag{7.30}$$

and the corresponding phase can be obtained as

$$\varphi = -\sin^{-1}\left(\frac{0\ \mathrm{mV}}{10\ \mathrm{mV}}\right) = 0° \tag{7.31}$$

The magnitude for Figure 7.8(b) can be obtained as

$$|Z| = \frac{10\ \mathrm{mV}}{0.275\ \mathrm{mA/cm^2}} = 36.4\ \Omega\mathrm{cm}^2 \tag{7.32}$$

and the corresponding phase can be obtained as

$$\varphi = -\sin^{-1}\left(\frac{7.07\ \mathrm{mV}}{10\ \mathrm{mV}}\right) = -45.0° \tag{7.33}$$

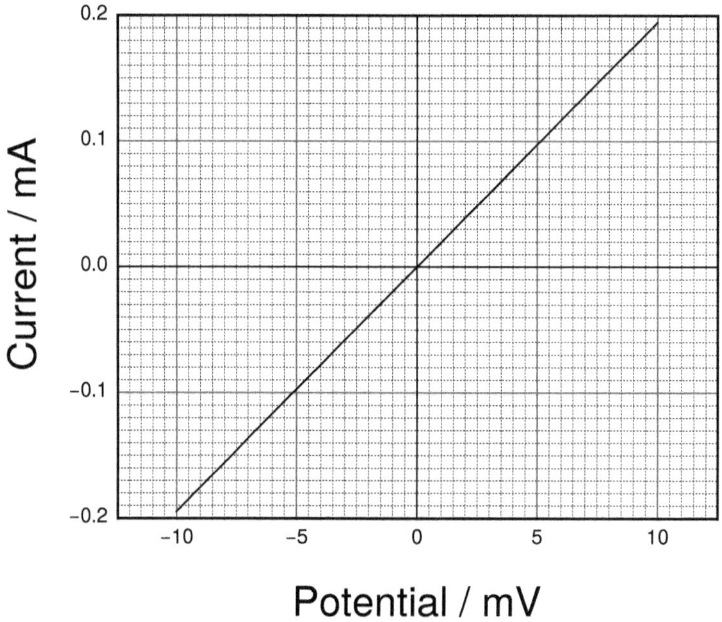

(a)

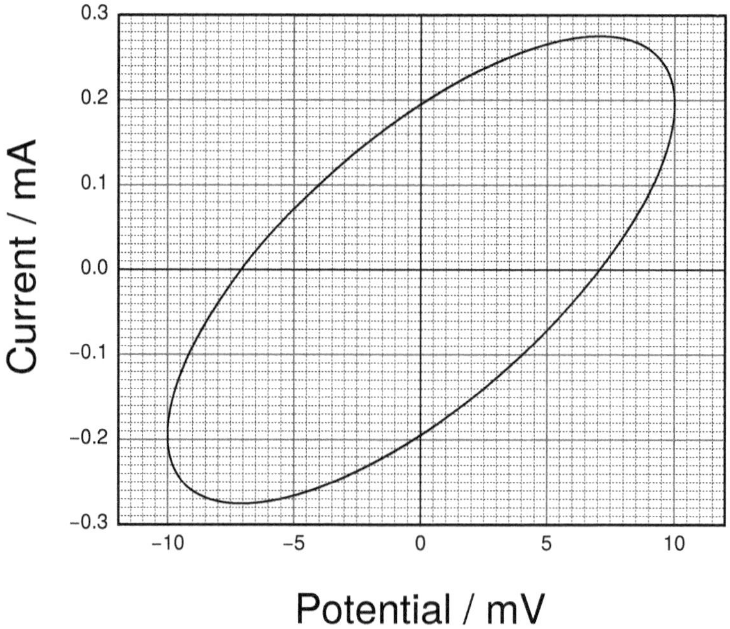

(b)

Figure 7.8: Lissajous plots for Problem 7.7. Taken from Orazem and Tribollet[1] with permission of John Wiley & Sons.

Chapter 8

Experimental Design

Selection of experimental parameters, such as frequency range and perturbation amplitude, are constrained by the system under study. The problems presented in this chapter explore guidelines for the design of experimental cells, for selection of appropriate impedance parameters, and for selection of appropriate instrument controls. Problems 8.1–8.7 are based on Orazem and Tribollet.[1]

Problems

8.1 Estimate the effect an ohmic resistance has on the maximum potential perturbation amplitude that can be applied to an electrochemical system while satisfying the constraint that ΔV is smaller than $0.2/b$.

Solution 8.1. Following

$$Z = R_e + \frac{1}{\frac{1}{Z_F} + j\omega C_{dl}} \tag{8.1}$$

the presence of an ohmic resistance will reduce the portion of the potential perturbation experienced at the electrode–electrolyte interface. The potential drop across the electrolyte is given by iR_e, and the potential drop across the electrode–electrolyte interface η_s is given by $i/(1/Z_F + j\omega C_{dl})$. At large frequencies, the interfacial impedance tends toward zero and the linear ohmic resistance dominates. At low frequency, the capacitive charging has negligible effect and the interfacial impedance is dominated by the faradaic contribution Z_F. The guidelines for the potential perturbation can therefore be determined by the behavior at low frequency. A value $b_a\Delta V \leq 0.2$ can be achieved by setting

$$b_a \Delta U \frac{R_t}{R_e + R_t} \leq 0.2 \tag{8.2}$$

Thus, the influence of ohmic resistance is to increase the allowable magnitude for the potential perturbation.

8.2 Estimate the maximum amplitude one should use for a potential perturbation for a system under Tafel kinetics with:

(a) A Tafel slope of 60 mV/decade and negligible ohmic resistance. Keep in mind the relationship between Tafel slope β and Tafel constant b given in

$$b = \frac{2.303 \times 10^3}{\beta} \tag{8.3}$$

(b) A Tafel slope of 120 mV/decade and negligible ohmic resistance.

(c) A Tafel slope of 60 mV/decade, an exchange current density equal to 1 mA/cm^2, an applied potential of 100 mV, and an ohmic resistance of 10 Ωcm^2.

(d) A Tafel slope of 60 mV/decade, an exchange current density equal to 1 mA/cm^2, an applied potential of 200 mV, and an ohmic resistance of 10 Ωcm^2.

Solution 8.2.

(a) Following equation (8.3) and given the guideline that the desired potential perturbation is

$$\Delta V = \frac{0.2}{b} = \frac{0.2\beta}{2.303 \times 10^3} \tag{8.4}$$

the maximum amplitude one should use for a potential perturbation for a system with negligible ohmic resistance and under Tafel kinetics with $\beta = 60$ mV/decade is 0.0052 V or 5.2 mV.

(b) Following equations (8.3) and (8.4), the maximum amplitude one should use for a potential perturbation for a system with negligible ohmic resistance and under Tafel kinetics with $\beta = 120$ mV/decade is 0.0104 V or 10.4 mV.

(c) At low frequency,

$$\overline{V} = \overline{U}\frac{R_t}{R_t + R_e} \tag{8.5}$$

where

$$R_t = \frac{1}{bK_0 \exp(b\overline{V})} \tag{8.6}$$

Introduction of equation (8.6) into equation (8.5) yields

$$\overline{V} = \frac{\overline{U}}{1 + R_e b K_0 \exp(b\overline{V})} \tag{8.7}$$

which may be expressed as

$$\overline{V} = \frac{1}{b} \ln \left(\left(\frac{\overline{U}}{\overline{V}} - 1 \right) \frac{1}{bR_eK_0} \right) \tag{8.8}$$

The solution to equation (8.8) may be obtained by trial and error or by use of Newton's method in which

$$\overline{V}_{n+1} = \overline{V}_n - \frac{f(\overline{V}_n)}{f'(\overline{V}_n)} \tag{8.9}$$

where

$$f(\overline{V}_n) = \overline{V}_n - \frac{1}{b} \ln \left(\left(\frac{\overline{U}}{\overline{V}_n} - 1 \right) \frac{1}{bR_eK_0} \right) = 0 \tag{8.10}$$

and

$$f'(\overline{V}_n) = 1 + \frac{\overline{U}}{b\overline{V}_n^2 \left(\overline{U}/\overline{V}_n - 1 \right)} \tag{8.11}$$

For $K_0 = 1 \times 10^{-3}$ A/cm^2, $R_e = 10\ \Omega\text{cm}^2$, $\beta = 60$ mV/decade ($b = 38.4$ V^{-1}), and $\overline{U} = 0.1$ V, iterative solution of equation (8.9) yields $V = 3.79 \times 10^{-2}$ V, $R_t = 6.09\ \Omega\text{cm}^2$, and, from

$$\Delta U = \frac{0.2}{b} \left(\frac{R_e + R_t}{R_e} \right) \tag{8.12}$$

$\Delta U = 1.38 \times 10^{-2}$ V (13.8 mV).

(d) For $K_0 = 1 \times 10^{-3}$ A/cm^2, $R_e = 10\ \Omega\text{cm}^2$, $\beta = 60$ mV/decade ($b = 38.4$ V^{-1}), and $\overline{U} = 0.2$ V, iterative solution of equation (8.9) yields $V = 5.21 \times 10^{-2}$ V, $R_t = 3.52\ \Omega\text{cm}^2$, and $\Delta U = 2.00 \times 10^{-2}$ V (20.0 mV).

8.3 Estimate the error in the real part of the impedance associated with a 0.22% error in the fundamental current response.

Solution 8.3. The error in the impedance is seen at low frequency, where the faradaic current dominates. The charge transfer resistance for a single-step potential-dependent electrochemical reaction is given by

$$R_t = \frac{1}{b\bar{i}} \tag{8.13}$$

where

$$\bar{i} = i_0(1 + \varepsilon_i) \tag{8.14}$$

The error in R_t associated with an error ε_i in the steady-state current can be expressed as

$$\varepsilon_{R_t} = \frac{1/(bi_0(1+\varepsilon_i)) - 1/(bi_0)}{1/(bi_0)} \tag{8.15}$$

or

$$\varepsilon_{R_t} = \frac{1}{(1+\varepsilon_i)} - 1 = \varepsilon_i \tag{8.16}$$

Thus, an error of 0.22% in the fundamental current response yields an error of 0.22% in the charge-transfer resistance.

8.4 Should the same potential perturbation amplitude be applied for all parts of the polarization curve? Give examples to demonstrate your answer.

Solution 8.4. The potential perturbation associated with a given error will change for different parts of the polarization curve. For simple Tafel kinetics, the value of R_t will change with applied potential and, for a given ohmic resistance R_e, the perturbation amplitude given by equation (8.12) will also change. A very small perturbation amplitude may be needed at potentials near the potential at which a metal passivates.

8.5 Will a spatial distribution of capacitance lead to a nonlinear current response to a sinusoidal potential input?

Solution 8.5. A spatial distribution of capacitance will not lead to a nonlinear current response to a sinusoidal potential input because the interfacial potential tends toward zero at high frequency where the charging current dominates.

8.6 Reproduce the results presented in Figures 8.1 and 8.2. This problem requires use of a spreadsheet program such as Microsoft Excel® or a computational programming environment such as MATLAB®.

Solution 8.6. The Fourier analysis requires calculation of complex potential and current following equations (7.4)–(7.7). The faradaic current used in Problem 7.1 was given by equation (7.2). The potential was given by equation (7.1), and the corresponding charging current was given by equation (7.3).

This problem could be solved using a spreadsheet or by writing a script in a programming language such as MATLAB.® The MATLAB® code used to generate Figures 8.1 and 8.2 is presented in Appendix B. The input for the code is presented in Table 8.1.

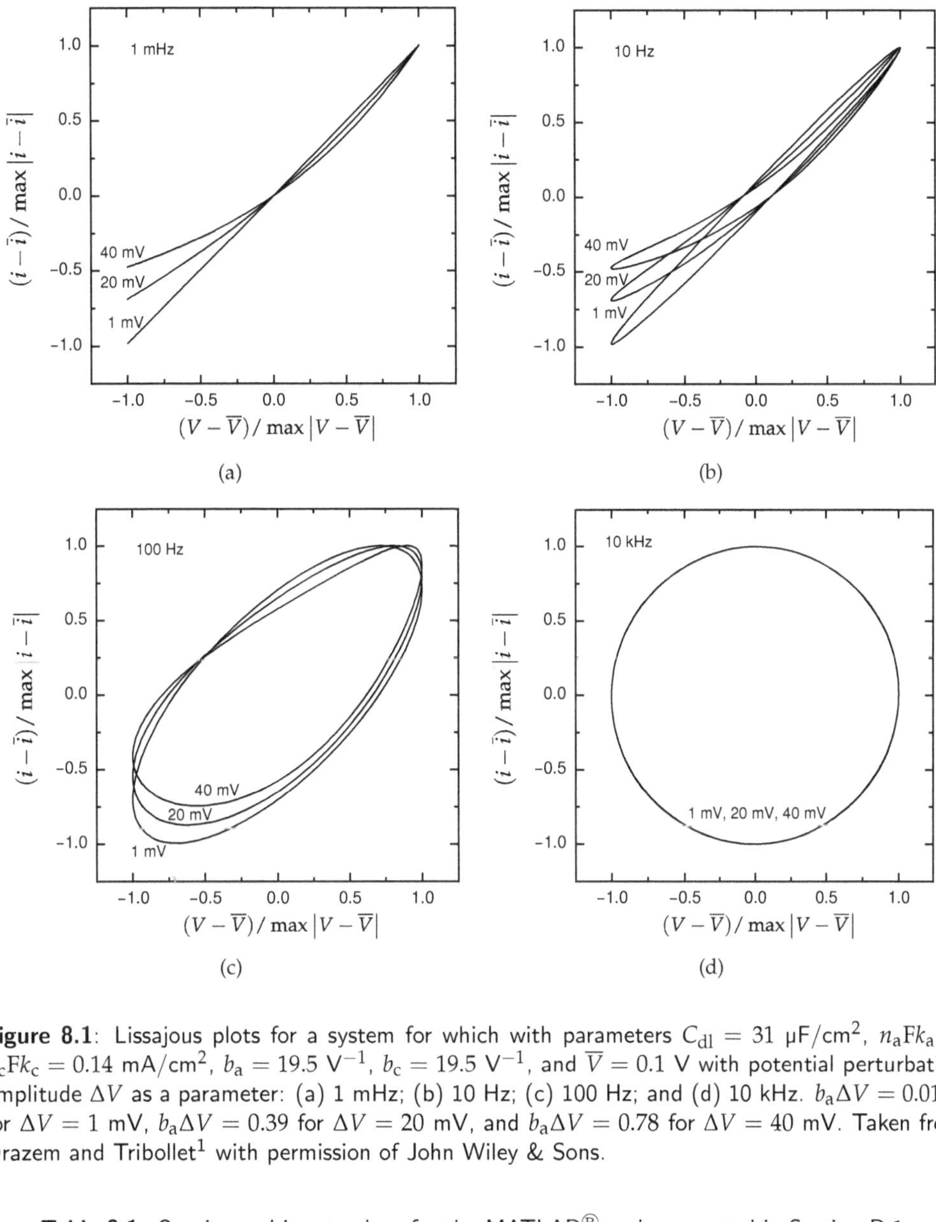

Figure 8.1: Lissajous plots for a system for which with parameters $C_{\mathrm{dl}} = 31\ \mu\mathrm{F/cm^2}$, $n_{\mathrm{a}}\mathrm{F}k_{\mathrm{a}} = n_{\mathrm{c}}\mathrm{F}k_{\mathrm{c}} = 0.14\ \mathrm{mA/cm^2}$, $b_{\mathrm{a}} = 19.5\ \mathrm{V^{-1}}$, $b_{\mathrm{c}} = 19.5\ \mathrm{V^{-1}}$, and $\overline{V} = 0.1$ V with potential perturbation amplitude ΔV as a parameter: (a) 1 mHz; (b) 10 Hz; (c) 100 Hz; and (d) 10 kHz. $b_{\mathrm{a}}\Delta V = 0.0195$ for $\Delta V = 1$ mV, $b_{\mathrm{a}}\Delta V = 0.39$ for $\Delta V = 20$ mV, and $b_{\mathrm{a}}\Delta V = 0.78$ for $\Delta V = 40$ mV. Taken from Orazem and Tribollet[1] with permission of John Wiley & Sons.

Table 8.1: Queries and input values for the MATLAB® code presented in Section B.1.

Query	Response
Voltage perturbation in mV	1, 20, or 40
Steady-state potential in mV	100
Capacity in $\mu\mathrm{F/cm^2}$	31
Anodic rate constant in $\mathrm{mA/cm^2}$	0.14
Cathodic rate constant in $\mathrm{mA/cm^2}$	0.14
Anodic transfer coefficient b_{a} in $\mathrm{V^{-1}}$	19.5
Cathodic transfer coefficient b_{c} in $\mathrm{V^{-1}}$	19.5

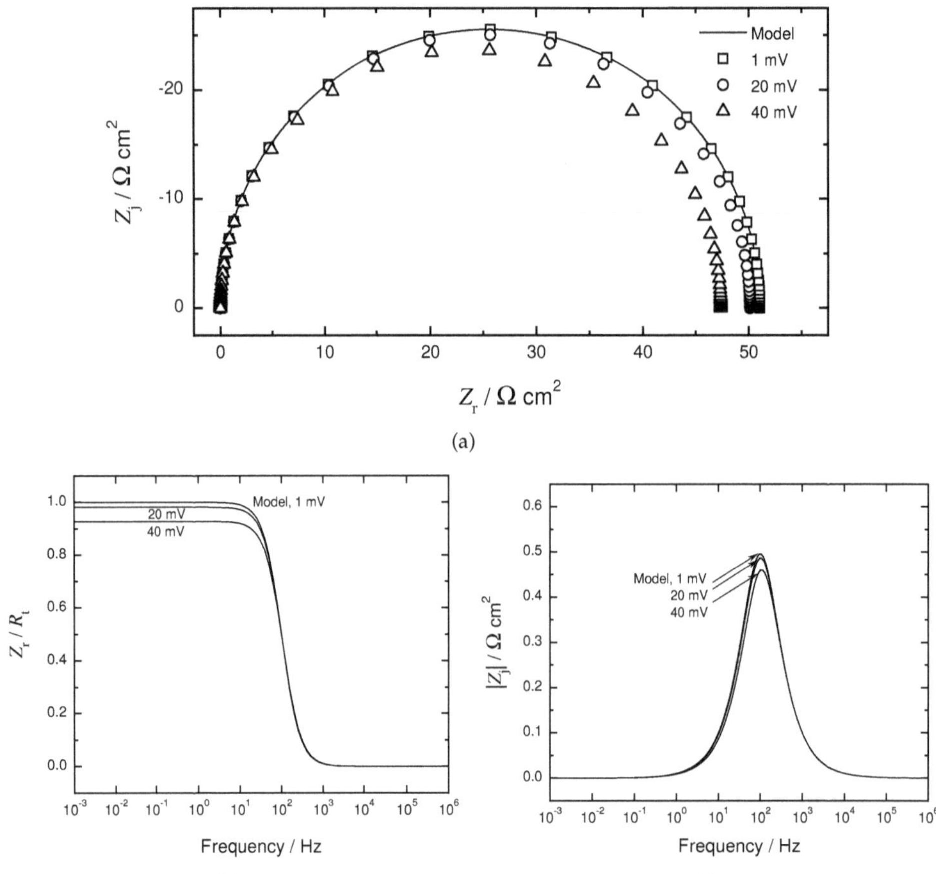

Figure 8.2: Impedance results obtained for the time-domain results presented in Figure 8.1 by use of a Fourier analysis with potential perturbation amplitude ΔV as a parameter: (a) Nyquist representation; (b) real part of the impedance as a function of frequency; and (c) imaginary part of the impedance as a function of frequency. Taken from Orazem and Tribollet[1] with permission of John Wiley & Sons.

8.7 Researchers have reported that, for impedance measurements on human skin under fixed-amplitude galvanostatic modulation, the measurement caused significant changes in skin properties. The magnitude of the skin impedance varied from about 10 Ωcm^2 at high frequency to 100 $\text{k}\Omega\text{cm}^2$ at low frequency. The perturbation amplitude was 0.1 mA on an exposed skin sample of 1 cm^2 area. Explain the reasons for their observation and suggest an improved experimental protocol.

Solution 8.7. The potential swing associated with the current perturbation is given by

$$\Delta V = \Delta I |Z(f)| \tag{8.17}$$

At high frequency, the potential swing associated with a current perturbation of 1×10^{-4} A at the high frequency impedance of 10 Ω is 1×10^{-3} V or 1 mV. At low frequency, where the impedance is equal to 1×10^5 Ω, the potential swing is 10 V. This large applied potential is sufficient to disrupt cell membranes leading to large changes in skin properties. For systems showing large changes in impedance with changes in frequency, the amplitude of current perturbation should change with frequency to achieve a smaller potential swing. A method for variable-amplitude galvanostatically modulated impedance spectroscopy was described by Wojcik et al.[19,20]

8.8 Accuracy contour plots provide a useful way to understand the performance for impedance instrumentation.[21–23] A generic accuracy contour plot is shown in Figure 8.3. The absolute high-impedance limits of a particular potentiostat and associated cables can be obtained by measuring the impedance in which the wires are not connected, simulating an infinite cell impedance. The result yields the effective capacitance of the leads and instrumentation. The line "B" is based on the wire capacitance. The absolute low-impedance limits can be obtained by measuring the impedance in which the wires are shorted, simulating a zero cell impedance. The result yields the effective resistance and inductance of the leads and instrumentation. The line "D" is based on the wire inductance. Upper and lower bounds for the measurable impedance may be obtained by accounting for the minimum current resolution and the largest measurable current. The line "A" is based on the minimum current resolution, and the line "E" is based on the maximum measurable current. The line "C" corresponds to the maximum frequency capability of the instrument.

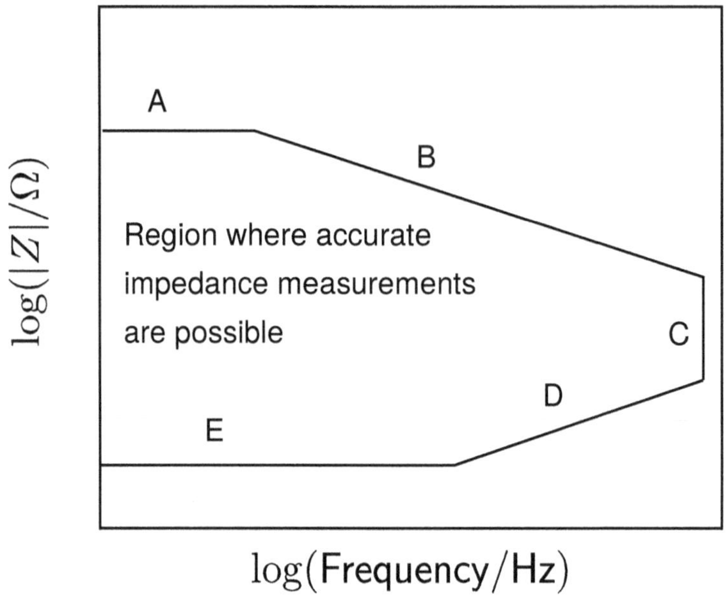

Figure 8.3: Generic accuracy contour plot for impedance measurements.

(a) Under the assumptions that the capacitance of the wires obtained using the open lead measurement is 1×10^{-14} F, the resistance and inductance of the wires obtained using the shorted leads measurement are 1×10^{-6} Ω and 1×10^{-8} H, respectively, the maximum measurable frequency is 1 MHz, and the minimum current resolution is 1×10^{-16} A, plot the absolute limits for the accuracy contour plot shown in Figure 8.3. Assume that the potential perturbation is 10 mV. Note that measurement outside these ranges is not possible.

(b) Estimate the boundary corresponding to impedance measurements that have a minimum accuracy in modulus of 1%, and draw these lines on the figure produced for Problem 8.8.a. Assume that the maximum measurable current is 0.6 A. Note that corrections to lines "B" and "D" result from a desire to minimize systemic errors and corrections to line "A" result from a desire to reduce stochastic errors.

Solution 8.8.

(a) Line "A" is obtained by estimating the maximum measurable impedance as

$$Z_{max} = \frac{\Delta V}{I_{min}} = \frac{0.01\ \text{V}}{10^{-16}\ \text{A}} = 10^{14}\ \Omega \tag{8.18}$$

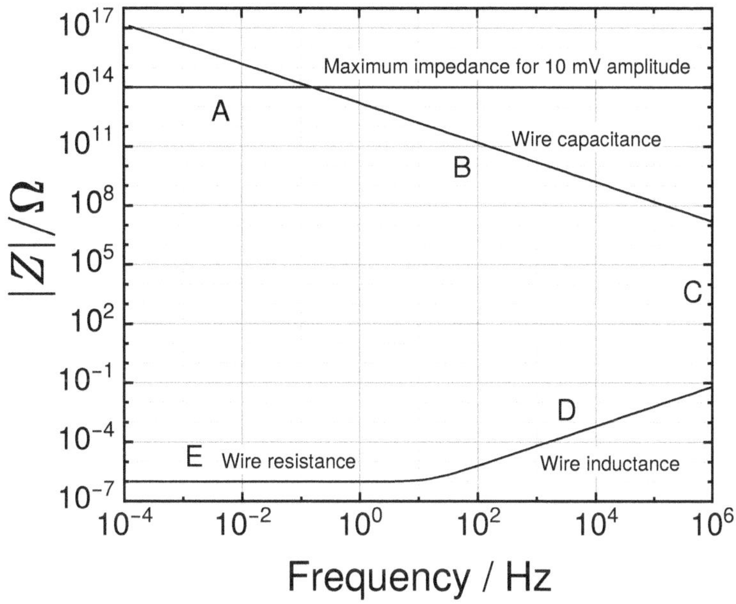

Figure 8.4: Accuracy contour plot for impedance measurements based on parameters given in Problem 8.8.a.

Line "B" can be obtained from the open-lead capacitance. Thus,

$$Z_C = \frac{1}{j\omega C_{wire}} \tag{8.19}$$

The magnitude can be calculated as

$$|Z_C| = \frac{1}{2\pi f C_{wire}} = \frac{1}{2\pi f 10^{-14}\text{ F}} \tag{8.20}$$

The line "C" is simply a vertical line at $f = 1$ MHz. The lines "E" and "D" are obtained as

$$Z_L = R_{wire} + j\omega L_{wire} \tag{8.21}$$

or

$$|Z_L| = R_{wire} + 2\pi f L_{wire} = 10^{-6}\ \Omega + (2\pi f 10^{-8})\ \Omega \tag{8.22}$$

These lines are given in Figure 8.4. Note that capacitances less than C_{wire} cannot be measured. As indicated by equation (8.18), larger impedances may be measured by increasing the potential perturbation. The increase of potential perturbation, however, is limited by the need to maintain linearity. In addition, available frequency range is constrained by the capacitance and inductance of the wires.

(b) The region for which the impedance may be measured with errors of less than 1% is smaller than the region outlined in Figure 8.4. The measured impedance can be

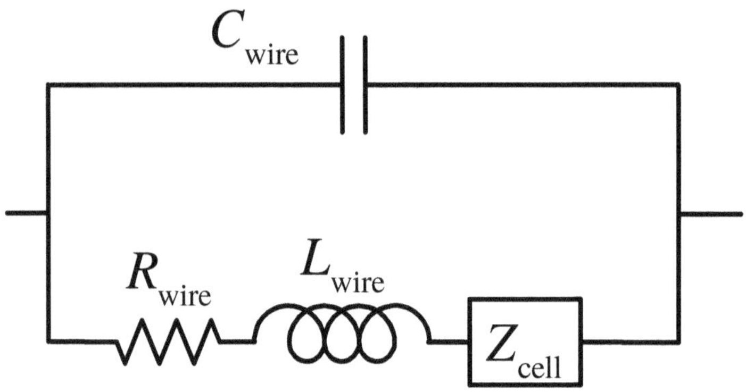

Figure 8.5: Circuit representation showing the influence of wire capacitance, resistance, and inductance on the measured impedance.

represented by the circuit shown in Figure 8.5, showing that a systemic error is caused by the capacitance, resistance, and inductance of the leads. Figure 8.5 provides guidance for adjusting the lines corresponding to the limits of the influence of wire capacitance, resistance, and inductance on the measured impedance.

The line "B" may be adjusted by use of a capacitance that is 100 times larger than the wire capacitance. Thus,

$$|Z_{C,1\%}| = \frac{1}{2\pi f 100 C_{wire}} = \frac{1}{2\pi f 10^{-12}}\ \Omega \tag{8.23}$$

Similarly, lines "E" and "D" can be adjusted following

$$|Z_{L,1\%}| = 100 R_{wire} + 2\pi f 100 L_{wire} = 10^{-4}\ \Omega + (2\pi f 10^{-6})\ \Omega \tag{8.24}$$

The adjustment of line "A" is associated with concern with the noise of the measurement:

$$Z_{max,1\%} = \frac{\Delta V}{100 I_{min}} = \frac{0.01\ \text{V}}{10^{-14}\ \text{A}} = 10^{12}\ \Omega \tag{8.25}$$

The smallest measurable impedance is associated with the maximum measurable current, and this impedance may be larger than the resistance of the leads. The smallest measurable impedance is given as

$$Z_{min,1\%} = \frac{\Delta V}{I_{max}} = \frac{0.01\ \text{V}}{0.6\ \text{A}} = 0.0167\ \Omega \tag{8.26}$$

The boundary resulting from equations (8.23) to (8.26) is presented in Figure 8.6.

It is important to note that the accuracy of current measurement is greatest at the maximum of the current range; thus, the value of $Z_{min,1\%}$ in equation (8.26) does not need to be adjusted by a factor of 100. In contrast, the maximum error in current measurement corresponds to the resolution of the current, which in the

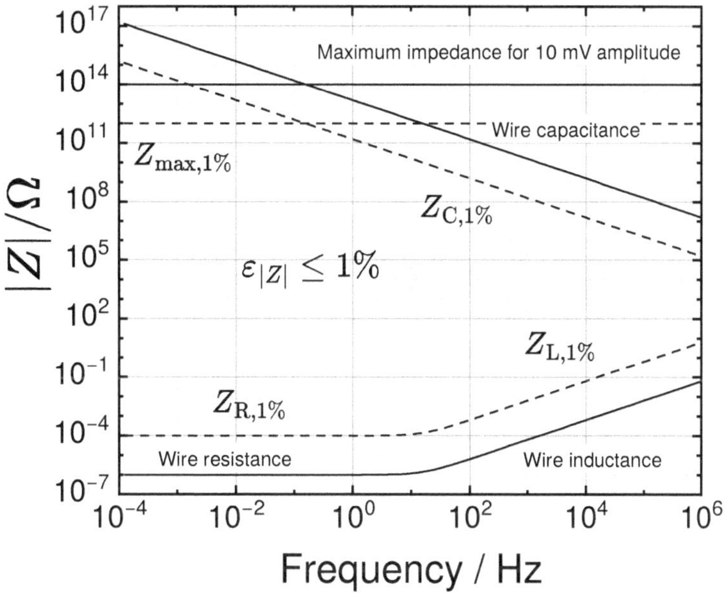

Figure 8.6: Accuracy contour plot showing the region for which the measured modulus of the impedance will have an error less than or equal to 1%.

present case is assumed to be 1 aA (1 attoampere or 10^{-18} A). A current 100 times larger can be expected to have an uncertainty of 1%.

The region for which the error in impedance magnitude, i.e., $\varepsilon_{|Z|}$ is less than 1 % can be shifted up or down by changing the perturbation amplitude. Measurement of smaller impedances can be achieved by use of a smaller potential perturbation, just as use of a larger potential perturbation gives access to larger impedances. More commonly, measurements for small impedances are performed under galvanostatic modulation. Such adjustments have hard limits indicated by the lines corresponding to the lead capacitance (line "B" in Figure 8.5) and the lead resistance and inductance (lines "E" and "D" in Figure 8.5). For an accuracy consistent with $\varepsilon_{|Z|} \leq 1$, the corresponding limits are indicated as $Z_{C,1\%}$ and $Z_{L,1\%}$ in Figure 8.6. In addition, a decrease of potential perturbation amplitude may increase stochastic errors, and an increase of potential perturbation amplitude may cause a nonlinear system response.

The accuracy contour plot shown in Figure 8.6 will be affected by system changes such as the use of longer leads to connect the cell to the instrumentation. The wire capacitance, inductance, and resistance scale with wire length, thus reducing the size of the $\varepsilon_{|Z|} \leq 1\%$ region.

8.9 Calculate the impedance for a system described by a parallel combination of a resistor $R_{\text{cell}} = 1 \times 10^{10}\ \Omega$ and a capacitor $C_{\text{cell}} = 1 \times 10^{-13}$ F. Plot the cell impedance and the measured impedance on the figure produced for Problem 8.8.b. Estimate the error as a function of frequency.

Solution 8.9. The impedance for a system described by a parallel combination of a resistor R_{cell} and capacitor C_{cell} can be expressed as

$$Z_{\text{cell}} = \frac{R_{\text{cell}}}{1 + \text{j}2\pi f R_{\text{cell}} C_{\text{cell}}} \tag{8.27}$$

and the corresponding magnitude may be expressed as

$$|Z_{\text{cell}}| = \sqrt{Z_{\text{cell,r}}^2 + Z_{\text{cell,j}}^2} \tag{8.28}$$

The parameters given in Problem 8.9 may be associated with the geometric response of a highly resistive lubrication oil, for which the resistance is large and the capacitance is small. The measured impedance, represented by the circuit given in Figure 8.5, is given by

$$Z_{\text{meas}} = \frac{(Z_{\text{L}} + Z_{\text{cell}})\, Z_{\text{C}}}{(Z_{\text{L}} + Z_{\text{cell}}) + Z_{\text{C}}} \tag{8.29}$$

where Z_{L} is given by equation (8.21) and Z_{C} is given by equation (8.19). The results are shown in Figure 8.7(a). At low frequencies, the impedance is well below the line indicated as $Z_{\text{max,1\%}}$, but, at frequencies larger than 10 Hz, the impedance falls between $Z_{\text{C,1\%}}$ and Z_{C}. The line and symbols seem to be superposed, but the calculated error shown in Figure 8.7(b) approaches a −9.1% error at frequencies above 1 kHz.

This calculation shows the value of the accuracy contour plot. Experimental measurements that extend beyond the $\varepsilon_{|Z|} \leq 1\%$ region are not obviously corrupted by the electric response of the instrument and leads. Knowledge of the instrumentation limits can prevent mistakes in the lab.

8.10 Calculate the impedance for a system described by a parallel combination of a resistor $R_{\text{cell}} = 1 \times 10^{10}\ \Omega$ and a capacitor $C_{\text{cell}} = 1 \times 10^{-15}$ F. Plot the cell impedance and the measured impedance on the figure produced for Problem 8.8.b. Estimate the error as a function of frequency.

Solution 8.10. This problem is similar to that presented in Problem 8.9 with a smaller value for the cell capacitance. The comparison of the calculated impedance to the accuracy contour plot for the parameters given in Problem 8.10 is presented in Figure 8.8(a), in which the magnitude of the impedance represented by equation (8.29)

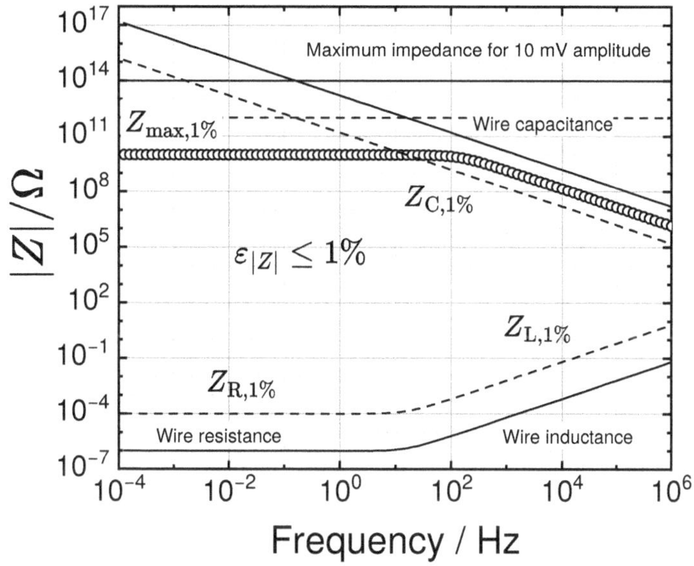

(a)

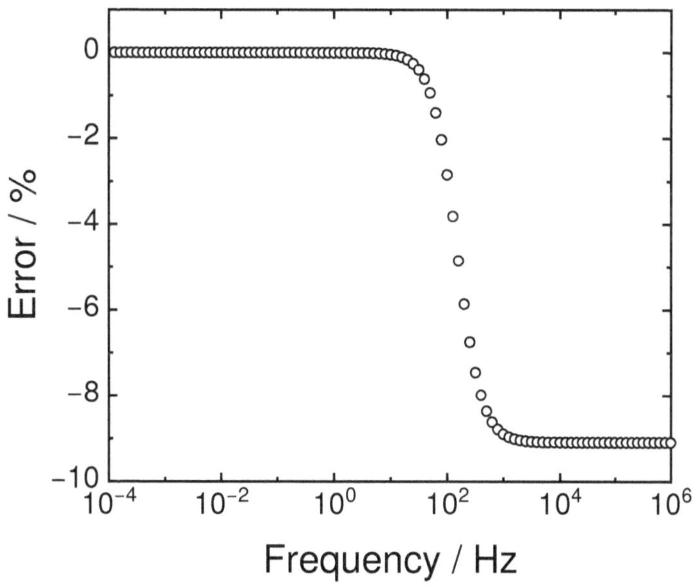

(b)

Figure 8.7: Comparison of the calculated impedance to the accuracy contour plot for the parameters given in Problem 8.9: (a) accuracy contour plot, in which the magnitude of the impedance represented by equation (8.29) is given as open circles and the cell impedance represented by equation (8.27) is represented by a solid line, and (b) error in the measured impedance given by equation (8.29) compared to the cell impedance given by equation (8.27).

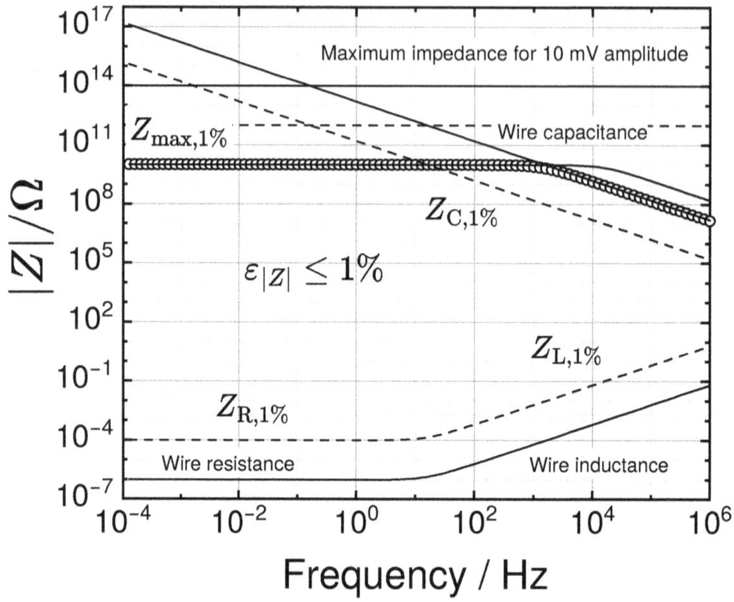

(a)

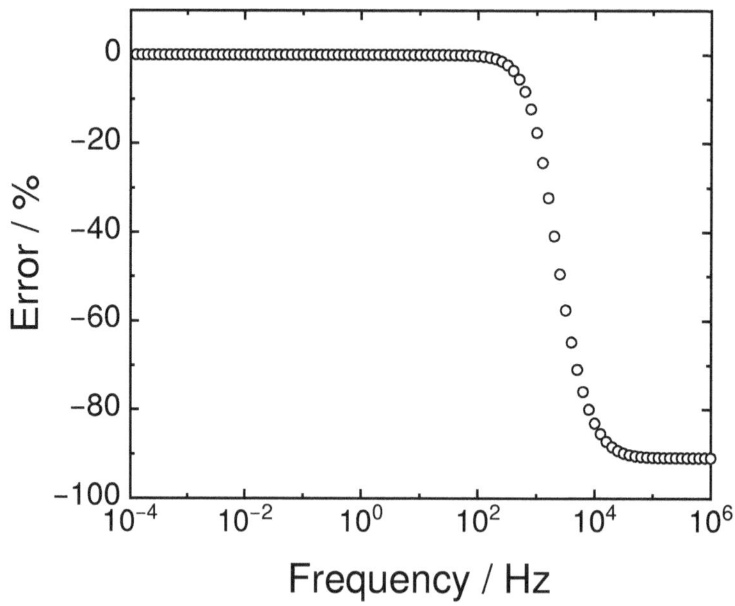

(b)

Figure 8.8: Comparison of the calculated impedance to the accuracy contour plot for the parameters given in Problem 8.10: (a) accuracy contour plot, in which the magnitude of the impedance represented by equation (8.29) is given as open circles and the cell impedance represented by equation (8.27) is represented by a solid line, and (b) error in the measured impedance given by equation (8.29) compared to the cell impedance given by equation (8.27).

is given as open circles and the cell impedance represented by equation (8.27) is represented by a solid line. In Figure 8.8(a), the cell impedance, representing the desired measurement, extends beyond the line associated with the capacitance of the wires; whereas, the measured impedance follows the wire capacitance line. The error, shown in Figure 8.8(b), reaches a value of 91% at frequencies above 10 kHz.

8.11 Calculate the impedance for a system described by a parallel combination of a resistor $R_{\text{cell}} = 1\ \Omega$ and a capacitor $C_{\text{cell}} = 1 \times 10^{-3}$ F. Plot the cell impedance and the measured impedance on the figure produced for Problem 8.8.b. Estimate the error as a function of frequency.

Solution 8.11. This problem is also similar to that presented in Problem 8.9, with a much smaller value for the cell resistance and a much larger value for the capacitance. These parameters are typical of an energy system such as a battery or fuel cell in which the impedance is small, and the large capacitance reflects a large surface area.

The comparison of the calculated impedance to the accuracy contour plot for the parameters given in Problem 8.10 is presented in Figure 8.9(a), in which the magnitude of the impedance represented by equation (8.29) is given as open circles and the cell impedance represented by equation (8.27) is represented by a solid line. In Figure 8.9(a), the cell impedance, representing the desired measurement, extends beyond the line associated with the inductance of the wires; whereas, the measured impedance follows the wire inductance line. The error is shown in Figure 8.9(b). The error reached a value of almost 40,000% at a frequency of 1 MHz.

The effect of the wire inductance on the measured impedance is shown in Figure 8.10. The inductive feature seen for frequencies larger than 1 kHz is typical of energy systems. Using a measurement model analysis, Roy and Orazem[24] showed that, for impedance measurements on PEM fuel cells, an inductive artifact existed for frequencies larger than 1 kHz and that the artifact extended into the negative imaginary impedance quadrant. An explanation is readily seen in Figure 8.8(a), where the lead inductance is seen to cancel the contribution of the cell capacitance for half a frequency decade before the measured impedance follows the inductance of the wires.

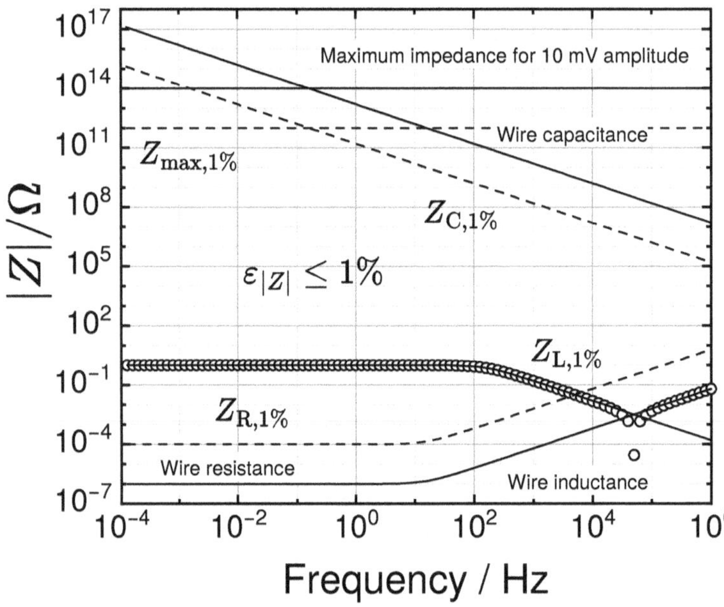

(a)

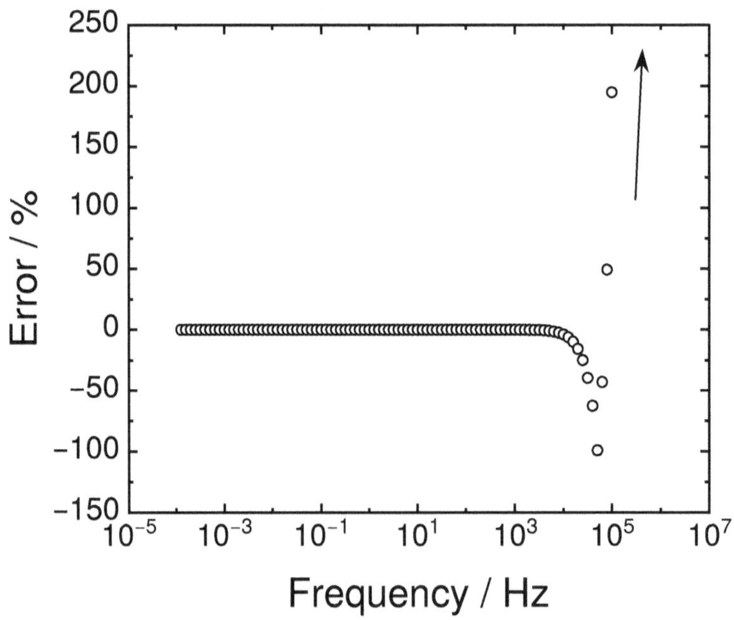

(b)

Figure 8.9: Comparison of the calculated impedance to the accuracy contour plot for the parameters given in Problem 8.11: (a) accuracy contour plot, in which the magnitude of the impedance represented by equation (8.29) is given as open circles and the cell impedance represented by equation (8.27) is represented by a solid line, and (b) error in the measured impedance given by equation (8.29) compared to the cell impedance given by equation (8.27).

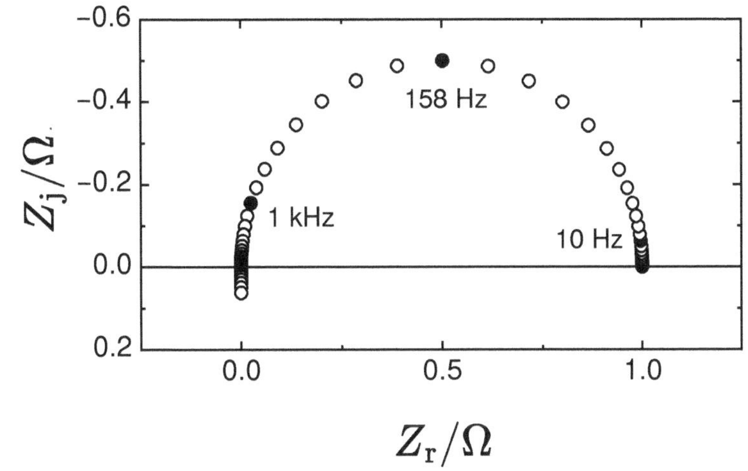

Figure 8.10: Calculated measured impedance in Nyquist format obtained from equation (8.29) for the parameters given in Problem 8.11. The inductive feature evident at high frequency is due to the inductance of the leads.

8.12 The accuracy contour plot presented in Problem 8.8 is based on using a potentiostatic modulation with a potential perturbation of 10 mV. Show that line E in Figure 8.3 can be lowered by using a smaller potential perturbation or by using galvanostatic modulation.

Solution 8.12. Line E in Figure 8.3 was obtained from equation (8.26). Use of a smaller potential perturbation will lower the value of the minimum measurable impedance. For example, use of a $\Delta V = 0.001$ V will reduce the minimum measurable impedance from $Z_{\text{min,1\%}} = 0.0167\ \Omega$ to $Z_{\text{min,1\%}} = 0.00167\ \Omega$. Under galvanostatic modulation, the issue will be the resolution of the potential measurement. Under the assumption that the potential can be resolved to 1 μV, the impedance under galvanostatic modulation with a 0.6 A perturbation can be

$$Z_{\text{min}} = \frac{\Delta V}{\Delta I} = \frac{1 \times 10^{-6}\ \text{V}}{0.6\ \text{A}} = 1.67 \times 10^{-6}\ \Omega \tag{8.30}$$

approaching the value of R_{wire} provided in Problem 8.8.a. As the accuracy of the potential measurement is controlled in this case by the potential measurement, the boundary of the $\varepsilon_{|Z|} < 1\%$ domain can be expressed as

$$Z_{\text{min}} = \frac{100\Delta V}{\Delta I} = \frac{1 \times 10^{-4}\ \text{V}}{0.6\ \text{A}} = 1.67 \times 10^{-4}\ \Omega \tag{8.31}$$

The lower bound of the accuracy contour plot can, therefore, be lowered by using galvanostatic modulation.

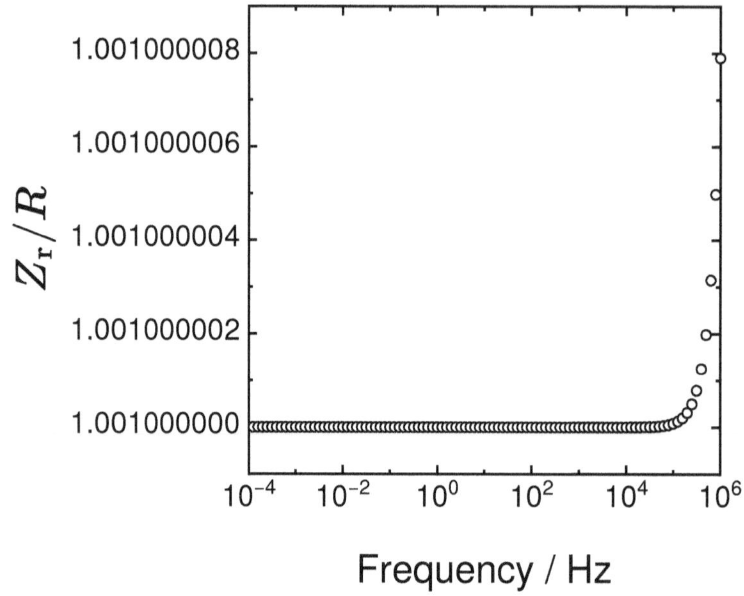

Figure 8.11: Calculated real part of the impedance scaled by resistor value obtained following equation (8.29) for the parameters given in Problem 8.13. The increase in resistance evident at high frequency is caused by the contribution of the lead capacitance as Z_C is no longer negligible at high frequency.

8.13 For the accuracy contour plot presented in Problem 8.8, calculate the measured impedance for a resistor with a value of 10^{-3} Ω. Plot the resulting real and imaginary parts of the impedance as functions of frequency.

Solution 8.13. The measured impedance is given by equation (8.29), where Z_L is given by equation (8.21), Z_C is given by equation (8.19), and

$$Z_{cell} = R = 10^{-3}\ \Omega \tag{8.32}$$

As the impedance $Z_C \gg R$, the measured impedance is given by

$$Z_{meas} = R_{wire} + R + j2\pi f L_{wire} \tag{8.33}$$

The real part of the impedance is shown in Figure 8.11 as a function of frequency. The increase in resistance evident at high frequency is caused by the contribution of the lead capacitance as Z_C is no longer negligible at high frequency, and the assumption used to derive equation (8.33) is not valid at high frequency. The influence of the resistance of the wires is evident in the value of the measured impedance, $Z_r/R = 1.001$.

The imaginary part of the impedance is shown in Figure 8.12. The increase in imag-

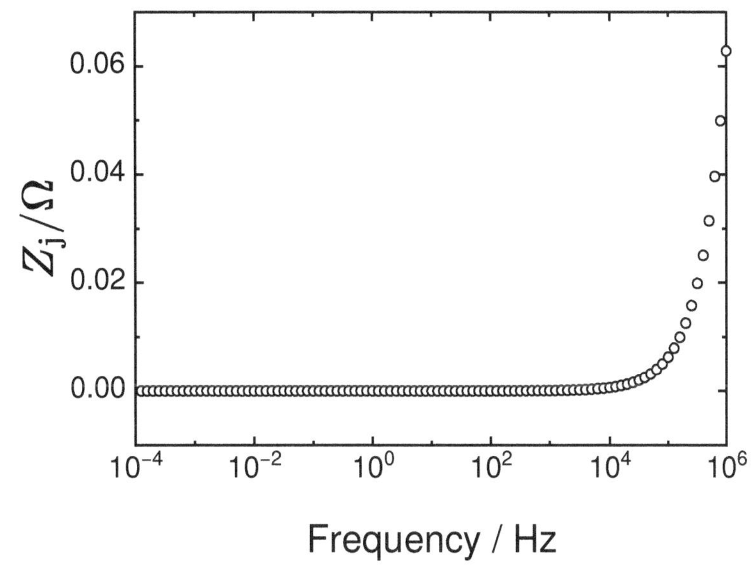

Figure 8.12: Calculated imaginary part of the impedance following equation (8.29) for the parameters given in Problem 8.13. The increase in imaginary impedance evident at high frequency is due to the inductance of the wires.

inary impedance evident at high frequency is due to the inductance of the wires. The influence of wire inductance may be seen more vividly by plotting the imaginary part of the impedance on a logarithmic scale. The results are shown in Figure 8.13. The horizontal line in Figure 8.13 corresponds to the magnitude of the impedance of the resistor.

8.14 The lead resistance, inductance, and capacitance are influenced by the type of connections that are made to the system or device being measured. Consider the impedance of a ZnO layer that acts as a resistance with value $R_{\text{ZnO}} = 1.34\ \Omega\text{cm}^2$ and an area $A_{\text{ZnO}} = 0.04\ \text{cm}^2$. The apparent values for lead properties were $C_{\text{wire}} = 3 \times 10^{-10}$ F, $R_{\text{wire}} = 7.3 \times 10^{-2}\ \Omega$, and $L_{\text{wire}} = 5 \times 10^{-7}$ H.

(a) Plot the measured ZnO impedance in the corresponding accuracy contour plot.

(b) Plot the percent error in the measured ZnO impedance as a function of frequency.

(c) Plot the real and imaginary parts of the measured ZnO impedance as functions of frequency.

(d) Plot the measured ZnO impedance in Nyquist format.

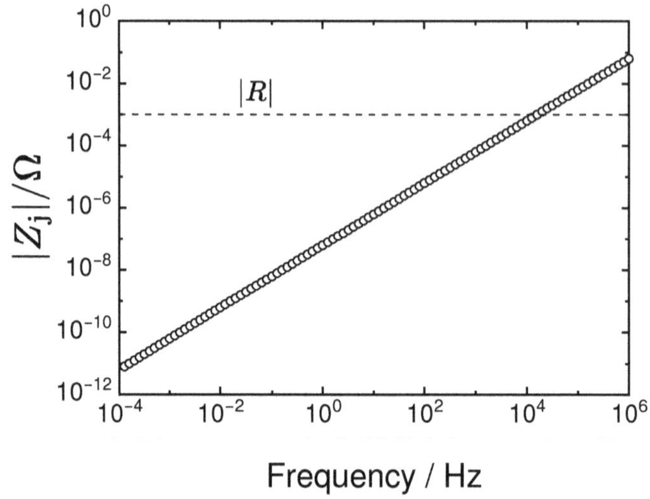

Figure 8.13: Calculated imaginary part of the impedance following equation (8.29) for the parameters given in Problem 8.13.

Solution 8.14.

(a) The measured impedance is given by equation (8.29), where Z_L is given by equation (8.21), Z_C is given by equation (8.19), and

$$Z_{cell} = Z_{ZnO} = \frac{R_{ZnO}}{A_{ZnO}} = 33.5\ \Omega \tag{8.34}$$

The corresponding accuracy contour plot is shown in Figure 8.14. The zone for which $\varepsilon_{|Z|} \leq 1\%$ is much smaller than that shown in Figure 8.6, but the impedance seems to be well within the lines for the lead capacitance, resistance, and inductance.

(b) The percent error for the impedance measurement is shown in Figure 8.15. The error at low frequency of about 0.2% can be attributed to the lead resistance. The increase in error at high frequency can be attributed to the lead capacitance. The lead inductance does not affect the real part of the impedance and, therefore, does not influence the error in the measurement of the ZnO resistance.

(c) The real and imaginary parts of the impedance are presented in Figure 8.16. The value of the real part of the impedance increases from 33.57 Ω at low frequencies to 33.83 Ω at 10^6 Hz. These values may be compared to the input resistance value of 33.50 Ω. The increase in the imaginary part of the impedance for frequencies greater than 10^4 Hz can be attributed to the influence of the lead inductance, even though the synthetic data, as shown in Figure 8.14, fall within the bounds formed by the lead inductance and capacitance.

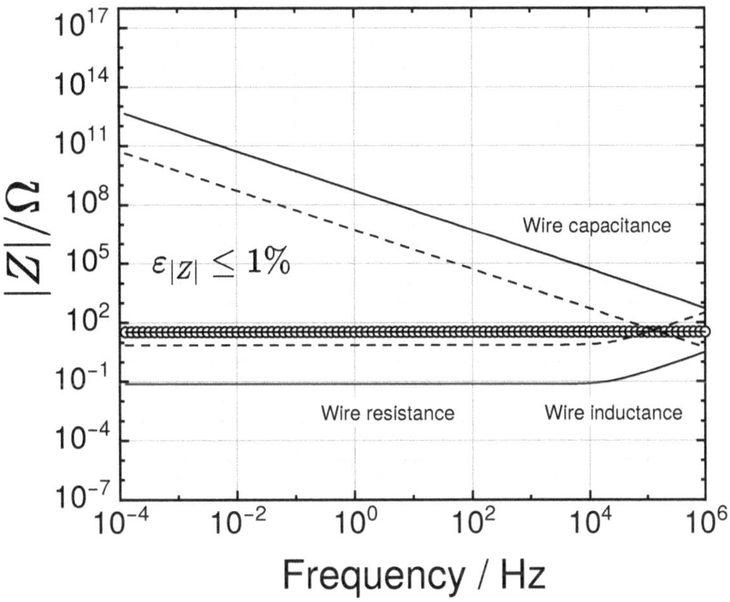

Figure 8.14: Accuracy contour plot for Problem 8.14 showing the region for which the measured modulus of the impedance will have an error less than or equal to 1%. The calculated impedance for the ZnO device is presented as circles.

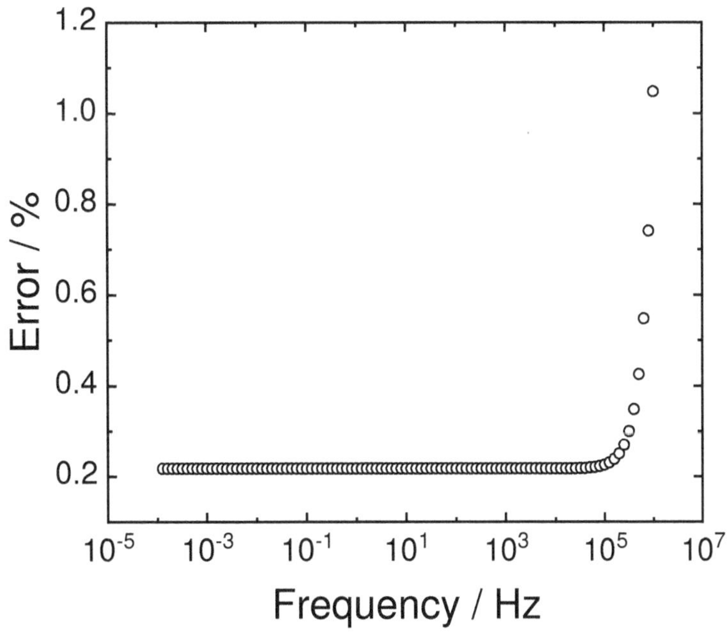

Figure 8.15: The percent error for the measurement of the resistance of the ZnO device described in Problem 8.14. The error at low frequency of about 0.2% can be attributed to the lead resistance. The increase in error at high frequency can be attributed to the lead capacitance.

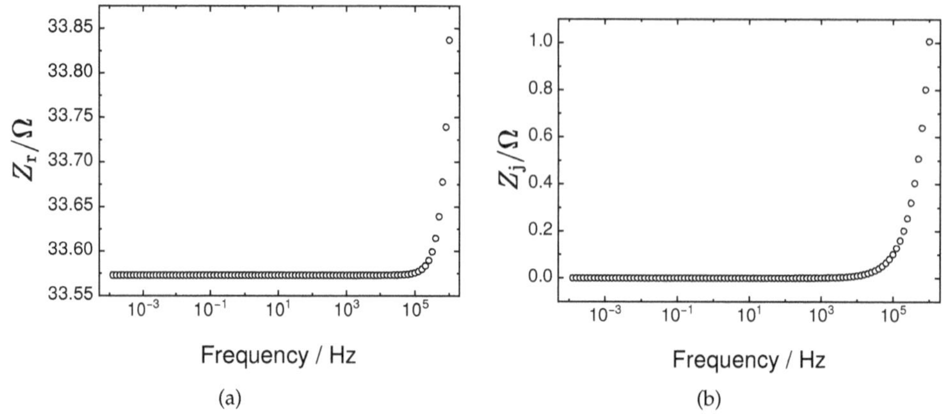

Figure 8.16: The calculated impedance for the ZnO device described in Problem 8.14: (a) real part of the impedance and (b) imaginary part of the impedance.

(d) The Nyquist plot is presented in Figure 8.17. As shown in Figure 8.16, the real part of the impedance behaves as a resistor for frequencies less than 100 kHz, and the imaginary part of the impedance becomes substantially greater than zero for frequencies greater than 10 kHz. The curvature of the high-frequency inductive feature shown in Figure 8.17 can be attributed to the combined effects of the lead inductance and capacitance. In contrast, the results presented in Figure 8.10, with a high-frequency inductive feature that resembles a vertical line, can be attributed to the influence of the lead inductance.

8.15 The data presented in Figure 8.18 were taken from Makharia et al.[25] for a pure-hydrogen-fueled polymer electrolyte fuel cell at a relatively low current density of 0.03 A/cm^2. Similar results were seen for higher current densities. Inductive features are seen at both high and low frequencies. Provide a suitable hypothesis for both the high-frequency and low-frequency inductive loops.

Solution 8.15. The high-frequency inductive feature may be attributed to the inductance of the leads. This hypothesis is supported by the relatively low value of the impedance for frequencies higher than 1 kHz. A similar behavior is shown in Figure 8.10. The low-frequency inductive feature may possibly be associated with the chemistry of the fuel cell. A similar behavior in PEM fuel cells has been attributed to formation of oxides on the Pt catalyst.[26–28]

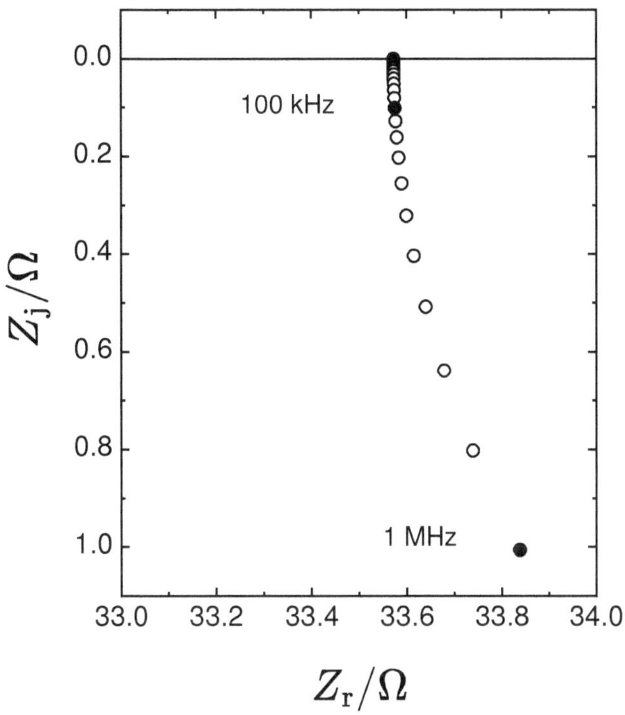

Figure 8.17: The calculated impedance for the ZnO device described in Problem 8.14 in Nyquist format.

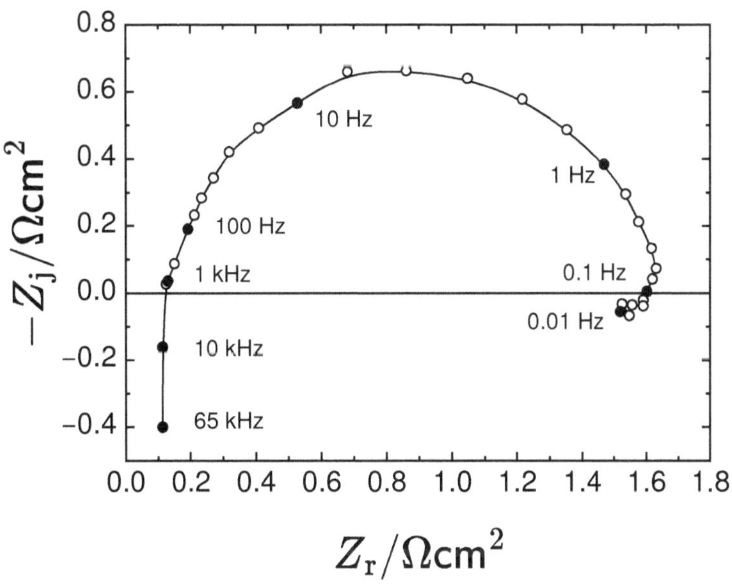

Figure 8.18: Impedance data for a pure-hydrogen-fueled polymer electrolyte fuel cell in Nyquist format. The data were taken from Makharia et al.[25] The current density was 0.03 A/cm^2.

8.16 Please answer the following in a few sentences each.

(a) Explain why the maximum useful frequency for low-impedance systems such as batteries and fuel cells is on the order of 1 kHz, even when using impedance instrumentation that has a maximum frequency of 1 MHz.

(b) Why are long cables not recommended for connecting impedance instrumentation to electrochemical systems?

(c) Superposition in a Nyquist plot of dissimilar impedance data collected at different elapsed times has been used to suggest that the impedance model should be the same for each data set, although model parameters may be dependent on elapsed time. Provide the dimensionless expressions used for the real and imaginary impedance to achieve this superposition.

Solution 8.16.

(a) The measured impedance for low-impedance systems may include the contribution of the resistance and inductance of the leads used to connect the instrument to the system under study.

(b) The capacitance, resistance, and inductance of the cable is proportional to cable length. Use of long cables to connect the instrument to the system under investigation will reduce the range of impedance that may be accurately measured.

(c) The normalized real part of the impedance may be expressed as

$$Z_{\mathrm{r,norm}} = \frac{Z_{\mathrm{r}} - R_{\mathrm{e}}}{Z_{\mathrm{r}}(0) - R_{\mathrm{e}}} \tag{8.35}$$

and the normalized imaginary part of the impedance may be expressed as

$$Z_{\mathrm{j,norm}} = \frac{Z_{\mathrm{j}}}{Z_{\mathrm{r}}(0) - R_{\mathrm{e}}} \tag{8.36}$$

Equations (8.35) and (8.36) are useful for graphical interpretation of impedance data.

Chapter 9

Equivalent Circuit Analogs

Models for the impedance response may be developed from proposed hypotheses involving reaction sequences, mass transfer, and physical phenomena. These models can often be expressed in the mathematical formalism of electrical circuits. Electrical circuits can also be used to construct a framework for accounting for the phenomena that influence the impedance response of electrochemical systems. The problems presented in this chapter explore use of electrical circuits to create a framework for the system impedance response. The role of reaction mechanisms and transport are treated solely as a lumped faradaic impedance Z_{F}. Subsequent chapters explore the development of expressions for Z_{F}. Problems 9.1–9.9 are taken from Orazem and Tribollet.[1]

Problems

9.1 Consider the schematic representation of a coated electrode presented in Figure 9.1. Develop the corresponding equivalent electrical circuit.

Solution 9.1. The solution to Problem 9.5 identifies conditions under which the ohmic resistance of pores can be neglected. Under the assumption that the thin dielectric is sufficiently thin that the ohmic resistance in the respective pores can be neglected, the circuit corresponding to the system shown schematically in Figure 9.1 is presented in Figure 9.2. In this case, the two capacitances $C_{\mathrm{L},1}$ and C_{dl} are in parallel; thus, only an effective capacitance $C_{\mathrm{L},1} + C_{\mathrm{dl}}$ can be seen in parallel to the Faradaic impedance Z_{F}.

9.2 Is the circuit developed for Problem 9.1 mathematically equivalent to that developed for Figure 9.3?

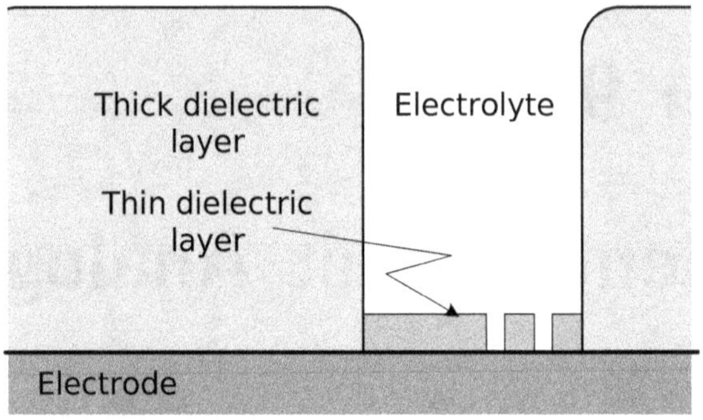

Figure 9.1: Schematic representation for the electrode coated by a thick dielectric layer with pores exposing a thin dielectric layer and the bare electrode as discussed in Problem 9.1. Taken from Orazem and Tribollet[1] with permission of John Wiley & Sons.

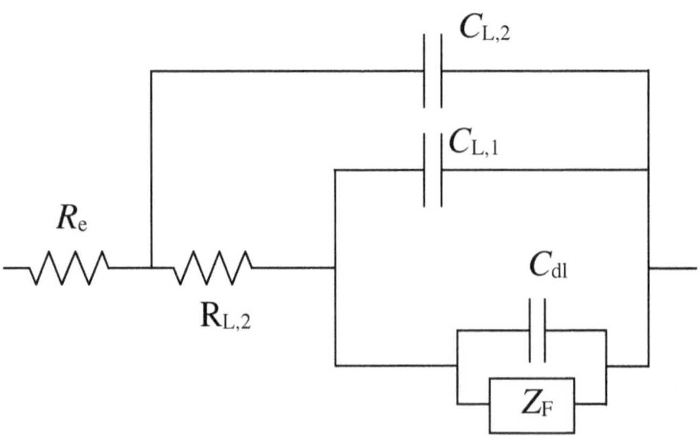

Figure 9.2: Electrical circuit corresponding to the system shown schematically in Figure 9.1.

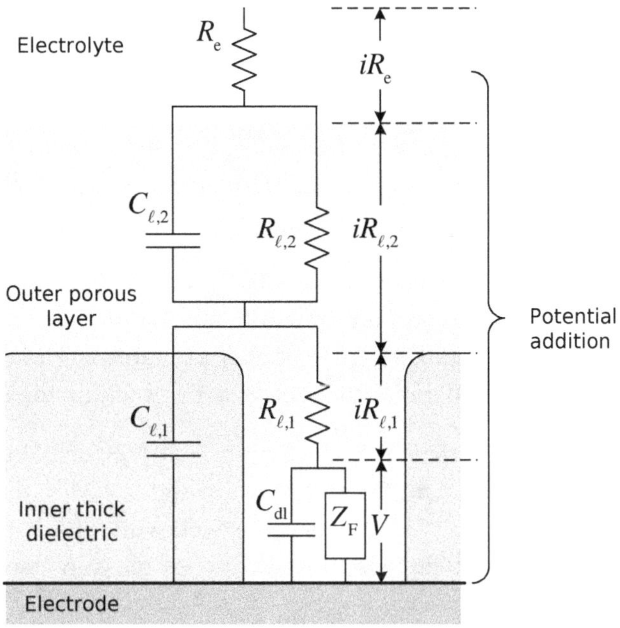

Figure 9.3: Equivalent electrical circuit of the impedance for an electrode coated by a porous layer covering a thick dielectric layer with pores exposing the electrode to the electrolyte. Taken from Orazem and Tribollet[1] with permission of John Wiley & Sons.

Solution 9.2. No, the two circuits are not mathematically equivalent. The circuit shown in Figure 9.2 has three capacitances, two ohmic resistances, and Z_F. The circuit shown in Figure 9.3 has three capacitances, three ohmic resistances, and Z_F. The two circuits would be mathematically equivalent if an ohmic resistance is added to Figure 9.2 to account for the pore resistance of the thin dielectric layer. As indicated in Problem 9.5, the pore resistance may be neglected for a film that is sufficiently thin.

9.3 Estimate the characteristic frequency for the impedance response of the outer porous layer of the system shown in Figure 9.3 for the following materials:

(a) Salt film with effective conductivity $\kappa = 10^{-4}$ S/cm

(b) Polymer coating with effective conductivity $\kappa = 10^{-8}$ S/cm

(c) Epoxy coating with effective conductivity $\kappa = 10^{-9}$ S/cm

Solution 9.3. The characteristic frequency is given by

$$f_{\ell 2} = \frac{1}{2\pi\varepsilon_{\ell 2}\varepsilon_0\rho_{\ell 2}} \tag{9.1}$$

where $\varepsilon_{\ell 2}$ is the dielectric constant of the film, $\varepsilon_0 = 8.8542 \times 10^{-14}$ F/cm is the permittivity of vacuum, and $\rho_{\ell 2} = 1/\kappa_{\ell 2}$ is the resistivity of the film in units of Ωcm.

(a) Salt film with effective conductivity $\kappa = 10^{-4}$ S/cm: Under the assumption that the dielectric constant for a salt film is $\varepsilon = 40$, the characteristic frequency is $f_c = 4.5$ MHz. This frequency is much higher than that used for typical electrochemical impedance spectroscopy measurements.

(b) Polymer coating with effective conductivity $\kappa = 10^{-8}$ S/cm: Under the assumption that the dielectric constant for a polymer coating is $\varepsilon = 3$, the characteristic frequency is $f_c = 6$ kHz. This frequency is within the range of that used for typical electrochemical impedance spectroscopy measurements.

(c) Epoxy coating with effective conductivity $\kappa = 10^{-9}$ S/cm: Under the assumption that the dielectric constant for an epoxy coating is $\varepsilon = 3$, the characteristic frequency is $f_c = 600$ Hz. This frequency is well within the range of that used for typical electrochemical impedance spectroscopy measurements.

9.4 Show that the characteristic frequency for the impedance of the outer porous layer of the system shown in Figure 9.3 is independent of film thickness. Show the influence of layer thickness on the resistance of the layer.

Solution 9.4. The characteristic frequency for the film is given as

$$f_{\ell 2} = \frac{1}{2\pi R_{\ell 2} C_{\ell 2}} \tag{9.2}$$

where

$$R_{\ell 2} = \delta_{\ell 2}/\kappa_{\ell 2} = \delta_{\ell 2}\rho_{\ell 2} \tag{9.3}$$

and

$$C_{\ell 2} = \varepsilon_{\ell 2}\varepsilon_0/\delta_{\ell 2} \tag{9.4}$$

Thus,

$$f_{\ell 2} = \frac{1}{2\pi\varepsilon_{\ell 2}\varepsilon_0\rho_{\ell 2}} \tag{9.5}$$

which is independent of film thickness. The layer resistance, however, is given by equation (9.3), which is proportional to film thickness.

9.5 Devise a criterion for deciding whether a dielectric film partially covering an electrode should be considered to be thin, as shown in Figure 9.4, or thick, as shown in Figure 9.5.

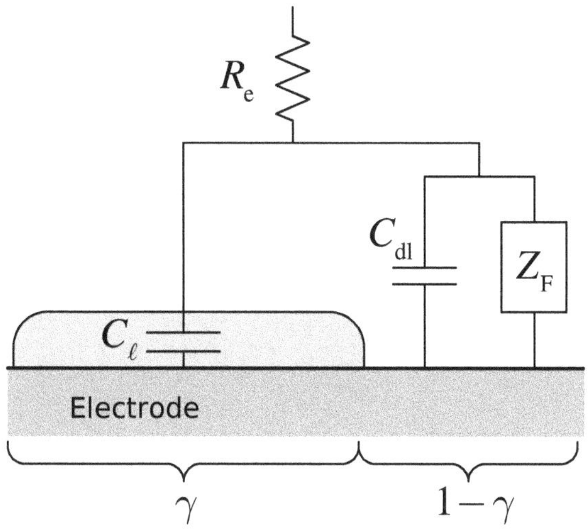

Figure 9.4: Equivalent electrical circuit for a partially blocked surface. Taken from Orazem and Tribollet[1] with permission of John Wiley & Sons.

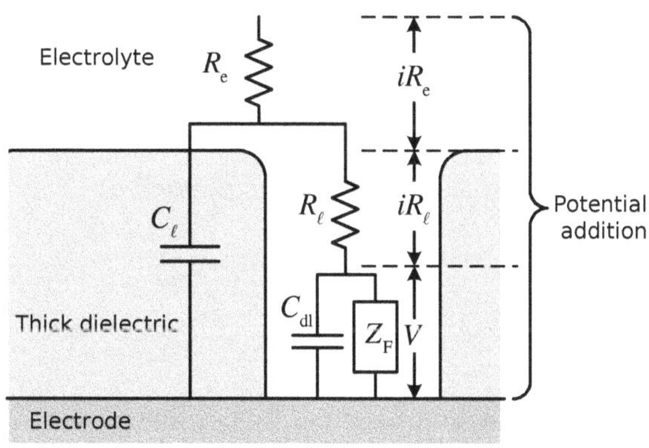

Figure 9.5: Equivalent electrical circuit of the impedance for an electrode coated by a thick dielectric layer with pores exposing the electrode to the electrolyte. Taken from Orazem and Tribollet[1] with permission of John Wiley & Sons.

Solution 9.5. The criterion is based on the question of whether the pore resistance can be considered negligible as compared to the interfacial impedance. The interfacial impedance is smallest at high frequency, therefore, the pore resistance, expressed as

$$R_{\text{pore}} = \rho_{\text{pore}}\delta_{\text{pore}} \tag{9.6}$$

should be less than

$$Z_{\text{C}} = \frac{1}{\omega C_{\text{dl}}} \tag{9.7}$$

The criterion is, therefore, that $R_{\text{pore}}/Z_{\text{C}} \ll 1$, or that

$$\delta_{\text{pore}} \ll \frac{1}{\omega C_{\text{dl}}\rho_{\text{pore}}} \tag{9.8}$$

or

$$\delta_{\text{pore}} = \frac{\xi_{\text{pore}}}{\omega C_{\text{dl}}\rho_{\text{pore}}} \tag{9.9}$$

where ξ_{pore} is a number much less than unity. The film thickness that corresponds to a thin film depends on the electrolyte resistivity, the double-layer capacitance, and the value of the highest frequency measured. For $\xi_{\text{pore}} = 0.001$, $\rho_{\text{pore}} = 50\ \Omega\text{cm}$, $C_{\text{dl}} = 20\ \mu\text{F/cm}^2$, and a maximum frequency of 100 kHz ($\omega = 1.6 \times 10^4\ \text{s}^{-1}$), the maximum film thickness is 16 nm.

9.6 Consider that a thin 10 nm thick dielectric film of α-Fe_2O_3– (hematite) covers a metal electrode and that the fractional coverage of the electrode is $\gamma = 0.7$. Estimate the capacitance of the system.

Solution 9.6. The capacitance can be expressed as

$$C_{\text{eff}} = \gamma C_{\ell} + (1-\gamma)C_{\text{dl}} \tag{9.10}$$

where

$$C_{\ell} = \frac{\varepsilon\varepsilon_0}{\delta} \tag{9.11}$$

The dielectric constant for hematite is $\varepsilon = 12$, thus, $C_{\ell} = 1\ \mu\text{F/cm}^2$. Under the assumption that the double-layer capacity is $C_{\text{dl}} = 20\ \mu\text{F/cm}^2$, $C_{\text{eff}} = 6.7\ \mu\text{F/cm}^2$.

9.7 How is the circuit of Figure 9.3 affected by a nonuniform thickness of the outer salt layer? Can the nonuniformity of the salt-film layer thickness be identified in the impedance response?

Solution 9.7. The part of the circuit shown in Figure 9.3 associated with the salt film could be expressed as the parallel combination of resistors and capacitors. These circuit elements could not be distinguished from an effective resistance

$$R_{\text{eff}} = \left(\frac{1}{R_{\ell 2,1}} + \frac{1}{R_{\ell 2,2}} + \ldots \right)^{-1} \tag{9.12}$$

in parallel with an effective capacitance

$$C_{\text{eff}} = C_{\ell 2,1} + C_{\ell 2,2} + \ldots \tag{9.13}$$

Thus, impedance cannot be used to determine if the outer salt film has a uniform thickness.

9.8 Under the assumption that Figure 7.6 properly describes a partially blocked electrode, will the associated equivalent circuit account for the influence of nonuniform current and potential distributions associated with this geometry?

Solution 9.8. No, the circuit does not account for the influence of geometry-induced nonuniform current and potential distributions. These effects require numerical solutions of the appropriate governing differential equations.

9.9 If $Z_{\text{a}} \gg Z_{\text{c}}$,

$$Z = R_{\text{e}} + \frac{1}{\dfrac{1}{Z_{\text{a}}} + \dfrac{1}{Z_{\text{c}}} + j\omega C_{\text{dl}}} \tag{9.14}$$

for corrosion at the open-circuit potential can be expressed as

$$Z = R_{\text{e}} + \frac{Z_{\text{c}}}{1 + j\omega C_{\text{dl}} Z_{\text{c}}} \tag{9.15}$$

Conversely, if $Z_{\text{c}} \gg Z_{\text{a}}$, equation (9.14) can be expressed as

$$Z = R_{\text{e}} + \frac{Z_{\text{a}}}{1 + j\omega C_{\text{dl}} Z_{\text{a}}} \tag{9.16}$$

Verify the results presented in Table 9.1.

Solution 9.9. If both reactions are kinetically controlled, then $Z_{\text{c}} = R_{\text{t,c}}$ and

Table 9.1: Results of experiments performed to determine if the simplifications to equation (9.14) expressed as equations (9.15) and (9.16) are justified. Experiments are performed at V_{corr}, $V_{corr}+\Delta V$, and $V_{corr}-\Delta V$. Typical values of ΔV are between 10 and 50 mV. Results are presented in terms of the expected change in the impedance as compared to the value measured at the open-circuit potential.

Assumption	$V_{corr}+\Delta V$	$V_{corr}-\Delta V$
$Z_c \gg Z_a$	Decrease $\Downarrow$	Increase $\Uparrow$
$Z_c \ll Z_a$	Increase $\Uparrow$	Decrease $\Downarrow$
$Z_c \approx Z_a$	Decrease $\Downarrow$	Decrease $\Downarrow$
$Z_a \neq f(V)$ & $Z_a \gg Z_c$	Increase $\Uparrow$	Decrease $\Downarrow$
$Z_a \neq f(V)$ & $Z_a \ll Z_c$	Unchanged $\Leftrightarrow$	Unchanged $\Leftrightarrow$

$Z_a = R_{t,a}$. The cathodic charge-transfer resistance may be expressed as

$$R_{t,c} = \frac{1}{b_c K_c \exp\left(-b_c\left(\overline{V}-V_{corr}\right)\right)} \tag{9.17}$$

As potential $\overline{V}$ increases, $R_{t,c}$ increases. Conversely, as $\overline{V}$ decreases, $R_{t,c}$ decreases. The anodic charge-transfer resistance may be expressed as

$$R_{t,a} = \frac{1}{b_a K_a \exp\left(b_a\left(\overline{V}-V_{corr}\right)\right)} \tag{9.18}$$

As $\overline{V}$ increases, $R_{t,a}$ decreases, and, as $\overline{V}$ decreases, $R_{t,c}$ increases. For $R_c \gg R_a$, equation (9.16) applies, and the impedance decreases at potentials larger than the open-circuit potential and the impedance increases at potentials smaller than the open-circuit potential. A similar argument applies for each of the entries in Table 9.1.

9.10 Orazem et al.[29] present a polarization curve, shown in Figure 9.6, that they attributed to corrosion of steel $Fe \rightarrow Fe^{2+} + 2e^-$, reduction of dissolved oxygen, $O_2 + 2H_2O + 4e^- \rightarrow 4OH^-$, and hydrogen evolution $H_2O + e^- \rightarrow \frac{1}{2}H_2 + OH^-$.

(a) Derive expressions for current density that could be used to reproduce Figure 9.6. Estimate the concentration of oxygen at the electrode surface in terms of a mass-transfer-limited current density.

(b) Develop an electrical circuit framework that describes the expected impedance response at the open circuit condition. Use a general expression $Z_{F,k}$ for the faradaic impedance of reaction k.

(c) Show the circuit that should apply at an applied potential of −400 mV referenced to the Cu|CuSO$_4$ electrode.

(d) Show the circuit that should apply at an applied potential of −1200 mV referenced to the Cu|CuSO$_4$ electrode.

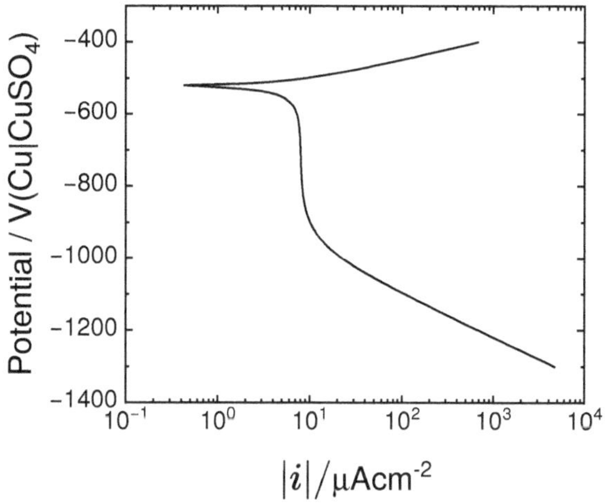

Figure 9.6: Polarization curve for corrosion of a steel pipe adapted from Orazem et al.[29]

Solution 9.10.

(a) The current density associated with the corrosion reaction may be expressed as

$$i_{\mathrm{Fe}} = K_{\mathrm{Fe}}^{*} \exp\left(b_{\mathrm{Fe}}\left(V - V_{\mathrm{Fe}}^{\circ}\right)\right) \tag{9.19}$$

where K_{Fe}^{*} is a rate constant with units of current density, $b_{\mathrm{Fe}} = 2,303/\beta_{\mathrm{Fe}}$, β_{Fe} is the Tafel slope with units of mV/decade, and V_{Fe}° is the equilibrium potential for the corrosion reaction. Similarly, the current density for the hydrogen evolution reaction may be expressed as

$$i_{\mathrm{H_2}} = -K_{\mathrm{H_2}}^{*} \exp\left(-b_{\mathrm{H_2}}\left(V - V_{\mathrm{H_2}}^{\circ}\right)\right) \tag{9.20}$$

The current density for oxygen reduction depends on the concentration of oxygen at the electrode surface. Thus,

$$i_{\mathrm{O_2}} = -K_{\mathrm{O_2}}^{*} c_{\mathrm{O_2}}(0) \exp\left(-b_{\mathrm{O_2}}\left(V - V_{\mathrm{O_2}}^{\circ}\right)\right) \tag{9.21}$$

Following the solution for Problem 5.4,

$$\frac{c_{\mathrm{O_2}}(0)}{c_{\mathrm{O_2}}(\infty)} = 1 - \frac{i_{\mathrm{O_2}}}{i_{\mathrm{lim,O_2}}} \tag{9.22}$$

where both $i_{\mathrm{O_2}}$ and $i_{\mathrm{lim,O_2}}$ are negative. Thus, equation (9.21) may be expressed as

$$i_{\mathrm{O_2}} = -K_{\mathrm{O_2}}^{*} c_{\mathrm{O_2}}(\infty) \left(1 - \frac{i_{\mathrm{O_2}}}{i_{\mathrm{lim,O_2}}}\right) \exp\left(-b_{\mathrm{O_2}}\left(V - V_{\mathrm{O_2}}^{\circ}\right)\right) \tag{9.23}$$

or

$$i_{O_2} = \frac{-K^*_{O_2} c_{O_2}(\infty) \exp\left(-b_{O_2}\left(V - V^\circ_{O_2}\right)\right)}{1 - \dfrac{K^*_{O_2} c_{O_2}(\infty)}{i_{\text{lim},O_2}} \exp\left(-b_{O_2}\left(V - V^\circ_{O_2}\right)\right)} \tag{9.24}$$

The total current density may be expressed as

$$i = i_{\text{Fe}} + i_{O_2} + i_{H_2} \tag{9.25}$$

While equation (9.19) is written in terms of three parameters, only two parameters are needed to define the value of the current density. Thus,

$$i_{\text{Fe}} = K_{\text{Fe}} \exp\left(b_{\text{Fe}} V\right) \tag{9.26}$$

$$i_{H_2} = -K_{H_2} \exp\left(-b_{H_2} V\right) \tag{9.27}$$

and

$$i_{O_2} = \frac{-K_{O_2} c_{O_2}(\infty) \exp\left(-b_{O_2} V\right)}{1 - \dfrac{K_{O_2} c_{O_2}(\infty)}{i_{\text{lim},O_2}} \exp\left(-b_{O_2} V\right)} \tag{9.28}$$

where $K_{\text{Fe}} = K^*_{\text{Fe}} \exp(-b_{\text{Fe}} V^\circ_{\text{Fe}})$ and $K_{H_2} = K^*_{H_2} \exp(b_{H_2} V^\circ_{H_2})$. As the bulk concentration of oxygen is independent of applied potential, the parameter $K_{O_2} c_{O_2}(\infty)$ may be treated as a lumped constant.

It should be noted that equation (9.28) is an expression of the Koutecky–Levich formulation,[30,31] in which

$$i_{k,O_2} = -K_{O_2} c_{O_2}(\infty) \exp\left(-b_{O_2} V\right) \tag{9.29}$$

and

$$i_{O_2} = \frac{i_{k,O_2}}{1 - \dfrac{i_{k,O_2}}{i_{\text{lim},O_2}}} \tag{9.30}$$

Equation (9.30) may be written in the usual Koutecky–Levich manner as

$$\frac{1}{i_{O_2}} = \frac{1}{i_{k,O_2}} + \frac{1}{i_{\text{lim},O_2}} \tag{9.31}$$

The calculated current density is compared in Figure 9.7 to values extracted from Figure 9.6. Model parameters are given in the figure caption.

(b) The general circuit for reactions on a bare electrode, shown in Figure 9.8(a), accounts for the contribution of faradaic impedances for corrosion, oxygen reduction, and hydrogen evolution reactions. As shown in Figure 9.7, the current density at the open-circuit potential associated with hydrogen evolution may be expected to be very small and the corrosion reaction is balanced by oxygen reduction. A small current density is associated with a large impedance; thus, the circuit associated with open-circuit condition is given in Figure 9.8(b).

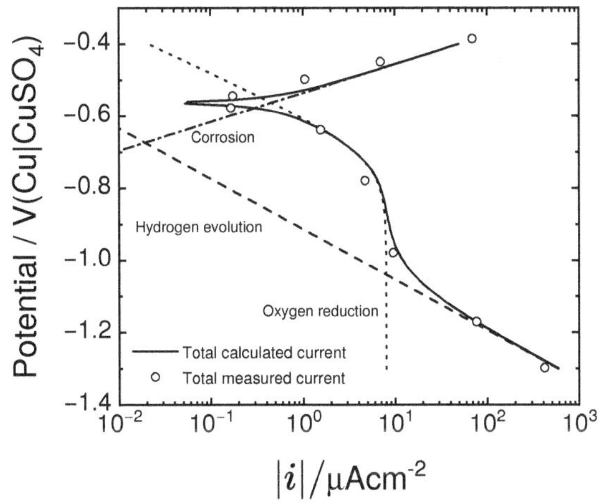

Figure 9.7: Calculated polarization curves corresponding to equations (9.25)–(9.28). Circles represent points extracted from Figure 9.6. The parameters used were $K_{Fe} = 5 \times 10^6$ μA/cm², $\beta_{Fe} =$ 80 mV/decade, $K_{O_2}c_{O_2}(\infty) = 1 \times 10^{-5}$ μA/cm², $\beta_{O_2} = -120$ mV/decade, $i_{\text{lim},O_2} = -8$ μA/cm², $K_{H_2} = 3 \times 10^{-7}$ μA/cm², and $\beta_{H_2} = -140$ mV/decade.

(c) At the anodic potential of −400 mV(Cu|CuSO$_4$), the contribution of the oxygen reduction current is negligible, and the corresponding circuit may be given by Figure 9.8(c).

(d) At the cathodic potential of −1,200 mV(Cu|CuSO$_4$), the dominant reaction is hydrogen evolution, and the corresponding circuit may be given by Figure 9.8(d).

9.11 Hirschorn et al.[32] discuss the role of a surface distribution of capacitance, charge-transfer resistance, and ohmic resistance on time-constant dispersion.

(a) Develop an expression for the impedance response for the circuit described in Figure 9.9(a). You may assume that each element, i.e., $R_{e,i}$, R_i, and C_i, has a unique value.

(b) Is a distributed time-constant response possible in the absence of an ohmic resistance, shown schematically in Figure 9.9(b)?

(c) Develop expressions for the effective resistance and capacitance associated with Figure 9.9(b).

Solution 9.11.

(a) The impedance of the circuit shown in Figure 9.9(a) may be described as the parallel combination of the impedance associated with individual elements, i.e.,

$$Z_i = R_{e,i} + \frac{R_i}{1 + j\omega R_i C_i} \tag{9.32}$$

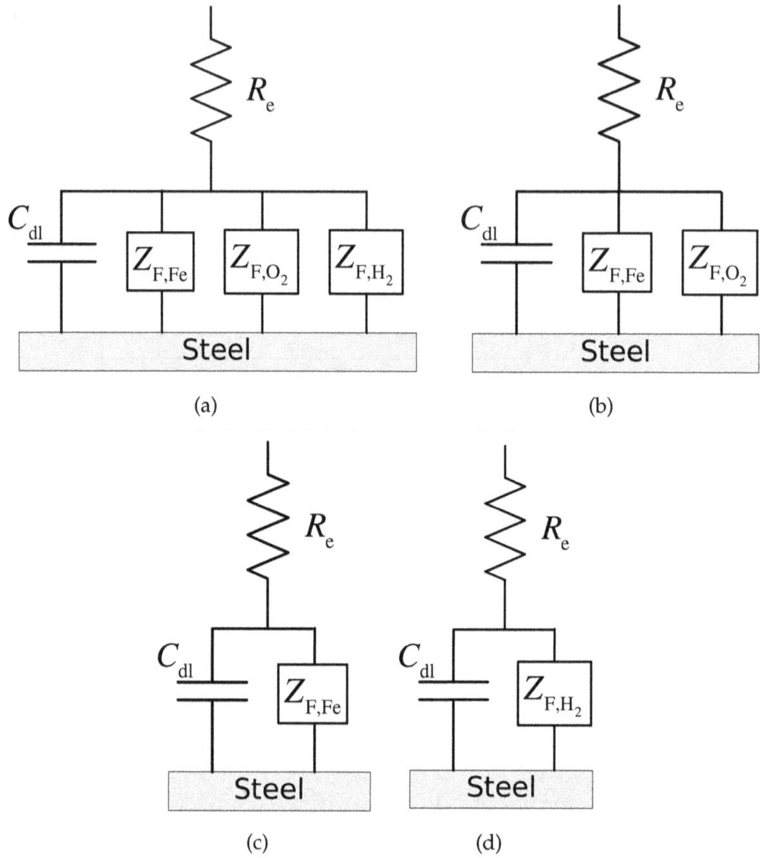

Figure 9.8: Electrical circuit frameworks for corrosion, oxygen reduction, and hydrogen evolution reactions on a bare steel electrode: (a) circuit showing influence of all three reactions; (b) circuit corresponding to the corrosion potential; (c) circuit corresponding to an anodic potential (-400 mV(Cu|$CuSO_4$); and (d) circuit corresponding to a cathodic potential ($-1,200$ mV(Cu|$CuSO_4$).

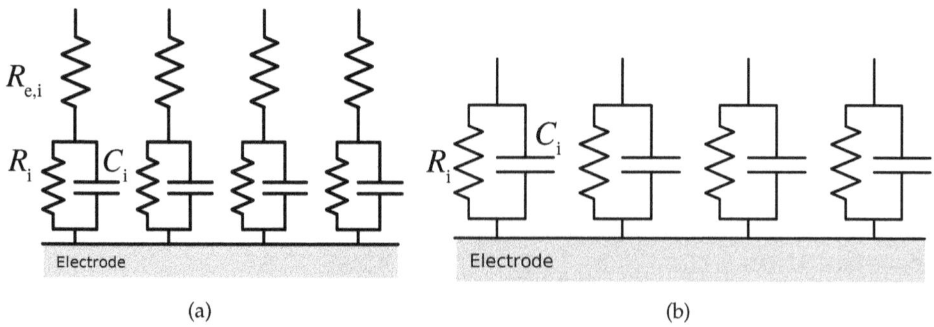

Figure 9.9: Schematic representation of a surface distribution of time constants: (a) leading to distributed time-constant behavior and (b) distribution in the limit that $R_e = 0$. Taken from Hirschorn et al.[32]

This situation corresponds to a sum of current contributions. Thus,

$$Z = \frac{1}{\sum_i \frac{1}{Z_i}} \tag{9.33}$$

Each of the impedance elements can be described as having two time constants, R_iC_i and R_eC_i. Thus, equation (9.33) represents a distribution of time constants.

(b) Interestingly, the time-constant distribution evident in equation (9.33) disappears in the absence of an ohmic resistance, as shown in Figure 9.9(b).

(c) Following equation (9.33), the effective resistance associated with a parallel combination of resistors shown in Figure 9.9(b) is given by

$$R_{eff} = \frac{1}{\sum_i \frac{1}{R_i}} \tag{9.34}$$

As the impedance associated with a capacitor is

$$Z_C = \frac{1}{j\omega C} \tag{9.35}$$

the effective capacitance associated with the parallel combination of capacitors shown in Figure 9.9(b) is given by

$$\frac{1}{j\omega C_{eff}} = \frac{1}{\sum_i j\omega C_i} \tag{9.36}$$

or

$$C_{eff} = \sum_i C_i \tag{9.37}$$

The circuit shown in Figure 9.9(b) may be drawn as the parallel combination of a single resistor and a single capacitor with values R_{eff} and C_{eff}, respectively.

9.12 We wish to study corrosion by measuring impedance at the open-circuit potential.

(a) Consider that, in addition to making impedance measurements at the open-circuit potential, we perform impedance measurements at potentials that are ± 10 mV from the open-circuit potential. What results will support using only the anodic faradaic impedance in the interpretation model?

(b) If only the anodic faradaic impedance can be extracted from the impedance data, does this mean that the cathodic current is equal to zero at open circuit?

Solution 9.12. At the open-circuit potential, the total current is equal to zero, and $i_a + i_c = 0$, where i_a is the anodic current and i_c is the cathodic current. The interpretation of impedance begins, then, with the recognition that the circuit includes the parallel contributions of the anodic impedance, the cathodic impedance, and the capacitance.

(a) Neglect of the cathodic faradaic impedance in the interpretation model described above would be supported by an impedance that is smaller at the OCP $+ \Delta V$ and larger at OCP $- \Delta V$.

(b) The statement that only the anodic impedance can be extracted from the impedance data suggests that the cathodic impedance is much larger than the anodic impedance. This does not imply that the cathodic current is equal to zero. The condition of an open-circuit potential requires that $i_a + i_c = 0$.

Chapter 10

Kinetic Models

The electrical circuits developed in Chapter 9 made use of boxes and undefined transfer functions Z_F to account for the impedance associated with interfacial reactions. In some cases, the interfacial impedance may be described in terms of circuit elements such as resistors and capacitors, but the nature of the impedance response depends on the proposed reaction mechanism. The objective of problems presented in this chapter is to explore the relationship between proposed reaction mechanisms and the interfacial impedance response. Problems 10.1–10.11 are based on problems presented by Orazem and Tribollet.[1]

Problems

10.1 Develop an expression for the faradaic impedance for the reaction

$$\mathrm{Ag} \rightarrow \mathrm{Ag}^{+} + \mathrm{e}^{-} \tag{10.1}$$

Solution 10.1. This problem involves the dissolution of silver in an aqueous medium. The faradaic current associated with this reaction can be expressed in terms of Tafel kinetics as

$$i_{\mathrm{Ag}} = K_{\mathrm{Ag}} \exp\left(b_{\mathrm{Ag}} V\right) \tag{10.2}$$

The current density given in equation (10.2) is a function only of potential. The steady current density increases exponentially as the potential $\overline{V}$ becomes more positive.

The phasor corresponding to the faradaic current response to a potential perturbation can be expressed as

$$\widetilde{i}_{\mathrm{Ag}} = \frac{\mathrm{d}i_{\mathrm{Ag}}}{\mathrm{d}V}\widetilde{V} = K_{\mathrm{Ag}} \exp\left(b_{\mathrm{Ag}}\overline{V}\right) b_{\mathrm{Ag}}\widetilde{V} \tag{10.3}$$

where $\widetilde{V}$ represents the potential phasor. The charge-transfer resistance for this reaction can be expressed as

$$R_{\mathrm{t,Ag}} = \frac{1}{K_{\mathrm{Ag}} \exp\left(b_{\mathrm{Ag}} \overline{V}\right) b_{\mathrm{Ag}}} = \frac{1}{\overline{i}_{\mathrm{Ag}} b_{\mathrm{Ag}}} \tag{10.4}$$

The faradaic impedance is obtained from

$$Z_{\mathrm{F,Ag}} = \frac{\widetilde{V}}{\widetilde{i}_{\mathrm{Ag}}} = R_{\mathrm{t,Ag}} \tag{10.5}$$

Equation (10.5) may be inserted in an electrical circuit framework such as those developed in Chapter 9.

10.2 Develop an expression for the faradaic impedance for the reaction

$$\mathrm{Ag}^{+} + \mathrm{e}^{-} \rightarrow \mathrm{Ag} \tag{10.6}$$

in which the concentration of Ag^{+} is influenced by mass transfer. Use a spreadsheet program to plot the current density as a function of potential. Use the same estimated parameters to plot the impedance response at a potential corresponding to 1/4, 1/2, and 3/4 of the mass-transfer-limited current density.

Solution 10.2. This problem is the reverse reaction of that presented in Problem 10.1. It looks similar, but now the reactant concentration is influenced by mass transfer. The current density is given by

$$i_{\mathrm{Ag}^{+}} = -K_{\mathrm{Ag}^{+}} c_{\mathrm{Ag}^{+}}(0) \exp\left(-b_{\mathrm{Ag}^{+}} V\right) \tag{10.7}$$

The current density is a function of both the concentration of species Ag^{+} at the electrode surface and the potential difference between the electrode and the solution adjacent to the electrode. The current density phasor is given by the solution to

$$\widetilde{i}_{\mathrm{Ag}^{+}} = \left(\frac{\partial i_{\mathrm{Ag}^{+}}}{\partial V}\right)_{c_{\mathrm{Ag}^{+}}(0)} \widetilde{V} + \left(\frac{\partial i_{\mathrm{Ag}^{+}}}{\partial c_{\mathrm{Ag}^{+}}(0)}\right)_{V} \widetilde{c}_{\mathrm{Ag}^{+}}(0) \tag{10.8}$$

Thus,

$$\widetilde{i}_{\mathrm{Ag}^{+}} = K_{\mathrm{Ag}^{+}} b_{\mathrm{Ag}^{+}} \overline{c}_{\mathrm{Ag}^{+}}(0) \exp\left(-b_{\mathrm{Ag}^{+}} \overline{V}\right) \widetilde{V} - K_{\mathrm{Ag}^{+}} \exp\left(-b_{\mathrm{Ag}^{+}} \overline{V}\right) \widetilde{c}_{\mathrm{Ag}^{+}}(0) \tag{10.9}$$

where $\widetilde{c}_{\mathrm{Ag}^{+}}(0)$ represents the concentration phasor for species Ag^{+} evaluated at the electrode surface.

The current density may be related to the flux of reactant to the surface according to

$$\bar{i}_{\mathrm{Ag}^+} = -\mathrm{F}D_{\mathrm{Ag}^+} \left. \frac{\mathrm{d}c_{\mathrm{Ag}^+}}{\mathrm{d}y} \right|_{y=0} \tag{10.10}$$

which can be expressed in terms of phasors as

$$\widetilde{i}_{\mathrm{Ag}^+} = -\mathrm{F}D_{\mathrm{Ag}^+} \left. \frac{\mathrm{d}\widetilde{c}_{\mathrm{Ag}^+}}{\mathrm{d}y} \right|_{y=0} \tag{10.11}$$

Equation (10.9) may be written

$$1 = \frac{1}{R_{\mathrm{t,Ag}^+}} \frac{\widetilde{V}}{\widetilde{i}_{\mathrm{Ag}^+}} - K_{\mathrm{Ag}^+} \exp\left(-b_{\mathrm{Ag}^+}\overline{V}\right) \frac{\widetilde{c}_{\mathrm{Ag}^+}(0)}{\widetilde{i}_{\mathrm{Ag}^+}} \tag{10.12}$$

where the charge-transfer resistance for reaction (10.6) is

$$R_{\mathrm{t,Ag}^+} = \frac{1}{K_{\mathrm{Ag}^+} b_{\mathrm{Ag}^+} \overline{c}_{\mathrm{Ag}^+}(0) \exp\left(-b_{\mathrm{Ag}^+}\overline{V}\right)} \tag{10.13}$$

Introduction of equation (10.11) yields

$$Z_{\mathrm{F,Ag}^+} = R_{\mathrm{t,Ag}^+} + Z_{\mathrm{D,Ag}^+} \tag{10.14}$$

where the diffusion impedance is given by

$$Z_{\mathrm{D,Ag}^+} = \frac{1}{b_{\mathrm{Ag}^+} \mathrm{F} D_{\mathrm{Ag}^+} \overline{c}_{\mathrm{Ag}^+}(0)} \left(\frac{-\widetilde{c}_{\mathrm{Ag}^+}(0)}{\left. \frac{\mathrm{d}\widetilde{c}_{\mathrm{Ag}^+}}{\mathrm{d}y} \right|_{y=0}} \right) \tag{10.15}$$

The diffusion impedance may be written in terms of a dimensionless position $\xi = y/\delta_{\mathrm{Ag}^+}$ and the dimensionless concentration phasor $\theta_{\mathrm{Ag}^+} = \widetilde{c}_{\mathrm{Ag}^+}/\widetilde{c}_{\mathrm{Ag}^+}(0)$ as

$$Z_{\mathrm{D,Ag}^+} = \frac{\delta_{\mathrm{Ag}^+}}{\mathrm{F}D_{\mathrm{Ag}^+}\overline{c}_{\mathrm{Ag}^+}(0)} \frac{1}{b_{\mathrm{Ag}^+}} \left(-\frac{1}{\theta'_{\mathrm{Ag}^+}(0)} \right) \tag{10.16}$$

The diffusion resistance may be expressed as

$$R_{\mathrm{d,Ag}^+} = \frac{\delta_{\mathrm{Ag}^+}}{\mathrm{F}D_{\mathrm{Ag}^+}\overline{c}_{\mathrm{Ag}^+}(0)} \frac{1}{b_{\mathrm{Ag}^+}} \tag{10.17}$$

and equation (10.14) may be expressed conveniently as

$$Z_{\mathrm{F,Ag}^+} = R_{\mathrm{t,Ag}^+} + R_{\mathrm{d,Ag}^+} \left(-\frac{1}{\theta'_{\mathrm{Ag}^+}(0)} \right) \tag{10.18}$$

Expressions for $-1/\theta'_{\mathrm{i}}(0)$ are developed in Chapter 11.

Table 10.1: Parameters employed for the simulations presented in the solution to Problem 10.2.

Parameter	Value	Units
$c_i(\infty)$	10^{-4}	mol/cm^3
δ_i	4×10^{-3}	cm
D_i	10^{-5}	cm^2/s
b_c	19.5	V^{-1}
K_c	10^{-2}	A cm/mol
C	2×10^{-5}	F/cm^2

The second part of the problem involves using a spreadsheet program to calculate a polarization curve and associated impedance. This is most conveniently done using the application of the Koutecky–Levich formulation to impedance presented in Tran et al.[33] and in Section 10.2.2 of Orazem and Tribollet.[1] The development is presented here for a generalized cathodic reaction. The current may be expressed as

$$\frac{1}{\bar{i}} = \frac{1}{\bar{i}_k} + \frac{1}{\bar{i}_{lim}} \tag{10.19}$$

where

$$\bar{i}_{lim} = -\frac{nFD_ic_i(\infty)}{\delta_i} \tag{10.20}$$

is the mass-transfer-limited current density and

$$\bar{i}_k = -K_c c_i(\infty)\exp(-b_c\bar{V}) \tag{10.21}$$

is the kinetic current based on the bulk concentration. The parameters used in the calculations are presented in Table 10.1. The resulting polarization curve is presented in Figure 10.1. The mass-transfer-limited current density, calculated to have a value of -0.0241 A/cm^2, is approached at negative potentials.

The next step is calculation of the charge-transfer and diffusion resistances. The charge-transfer resistance is a function of potential and can be obtained from

$$R_t = \frac{1}{b_c\bar{i}_{lim}}\left(1 + \frac{\bar{i}_{lim}}{\bar{i}_k}\right) \tag{10.22}$$

The diffusion resistance is obtained from

$$R_d = \frac{1}{b_c\bar{i}_{lim}}\left(1 + \frac{\bar{i}_k}{\bar{i}_{lim}}\right) \tag{10.23}$$

The results are presented in Figure 10.2. The value of the charge-transfer resistance decreases as potential becomes more negative, as is expected for a cathodic reaction. The value of the diffusion resistance decreases as potential becomes more positive, as is

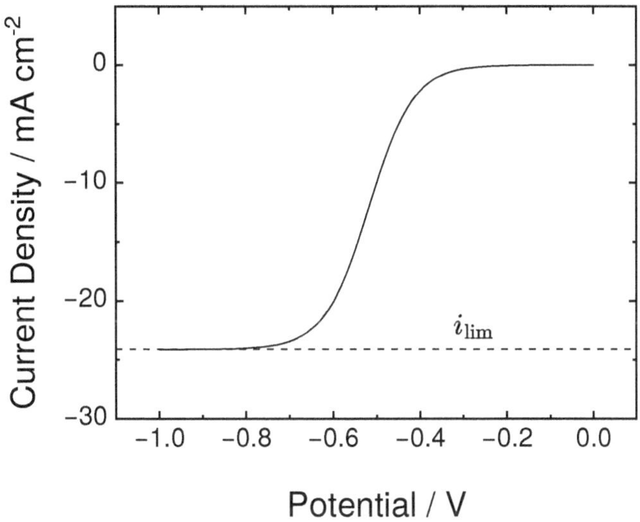

Figure 10.1: Polarization curve generated for the solution to Problem 10.2 using the parameters presented in Table 10.1.

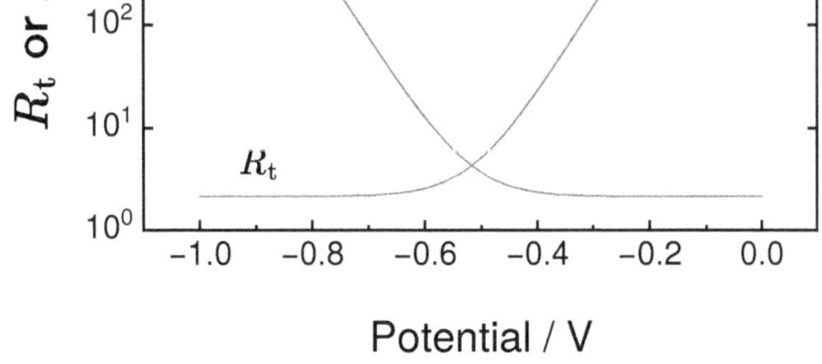

Figure 10.2: Charge-transfer and diffusion resistances for the solution to Problem 10.2 using equations (10.22) and (10.23), respectively. Parameters are presented in Table 10.1.

Table 10.2: Values of $\bar{i}_k/\bar{i}_{lim}$ corresponding to specified values of $\bar{i}/\bar{i}_{lim}$. Calculated values for charge-transfer and diffusion resistances are also presented.

$\bar{i}/\bar{i}_{lim}$	0.25	0.5	0.75
$\bar{i}_k/\bar{i}_{lim}$	1/3	1	3
R_t, Ωcm^2	8.5039	4.2519	2.8346
R_d, Ωcm^2	2.8346	4.2519	8.5039

expected for conditions under which mass transfer plays a decreased role. The limiting values for these parameters are equal, i.e.,

$$\lim_{V\to-\infty} R_t = \lim_{V\to+\infty} R_d \tag{10.24}$$

The relationship between $\bar{i}_k/\bar{i}_{lim}$ and the specified values of $\bar{i}/\bar{i}_{lim}$ can be found from equation (10.19) to be

$$\frac{\bar{i}_k}{\bar{i}_{lim}} = \frac{\bar{i}/\bar{i}_{lim}}{1-\bar{i}/\bar{i}_{lim}} \tag{10.25}$$

The resulting values for $\bar{i}_k/\bar{i}_{lim}$, R_t, and R_d are presented in Table 10.2. Interestingly, $\bar{i}_k = \bar{i}_{lim}$ when $\bar{i}/\bar{i}_{lim} = 1/2$, and $R_t = R_d$. Values of R_t and R_d can be obtained from equations (10.22) and (10.23), respectively, for specified values of i_k/i_{lim}.

Calculation of impedance requires a model for the dimensionless diffusion impedance. Under the assumption of diffusion through a film,

$$-\frac{1}{\theta'(0)} = \frac{\tanh\sqrt{j\omega\delta_i^2/D_i}}{\sqrt{j\omega\delta_i^2/D_i}} \tag{10.26}$$

The calculation also requires a value for capacitance, shown in Table 10.1 to be $C = 20\ \mu F/cm^2$. The resulting impedance is shown in Figure 10.3. The high-frequency loop corresponds to the reaction kinetics and has a characteristic frequency $f_c = 1/(2\pi R_t C)$ that depends on the value of $\bar{i}/\bar{i}_{lim}$. The low-frequency loop corresponds to diffusion has a characteristic frequency $f_c = 2.5D_i/(2\pi\delta_i^2)$. As $\bar{i}/\bar{i}_{lim}$ increases, the diffusion resistance increases and the charge-transfer resistance decreases. As $\bar{i}/\bar{i}_{lim}$ decreases, the diffusion resistance decreases and the charge-transfer resistance increases.

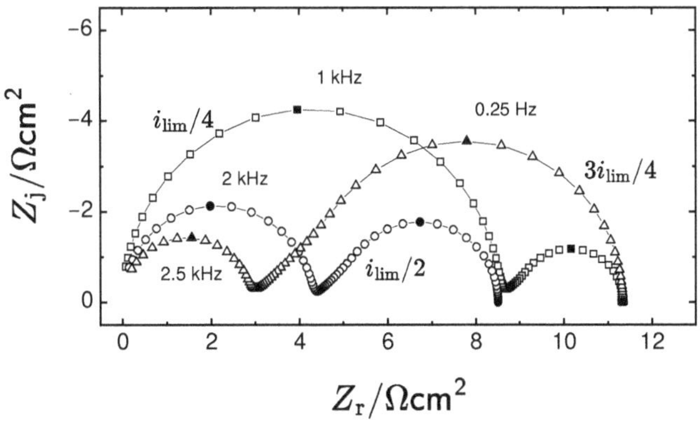

Figure 10.3: Impedance spectra generated for the solution to Problem 10.2 using the parameters presented in Table 10.1 with current density as a parameter. Values are presented for current densities that are 1/4, 1/2, and 3/4 of the limiting current density.

10.3 Should the Stern–Geary relation expressed as

$$\bar{i}_{\rm corr} = \frac{-\beta_c \beta_a}{2.303 \times 10^3 R_p (\beta_a - \beta_c)} \tag{10.27}$$

be used to assess the corrosion rate of magnesium? We may assume that corrosion of magnesium proceeds according to a two-step reaction sequence in which

$$\text{Mg} \xrightarrow{k_1} \text{Mg}^+_{\rm ads} + \text{e}^- \tag{10.28}$$

involves production of an adsorbed reaction intermediate (Mg^+_{ads}), which reacts further to form the divalent Mg^{2+} ion, i.e.,

$$\text{Mg}^+_{\rm ads} \underset{k_{22}}{\overset{k_2}{\rightleftarrows}} \text{Mg}^{2+} + \text{e}^- \tag{10.29}$$

The product Mg^{2+} diffuses through a porous layer of $Mg(OH)_2$ that has a thickness of δ. If equation (10.27) is not appropriate, how should the corrosion rate be estimated?

Solution 10.3. The Stern–Geary relationship applies for impedance measurements under open-circuit situations for which the anodic and cathodic reactions may be expressed by a Tafel equation. For example, the current density for hydrogen evolution may be expressed as

$$\bar{i}_{H_2} = -K_{H_2} \exp\left(-\frac{2.303 \times 10^3 \overline{V}}{\beta_{H_2}}\right) \tag{10.30}$$

Equation (10.27) can apply for cases where the current is independent of potential, for example, where $\beta = \infty$, but it strictly does not apply for more complex mechanisms that involve multiple reactions, the influence of concentration, and surface coverage by adsorbed intermediates. For such cases, the current density for reaction (10.28) provides a value for the corrosion current density, i.e.,

$$\bar{i}_{\text{corr}} = \frac{\beta_{\text{Mg}}}{2.303 \times 10^3 R_{\text{t,Mg}}} \tag{10.31}$$

where Mg refers to the metal dissolution reaction (10.28). The difficulty with applying equation (10.31) or, indeed, equation (10.27), is that both the Tafel slope β_{Mg} and the charge-transfer resistance $R_{\text{t,Mg}}$ are required. The charge-transfer resistance may be obtained from regression of a model to the impedance data, and the Tafel slope may be obtained from the polarization curve. In the case of magnesium, determination of the Tafel slope for reaction (10.28) from the anodic polarization curve is confounded by the other reactions that take place. In some cases the Tafel slope may be more easily obtained for the cathodic reaction. At open circuit, the current for the anodic and cathodic reactions cancel such that $i_{\text{corr}} = i_{\text{a}} = -i_{\text{c}}$. If the cathodic reaction is given by reaction (10.30), the corrosion current may be obtained from

$$\bar{i}_{\text{corr}} = -\frac{\beta_{\text{H}_2}}{2.303 \times 10^3 R_{\text{t,H}_2}} \tag{10.32}$$

where β_{H_2} is obtained from the slope of the cathodic part of the polarization curve and $R_{\text{t,H}_2}$ is obtained by regression of impedance data.

10.4 Consider the reaction sequence

$$\text{M} + \text{A} \rightarrow \text{MA}^+_{\text{ads}} + \text{e}^- \tag{10.33}$$

where MA^+_{ads} is an adsorbed intermediate that reacts according to

$$\text{MA}^+_{\text{ads}} + \text{A} \rightarrow \text{MA}_2^{2+} + \text{e}^- \tag{10.34}$$

Derive the faradaic impedance taking into account the mass-transfer limitation to the species A.

Solution 10.4. The expression for current density for reaction (10.33) can be expressed as

$$i_1 = K_1 c_{\text{A}}(0)(1-\gamma)\exp(b_1 V) \tag{10.35}$$

The current density for reaction (10.34) can be expressed as

$$i_2 = K_2 c_{\text{A}}(0)\gamma \exp(b_2 V) \tag{10.36}$$

The other two equations needed to solve for the impedance response are a material balance on the adsorbed species, given as

$$\Gamma\frac{\partial\gamma}{\partial t} = \frac{i_1}{\mathrm{F}} - \frac{i_2}{\mathrm{F}} \tag{10.37}$$

and the relationship between current and diffusive flux, given as

$$D_\mathrm{A}\left.\frac{\partial c_\mathrm{A}}{\partial y}\right|_{y=0} = \frac{i_1}{\mathrm{F}} + \frac{i_2}{\mathrm{F}} \tag{10.38}$$

The diffusion term includes the sum of i_1 and i_2 because the flux of species A contributes to both reactions.

The current density phasor for reaction (10.33) is expressed as

$$\begin{aligned}\widetilde{i}_1 &= K_1\bar{c}_\mathrm{A}(0)(1-\bar{\gamma})b_1\exp\left(b_1\overline{V}\right)\widetilde{V} - K_1\bar{c}_\mathrm{A}(0)\exp\left(b_1\overline{V}\right)\widetilde{\gamma}\\ &\quad +K_1(1-\bar{\gamma})\exp\left(b_1\overline{V}\right)\widetilde{c}_\mathrm{A}(0)\end{aligned} \tag{10.39}$$

and the current density phasor for reaction (10.34) is expressed as

$$\begin{aligned}\widetilde{i}_2 &= K_2\bar{c}_\mathrm{A}(0)\bar{\gamma}b_2\exp\left(b_2\overline{V}\right)\widetilde{V} + K_2\bar{c}_\mathrm{A}(0)\exp\left(b_2\overline{V}\right)\widetilde{\gamma}\\ &\quad +K_2\bar{\gamma}\exp\left(b_2\overline{V}\right)\widetilde{c}_\mathrm{A}(0)\end{aligned} \tag{10.40}$$

Equations (10.39) and (10.40) may be expressed in compact notation as

$$\widetilde{i}_1 = \frac{1}{R_{t,1}}\widetilde{V} - B_1\widetilde{\gamma} + C_1\widetilde{c}_\mathrm{A}(0) \tag{10.41}$$

and

$$\widetilde{i}_2 = \frac{1}{R_{t,2}}\widetilde{V} + B_2\widetilde{\gamma} + C_2\widetilde{c}_\mathrm{A}(0) \tag{10.42}$$

respectively. The total current density phasor may be expressed as

$$\widetilde{i} = \left(\frac{1}{R_{t,1}} + \frac{1}{R_{t,2}}\right)\widetilde{V} + (B_2 - B_1)\,\widetilde{\gamma} + (C_1 + C_2)\,\widetilde{c}_\mathrm{A}(0) \tag{10.43}$$

or

$$1 = \left(\frac{1}{R_{t,1}} + \frac{1}{R_{t,2}}\right)Z_\mathrm{F} + (B_2 - B_1)\,\frac{\widetilde{\gamma}}{\widetilde{i}} + (C_1 + C_2)\,\frac{\widetilde{c}_\mathrm{A}(0)}{\widetilde{i}} \tag{10.44}$$

where $Z_\mathrm{F} = \widetilde{V}/\widetilde{i}$.

Upon introduction of $\gamma = \bar{\gamma} + \mathrm{Re}\{\widetilde{\gamma}\exp(\mathrm{j}\omega t)\}$, equation (10.37) may be expressed as

$$\mathrm{j}\omega\Gamma\mathrm{F}\widetilde{\gamma} = \widetilde{i}_1 - \widetilde{i}_2 \tag{10.45}$$

or, in terms of the notation introduced in equations (10.41) and (10.42),

$$\mathrm{j}\omega\Gamma\mathrm{F}\widetilde{\gamma} = \left(\frac{1}{R_{t,1}} - \frac{1}{R_{t,2}}\right)\widetilde{V} - (B_1 + B_2)\,\widetilde{\gamma} + (C_1 - C_2)\,\widetilde{c}_\mathrm{A}(0) \tag{10.46}$$

Thus,

$$\widetilde{\gamma}=\left(\frac{1}{R_{t,1}}-\frac{1}{R_{t,2}}\right)\frac{1}{j\omega\Gamma F+B_1+B_2}\widetilde{V}+\frac{C_1-C_2}{j\omega\Gamma F+B_1+B_2}\widetilde{c}_A(0) \tag{10.47}$$

Equation (10.38) may be expressed as

$$D_A\left.\frac{d\widetilde{c}_A}{dy}\right|_{y=0}=\frac{\widetilde{i}}{F} \tag{10.48}$$

Upon introduction of equations (10.47) and (10.48), equation (10.44) may be expressed as

$$\begin{aligned}1 = &\left(\left(\frac{1}{R_{t,1}}+\frac{1}{R_{t,2}}\right)+\left(\frac{1}{R_{t,1}}-\frac{1}{R_{t,2}}\right)\frac{B_2-B_1}{j\omega\Gamma F+B_1+B_2}\right)Z_F \\ &+\left((C_1+C_2)+\frac{(B_2-B_1)(C_1-C_2)}{j\omega\Gamma F+B_1+B_2}\right)\frac{1}{FD_A}\frac{\widetilde{c}_A(0)}{\left.\frac{d\widetilde{c}_A}{dy}\right|_{y=0}}\end{aligned} \tag{10.49}$$

and the faradaic impedance is obtained as

$$Z_F=\left[\frac{1-\left((C_1+C_2)+\frac{(B_2-B_1)(C_1-C_2)}{j\omega\Gamma F+B_1+B_2}\right)\frac{\widetilde{c}_A(0)}{FD_A\left.\frac{d\widetilde{c}_A}{dy}\right|_{y=0}}}{\left(\frac{1}{R_{t,1}}+\frac{1}{R_{t,2}}\right)+\left(\frac{1}{R_{t,1}}-\frac{1}{R_{t,2}}\right)\frac{1}{j\omega\Gamma F+B_1+B_2}}\right] \tag{10.50}$$

The faradaic impedance may be inserted into the appropriate electrical circuit framework such as developed in Chapter 9 to obtain the impedance for the system.

10.5 Derive the faradaic impedance for the anodic dissolution of copper at low overpotential in a chloride medium, where the reactions proceed according to

$$Cu+Cl^-\rightleftarrows CuCl_{ads}+e^- \tag{10.51}$$

where $CuCl_{ads}$ is an adsorbed intermediate that reacts with chloride ions to form $CuCl_2^-$, i.e.,

$$CuCl_{ads}+Cl^-\rightleftarrows CuCl_2^- \tag{10.52}$$

The mass-transfer limitation is due only to $CuCl_2^-$.

Solution 10.5. The electrochemical reaction may be expressed as

$$Cu+Cl^-\underset{K_{-1}}{\overset{K_1}{\rightleftarrows}} CuCl_{ads}+e^- \tag{10.53}$$

where K_i is a lumped-parameter rate constant including the interfacial equilibrium potential for reaction with index i and given as

$$K_i = nFk_i \exp(-b_i V_{0,i}) \tag{10.54}$$

where $V_{0,i}$ is the interfacial equilibrium potential. The chemical reaction may be expressed as

$$CuCl_{ads} + Cl^- \underset{k_b}{\overset{k_f}{\rightleftarrows}} CuCl_2^- \tag{10.55}$$

where k_f and k_b are the forward and backward rate constants, respectively. The total faradic current can be expressed as

$$i_{F,Cu} = K_1 c_{Cl^-}(0)(1-\gamma)\exp(bV) - K_{-1}\Gamma\gamma\exp(-bV) \tag{10.56}$$

A material balance is expressed as a function of the fraction of surface coverage by the adsorbed species $CuCl_{ads}$ as

$$\begin{aligned} F\Gamma\frac{d\gamma}{dt} &= K_1 c_{Cl^-}(0)(1-\gamma)\exp(bV) - K_{-1}\Gamma\gamma\exp(-bV) \\ &\quad -k_f F\Gamma\gamma c_{Cl^-}(0) + k_b F(1-\gamma)c_{CuCl_2^-}(0) \end{aligned} \tag{10.57}$$

where $c_{Cl^-}(0)$ and $c_{CuCl_2^-}(0)$ represent concentrations at the electrode surface.

As the mass transfer limitation can be attributed to $CuCl_2^-$, $c_{Cl^-}(0) = c_{Cl^-}(\infty)$ in equations (10.56) and (10.57). The steady flux of $CuCl_2^-$ is related to the current density by

$$\frac{\bar{i}_{F,Cu}}{F} = D_{CuCl_2^-}\frac{\bar{c}_{CuCl_2^-}(0)}{\delta_{CuCl_2^-}} \tag{10.58}$$

where $\delta_{CuCl_2^-}$ is the diffusion layer thickness of $CuCl_2^-$. At steady state, $d\gamma/dt = 0$; thus, combination of equations (10.57) and (10.58) yields

$$\bar{i}_{F,Cu} = k_f\Gamma\bar{\gamma}\, c_{Cl^-}(\infty) - k_b(1-\gamma)\frac{\bar{i}_F\delta_{CuCl_2^-}}{FD_{CuCl_2^-}} \tag{10.59}$$

In reference 34, the assumption was made that $\gamma \ll 1$. With this assumption, equation (10.56) can be expressed as

$$i_{F,Cu} = K_1 c_{Cl^-}(\infty)\exp(bV) - K_{-1}\Gamma\bar{\gamma}\exp(-bV) \tag{10.60}$$

and equation (10.59) can be expressed as

$$\bar{i}_{F,Cu} = k_f\Gamma\bar{\gamma}\, c_{Cl^-}(\infty) - k_b\frac{\bar{i}_F\delta_{CuCl_2^-}}{FD_{CuCl_2^-}} \tag{10.61}$$

From equation (10.61), the steady-state surface coverage can be expressed as

$$\overline{\gamma} = \frac{\overline{i}_{\mathrm{F,Cu}}\left(1 + k_{\mathrm{b}}\dfrac{\delta_{\mathrm{CuCl}_2^-}}{FD_{\mathrm{CuCl}_2^-}}\right)}{k_{\mathrm{f}}\Gamma c_{\mathrm{Cl}^-}(\infty)} \tag{10.62}$$

Insertion of equation (10.62) in equation (10.56) yields

$$\frac{1}{\overline{i}_{\mathrm{F,Cu}}} = \frac{1}{K_1\overline{c}_{\mathrm{Cl}^-}(\infty)\exp(b\overline{V})} + \frac{K_{-1}\exp(-2b\overline{V})}{FK_1k_{\mathrm{f}}\left(\overline{c}_{\mathrm{Cl}^-}(\infty)\right)^2\exp(b\overline{V})}\left(1 + \frac{k_{\mathrm{b}}\delta_{\mathrm{CuCl}_2^-}}{FD_{\mathrm{CuCl}_2^-}}\right) \tag{10.63}$$

By elimination of the faradaic current in equation (10.62), the steady-state surface concentration may be expressed as

$$\Gamma\overline{\gamma} = \frac{K_1\overline{c}_{\mathrm{Cl}^-}(\infty)\exp(b\overline{V})\left(1 + \dfrac{k_{\mathrm{b}}\delta_{\mathrm{CuCl}_2^-}}{FD_{\mathrm{CuCl}_2^-}}\right)}{k_{\mathrm{f}}\overline{c}_{\mathrm{Cl}^-}(\infty) + K_{-1}\exp(-b\overline{V})\left(1 + \dfrac{k_{\mathrm{b}}\delta_{\mathrm{CuCl}_2^-}}{FD_{\mathrm{CuCl}_2^-}}\right)} \tag{10.64}$$

Under a sinusoidal perturbation and under the assumptions used to generate equations (10.63) and (10.64), equation (10.60) may be expressed as

$$\widetilde{i}_{\mathrm{F,Cu}} = \frac{\widetilde{V}}{R_{\mathrm{t}}} - K_{-1}\Gamma\exp(-b\overline{V})\widetilde{\gamma} \tag{10.65}$$

where

$$\frac{1}{R_{\mathrm{t}}} = K_1\overline{c}_{\mathrm{Cl}^-}(\infty)b\exp\left(b\overline{V}\right) + K_{-1}\Gamma\overline{\gamma}b\exp\left(-b\overline{V}\right) \tag{10.66}$$

Equation (10.57) can be expressed as

$$\mathrm{j}\omega\mathrm{F}\Gamma\widetilde{\gamma} = \frac{\widetilde{V}}{R_{\mathrm{t}}} - \left(K_{-1}\exp(-b\overline{V}) + k_{\mathrm{f}}\mathrm{F}c_{\mathrm{Cl}^-}(\infty)\right)\Gamma\widetilde{\gamma} + k_{\mathrm{b}}\mathrm{F}\widetilde{c}_{\mathrm{CuCl}_2^-}(0) \tag{10.67}$$

Equations (10.65) and (10.67) are written in terms of four phasors: $\widetilde{V}, \widetilde{i}, \widetilde{\gamma}$, and $\widetilde{c}_{\mathrm{CuCl}_2^-}(0)$.

An additional equation is needed to allow calculation of transfer functions. Equation (10.58) may be expressed as

$$D_{\mathrm{CuCl}_2^-}\left.\frac{\mathrm{d}\widetilde{c}_{\mathrm{CuCl}_2^-}}{\mathrm{d}y}\right|_{y=0} = k_{\mathrm{f}}c_{\mathrm{Cl}^-}(\infty)\Gamma\widetilde{\gamma} - k_{\mathrm{b}}\widetilde{c}_{\mathrm{CuCl}_2^-}(0) \tag{10.68}$$

Equation (10.68) is written in terms of reaction (10.55) because, under non-steady-state conditions, the rates of reactions (10.53) and (10.55) are not equal. Following the development in Chapter 11 of Orazem and Tribollet,[1]

$$\frac{\widetilde{c}_{\mathrm{CuCl}_2^-}(0)}{\left.\dfrac{\mathrm{d}\widetilde{c}_{\mathrm{CuCl}_2^-}}{\mathrm{d}y}\right|_{y=0}} = \delta_{\mathrm{CuCl}_2^-}\left(\frac{-1}{\theta'_{\mathrm{CuCl}_2^-}(0)}\right) \tag{10.69}$$

Equations (10.65), (10.67), (10.68), and (10.69) constitute a set of four equations with four unknowns,

$$Y_F = \frac{1}{Z_F} = \frac{\widetilde{i}_{F,Cu}}{\widetilde{V}} \tag{10.70}$$

$$Z_\gamma = \frac{\widetilde{\gamma}}{\widetilde{V}} \tag{10.71}$$

$$Z_c = \frac{\widetilde{c}_{CuCl_2^-}}{\widetilde{V}} \tag{10.72}$$

$$Z_d = \frac{\left.\dfrac{d\widetilde{c}_{CuCl_2^-}}{dy}\right|_{y=0}}{\widetilde{V}} \tag{10.73}$$

where equation (10.70) represents the faradaic admittance and equations (10.71)–(10.73) are also transfer functions. Analytic expressions for the transfer functions are possible, but it could be easier to solve the problem numerically for specified conditions.

10.6 Derive the faradaic impedance for corrosion of magnesium taking into account the following reactions:

$$Mg \xrightarrow{k_1} Mg^+_{ads} + e^- \tag{10.74}$$

involves production of an adsorbed reaction intermediate (Mg^+_{ads}), which reacts further to form the divalent Mg^{2+} ion, i.e.,

$$Mg^+_{ads} \underset{k_{22}}{\overset{k_2}{\rightleftarrows}} Mg^{2+} + e^- \tag{10.75}$$

The product Mg^{2+} diffuses through a porous layer of $Mg(OH)_2$ that has a thickness of δ. The adsorbed species Mg^+_{ads} may react further with water, i.e.,

$$2Mg^+_{ads} + 2H_2O \rightarrow 2Mg^{2+} + 2OH^- + H_2 \tag{10.76}$$

Reaction (10.76) is a chemical reaction that results in hydrogen evolution at potentials associated with Mg dissolution. Due to the anomalous production of hydrogen at anodic rather than cathodic potentials, this reaction was termed the Negative Difference Effect (NDE).

Solution 10.6. Each reaction has a normalized rate constant K_i corresponding to its rate constant k_i by

$$K_i = nFk_i \exp\left(b_i(V - V_{0,i})\right) \tag{10.77}$$

where Γ is the maximum number of sites per surface unit that can be occupied by the adsorbate Mg^+_{ads}. Moreover, the chemical reaction (10.76) is assumed to be independent of potential with a rate constant k_3. The mass and charge balances are expressed as a function of the fraction of the surface coverage by the adsorbed species γ as

$$\mathrm{F}\Gamma\frac{\mathrm{d}\gamma}{\mathrm{d}t} = K_1(1-\gamma) - K_2\Gamma\gamma + K_{22}c_{\mathrm{Mg}^{2+}}(0)(1-\gamma) - k_3\mathrm{F}\Gamma\gamma \tag{10.78}$$

where Γ is the maximum number of sites per surface unit that can be occupied by the adsorbed $\mathrm{Mg}^+_{\mathrm{ads}}$. A material balance on Mg^{2+} under the steady-state condition yields

$$\mathrm{F}\frac{D_{\mathrm{Mg}^{2+}}\overline{c}_{\mathrm{Mg}^{2+}}(0)}{\delta_{\mathrm{Mg}^{2+}}} = K_2\Gamma\overline{\gamma} - K_{22}\overline{c}_{\mathrm{Mg}^{2+}}(0)(1-\overline{\gamma}) + k_3\mathrm{F}\Gamma\overline{\gamma} \tag{10.79}$$

where $\delta_{\mathrm{Mg}^{2+}}$ is the thickness of the Nernst diffusion layer and $\overline{c}_{\mathrm{Mg}^{2+}}(0)$ is the concentration of the Mg^{2+} ion at the electrode interface. The total faradic current density $i_{\mathrm{F,Mg}}$ can be expressed as

$$i_{\mathrm{F,Mg}} = K_1(1-\overline{\gamma}) + K_2\Gamma\overline{\gamma} - K_{22}\overline{c}_{\mathrm{Mg}^{2+}}(0) \tag{10.80}$$

At steady state, $\overline{\gamma}$ and $\overline{c}_{\mathrm{Mg}^{2+}}(0)$ are thus given by

$$\overline{\gamma} = \frac{K_1\left(\dfrac{\mathrm{F}D_{\mathrm{Mg}^{2+}}}{\delta_{\mathrm{Mg}^{2+}}} + K_{22}\right)}{K_1\left(\dfrac{D_{\mathrm{Mg}^{2+}}}{\delta_{\mathrm{Mg}^{2+}}} + K_{22}\right) + \Gamma(K_2 + k_3\mathrm{F})\dfrac{\mathrm{F}D_{\mathrm{Mg}^{2+}}}{\delta_{\mathrm{Mg}^{2+}}}} \tag{10.81}$$

and

$$\overline{c}_{\mathrm{Mg}^{2+}}(0) = \frac{\overline{\gamma}\Gamma(K_2 + k_3\mathrm{F})}{\dfrac{\mathrm{F}D_{\mathrm{Mg}^{2+}}}{\delta_{\mathrm{Mg}^{2+}}} + K_{22}} \tag{10.82}$$

respectively. The faradic impedance Z_{F} is thus calculated by linearizing the equations (10.78), (10.79), and (10.80) for small sine wave perturbations to obtain

$$\begin{aligned}(\mathrm{F}\Gamma\mathrm{j}\omega + K_1 + \Gamma(K_2 + k_3\mathrm{F}))\,\widetilde{\gamma} = \\ \left[(1-\gamma)\,K_1 b_1 - \Gamma\gamma K_2 b_2 - c_{\mathrm{Mg}^{2+}}(0) K_{22} b_{22}\right]\widetilde{V} + K_{22}\widetilde{c}_{\mathrm{Mg}^{2+}}(0)\end{aligned} \tag{10.83}$$

and

$$\begin{aligned}\widetilde{i}_{\mathrm{F,Mg}} &= (\Gamma K_2 - K_1)\,\widetilde{\gamma} \\ &+ \left[(1-\gamma)\,K_1 b_1 + \Gamma\gamma K_2 b_2 + c_{\mathrm{Mg}^{2+}}(0) K_{22} b_{22}\right]\widetilde{V} - K_{22}\widetilde{c}_{\mathrm{Mg}^{2+}}(0)\end{aligned} \tag{10.84}$$

Since the specie Mg^{2+} diffuses toward the electrode surface, the resulting concentration perturbation $\widetilde{c}_{\mathrm{Mg}^{2+}}(0)$ is obtained from the finite-length diffusion impedance, represented by $-1/\theta'_{\mathrm{Mg}^{2+}}(0)$ and developed in Chapter 11 of Orazem and Tribollet.[1]

The resulting impedance response is given by

$$Z_{F,Mg} = \frac{\widetilde{V}}{\widetilde{i}_{F,Mg}} = \frac{1 + K_{22}\left(-1/\theta'_{Mg^{2+}}(0)\right)\left(1 - \dfrac{\Gamma K_2 - K_1}{F\Gamma j\omega + K_1 + \Gamma(K_2 + k_3 F)}\right)}{\left(\dfrac{(r_1 - r_2)(\Gamma K_2 - K_1)}{F\Gamma j\omega + K_1 + \Gamma(K_2 + k_3 F)} + (r_1 + r_2)\right)} \tag{10.85}$$

with $r_1 = (1 - \gamma) K_1 b_1$ and $r_2 = K_2 b_2 \Gamma \gamma + K_{22} b_{22} c_{Mg^{2+}}(0)$.

10.7 Perform the calculations needed to generate Figure 10.4 for a system in which a metal M dissolves through an adsorbed intermediate X^+ following

$$M \rightarrow X^+ + e^- \tag{10.86}$$

The adsorbed intermediate reacts in a second electrochemical step

$$X^+ \rightarrow P^{2+} + e^- \tag{10.87}$$

to form the final product P. As shown by Orazem and Tribollet,[1] the total steady-state current density is given by

$$\bar{i}_F = \bar{i}_M + \bar{i}_X = \frac{2K_M \exp(b_M \overline{V}) K_X \exp(b_X \overline{V})}{K_M \exp(b_M \overline{V}) + K_X \exp(b_X \overline{V})} \tag{10.88}$$

and the associated faradaic impedance is given as

$$\frac{1}{Z_F} = \frac{1}{R_t} + \frac{\left(K_X \exp\left(b_X \overline{V}\right) - K_M \exp\left(b_M \overline{V}\right)\right)\left(R_{t,M}{}^{-1} - R_{t,X}{}^{-1}\right)}{\Gamma F j\omega + \left(K_X \exp\left(b_X \overline{V}\right) + K_M \exp\left(b_M \overline{V}\right)\right)} \tag{10.89}$$

where

$$\frac{1}{R_t} = \frac{1}{R_{t,M}} + \frac{1}{R_{t,X}} \tag{10.90}$$

represents an effective charge-transfer resistance.

Solution 10.7. Adsorption of reaction intermediate X^+ is assumed to obey a Langmuir isotherm and is characterized by a surface coverage γ_X. The steady-state current density associated with reaction (10.86) is given by

$$\bar{i}_M = K_M (1 - \overline{\gamma}_X) \exp\left(b_M \overline{V}\right) \tag{10.91}$$

Similar mechanisms were proposed by Epelboin and Keddam[35] for calculating the impedance of iron dissolution through two steps involving an adsorbed FeOH intermediate and by Peter et al.[36] for the impedance model of the dissolution of aluminum in three consecutive steps with two adsorbed intermediates.

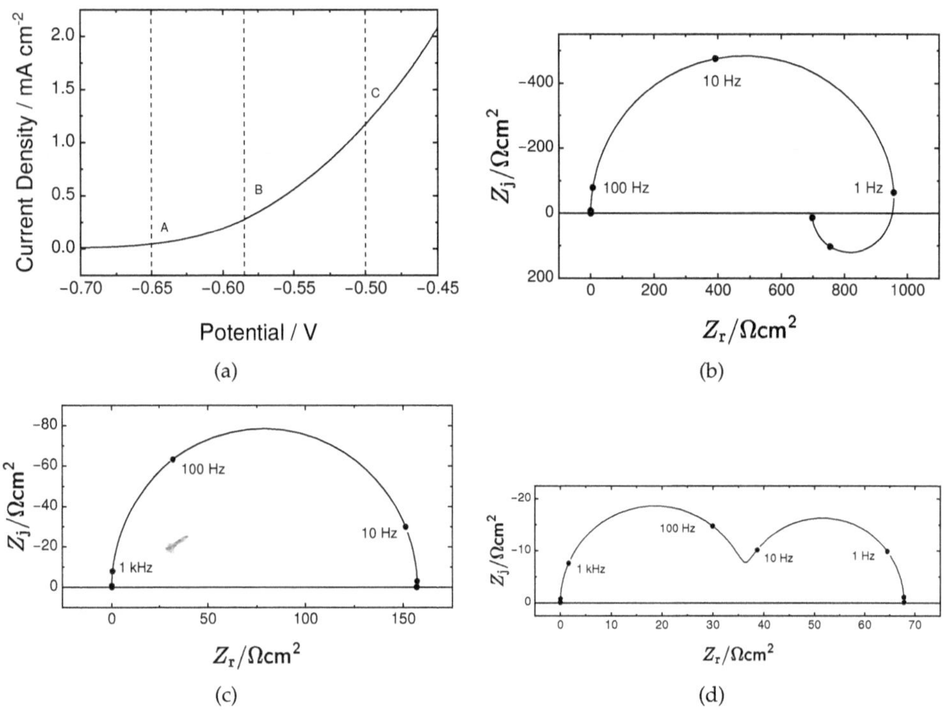

Figure 10.4: Calculated steady-state and impedance response for coupled reactions dependent on potential and surface coverage for Problem 10.7. (a) Simulated current-potential curve following equation (10.88) with the kinetic parameters $K_M = 4F$ A/cm^2, $b_M = 36$ V^{-1}, $K_X = 10^{-6}F$ A/cm^2, $b_X = 10$ V^{-1}, $\Gamma = 2 \times 10^{-9}$ mol/cm^2, and $C_{dl} = 20$ μF/cm^2. The points A, B, and C correspond to the simulated impedance; (b) at the point A ($V = -0.65$ V); (c) at the point B ($V = -0.585$ V); and (d) at the point C ($V = -0.50$ V). Taken from Orazem and Tribollet[1] with permission of John Wiley & Sons.

The steady-state current density associated with reaction (10.87) is given by

$$\bar{i}_X = K_X \bar{\gamma}_X \exp\left(b_X \overline{V}\right) \tag{10.92}$$

where the rate constant K_X includes the maximum surface concentration of the intermediate X, defined in equation (10.93) as Γ. The variation of the surface coverage by the intermediate X^+ is given by

$$\Gamma \frac{d\gamma_X}{dt} = \frac{i_M}{F} - \frac{i_X}{F} \tag{10.93}$$

Under a steady-state condition, $d\gamma_X/dt = 0$ and $\bar{i}_M = \bar{i}_X$. By using the corresponding equations (10.91) and (10.92), an expression for the steady-state surface coverage $\bar{\gamma}_X$ is obtained as

$$\bar{\gamma}_X = \frac{K_M \exp\left(b_M \overline{V}\right)}{K_M \exp\left(b_M \overline{V}\right) + K_X \exp\left(b_X \overline{V}\right)} \tag{10.94}$$

Examination of equation (10.94) reveals that, if $K_M \exp\left(b_M \overline{V}\right) \gg K_X \exp\left(b_X \overline{V}\right)$, then $\bar{\gamma}_X \rightarrow 1$. If $K_M \exp\left(b_M \overline{V}\right) \ll K_X \exp\left(b_X \overline{V}\right)$, then $\bar{\gamma}_X \rightarrow 0$. An expression for the total steady-state current density, given by

$$\bar{i}_F = \bar{i}_M + \bar{i}_X = \frac{2K_M \exp(b_M \overline{V}) K_X \exp(b_X \overline{V})}{K_M \exp(b_M \overline{V}) + K_X \exp(b_X \overline{V})} \tag{10.95}$$

can be found from equations (10.91), (10.87), and (10.94). Equation (10.95) is the same as equation (10.88).

The current density phasors for reactions (10.86) and (10.87) are given, respectively, by

$$\tilde{i}_M = \frac{1}{R_{t,M}} \tilde{V} - K_M \exp\left(b_M \overline{V}\right) \tilde{\gamma}_X \tag{10.96}$$

and

$$\tilde{i}_X = \frac{1}{R_{t,X}} \tilde{V} + K_X \exp\left(b_X \overline{V}\right) \tilde{\gamma}_X \tag{10.97}$$

where the charge-transfer resistances are defined by

$$R_{t,M} - \frac{1}{K_M \left(1 - \bar{\gamma}_X\right) b_M \exp\left(b_M \overline{V}\right)} \tag{10.98}$$

and

$$R_{t,X} = \frac{1}{K_X \bar{\gamma}_X b_X \exp\left(b_X \overline{V}\right)} \tag{10.99}$$

respectively. According to equation (10.93), the surface coverage phasor can be expressed as

$$\Gamma F\, j\omega \tilde{\gamma}_X = \left(\frac{1}{R_{t,M}} - \frac{1}{R_{t,X}}\right) \tilde{V} - \left(K_X \exp\left(b_X \overline{V}\right) + K_M \exp\left(b_M \overline{V}\right)\right) \tilde{\gamma}_X \tag{10.100}$$

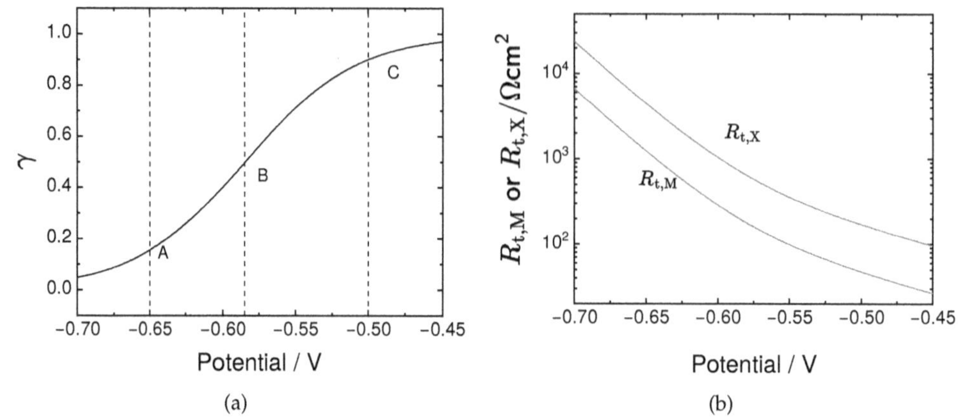

Figure 10.5: Calculated steady-state values as functions of applied potential obtained as the solution to Problem 10.7 for kinetic parameters $K_M = 4F\ A/cm^2$, $b_M = 36\ V^{-1}$, $K_X = 10^{-6}F\ A/cm^2$, $b_X = 10\ V^{-1}$, $\Gamma = 2 \times 10^{-9}$ mol/cm^2, and $C_{dl} = 20\ \mu F/cm^2$. (a) fractional coverage of the electrode and (b) charge-transfer resistances for the electrochemical reactions (10.86) and (10.87).

yielding

$$\widetilde{\gamma}_X = \left(\frac{\dfrac{1}{R_{t,M}} - \dfrac{1}{R_{t,X}}}{\Gamma Fj\omega + \left(K_X \exp\left(b_X\overline{V}\right) + K_M \exp\left(b_M\overline{V}\right)\right)} \right) \widetilde{V} \tag{10.101}$$

The net faradaic current density phasor, given by the sum of contributions from reactions (10.86) and (10.87), is a function of γ_X and V; thus,

$$\begin{aligned} \widetilde{i}_F &= \widetilde{i}_X + \widetilde{i}_M \\ &= \left(\frac{1}{R_{t,M}} + \frac{1}{R_{t,X}}\right)\widetilde{V} + \left(K_X \exp\left(b_X\overline{V}\right) - K_M \exp\left(b_M\overline{V}\right)\right)\widetilde{\gamma}_X \end{aligned} \tag{10.102}$$

The expression of the faradaic admittance, i.e., $Y_F = 1/Z_F$, is

$$Y_F = \frac{1}{Z_F} = \frac{\widetilde{i}_F}{\widetilde{V}} \tag{10.103}$$

or

$$\frac{1}{Z_F} = \frac{1}{R_t} + \frac{\left(K_X \exp\left(b_X\overline{V}\right) - K_M \exp\left(b_M\overline{V}\right)\right)\left({R_{t,M}}^{-1} - {R_{t,X}}^{-1}\right)}{\Gamma Fj\omega + \left(K_X \exp\left(b_X\overline{V}\right) + K_M \exp\left(b_M\overline{V}\right)\right)} \tag{10.104}$$

where

$$\frac{1}{R_t} = \frac{1}{R_{t,M}} + \frac{1}{R_{t,X}} \tag{10.105}$$

which are the same as equations (10.89) and (10.90).

The calculated current density agrees with Figure 10.4(a). Some insight can be found from the fractional coverage of the surface given in Figure 10.5(a). The fractional coverage shows an inflection point at a potential of −0.585 V. The semilogarithmic plot of charge-transfer resistances has curvature due to the influence of the potential-dependent coverage of the surface.

The calculated impedance values are the same as are shown in Figures 10.4(b), 10.4(c), and 10.4(d). The impedance show a transition from a high-frequency capacitive loop with a low-frequency inductive loop at −0.65 V (Figure 10.4(b)) to two capacitive loops at −0.50 V (Figure 10.4(d)). The impedance decreased as the potential increased, in agreement with the increase in current with potential as shown in Figure 10.4(a). A single capacitive loop is observed in Figure 10.4(c) at a potential of −0.585 V. This potential coincides with the inflection of the fractional coverage of the surface shown in Figure 10.5(a).

10.8 Explain why K_M^*, defined in

$$i_M = K_M^* \exp\left(b_M\left(V - V_{0,M}\right)\right) \tag{10.106}$$

has a value that is independent of the reference electrode.

Solution 10.8. The term K_M^* can be expressed as

$$K_M^* = \frac{i_M}{\exp\left(b_M\left(V \quad V_{0,M}\right)\right)} \tag{10.107}$$

The potential $V = \Phi_m - \Phi_0$ can be expressed in terms of reference electrode as

$$V = \Phi_m - \Phi_{ref} - \left(\Phi_0 - \Phi_{ref}\right) \tag{10.108}$$

where the reference electrode potentials cancel. Thus, K_M^* is independent of the reference electrode.

While models are typically developed in terms of a local potential difference as expressed in equation (10.108), experimental data may be available, as is the case in Problem 9.10, in terms of potentials $V = \Phi_m - \Phi_{ref}$ and $V^\circ = \Phi_m^\circ - \Phi_{ref}$. Again, the reference electrode potentials cancel, leaving K_M^* independent of the arbitrary choice of a reference electrode. The value of

$$K_M = K_M^* \exp\left(-b_M V_{0,M}\right) \tag{10.109}$$

does depend on the arbitrary choice of reference electrode.

10.9 If $V_{ref1} - V_{ref2} = 0.4$ V, give the expression of $K_{M,ref2}$ with respect to $K_{M,ref1}$, where $K_{M,refi}$ is the value corresponding to an experiment performed with a given reference electrode. Show that the value of the charge-transfer resistance R_t is independent of the reference electrode used.

Solution 10.9. Following equation (10.109),

$$K_{\mathrm{M,ref1}} = K_{\mathrm{M}}^{*} \exp\left(-b_{\mathrm{M}}\left(\Phi_{\mathrm{m}}^{\circ} - \Phi_{\mathrm{ref1}}\right)\right) \tag{10.110}$$

Thus

$$\frac{K_{\mathrm{M,ref1}}}{K_{\mathrm{M,ref2}}} = \exp\left(b_{\mathrm{M}}\left(\Phi_{\mathrm{ref1}} - \Phi_{\mathrm{ref2}}\right)\right) \tag{10.111}$$

The result depends on the value of b_{M}. For a Tafel slope of 120 mV/decade, $b_{\mathrm{M}} = 19.2\ \mathrm{V}^{-1}$ and $K_{\mathrm{M,ref1}}/K_{\mathrm{M,ref2}} = 2.2 \times 10^{3}$. For a Tafel slope of 60 mV/decade, $b_{\mathrm{M}} = 38.4\ \mathrm{V}^{-1}$ and $K_{\mathrm{M,ref1}}/K_{\mathrm{M,ref2}} = 4.7 \times 10^{6}$.

The charge-transfer resistance may be expressed as

$$R_{\mathrm{t,M}} = \frac{1}{K_{\mathrm{M}}^{*} \exp\left(b_{\mathrm{M}}\left(V - V_{0,\mathrm{M}}\right)\right)} \tag{10.112}$$

Given that both K_{M}^{*} and $(V - V_{0,\mathrm{M}})$ are independent of the reference electrode, $R_{\mathrm{t,M}}$ is independent of the reference electrode.

10.10 The oxidation of hydrogen peroxide on a platinum disk electrode may be expressed as

$$H_2O_2 \rightarrow O_2 + 2H^{+} + 2e^{-} \tag{10.113}$$

(a) Develop an expression for the faradaic current density phasor $\widetilde{i}_{\mathrm{F}}$ in terms of potential and concentration phasors $\widetilde{V}$ and $\widetilde{c}_{\mathrm{H_2O_2}}(0)$, respectively.

(b) Develop an expression for the faradaic impedance in terms of the charge-transfer resistance and the diffusion impedance. Write the diffusion impedance in terms of the dimensionless diffusion impedance $-1/\theta_{\mathrm{i}}'$ and the diffusion resistance, providing a clear relationship to kinetic parameters.

(c) Show how knowledge of the charge-transfer resistance may be used to extract the peroxide oxidation current density.

Solution 10.10.

(a) The current density for reaction (10.113) may be expressed as

$$i_{\mathrm{H_2O_2}} = K_{\mathrm{H_2O_2}} c_{\mathrm{H_2O_2}}(0) \exp\left(b_{\mathrm{H_2O_2}} V\right) \tag{10.114}$$

The current density phasor is given in the first terms of a Taylor series expansion as

$$\widetilde{i}_{\mathrm{H_2O_2}} = \left.\frac{\partial i_{\mathrm{H_2O_2}}}{\partial V}\right|_{c_{\mathrm{H_2O_2}}(0)} \widetilde{V} + \left.\frac{\partial i_{\mathrm{H_2O_2}}}{\partial c_{\mathrm{H_2O_2}}(0)}\right|_{V} \widetilde{c}_{\mathrm{H_2O_2}}(0) \tag{10.115}$$

or

$$\begin{aligned} \widetilde{i}_{H_2O_2} &= K_{H_2O_2} b_{H_2O_2} c_{H_2O_2}(0) \exp\left(b_{H_2O_2}\overline{V}\right) \widetilde{V} \\ &\quad + K_{H_2O_2} \exp\left(b_{H_2O_2}\overline{V}\right) \widetilde{c}_{H_2O_2}(0) \end{aligned} \tag{10.116}$$

Equation (10.116) may be written as

$$\widetilde{i}_{H_2O_2} = \frac{1}{R_{t,H_2O_2}} \widetilde{V} + K_{H_2O_2} \exp\left(b_{H_2O_2}\overline{V}\right) \widetilde{c}_{H_2O_2}(0) \tag{10.117}$$

where

$$R_{t,H_2O_2} = \frac{1}{K_{H_2O_2} b_{H_2O_2} c_{H_2O_2}(0) \exp\left(b_{H_2O_2}\overline{V}\right)} \tag{10.118}$$

(b) A transfer function is defined as the ratio of two phasors; whereas, equation (10.117) is written in terms of three phasors. The current may be written in terms of concentration through Fick's first law as

$$i_{H_2O_2} = 2FD_{H_2O_2} \left.\frac{dc_{H_2O_2}}{dy}\right|_{y=0} \tag{10.119}$$

which can be written in terms of phasor quantities as

$$\widetilde{i}_{H_2O_2} = 2FD_{H_2O_2} \left.\frac{d\widetilde{c}_{H_2O_2}}{dy}\right|_{y=0} \tag{10.120}$$

Equation (10.117) may be combined with equation (10.120) to yield

$$1 = \frac{1}{R_{t,H_2O_2}} \frac{\widetilde{V}}{\widetilde{i}_{H_2O_2}} + \frac{K_{H_2O_2} \exp\left(b_{H_2O_2}\overline{V}\right)}{2FD_{H_2O_2}} \frac{\widetilde{c}_{H_2O_2}(0)}{\left.\frac{d\widetilde{c}_{H_2O_2}}{dy}\right|_{y=0}} \tag{10.121}$$

where $\widetilde{V}/\widetilde{i}_{H_2O_2} = Z_{F,H_2O_2}$. The faradaic impedance can be found to be

$$Z_{F,H_2O_2} = R_{t,H_2O_2} - \frac{R_{t,H_2O_2} K_{H_2O_2} \exp\left(b_{H_2O_2}\overline{V}\right)}{2FD_{H_2O_2}} \frac{\widetilde{c}_{H_2O_2}(0)}{\left.\frac{d\widetilde{c}_{H_2O_2}}{dy}\right|_{y=0}} \tag{10.122}$$

For $\theta_{H_2O_2} = \widetilde{c}_{H_2O_2}/\widetilde{c}_{H_2O_2}(0)$ and $\xi = y/\delta_{H_2O_2}$,

$$\frac{\widetilde{c}_{H_2O_2}(0)}{\left.\frac{d\widetilde{c}_{H_2O_2}}{dy}\right|_{y=0}} = \delta_{H_2O_2} \frac{1}{\theta'_{H_2O_2}(0)} \tag{10.123}$$

where $\theta'_{H_2O_2} = d\theta_{H_2O_2}/d\xi$. Equation (10.122) may be expressed as

$$Z_{F,H_2O_2} = R_{t,H_2O_2} + R_{d,H_2O_2} \frac{-1}{\theta'_{H_2O_2}(0)} \tag{10.124}$$

where

$$R_{d,H_2O_2} = \frac{\delta_{H_2O_2}}{2Fb_{H_2O_2} D_{H_2O_2} c_{H_2O_2}(0)} \tag{10.125}$$

(c) The charge-transfer resistance defined by equation (10.118) may be expressed in terms of the hydrogen peroxide oxidation current density as

$$R_{t,H_2O_2} = \frac{1}{b_{H_2O_2}\bar{i}_{H_2O_2}} \tag{10.126}$$

Equation (10.126) may be expressed in terms of the Tafel slope as

$$R_{t,H_2O_2} = \frac{\beta_{H_2O_2}}{2,303\bar{i}_{H_2O_2}} \tag{10.127}$$

or

$$\bar{i}_{H_2O_2} = \frac{\beta_{H_2O_2}}{2,303R_{t,H_2O_2}} \tag{10.128}$$

Calculation of the hydrogen peroxide oxidation current density requires the regressed value of the charge-transfer resistance and the Tafel slope. The Tafel slope may be obtained from the steady-state polarization curve. See, for example, the solution to Problem 9.10.

10.11 Consider the cathodic reduction of oxygen on a Pt catalyst, e.g.,

$$O_2 + 2H^+ + 4e^- \rightarrow 2H_2O \tag{10.129}$$

Consider also that the Pt catalyst can be oxidized by the reversible reaction

$$Pt + H_2O \rightleftarrows PtO + 2H_2 + 2e^- \tag{10.130}$$

(a) Can reaction (10.129) yield a low-frequency inductive loop in the impedance response? You need not derive the associated faradaic impedance, but you should explain your reasoning.

(b) Find the steady-state fractional coverage by PtO, $\overline{\gamma}_{PtO}$.

(c) Develop an expression for the faradaic impedance response associated with reactions (10.129) and (10.130). You may assume that reaction (10.129) proceeds rapidly on the Pt catalyst but does not proceed on PtO. Thus, the effective rate constant for reaction (10.129) can be expressed as $K_{eff} = K_{O_2}(1 - \gamma_{PtO})$. In addition, mass-transfer limitations to hydrogen ions and water may be neglected, but the influence of mass transfer on the oxygen concentration may not be ignored.

Solution 10.11.

(a) Epelboin and Keddam[35] and Armstrong et al.[37] showed that low-frequency inductive features can be attributed to the coupling of electrochemical reactions through an adsorbed intermediate. In this case, two electrochemical reactions are proposed in which the rate constant for the electrochemical reduction of oxygen is influenced by the electrochemical formation of an oxidized platinum with lower reactivity. The two electrochemical reactions are coupled by the rate constant, rather than by the absorbed intermediate invoked by Epelboin and Keddam[35] and Armstrong et al.[37] The coupling of electrochemical reactions could, in principle, account for low-frequency inductive loops in the impedance response.

(b) The reactions with the reaction rate constant are given by

$$O_2 + 2H^+ + 4e^- \overset{k_{O_2}}{\rightarrow} 2H_2O \tag{10.131}$$

and

$$Pt + H_2O \underset{k_2}{\overset{k_1}{\rightleftarrows}} PtO + 2H_2 + 2e^- \tag{10.132}$$

where K_i is the lumped-parameter rate constant including the interfacial equilibrium potential for reaction with index i and given as

$$K_i = nFk_i \exp\left(-b_i V_{0,i}\right) \tag{10.133}$$

where $V_{0,i}$ is the interfacial equilibrium potential and k_i is the rate constant for electrochemical reaction with index i.

The steady-state current density corresponding to reaction (10.132) was given by

$$\bar{i}_{Pt} = K_1(1-\bar{\gamma})\exp\left(b_1\bar{V}\right) - K_2\Gamma\bar{\gamma}\exp\left(-b_2\bar{V}\right) \tag{10.134}$$

where Γ is the maximum surface coverage, and γ is the fractional surface coverage by PtO. The corresponding material balance for the PtO was expressed as

$$\Gamma\frac{\partial\gamma_{PtO}}{\partial t} = \frac{i_{Pt}}{2F} \tag{10.135}$$

Thus, the steady-state surface coverage of the platinum oxide is given as

$$\bar{\gamma} = \frac{K_1 b_1 \exp(b_1\bar{V})}{K_1 b_1 \exp(b_1\bar{V}) + K_2 b_2 \exp(-b_2\bar{V})} \tag{10.136}$$

(c) The steady-state current density corresponding to reaction (10.131) was given by

$$\bar{i}_{O_2} = -K_{eff}\bar{c}_{O_2}(0)\exp\left(-b_{O_2}\bar{V}\right) \tag{10.137}$$

Under a sinusoidal perturbation, the current density phasors for reaction (10.131) and (10.132) are given, respectively, by

$$\tilde{i}_{O_2} = b_{O_2}K_{eff}\bar{c}_{O_2}(0)\exp\left(-b_{O_2}\bar{V}\right)\tilde{V} - K_{eff}\exp\left(-b_{O_2}\bar{V}\right)\tilde{c}_{O_2}(0) \tag{10.138}$$

and

$$\tilde{i}_{Pt} = \left(\frac{1}{R_{t,1}} + \frac{1}{R_{t,2}}\right)\tilde{V} - \left(K_1\exp\left(b_1\bar{V}\right) + K_2\Gamma\exp\left(-b_2\bar{V}\right)\right)\tilde{\gamma} \tag{10.139}$$

where $R_{t,i}$ is the charge transfer resistance for each reaction and can be expressed, respectively, as

$$R_{t,\mathrm{O}_2} = \frac{1}{K_{\mathrm{eff}}\bar{c}_{\mathrm{O}_2}(0) b_{\mathrm{O}_2} \exp(-b_{\mathrm{O}_2}\bar{V})} \tag{10.140}$$

$$R_{t,1} = \frac{1}{K_1 b_1 (1-\overline{\gamma}) \exp(b_1 \overline{V})} \tag{10.141}$$

and

$$R_{t,2} = \frac{1}{K_2 b_2 \Gamma \overline{\gamma} \exp(b_2 \overline{V})} \tag{10.142}$$

Equation (10.135) can be expressed as

$$\mathrm{j}\omega\Gamma\widetilde{\gamma} = \frac{\widetilde{i}_{\mathrm{Pt}}}{2\mathrm{F}} \tag{10.143}$$

From equation (10.138), the faradic impedance for the reduction of the oxygen can be expressed as

$$Z_{\mathrm{F,O}_2} = R_{t,\mathrm{O}_2} + Z_{\mathrm{D,O}_2} \tag{10.144}$$

where

$$Z_{D,\mathrm{O}_2} = \frac{\delta_{\mathrm{O}_2}}{\bar{c}_{\mathrm{O}_2}(0) b_{\mathrm{O}_2} 2FD_{\mathrm{O}_2}} \frac{-1}{\theta'_{\mathrm{O}_2}(0)} \tag{10.145}$$

where $-1/\theta'_{\mathrm{O}_2}(0)$ is the dimensionless diffusion impedance for oxygen introduced in Chapter 11 of Orazem and Tribollet.[1] From equations (10.139) and (10.143), the faradaic impedance for the platinum dissolution can be expressed as

$$Z_{\mathrm{F,Pt}} = \frac{2\mathrm{j}\omega\mathrm{F}\Gamma\gamma + \left(K_1 \exp\left(b_1\overline{V}\right) + K_2\Gamma \exp\left(-b_2\overline{V}\right)\right)}{2\mathrm{j}\omega\mathrm{F}\Gamma\gamma\left(\frac{1}{R_{t,1}} + \frac{1}{R_{t,2}}\right)} \tag{10.146}$$

The total faradic impedance can be expressed as

$$Z_{\mathrm{F}} = \left(\frac{1}{Z_{\mathrm{F,O}_2}} + \frac{1}{Z_{\mathrm{F,Pt}}}\right)^{-1} \tag{10.147}$$

The faradaic impedance may be inserted into the appropriate electrical circuit framework such as developed in Chapter 9 to obtain the impedance for the system.

10.12 We wish to estimate values for the faradaic impedances for the reactions discussed in Problem 9.10. The reactions include corrosion of steel

$$Fe \rightarrow Fe^{2+} + 2e^- \tag{10.148}$$

reduction of dissolved oxygen

$$O_2 + 2H_2O + 4e^- \rightarrow 4OH^- \tag{10.149}$$

and hydrogen evolution

$$H_2O + e^- \rightarrow \frac{1}{2}H_2 + OH^- \tag{10.150}$$

Parameter values are given in Figure 9.7.

(a) Estimate values for the zero-frequency limit of the faradaic impedances for an applied potential of -0.4 V referenced to the $Cu/CuSO_4$ electrode.

(b) Estimate values for the zero-frequency limit of the faradaic impedances for an applied potential of -0.565 V referenced to the $Cu/CuSO_4$ electrode.

(c) Estimate values for the zero-frequency limit of the faradaic impedances for an applied potential of 1.2 V referenced to the $Cu/CuSO_4$ electrode.

Solution 10.12. The faradaic impedance for reaction (10.148) is given by

$$Z_{F,Fe} = R_{t,Fe} = \frac{1}{K_{Fe} b_{Fe} \exp\left(b_{Fe}\overline{V}\right)} \tag{10.151}$$

where V is referenced to the $Cu/CuSO_4$ electrode. The faradaic impedance for reaction (10.150) is given by

$$Z_{F,H_2} = R_{t,H_2} = \frac{1}{K_{H_2} b_{H_2} \exp\left(-b_{H_2}\overline{V}\right)} \tag{10.152}$$

The faradaic impedance for reaction (10.149) is given by

$$Z_{F,O_2} = R_{t,O_2} + R_{d,O_2}\left(\frac{-1}{\theta'_{O_2}(0)}\right) \tag{10.153}$$

Values for the charge-transfer resistances given in equations (10.151) and (10.152) may be readily calculated for a given potential using the parameters given in Figure 9.7.

Table 10.3: Calculated results for charge transfer resistances and diffusion impedances for reactions (10.148), (10.149), and (10.150).

Potential, V(Cu/CuSO$_4$)	$R_{t,Fe}$, Ωcm^2	R_{t,H_2}, Ωcm^2	R_{t,O_2}, Ωcm^2	R_{d,O_2}, Ωcm^2
−0.4	700	2.8×10^8	2.4×10^6	6.5×10^3
−0.565	8.0×10^4	1.9×10^7	1.1×10^5	6.5×10^3
−1.2	7.0×10^{12}	540	6.5×10^3	8.2×10^7

The values for R_{t,O_2} and R_{d,O_2} require a little more effort. The charge-transfer resistance may be expressed as

$$R_{t,O_2} = R_{t_lim,O_2}\left(1 + \frac{i_{lim,O_2}}{i_{k,O_2}}\right) \tag{10.154}$$

where, following Tran et al.,[33]

$$i_{k,O_2} = -K_{O_2}\bar{c}_{O_2}(\infty)\exp\left(-b_{O_2}\overline{V}\right) \tag{10.155}$$

and

$$R_{t_{lim},O_2} = \frac{1}{-b_{O_2}\bar{i}_{lim,O_2}} \tag{10.156}$$

A lumped value of $K_{O_2}\bar{c}_{O_2}(\infty)$ and a value for i_{lim,O_2} were defined in the solution to Problem 9.10. From these values, $R_{t_{lim},O_2} = 6.5 \times 10^3\ \Omega cm^2$. Thus, values may readily be obtained from equation (10.154) for R_{t,O_2} and

$$R_{d,O_2} = R_{t_{lim},O_2}\left(1 + \frac{i_{k,O_2}}{i_{lim,O_2}}\right) \tag{10.157}$$

In the limit that frequency approaches zero, the dimensionless diffusion impedance may be assumed to approach unity. Thus,

$$\lim_{\omega \to 0} Z_{F,O_2}(0) = R_{t,O_2} + R_{d,O_2} \tag{10.158}$$

Calculated values of charge-transfer and diffusion resistances are presented in Table 10.3 for potentials of −0.4, −0.565, and −1.2 V(Cu/CuSO$_4$), respectively.

(a) The potential −0.4 V(Cu/CuSO$_4$) is in the anodic regime, where the corrosion current may be expected to dominate. The faradaic impedance for the corrosion reaction $Z_{F,Fe} = 0.7\ k\Omega cm^2$ is much smaller than the faradaic impedance for the hydrogen evolution $Z_{F,H_2} = 2.8 \times 10^5\ k\Omega cm^2$ and oxygen reduction reactions $Z_{F,O_2} = 2.4 \times 10^3\ k\Omega cm^2$. This result confirms the suitability of Figure 9.8(c). Only the faradaic impedance of the corrosion reaction needs to be included in the circuit analysis.

(b) The potential −0.565 V(Cu/CuSO$_4$) is close to the open-circuit or corrosion potential. The faradaic impedance for hydrogen evolution, $Z_{F,H_2} = 1.9 \times 10^4\ k\Omega cm^2$, is substantially larger than the faradaic impedance for the corrosion reaction $Z_{F,Fe} = 80$ kΩcm^2 and the low-frequency faradaic impedance given by equation (10.158) for the oxygen reduction reaction, $Z_{F,O_2} = 115$ kΩcm^2. This result confirms the suitability of Figure 9.8(b). The faradaic impedances of both the corrosion and oxygen reduction reactions need to be included in the circuit analysis. The contribution of hydrogen evolution may be neglected.

(c) The potential -1.2 V(Cu/CuSO$_4$) is in the cathodic regime where hydrogen evolution becomes evident. The faradaic impedance for the hydrogen evolution is $Z_{F,H_2} = 0.54$ kΩcm^2. This value is much smaller than the impedance for corrosion reaction, $Z_{F,Fe} = 7 \times 10^9$ kΩcm^2. The faradaic impedance for the oxygen reduction reaction varies from $Z_{F,O_2} = 6.5$ kΩcm^2 at high frequency to $Z_{F,O_2} = 8.2 \times 10^4$ kΩcm^2 at low frequency. This result confirms the suitability of Figure 9.8(d). Only the faradaic impedance of the hydrogen evolution reaction needs to be included in the circuit analysis.

10.13 Calculate and plot the impedance response expected for Problem 10.12 at the specified potentials. You may assume that diffusion through a stagnant layer applies with $\delta = 6$ µm and $D_i = 10\ ^5$ cm^2/s. Kinetic and diffusion parameter values are those obtained in Problem 10.12. Assume an ohmic resistance $R_e = 10$ Ωcm^2 and a capacitance $C_{dl} = 20$ µF/cm^2.
(a) An applied potential of -0.4 V(Cu/CuSO$_4$).
(b) An applied potential of -0.565 V (Cu/CuSO$_4$).
(c) An applied potential of -1.2 V (Cu/CuSO$_4$).

Solution 10.13. The impedance may be written as

$$Z = R_e + \frac{1}{\dfrac{1}{R_{t,Fe}} + \dfrac{1}{R_{t,H_2}} + \dfrac{1}{R_{t,O_2} + R_{d,O_2}\left(\dfrac{-1}{\theta'(0)}\right)} + j\omega C_{dl}} \tag{10.159}$$

The dimensionless diffusion impedance is given in Figure 10.6 for a stagnant layer with thickness $\delta = 6$ µm and species diffusivity $D_i = 10^{-5}$ cm^2/s. The characteristic frequency is given by

$$f_c = \frac{2.5 D_i}{2\pi\delta^2} = 11 \text{ Hz} \tag{10.160}$$

(a) The system impedance at an applied potential of -0.4 V(Cu/CuSO$_4$) is presented in Figure 10.7. The characteristic frequency is on the order of 11.5 Hz, which yields an effective charge-transfer resistance of

$$R_{t,eff} = \frac{1}{2\pi f_c C_{dl}} = 690 \ \Omega\text{cm}^2 \tag{10.161}$$

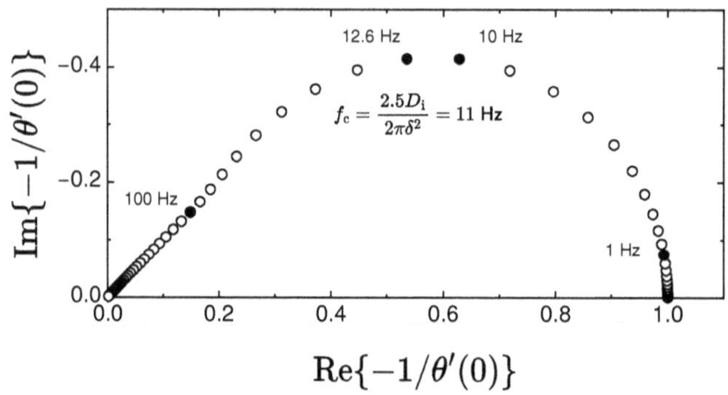

Figure 10.6: Dimensionless diffusion impedance for a stagnant layer with thickness $\delta = 6$ µm and species diffusivity $D_i = 10^{-5}$ cm^2/s as discussed in Problem 10.13.

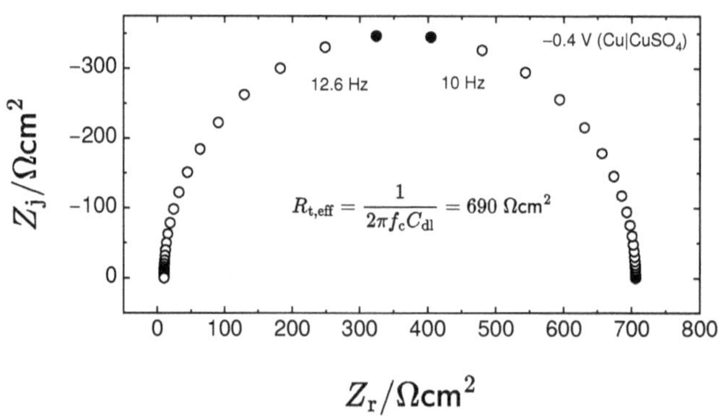

Figure 10.7: The system impedance at an applied potential of -0.4 V(Cu/$CuSO_4$).

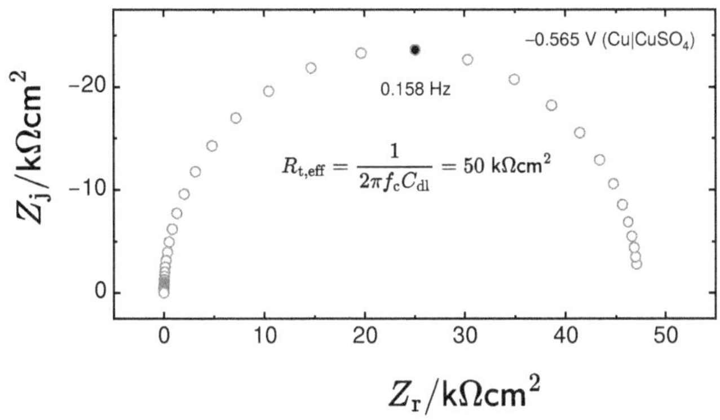

Figure 10.8: The system impedance at an applied potential of −0.565 V(Cu/$CuSO_4$).

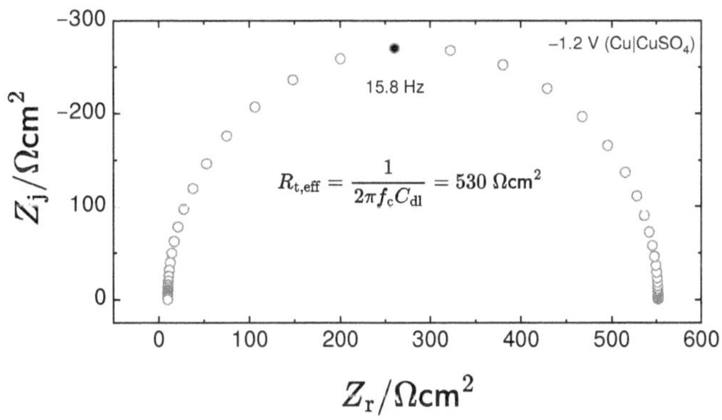

Figure 10.9: The system impedance at an applied potential of −1.2 V(Cu/$CuSO_4$).

The value is only slightly lower than the charge-transfer resistance of the corrosion reaction $R_{t,Fe} = 700\ \Omega\text{cm}^2$, suggesting that the oxygen reduction reaction plays a minor role at this potential.

(b) The system impedance at an applied potential of −0.565 V(Cu/$CuSO_4$) is presented in Figure 10.8. The characteristic frequency is on the order of 0.170 Hz, which yields an effective charge-transfer resistance of $R_{t,eff} = 50\ \text{k}\Omega\text{cm}^2$. The value is smaller than that of the corrosion reaction $R_{t,Fe} = 80\ \text{k}\Omega\text{cm}^2$, suggesting that the oxygen reduction reaction plays a significant role at this potential.

(c) The system impedance at an applied potential of −1.2 V(Cu/$CuSO_4$) is presented in Figure 10.9. The characteristic frequency is on the order of 15 Hz, which yields an effective charge-transfer resistance of $R_{t,eff} = 530\ \Omega\text{cm}^2$, which is on the order of the charge-transfer resistance of the hydrogen evolution reaction $R_{t,H_2} = 540\ \Omega\text{cm}^2$. These results suggest that the oxygen reduction reaction is not visible in the impedance response at this potential.

10.14 Using the steps given below, develop a model for the faradaic impedance for the reaction $Cu + Cl^- \rightarrow CuCl + e^-$. We wish to consider only early times such that the CuCl film has not grown to the thickness that would influence the diffusion of chloride ions.

(a) Find an expression for the faradaic current.

(b) Find an expression for the faradaic current phasor in terms of potential and chloride concentration phasors.

(c) Find an expression for the faradaic current in terms of the gradient of chloride concentration.

(d) Find an expression for the faradaic current phasor resulting from part (c).

(e) Get an expression for the faradaic impedance in terms of the charge-transfer and diffusion resistances and the dimensionless diffusion impedance.

(f) Find an expression for the charge-transfer resistance.

(g) Find an expression for the diffusion resistance.

(h) What is the relationship between the charge-transfer resistance and the Koutecky–Levich parameters?

(i) Please define i_k and i_{lim} used in part (h).

(j) What is the relationship between the diffusion resistance and the Koutecky–Levich parameters?

Solution 10.14.

(a) The faradaic current density may be expressed as

$$i_F = K_a c_{Cl^-}(0) \exp\left(b_a V\right) \tag{10.162}$$

where K_a includes the equilibrium potential for the reaction.

(b) The faradaic current may be expressed in terms of a steady state term $\bar{i}_F$ and a phasor $\widetilde{i}_F$ as

$$i_F = \bar{i}_F + \text{Re}\left\{\widetilde{i}_F \exp(j\omega t)\right\} \tag{10.163}$$

The faradaic current density phasor may be expressed in terms of potential and concentration phasors as

$$\widetilde{i}_F = K_a b_a \bar{c}_{Cl^-}(0) \exp\left(b_a \overline{V}\right) \widetilde{V} + K_a \exp\left(b_a \overline{V}\right) \widetilde{c}_{Cl^-}(0) \tag{10.164}$$

Equation (10.164) is written in terms of three phasors. An expression in terms of two phasors is needed to obtain a transfer function, for example, an impedance.

(c) The faradaic current density may be expressed in terms of a concentration gradient as

$$i_F = \frac{nFD_{Cl^-}}{s_{Cl^-}} \left.\frac{dc_{Cl^-}}{dy}\right|_{y=0} \tag{10.165}$$

where the number of electrons transferred $n = 1$ and the stoichiometric coefficient for Cl^- is $s_{Cl^-} = 1$.

(d) The concentration may be expressed as

$$c_{Cl^-} = \bar{c}_{Cl^-} + \mathrm{Re}\left\{\tilde{c}_{Cl^-}\exp(j\omega t)\right\} \tag{10.166}$$

Introduction of equations (10.163) and (10.166) into equation (10.165) yields

$$\bar{i}_F + \mathrm{Re}\left\{\tilde{i}_F\exp(j\omega t)\right\} = FD_{Cl^-}\left.\frac{d\bar{c}_{Cl^-}}{dy}\right|_{y=0} + FD_{Cl^-}\mathrm{Re}\left\{\left.\frac{d\tilde{c}_{Cl^-}}{dy}\right|_{y=0}\exp(j\omega t)\right\} \tag{10.167}$$

As equation (10.165) is satisfied at steady state, equation (10.167) may be expressed as

$$\tilde{i}_F = FD_{Cl^-}\left.\frac{d\tilde{c}_{Cl^-}}{dy}\right|_{y=0} \tag{10.168}$$

Combination of equations (10.164) and (10.168) allows development of an expression for the faradaic impedance.

(e) Equation (10.164) may be rearranged as

$$1 = \frac{1}{R_t}Z_F + K_a\exp\left(b_a\overline{V}\right)\frac{\tilde{c}_{Cl^-}(0)}{\tilde{i}_F} \tag{10.169}$$

where

$$R_t = \frac{1}{K_a b_a \bar{c}_{Cl^-}(0)\exp\left(b_a\overline{V}\right)} \tag{10.170}$$

and

$$Z_F = \frac{\widetilde{V}}{\tilde{i}_F} \tag{10.171}$$

Introduction of equation (10.168) to equation (10.169) yields

$$1 = \frac{1}{R_t}Z_F + \frac{K_a\exp\left(b_a\overline{V}\right)}{FD_{Cl^-}}\frac{\tilde{c}_{Cl^-}(0)}{\left.\frac{d\tilde{c}_{Cl^-}}{dy}\right|_{y=0}} \tag{10.172}$$

Multiplication of both sides by R_t yields

$$Z_F = R_t - R_t\frac{K_a\exp\left(b_a\overline{V}\right)}{FD_{Cl^-}}\frac{\tilde{c}_{Cl^-}(0)}{\left.\frac{d\tilde{c}_{Cl^-}}{dy}\right|_{y=0}} \tag{10.173}$$

The dimensionless diffusion impedance is obtained by definition of a dimensionless concentration phasor

$$\theta_{Cl^-} = \frac{\tilde{c}_{Cl^-}}{\tilde{c}_{Cl^-}(0)} \tag{10.174}$$

and position

$$\xi = \frac{y}{\delta_{\mathrm{Cl}^-}} \tag{10.175}$$

Application of the chain rule yields

$$\frac{\widetilde{c}_{\mathrm{Cl}^-}(0)}{\left.\dfrac{\mathrm{d}\widetilde{c}_{\mathrm{Cl}^-}}{\mathrm{d}y}\right|_{y=0}} = \delta_{\mathrm{Cl}^-}\frac{1}{\theta'_{\mathrm{Cl}^-}(0)} \tag{10.176}$$

Thus, equation (10.173) may be written as

$$Z_{\mathrm{F}} = R_{\mathrm{t}} + \frac{\delta_{\mathrm{Cl}^-}}{b_{\mathrm{a}}\bar{c}_{\mathrm{Cl}^-}(0)\mathrm{F}D_{\mathrm{Cl}^-}}\frac{-1}{\theta'_{\mathrm{Cl}^-}(0)} \tag{10.177}$$

or

$$Z_{\mathrm{F}} = R_{\mathrm{t}} + R_{\mathrm{d}}\frac{-1}{\theta'_{\mathrm{Cl}^-}(0)} \tag{10.178}$$

where

$$R_{\mathrm{d}} = \frac{\delta_{\mathrm{Cl}^-}}{b_{\mathrm{a}}\bar{c}_{\mathrm{Cl}^-}(0)\mathrm{F}D_{\mathrm{Cl}^-}} \tag{10.179}$$

(f) The charge-transfer resistance is given by equation (10.170).

(g) The diffusion resistance is given by equation (10.179).

(h) The charge-transfer resistance may be expressed in terms of Koutecky–Levich variables as

$$R_{\mathrm{t}} = \frac{1}{b_{\mathrm{a}}\bar{i}_{\mathrm{lim}}}\left(1 + \frac{\bar{i}_{\mathrm{lim}}}{\bar{i}_{\mathrm{k}}}\right) \tag{10.180}$$

where definitions of i_{k} and i_{lim} are provided in part (i).

(i) The Koutecky–Levich equation is given by

$$\frac{1}{i} = \frac{1}{i_{\mathrm{k}}} + \frac{1}{i_{\mathrm{lim}}} \tag{10.181}$$

where, for the present problem

$$i_{\mathrm{k}} = K_{\mathrm{a}}c_{\mathrm{Cl}^-}(\infty)\exp\left(b_{\mathrm{a}}V\right) \tag{10.182}$$

and

$$i_{\mathrm{lim}} = \frac{\mathrm{F}D_{\mathrm{Cl}^-}c_{\mathrm{Cl}^-}(\infty)}{\delta_{\mathrm{Cl}^-}} \tag{10.183}$$

Equations (10.182) and (10.183) provide a convenient way to calculate the current for a system influenced by mass transfer.

(j) The diffusion resistance may be expressed in terms of Koutecky–Levich variables as

$$R_{\rm d} = \frac{1}{b_{\rm a}\bar{i}_{\rm lim}}\left(1 + \frac{\bar{i}_{\rm k}}{\bar{i}_{\rm lim}}\right) \tag{10.184}$$

where definitions of $i_{\rm k}$ and $i_{\rm lim}$ are provided in part (i).

10.15 Using the steps given below, develop a model for the faradaic impedance for the reaction $H_2O_2 \rightarrow 2H^+ + O_2 + 2e^-$.

(a) Find an expression for the faradaic current.

(b) Find an expression for the faradaic current phasor in terms of potential and chloride concentration phasors.

(c) Find an expression for the faradaic current in terms of the gradient of chloride concentration.

(d) Find an expression for the faradaic current phasor resulting from part (c).

(e) Get an expression for the faradaic impedance in terms of the charge-transfer and diffusion resistances and the dimensionless diffusion impedance.

(f) Find an expression for the charge-transfer resistance.

(g) Find an expression for the diffusion resistance.

(h) What is the relationship between the charge-transfer resistance and the Koutecky–Levich parameters?

(i) Please define $i_{\rm k}$ and $i_{\rm lim}$ used in part (h).

(j) What is the relationship between the diffusion resistance and the Koutecky–Levich parameters?

Solution 10.15.

(a) The faradaic current density may be expressed as

$$i_{\rm F} = K_{\rm a} c_{\rm H_2O_2}(0) \exp\left(b_{\rm a} V\right) \tag{10.185}$$

where $K_{\rm a}$ includes the equilibrium potential for the reaction.

(b) The faradaic current may be expressed in terms of a steady state term $\bar{i}_{\rm F}$ and a phasor $\tilde{i}_{\rm F}$ as

$$i_{\rm F} = \bar{i}_{\rm F} + {\rm Re}\left\{\tilde{i}_{\rm F} \exp(j\omega t)\right\} \tag{10.186}$$

The faradaic current density phasor may be expressed in terms of potential and concentration phasors as

$$\tilde{i}_{\rm F} = K_{\rm a} b_{\rm a} \bar{c}_{\rm H_2O_2}(0) \exp\left(b_{\rm a}\overline{V}\right)\widetilde{V} + K_{\rm a} \exp\left(b_{\rm a}\overline{V}\right)\tilde{c}_{\rm H_2O_2}(0) \tag{10.187}$$

Equation (10.187) is written in terms of three phasors. An expression in terms of two phasors is needed to obtain a transfer function, for example, an impedance.

(c) Following equation (10.165), the faradaic current density may be expressed in terms of a concentration gradient as

$$i_F = 2FD_{H_2O_2} \left. \frac{dc_{H_2O_2}}{dy} \right|_{y=0} \tag{10.188}$$

where the number of electrons transferred $n = 2$ and the stoichiometric coefficient for H_2O_2 is $s_{H_2O_2} = 1$.

(d) The concentration may be expressed as

$$c_{H_2O_2} = \overline{c}_{H_2O_2} + \text{Re}\left\{\widetilde{c}_{H_2O_2} \exp(j\omega t)\right\} \tag{10.189}$$

Following the development of equation (10.168), the introduction of equations (10.186) and (10.189) into equation (10.188) yields

$$\widetilde{i}_F = 2FD_{H_2O_2} \left. \frac{d\widetilde{c}_{H_2O_2}}{dy} \right|_{y=0} \tag{10.190}$$

Combination of equations (10.187) and (10.190) allows development of an expression for the faradaic impedance.

(e) Equation (10.187) may be rearranged as

$$1 = \frac{1}{R_t} Z_F + K_a \exp\left(b_a \overline{V}\right) \frac{\widetilde{c}_{H_2O_2}(0)}{\widetilde{i}_F} \tag{10.191}$$

where

$$R_t = \frac{1}{K_a b_a \overline{c}_{H_2O_2}(0) \exp\left(b_a \overline{V}\right)} \tag{10.192}$$

and Z_F is given by equation (10.171). Introduction of equation (10.190) to equation (10.191) yields

$$1 = \frac{1}{R_t} Z_F + \frac{K_a \exp\left(b_a \overline{V}\right)}{2FD_{H_2O_2}} \frac{\widetilde{c}_{H_2O_2}(0)}{\left. \frac{d\widetilde{c}_{H_2O_2}}{dy} \right|_{y=0}} \tag{10.193}$$

Multiplication of both sides by R_t yields

$$Z_F = R_t - R_t \frac{1}{2b_a FD_{H_2O_2} \overline{c}_{H_2O_2}(0)} \frac{\widetilde{c}_{H_2O_2}(0)}{\left. \frac{d\widetilde{c}_{H_2O_2}}{dy} \right|_{y=0}} \tag{10.194}$$

The dimensionless diffusion impedance is obtained by definition of a dimensionless concentration phasor

$$\theta_{H_2O_2} = \frac{\widetilde{c}_{H_2O_2}}{\widetilde{c}_{H_2O_2}(0)} \tag{10.195}$$

and position

$$\xi = \frac{y}{\delta_{H_2O_2}} \tag{10.196}$$

Application of the chain rule yields

$$\frac{\widetilde{c}_{H_2O_2}(0)}{\left.\frac{d\widetilde{c}_{H_2O_2}}{dy}\right|_{y=0}} = \delta_{H_2O_2}\frac{1}{\theta'_{H_2O_2}(0)} \tag{10.197}$$

Thus, equation (10.194) may be written as

$$Z_F = R_t + R_d\frac{-1}{\theta'_{H_2O_2}(0)} \tag{10.198}$$

where

$$R_d = \frac{\delta_{H_2O_2}}{2b_a\bar{c}_{H_2O_2}(0)FD_{H_2O_2}} \tag{10.199}$$

(f) The charge-transfer resistance is given by equation (10.192).

(g) The diffusion resistance is given by equation (10.199).

(h) The charge-transfer resistance may be expressed in terms of Koutecky–Levich variables as equation (10.180).

(i) The Koutecky–Levich equation is given by equation (10.181). For the present problem,

$$i_k = K_a c_{H_2O_2}(\infty)\exp(b_a V) \tag{10.200}$$

and

$$i_{lim} = \frac{FD_{H_2O_2}c_{H_2O_2}(\infty)}{\delta_{H_2O_2}} \tag{10.201}$$

Equations (10.200) and (10.201) provide a convenient way to calculate the current for a system influenced by mass transfer.

(j) The diffusion resistance may be expressed in terms of Koutecky–Levich variables using equation (10.184), where definitions of i_k and i_{lim} are provided in part (i).

Chapter 11

Diffusion Impedance

The development of kinetic models presented in Chapter 10 requires expressions for the concentrations of reacting species at the electrode surface. The development was expressed in terms of an inverse dimensionless concentration gradient given as $-1/\theta_i'(0)$. The problems presented in this chapter explore conditions and systems for which expressions for diffusion impedance, for example, $-1/\theta_i'(0)$, can be developed. Problems 11.1–11.12 are based on problems taken from Orazem and Tribollet.[1]

Problems

11.1 Starting with the material balance equation, verify the development of

$$\frac{-1}{\theta_i'(0)} = \frac{\tanh\left(\sqrt{jK_i}\right)}{\sqrt{jK_i}} \tag{11.1}$$

for diffusion with an infinite supply of electro-active species.

Solution 11.1. The general solution to

$$\frac{d^2\theta_i}{d\xi^2} - jK_i\theta_i = 0 \tag{11.2}$$

is

$$\theta_i = C_1 \exp\left(\xi\sqrt{jK_i}\right) - C_2 \exp\left(-\xi\sqrt{jK_i}\right) \tag{11.3}$$

Equation (11.1) corresponds to the diffusion impedance associated with a film in which the reacting species is replenished by a reservoir of infinite volume. The boundary conditions are

$$\begin{aligned} \theta_i &= 0 \quad \text{at} \quad \xi = 1 \\ \theta_i &= 1 \quad \text{at} \quad \xi = 0 \end{aligned} \tag{11.4}$$

Application of the boundary conditions yields

$$0 = C_1 \exp\left(\sqrt{jK_i}\right) - C_2 \exp\left(-\sqrt{jK_i}\right) \tag{11.5}$$

and

$$1 = C_1 - C_2 \tag{11.6}$$

or

$$C_2 = C_1 - 1 \tag{11.7}$$

A combination of equations (11.5) and (11.6) yields

$$\begin{aligned} C_1 &= \frac{\exp\left(-\sqrt{jK_i}\right)}{\exp\left(-\sqrt{jK_i}\right) - \exp\left(\sqrt{jK_i}\right)} \\ &= \frac{\exp\left(-\sqrt{jK_i}\right)}{2\sinh\left(-\sqrt{jK_i}\right)} \end{aligned} \tag{11.8}$$

and

$$C_2 = \frac{\exp\left(\sqrt{jK_i}\right)}{2\sinh\left(-\sqrt{jK_i}\right)} \tag{11.9}$$

Introduction to equation (11.3) yields

$$\begin{aligned} \theta_i &= \frac{\exp\left(-\sqrt{jK_i}\right)\exp\left(\xi\sqrt{jK_i}\right)}{2\sinh\left(-\sqrt{jK_i}\right)} - \frac{\exp\left(\sqrt{jK_i}\right)\exp\left(-\xi\sqrt{jK_i}\right)}{2\sinh\left(-\sqrt{jK_i}\right)} \\ &= \frac{\sinh\left((\xi-1)\sqrt{jK_i}\right)}{\sinh\left(-\sqrt{jK_i}\right)} \end{aligned} \tag{11.10}$$

The derivative of θ_i with respect to ξ yields

$$\frac{d\theta_i}{d\xi} = \frac{\cosh\left((\xi-1)\sqrt{jK_i}\right)\sqrt{jK_i}}{\sinh\left(-\sqrt{jK_i}\right)} \tag{11.11}$$

Given the odd character of the sinh function and the even character of the cosh function, upon evaluation at $\xi = 0$,

$$\frac{-1}{\left.\frac{d\theta_i}{d\xi}\right|_{\xi=0}} = \frac{-1}{\theta_i'(0)} = \frac{\tanh\left(\sqrt{jK_i}\right)}{\sqrt{jK_i}} \tag{11.12}$$

Equation (11.12) is the diffusion impedance for a film of finite thickness δ_f with exchange of electro-active species.

11.2 Starting with the material balance equation, verify the development of

$$\frac{-1}{\theta_i'(0)} = \frac{\coth\left(\sqrt{jK_i}\right)}{\sqrt{jK_i}} \tag{11.13}$$

for diffusion without exchange of electro-active species.

Solution 11.2. The general solution to equation (11.2) is given as equation (11.3). The boundary conditions for a diffusion layer of finite thickness without exchange of electro-active species are

$$\begin{aligned} \frac{d\theta_i}{d\xi} &= 0 \quad \text{at} \quad \xi = 1 \\ \theta_i &= 1 \quad \text{at} \quad \xi = 0 \end{aligned} \tag{11.14}$$

The derivative of equation (11.3) with respect to position is given as

$$\frac{d\theta_i}{d\xi} = C_1\sqrt{jK_i}\exp\left(\xi\sqrt{jK_i}\right) - C_2\sqrt{jK_i}\exp\left(-\xi\sqrt{jK_i}\right) \tag{11.15}$$

Application of the boundary conditions yields

$$0 = C_1\sqrt{jK_i}\exp\left(\sqrt{jK_i}\right) + C_2\sqrt{jK_i}\exp\left(-\sqrt{jK_i}\right) \tag{11.16}$$

and

$$1 = C_1 - C_2 \tag{11.17}$$

or

$$C_2 = C_1 - 1 \tag{11.18}$$

A combination of equations (11.16) and (11.17) yields

$$\begin{aligned} C_1 &= \frac{\exp\left(-\sqrt{jK_i}\right)}{\exp\left(-\sqrt{jK_i}\right) + \exp\left(\sqrt{jK_i}\right)} \\ &= \frac{\exp\left(-\sqrt{jK_i}\right)}{2\cosh\left(\sqrt{jK_i}\right)} \end{aligned} \tag{11.19}$$

and

$$C_2 = -\frac{\exp\left(\sqrt{jK_i}\right)}{2\cosh\left(\sqrt{jK_i}\right)} \tag{11.20}$$

Introduction to equation (11.3) yields

$$\begin{aligned} \theta_i &= \frac{\exp\left(-\sqrt{jK_i}\right)\exp\left(\xi\sqrt{jK_i}\right)}{2\cosh\left(-\sqrt{jK_i}\right)} + \frac{\exp\left(\sqrt{jK_i}\right)\exp\left(-\xi\sqrt{jK_i}\right)}{2\cosh\left(-\sqrt{jK_i}\right)} \\ &= \frac{\cosh\left((\xi-1)\sqrt{jK_i}\right)}{\cosh\left(-\sqrt{jK_i}\right)} \end{aligned} \tag{11.21}$$

The derivative of θ_i with respect to ξ yields

$$\frac{d\theta_i}{d\xi} = \frac{\sinh\left((\xi-1)\sqrt{jK_i}\right)\sqrt{jK_i}}{\cosh\left(-\sqrt{jK_i}\right)} \tag{11.22}$$

Given the odd character of the sinh function and the even character of the cosh function, upon evaluation at $\xi = 0$,

$$\frac{-1}{\left.\frac{\mathrm{d}\theta_\mathrm{i}}{\mathrm{d}\xi}\right|_{\xi=0}} = \frac{-1}{\theta_\mathrm{i}'(0)} = \frac{\coth\left(\sqrt{\mathrm{j}K_\mathrm{i}}\right)}{\sqrt{\mathrm{j}K_\mathrm{i}}} \tag{11.23}$$

Equation (11.23) is the diffusion impedance for a film of finite thickness δ_f without exchange of electro-active species.

11.3 Show that, for disk rotating in an incompressible fluid and under the assumption that c_i is independent of r, $-\nabla \cdot \mathrm{N_i} = 0$ gives rise to

$$v_\mathrm{y}\frac{\mathrm{d}\bar{c}_\mathrm{i}}{\mathrm{d}y} - D_\mathrm{i}\frac{\mathrm{d}^2\bar{c}_\mathrm{i}}{\mathrm{d}y^2} = 0 \tag{11.24}$$

and

$$\nabla \cdot \mathrm{N_i} \neq \frac{\mathrm{d}N_\mathrm{i}}{\mathrm{d}y} \tag{11.25}$$

This problem statement is important when solving the convective diffusion for a rotating disk in terms of fluxes rather than concentrations.

Solution 11.3. The flux components can be written in cylindrical coordinates as

$$N_\mathrm{r,i} = c_\mathrm{i}v_\mathrm{r} = c_\mathrm{i}rf(y) \tag{11.26}$$

$$N_{\theta,\mathrm{i}} = c_\mathrm{i}v_\theta = c_\mathrm{i}rg(y) \tag{11.27}$$

and

$$N_\mathrm{y,i} = -D_i\frac{\partial c_i}{\partial y} + c_\mathrm{i}v_\mathrm{y} = -D_i\frac{\partial c_i}{\partial y} + c_\mathrm{i}h(y) \tag{11.28}$$

where $f(y)$, $g(y)$, and $h(y)$ are functions of axial position. The flux in the radial and angular directions is driven by convection, and the flux in the axial direction includes both diffusion and convection contributions. The steady-state conservation equation, in the absence of homogeneous reactions, can be expressed as

$$\nabla \cdot \vec{N}_\mathrm{i} = \frac{1}{r}\frac{\partial}{\partial r}\left(rN_\mathrm{r,i}\right) + \frac{1}{r}\frac{\partial N_{\theta,\mathrm{i}}}{\partial \theta} + \frac{\partial N_\mathrm{y,i}}{\partial y} = 0 \tag{11.29}$$

Insertion of equations (11.26)–(11.28) yields

$$\nabla \cdot \vec{N}_\mathrm{i} = \frac{1}{r}\frac{\partial}{\partial r}\left(r^2 c_\mathrm{i} f(y)\right) - D_\mathrm{i}\frac{\partial^2 c_\mathrm{i}}{\partial y^2} + \frac{\partial}{\partial y}\left(c_\mathrm{i}h(y)\right) = 0 \tag{11.30}$$

or, as c_i is independent of r,

$$\nabla \cdot \vec{N}_{\mathrm{i}} = 2c_{\mathrm{i}}f(y) - D_{\mathrm{i}}\frac{\partial^2 c_{\mathrm{i}}}{\partial y^2} + \frac{\partial}{\partial y}\left(c_{\mathrm{i}}h(y)\right) = 0 \tag{11.31}$$

Equation (11.31) shows that the radial contribution to flux cannot be neglected when evaluating the divergence of the flux. Thus, equation (11.25) is verified.

Both the concentration and axial velocity are functions of axial position. Equation (11.30) may be written as

$$\nabla \cdot \vec{N}_{\mathrm{i}} = c_{\mathrm{i}}\frac{1}{r}\frac{\partial}{\partial r}\left(rv_{\mathrm{r}}\right) - D_{\mathrm{i}}\frac{\partial^2 c_{\mathrm{i}}}{\partial y^2} + c_{\mathrm{i}}\frac{\partial v_{\mathrm{y}}}{\partial y} + v_{\mathrm{y}}\frac{\partial c_{\mathrm{i}}}{\partial y} = 0 \tag{11.32}$$

For an incompressible fluid, $\nabla \cdot \vec{v} = 0$, and

$$\nabla \cdot \vec{v} = \frac{1}{r}\frac{\partial}{\partial r}\left(rv_{\mathrm{r}}\right) + \frac{\partial v_{\mathrm{y}}}{\partial y} = 0 \tag{11.33}$$

Thus,

$$\nabla \cdot \vec{N}_{\mathrm{i}} = c_{\mathrm{i}}\left(\frac{1}{r}\frac{\partial}{\partial r}\left(rv_{\mathrm{r}}\right) + \frac{\partial v_{\mathrm{y}}}{\partial y}\right) + -D_{\mathrm{i}}\frac{\partial^2 c_{\mathrm{i}}}{\partial y^2} + v_{\mathrm{y}}\frac{\partial c_{\mathrm{i}}}{\partial y} = 0 \tag{11.34}$$

Cancellation of terms in equation (11.34) yields equation (11.24).

11.4 Plot, on an impedance plane format, the impedance obtained for a Nernst stagnant diffusion layer and the impedance obtained for a rotating disk electrode under assumption of an infinite Schmidt number. Show that, while the behaviors of the two models at high and low frequencies are in agreement, the two models do not agree at intermediate frequencies.[38] Explain.

Solution 11.4. The dimensionless diffusion impedance for a Nernst stagnant diffusion layer may be expressed as

$$\frac{-1}{\theta_{\mathrm{i}}'(0)} = \frac{\tanh\left(\sqrt{\mathrm{j}K_{\mathrm{i}}}\right)}{\sqrt{\mathrm{j}K_{\mathrm{i}}}} \tag{11.35}$$

where the dimensionless frequency

$$K_{\mathrm{i}} = \frac{\omega\delta_{\mathrm{f}}^2}{D_{\mathrm{i}}} \tag{11.36}$$

is a function of the film thickness and diffusivity of the reacting species i, and

$$\theta_{\mathrm{i}}(\xi) = \widetilde{c}_{\mathrm{i}}/\widetilde{c}_{\mathrm{i}}(0) \tag{11.37}$$

is the dimensionless concentration phasor referenced to the value at the electrode surface. The dimensionless diffusion impedance is readily calculated as a function of K_{i} by use of equation (11.1).

The impedance of a rotating disk electrode under assumption of an infinite Schmidt number is a more challenging problem. The governing equation may be expressed as

$$j\omega\widetilde{c}_{\mathrm{i}} + v_{\mathrm{y}}\frac{\mathrm{d}\widetilde{c}_{\mathrm{i}}}{\mathrm{d}y} - D_{\mathrm{i}}\frac{\mathrm{d}^2\widetilde{c}_{\mathrm{i}}}{\mathrm{d}y^2} = 0 \tag{11.38}$$

Under the assumption that the normal velocity is limited to the first term of its expansion near the electrode surface

$$v_{\mathrm{y}} = -a\frac{\Omega^{3/2}}{\sqrt{\nu}}y^2 = -\alpha y^2 \tag{11.39}$$

The dimensionless frequency may be expressed as

$$K_{\mathrm{i}} = \frac{\omega}{\Omega}\left(\frac{9\nu}{\mathrm{a}^2 D_{\mathrm{i}}}\right)^{1/3} = \frac{\omega}{\Omega}\left(\frac{9}{\mathrm{a}^2}\right)^{1/3}\mathrm{Sc}_{\mathrm{i}}{}^{1/3} = \frac{\omega\delta_{\mathrm{i}}^2}{D_{\mathrm{i}}} \tag{11.40}$$

and the governing equation may be expressed as

$$\frac{\mathrm{d}^2\theta_{\mathrm{i}}}{\mathrm{d}\xi^2} + 3\xi^2\frac{\mathrm{d}\theta_{\mathrm{i}}}{\mathrm{d}\xi} - \mathrm{j}K_{\mathrm{i}}\theta_{\mathrm{i}} = 0 \tag{11.41}$$

where $\xi = y/\delta_{\mathrm{i}}$ and the diffusion layer thickness

$$\delta_{\mathrm{i}} = \left(\frac{3D_{\mathrm{i}}}{\alpha}\right)^{1/3} = \left(\frac{3}{\mathrm{a}}\right)^{1/3}\frac{1}{\mathrm{Sc}_{\mathrm{i}}{}^{1/3}}\sqrt{\frac{\nu}{\Omega}} \tag{11.42}$$

is a function of disk rotation speed Ω.

Only approximative analytic solutions are possible for equation (11.41). A MATLAB code is presented in Section B.5.1 that solves equation (11.41) subject to the boundary conditions that $\theta_{\mathrm{i}} = 1$ at $\xi = 0$ and $\theta_{\mathrm{i}} = 0$ at $\xi \to \infty$. The comparison of the impedance obtained from equations (11.1) and (11.41) is presented in Figure 11.1. The impedance of both the Nernst stagnant diffusion layer and the rotating disk electrode approach the Warburg impedance at high frequency, i.e.,

$$\lim_{\omega\to\infty}\frac{-1}{\theta_{\mathrm{i}}'(0)} = \frac{1}{\sqrt{\mathrm{j}K_{\mathrm{i}}}} \tag{11.43}$$

and both impedances approach unity at low frequency. The two models differ for dimensionless frequencies $1 < K_{\mathrm{i}} < 10$.

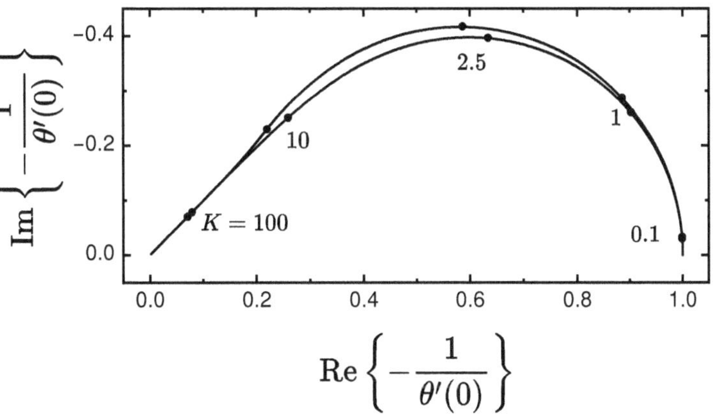

Figure 11.1: Impedance obtained for a Nernst stagnant diffusion layer (equation (11.1)) and the impedance obtained for a rotating disk electrode under assumption of an infinite Schmidt number (obtained from solution of equation (11.41)). Taken from Orazem and Tribollet[1] with permission of John Wiley & Sons.

11.5 It can be useful to have a general analytic expression for velocity that applies for regions both close to and far from the disk electrode. Velocity expansions appropriate for positions close to the disk are given as

$$H = -\mathrm{a}\zeta^2 + \frac{\zeta^3}{3} + \frac{\mathrm{b}}{6}\zeta^4 + \ldots \tag{11.44}$$

and

$$F = \mathrm{a}\zeta - \frac{\zeta^2}{2} - \frac{\mathrm{b}}{3}\zeta^3 + \ldots \tag{11.45}$$

for axial and radial velocities, respectively. For positions far from the disk electrode, the respective velocity expansions can be expressed as[39]

$$\begin{aligned} H &= -\alpha + \frac{2\mathrm{A}}{\alpha}\exp(-\alpha\zeta) - \frac{\mathrm{A}^2+\mathrm{B}^2}{2\alpha^3}\exp(-2\alpha\zeta) \\ &\quad + \frac{\mathrm{A}(\mathrm{A}^2+\mathrm{B}^2)}{6\alpha^5}\exp(-3\alpha\zeta) + \ldots \end{aligned} \tag{11.46}$$

and

$$\begin{aligned} F &= \mathrm{A}\exp(-\alpha\zeta) - \frac{\mathrm{A}^2+\mathrm{B}^2}{2\alpha^2}\exp(-2\alpha\zeta) \\ &\quad + \frac{\mathrm{A}(\mathrm{A}^2+\mathrm{B}^2)}{4\alpha^4}\exp(-3\alpha\zeta) + \ldots \end{aligned} \tag{11.47}$$

where $\alpha = 0.88447411$, $\mathrm{A} = 0.92486353$, and $\mathrm{B} = 1.20221175$.[40] Other constants and definitions are presented in the List of Symbols. Find an interpolation formula that applies for all values of ζ.

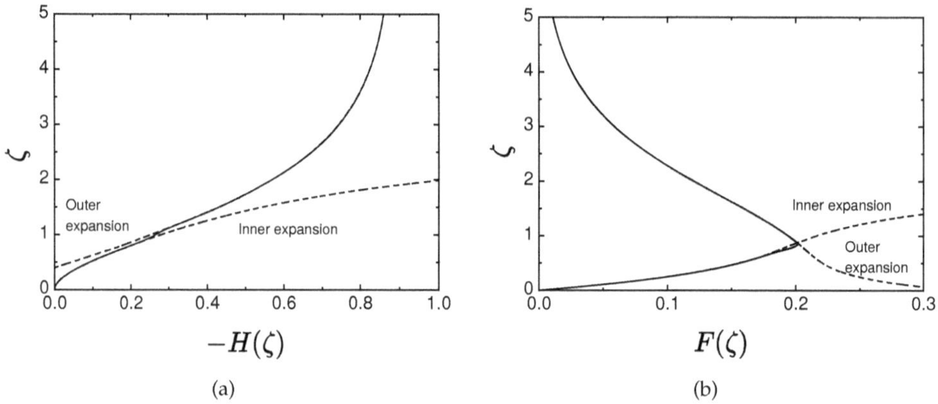

Figure 11.2: Interpolation formulas for rotating disk velocity following equations (11.48) and (11.49): (a) expansion for axial velocity $-H$ with $\zeta_0 = 1$ and $b = 25$; and (b) expansion for radial velocity F with $\zeta_0 = 0.8$ and $b = 25$. The interpolation formulas are presented as a solid line, and asymptotic expansions are presented as dashed lines.

Solution 11.5. The appropriate interpolation formula will not distort the velocity profile near the electrode, which is critical to accurate solutions for mass-transfer problems, and will approach the velocity far from the disk without discontinuities. One such formula is based on the Fermi–Dirac distribution function, which has a value of unity at low values of ζ and zero at large values of ζ. The expression is given as

$$f = \frac{1}{1 + \exp\left(b(\zeta - \zeta_0)\right)} \tag{11.48}$$

and the velocity profile can be expressed as

$$v = f v_{\text{near}} + (1 - f) v_{\text{far}} \tag{11.49}$$

The results are presented in Figure 11.2. As shown in Figure 11.3, the interpolated velocity presented in Figure 11.2 agrees well with the associated series expansion appropriate for the region close to the disk surface given as equations (11.44) and (11.45). In addition, as shown in Figure 11.2, the interpolated velocity is well behaved in the region far from the disk electrode $\zeta > 1$.

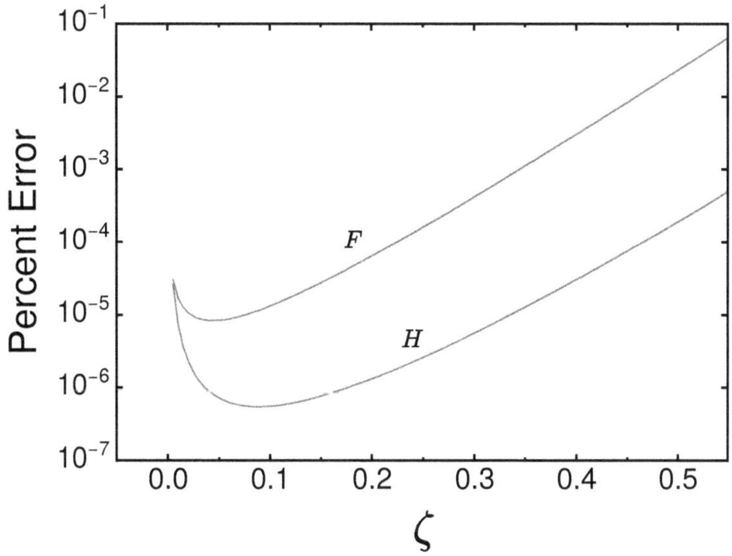

Figure 11.3: Deviation of interpolated velocity presented in Figure 11.2 from the respective series expansion appropriate for the region close to the disk surface, i.e., equations (11.44) and (11.45).

11.6 Starting with

$$-\frac{1}{\theta_i'(0)} = \frac{-1}{\theta_{i,0}'(0) + \theta_{i,1}'(0)\mathrm{Sc_i}^{-1/3} + \theta_{i,2}'(0)\mathrm{Sc_i}^{-2/3}} \tag{11.50}$$

derive

$$Z_{(0)} = -\frac{1}{\theta_{i,0}'(0)} \tag{11.51}$$

$$Z_{(1)} = \frac{\theta_{i,1}'(0)}{\left(\theta_{i,0}'(0)\right)^2} \tag{11.52}$$

and

$$Z_{(2)} = -\frac{1}{\theta_{i,0}'(0)}\left[\left(\frac{\theta_{i,1}'(0)}{\theta_{i,0}'(0)}\right)^2 - \frac{\theta_{i,2}'(0)}{\theta_{i,0}'(0)}\right] \tag{11.53}$$

as used in

$$\frac{-1}{\theta_i'(0)} = Z_{(0)} + \frac{Z_{(1)}}{\mathrm{Sc_i}^{1/3}} + \frac{Z_{(2)}}{\mathrm{Sc_i}^{2/3}} + \dots \tag{11.54}$$

for the rotating disk. Do these relationships apply as well for

$$-\frac{1}{\theta_i'(0)} = Z_{(0)} + \frac{Z_{(1)}}{\mathrm{Sc_i}^{1/3}} + \frac{Z_{(2)}}{\mathrm{Sc_i}^{2/3}} + \dots \tag{11.55}$$

as was developed for the impinging jet?

Solution 11.6. The diffusion impedance for a finite Schmidt number is obtained from

$$-\frac{1}{\theta_i'(0)} = \frac{-1}{\theta_{i,0}'(0) + \theta_{i,1}'(0)\mathrm{Sc_i}^{-1/3} + \theta_{i,2}'(0)\mathrm{Sc_i}^{-2/3}} \tag{11.56}$$

which can be expressed in compact form as

$$-\frac{1}{\theta_i'(0)} = \frac{-1}{a + bx^{-1/3} + cx^{-2/3}} \tag{11.57}$$

Multiply numerator and denominator by $a - bx^{-1/3} - cx^{-2/3}$ to obtain

$$-\frac{1}{\theta_i'(0)} = -\frac{a - bx^{-1/3} - cx^{-2/3}}{a^2 - b^2x^{-2/3}} \tag{11.58}$$

in which all terms of order x^{-1} or smaller were considered to be negligible. Multiply numerator and denominator by $a^2 + b^2x^{-2/3}$ to obtain

$$-\frac{1}{\theta_i'(0)} = -\frac{a^3 - a^2bx^{-1/3} + \left(ab^2 - a^2c\right)x^{-2/3}}{a^4} \tag{11.59}$$

where, again, all terms of order x^{-1} or smaller were considered to be negligible. Upon division by a^4,

$$-\frac{1}{\theta_i'(0)} = -\frac{1}{a} + \frac{b}{a^2}x^{-1/3} - \left(\frac{b^2}{a^3} - \frac{c}{a^2}\right)x^{-2/3} \tag{11.60}$$

Equation (11.60) is the same as equation (11.54). The relationship expressed as equation (11.56) applies for any system for which the expansion

$$\theta_i\left(\xi, \mathrm{Sc_i}, K\right) = \theta_{i,0}\left(\xi, K\right) + \frac{\theta_{i,1}\left(\xi, K\right)}{\mathrm{Sc_i}^{1/3}} + \frac{\theta_{i,2}\left(\xi, K\right)}{\mathrm{Sc_i}^{2/3}} + \ldots \tag{11.61}$$

applies. Therefore, the relationships expressed as equations (11.51), (11.52), and (11.53) apply for the submerged impinging jet as well as for the rotating disk.

11.7 Show that the line represented in Figure 11.4 is parallel to the line corresponding to the Levich equation for the mass-transfer-limited current density to a rotating disk.

Solution 11.7. For a rotating disk electrode, the generalized Levich equation yields a flux given by

$$J_{i,\lim} = 0.62045 D_i^{2/3} \Omega^{1/2} \nu^{-1/6} c_i(\infty) \tag{11.62}$$

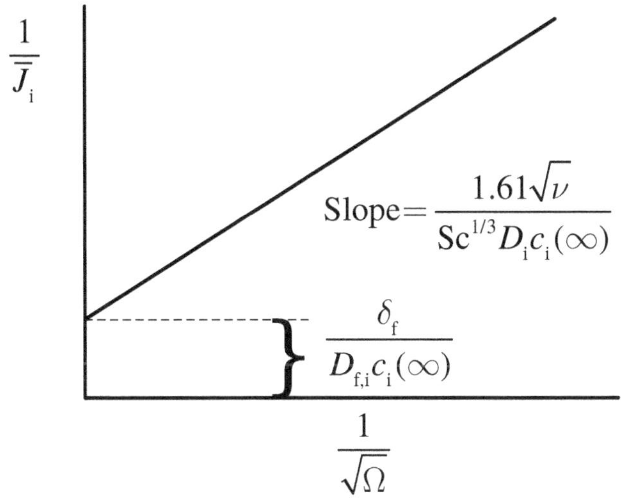

Figure 11.4: Schematic representation showing that a plot of $1/\overline{J}_i$ as a function of $\Omega^{-1/2}$ yields a straight line with an intercept of $1/J_{\Omega\to\infty,i} = \delta_f / D_{f,i} c_i(\infty)$. Taken from Orazem and Tribollet[1] with permission of John Wiley & Sons.

where Ω is the disk rotation speed in radian/s. The inverse of the flux is given by

$$\frac{1}{J_{i,\lim}} = \frac{1}{0.62045} D_i^{-2/3} \Omega^{-1/2} \nu^{1/6} \frac{1}{c_i(\infty)} \tag{11.63}$$

and the derivative with respect to the inverse of the square root of rotation speed is expressed as

$$\frac{d(1/J_{i,\lim})}{d\Omega^{-1/2}} = \frac{1}{0.62045} D_i^{-2/3} \nu^{1/6} \frac{1}{c_i(\infty)} \tag{11.64}$$

To place this in the form expressed in Figure 11.4,

$$\frac{d(1/J_{i,\lim})}{d\Omega^{-1/2}} = \frac{1.611 D_i^{1/3} \sqrt{\nu}}{\nu^{1/3} D_i c_i(\infty)} \tag{11.65}$$

Given that $Sc = \nu / D_i$,

$$\frac{d(1/J_{i,\lim})}{d\Omega^{-1/2}} = \frac{1.611 \sqrt{\nu}}{Sc^{1/3} D_i c_i(\infty)} \tag{11.66}$$

which is the slope given in Figure 11.4.

11.8 The equation governing mass transfer for an impinging jet can be expressed as

$$\frac{\partial c_i}{\partial t}+v_y\frac{\partial c_i}{\partial y}-D_i\frac{\partial^2 c_i}{\partial y^2}=0 \tag{11.67}$$

(a) Starting with $c_i = \bar{c}_i + \mathrm{Re}\left\{\tilde{c}_i e^{j\omega t}\right\}$, develop the equation we need to solve for $\tilde{c}_i$. You need not solve the equation.

(b) Identify the boundary conditions needed to solve the equation you develop.

(c) Explain the relationship between $\tilde{c}_i$ and the diffusion impedance. In other words, how would you use the solution to the equation you develop to obtain the diffusion impedance?

Solution 11.8. The submerged impinging jet geometry can be made to give uniform mass-transfer rates across a disk electrode within the stagnation region. The stagnation region is defined to be the region surrounding the stagnation point in which the axial velocity, given by

$$v_y=-\sqrt{a_{IJ}\nu}\,\phi\left(\eta\right) \tag{11.68}$$

is independent of radial position, and the radial velocity is given by

$$v_r=\frac{a_{IJ}r}{2}\frac{d\phi\left(\eta\right)}{d\eta} \tag{11.69}$$

where a_{IJ} is the hydrodynamic constant that is a function only of geometry and fluid velocity, r and y are the radial and axial positions, respectively, ν is the kinematic viscosity, and ϕ is the stream function that is given in terms of dimensionless axial position $\eta = y\sqrt{a_{IJ}/\nu}$ as[41]

$$\phi\left(\eta\right)=1.352\eta^2-\frac{1}{3}\eta^3+7.2888\times 10^{-3}\eta^6+\ldots \tag{11.70}$$

Esteban et al.[42] used ring electrodes to find that the stagnation region extends to a radial distance roughly equal to the inside radius of the nozzle.

(a) Under the assumption that the electrode is uniformly accessible, the equation governing mass transfer in the frequency domain is given by

$$\frac{d^2\theta_i}{d\xi^2}+\left(3\xi^2-\left(\frac{3}{1.352^4}\right)^{1/3}\frac{\xi^3}{\mathrm{Sc_i}^{1/3}}+\ldots\right)\frac{d\theta_i}{d\xi}-jK_i\theta_i=0 \tag{11.71}$$

where

$$K_i=\frac{\omega}{a_{IJ}}\left(\frac{9}{(1.352)^2}\right)^{1/3}\mathrm{Sc_i}^{1/3}=1.70123\frac{\omega\mathrm{Sc_i}^{1/3}}{a_{IJ}} \tag{11.72}$$

represents a dimensionless frequency, and $\xi = y/\delta_{\mathrm{i}}$ is a dimensionless position where

$$\delta_{\mathrm{i}} = \left(\frac{3}{1.352}\right)^{1/3} \frac{1}{\mathrm{Sc_i}^{1/3}} \sqrt{\frac{\nu}{\mathrm{a_{IJ}}}} = 1.180 \frac{1}{\mathrm{Sc_i}^{1/3}} \sqrt{\frac{\nu}{\mathrm{a_{IJ}}}} \tag{11.73}$$

is a characteristic length for transport of species i. Terms in ξ of order greater or equal to 6 were neglected in equation (11.71).

11.9 Consider a homogeneous reaction

$$\mathrm{AB} \underset{k_{\mathrm{b}}}{\overset{k_{\mathrm{f}}}{\rightleftarrows}} \mathrm{A^-} + \mathrm{B^+} \tag{11.74}$$

for which $\mathrm{B^+}$ is electro-active and is consumed at the electrode according to

$$\mathrm{B^+} + \mathrm{e^-} \rightarrow \mathrm{B} \tag{11.75}$$

In the presence of supporting electrolyte, the concentration of the anion $\mathrm{A^-}$ may be sufficiently large to be considered constant, making reaction (11.74) first order with respect to species AB and $\mathrm{B^+}$. Under the assumptions that the diffusion coefficients for AB and $\mathrm{B^+}$ are equal and that convection may be ignored, develop a solution for the steady-state distribution of concentration and reaction rate.

Solution 11.9. When the concentration of A^- may be considered invariant, the rate of production of species AB and $\mathrm{B^+}$ by reaction (11.74) may be expressed as

$$R_{\mathrm{B^+}} = -R_{\mathrm{AB}} = k_{\mathrm{f}} c_{\mathrm{AB}}(y) - k_{\mathrm{b}} c_{\mathrm{B^+}}(y) \tag{11.76}$$

Under the assumption of a steady state, the conservation equations for species AB and $\mathrm{B^+}$ may be expressed as

$$D \frac{\partial^2 c_{\mathrm{AB}}}{\partial y^2} - k_{\mathrm{f}} c_{\mathrm{AB}} + k_{\mathrm{b}} c_{\mathrm{B^+}} = 0 \tag{11.77}$$

and

$$D \frac{\partial^2 c_{\mathrm{B^+}}}{\partial y^2} + k_{\mathrm{f}} c_{\mathrm{AB}} - k_{\mathrm{b}} c_{\mathrm{B^+}} = 0 \tag{11.78}$$

respectively, where

$$D = D_{\mathrm{AB}} = D_{\mathrm{B^+}} \tag{11.79}$$

The sum of equations (11.104) and (11.105) yields

$$D \frac{\partial^2 \psi_1}{\partial y^2} = 0 \tag{11.80}$$

where

$$\psi_1 = c_{\mathrm{AB}} + c_{\mathrm{B^+}} \tag{11.81}$$

After algebraic manipulation, the difference between equations (11.77) and (11.78) may be expressed as

$$D\frac{\partial^2 \psi_2}{\partial y^2} - k\psi_2 = 0 \tag{11.82}$$

where

$$\psi_2 = c_{\mathrm{AB}} - \frac{c_{\mathrm{B^+}}}{K_{\mathrm{eq}}} \tag{11.83}$$

and where $k = k_{\mathrm{f}} + k_{\mathrm{b}}$ and $K_{\mathrm{eq}} = k_{\mathrm{f}}/k_{\mathrm{b}}$. The solution to equation (11.82) is

$$\psi_2 = C_1 \exp\left(\sqrt{k/D}y\right) + C_2 \exp\left(-\sqrt{k/D}y\right) \tag{11.84}$$

and the solution to equation (11.80) is

$$\psi_1 = C_3 y + C_4 \tag{11.85}$$

where C_1, C_2, C_3, and C_4 are constants obtained by the application of the boundary conditions.

Under the assumption that the system is operating at the mass-transfer-limited current, the boundary condition at $y = 0$ for the reacting species $\mathrm{B^+}$ may be expressed as

$$c_{\mathrm{B^+}}(0) = 0 \tag{11.86}$$

the flux of the non-reacting species is given by

$$\left.\frac{\mathrm{d}c_{\mathrm{AB}}}{\mathrm{d}y}\right|_{y=0} \tag{11.87}$$

Reaction (11.74) is assumed to be equilibrated at $y = \delta$, thus

$$c_{\mathrm{B^+}} = K_{\mathrm{eq}} c_{\mathrm{AB}} \tag{11.88}$$

and a value may be set for concentration $c_{\mathrm{AB}} + c_{\mathrm{B^+}}$

$$c_{\mathrm{AB}} + c_{\mathrm{B^+}} = c_{\mathrm{AB}}(\infty)(1 + K_{\mathrm{eq}}) \tag{11.89}$$

Boundary condition (11.86) at $y = 0$ may be expressed in terms of transformed variables as

$$\psi_1(0) = \psi_2(0) \tag{11.90}$$

and condition (11.87) yields

$$\left.\frac{d\psi_1}{dy}\right|_{y=0} = -K_{\mathrm{eq}} \left.\frac{d\psi_2}{dy}\right|_{y=0} \tag{11.91}$$

Equation (11.88) yields

$$\psi_2 = 0 \tag{11.92}$$

and equation (11.89) yields

$$\psi_1 = c_{\mathrm{AB}}(\infty)(1 + K_{\mathrm{eq}}) = c_0 \tag{11.93}$$

Application of equations (11.90)–(11.93) to equations (11.84) and (11.85) yields a set of algebraic equations

$$c_4 - C_1 - C_2 = 0 \tag{11.94}$$

$$C_3 + K_{eq}\sqrt{\frac{k}{D}}(C_1 - C_2) = 0 \tag{11.95}$$

$$C_1 \exp\left(\sqrt{k\delta^2/D}\right) + C_2 \exp\left(\sqrt{-k\delta^2/D}\right) \tag{11.96}$$

and

$$C_4 + C_3\delta = C_0 \tag{11.97}$$

Equations (11.94)–(11.97) are solved simultaneously.

The resulting values for C_1, C_2, C_3, and C_4 are given as

$$C_1 = \frac{c_0 \exp\left(\sqrt{k\delta^2/D}\right)}{\exp\left(2\sqrt{k\delta^2/D}\right)\left(K_{\mathrm{eq}}\sqrt{k\delta^2/D} + 1\right) + \left(K_{\mathrm{eq}}\sqrt{k\delta^2/D} - 1\right)} \tag{11.98}$$

$$C_2 = \frac{c_0\left(\cosh\sqrt{k\delta^2/D} + \sinh\sqrt{k\delta^2/D}\right)}{2\left(K_{\mathrm{eq}}\sqrt{k\delta^2/D}\cosh\sqrt{k\delta^2/D} + \sinh\sqrt{k\delta^2/D}\right)} \tag{11.99}$$

$$C_3 = \frac{c_0 K_{\mathrm{eq}}\sqrt{k/D}\cosh\sqrt{k\delta^2/D}}{\left(K_{\mathrm{eq}}\sqrt{k\delta^2/D}\cosh\sqrt{k\delta^2/D} + \sinh\sqrt{k\delta^2/D}\right)} \tag{11.100}$$

and

$$C_4 = \frac{c_0 \sinh\sqrt{k\delta^2/D}}{\left(K_{\mathrm{eq}}\sqrt{k\delta^2/D}\cosh\sqrt{k\delta^2/D} + \sinh\sqrt{k\delta^2/D}\right)} \tag{11.101}$$

respectively. The original variables may be expressed as

$$c_{\mathrm{B}^+} = \frac{K_{\mathrm{eq}}(\psi_1 - \psi_2)}{1 + K_{\mathrm{eq}}} \tag{11.102}$$

and

$$c_{\mathrm{AB}} = \frac{(\psi_1 + K_{\mathrm{eq}}\psi_2)}{1 + K_{\mathrm{eq}}} \tag{11.103}$$

The reaction rate may be obtained from equation (11.76).

11.10 Derive an expression for the dimensionless diffusion impedance for an electrochemical reaction influenced by a chemical reaction in an infinite domain as shown in Problem 11.9.

Solution 11.10. When the concentration of A^- may be considered invariant, the rate of production of species AB and B^+ by reaction (11.74) may be expressed as equation (11.76). The conservation equations for species AB and B^+ may be expressed as

$$\frac{\partial c_{AB}}{\partial t} = D\frac{\partial^2 c_{AB}}{\partial y^2} - k_f c_{AB} + k_b c_{B^+} \tag{11.104}$$

and

$$\frac{\partial c_{B^+}}{\partial t} = D\frac{\partial^2 c_{B^+}}{\partial y^2} + k_f c_{AB} - k_b c_{B^+} \tag{11.105}$$

respectively, where

$$D = D_{AB} = D_{B^+} \tag{11.106}$$

The sum of equations (11.104) and (11.105) yields

$$\frac{\partial}{\partial t}\left(c_{AB} + c_{B^+}\right) = D\frac{\partial^2}{\partial y^2}\left(c_{AB} + c_{B^+}\right) \tag{11.107}$$

After algebraic manipulation, the difference between equations (11.104) and (11.105) may be expressed as

$$\frac{\partial}{\partial t}\left(c_{AB} - \frac{c_{B^+}}{K_{eq}}\right) = D\frac{\partial^2}{\partial y^2}\left(c_{AB} - \frac{c_{B^+}}{K_{eq}}\right) - k\left(c_{AB} - \frac{c_{B^+}}{K_{eq}}\right) \tag{11.108}$$

where $k = k_f + k_b$ and $K_{eq} = k_f/k_b$. As equations (11.107) and (11.108) are linear, the solution for the corresponding diffusion impedance does not require a solution for the steady state.

Equations (11.107) and (11.108) may be expressed in frequency domain as

$$\frac{j\omega}{D}\left(\widetilde{c}_{AB} + \widetilde{c}_{B^+}\right) = \frac{\partial^2}{\partial y^2}\left(\widetilde{c}_{AB} + \widetilde{c}_{B^+}\right) \tag{11.109}$$

and

$$\frac{j\omega}{D}\left(\widetilde{c}_{AB} - \frac{\widetilde{c}_{B^+}}{K_{eq}}\right) = \frac{\partial^2}{\partial y^2}\left(\widetilde{c}_{AB} - \frac{\widetilde{c}_{B^+}}{K_{eq}}\right) - \frac{k}{D}\left(\widetilde{c}_{AB} - \frac{\widetilde{c}_{B^+}}{K_{eq}}\right) \tag{11.110}$$

respectively. The general solution to equations (11.109) and (11.110) may be expressed as

$$\left(\widetilde{c}_{AB} + \widetilde{c}_{B^+}\right) = C_1 \exp(-s_1 y) + C_2 \exp(s_1 y) \tag{11.111}$$

and

$$\left(\widetilde{c}_{AB} - \frac{\widetilde{c}_{B^+}}{K_{eq}}\right) = C_3 \exp(-s_2 y) + C_4 \exp(s_2 y) \tag{11.112}$$

respectively, where

$$s_1^2 = \frac{j\omega}{D} \tag{11.113}$$

and

$$s_2^2 = \frac{j\omega + k}{D} \tag{11.114}$$

The boundary conditions for diffusion through a film of finite thickness δ are:

$$\widetilde{c}_{AB}(\delta) = 0 \tag{11.115}$$

$$\widetilde{c}_{B^+}(\delta) = 0 \tag{11.116}$$

$$\widetilde{c}_{B^+}(0) = 1 \tag{11.117}$$

and

$$\left.\frac{d\widetilde{c}_{AB}}{dy}\right|_{y=0} = 0 \tag{11.118}$$

The corresponding expressions are

$$0 = C_1 \exp(-s_1 y) + C_2 \exp(s_1 y) + K_{eq} C_3 \exp(-s_2 y) + K_{eq} C_4 \exp(s_2 y) \tag{11.119}$$

$$0 = C_1 \exp(-s_1 y) + C_2 \exp(s_1 y) - C_3 \exp(-s_2 y) - C_4 \exp(s_2 y) \tag{11.120}$$

$$1 + \frac{1}{K_{eq}} = C_1 + C_2 - C_3 - C_4 \tag{11.121}$$

and

$$0 = -C_1 s_1 + C_2 s_1 - K_{eq} C_3 s_2 + K_{eq} C_4 s_2 \tag{11.122}$$

respectively.

The dimensionless diffusion impedance may be expressed as

$$-\frac{1}{\theta'_{B^+}} = -\frac{1}{\delta} \frac{\widetilde{c}_{B^+}(0)}{\left.\frac{d\widetilde{c}_{B^+}}{dy}\right|_{y=0}} \tag{11.123}$$

$$= \frac{1}{K_{eq}+1} \frac{\tanh\sqrt{(j\omega + k)\frac{\delta^2}{D}}}{\sqrt{(j\omega + k)\frac{\delta^2}{D}}} + \frac{K_{eq}}{K_{eq}+1} \frac{\tanh\sqrt{j\omega\frac{\delta^2}{D}}}{\sqrt{j\omega\frac{\delta^2}{D}}} \tag{11.124}$$

or

$$-\frac{1}{\theta'_{B^+}} = \frac{1}{K_{eq}+1} \frac{\tanh\sqrt{jK + k_{dim}}}{\sqrt{jK + k_{dim}}} + \frac{K_{eq}}{K_{eq}+1} \frac{\tanh\sqrt{jK}}{\sqrt{jK}} \tag{11.125}$$

where $K = \omega\delta^2/D$ and $k_{dim} = k\delta^2/D$. The first term on the right-hand side of equation (11.125) corresponds to a modified Gerischer impedance (see, for example, Boukamp and Bouwmeester[43]). The second term on the right-hand side of equation (11.125) corresponds to the diffusion impedance.

11.11 Use

$$\omega_c = 1.7783k \tag{11.126}$$

to estimate the thickness of the reaction zone for an electrochemical reaction influenced by a chemical reaction in an infinite domain.

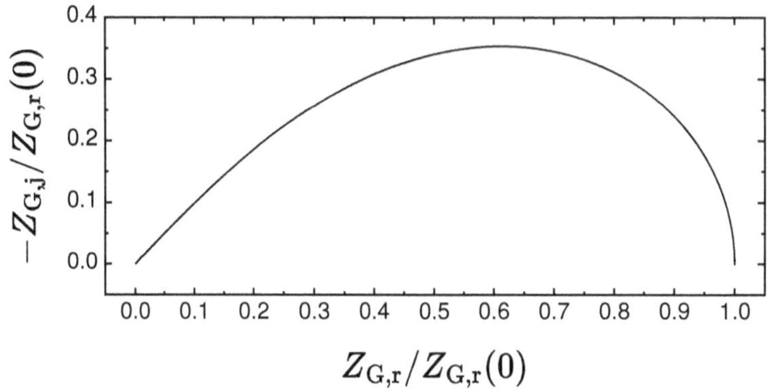

Figure 11.5: The normalized Gerischer impedance in Nyquist format.

Solution 11.11. At high frequencies,

$$-\frac{1}{\theta'_{B+}} = \frac{1}{K_{eq}+1}\frac{1}{\sqrt{jK+k_{dim}}} + \frac{K_{eq}}{K_{eq}+1}\frac{1}{\sqrt{jK}} \tag{11.127}$$

$$= \frac{1}{K_{eq}+1}Z_G + \frac{K_{eq}}{K_{eq}+1}Z_W \tag{11.128}$$

where Z_G is the Gerischer impedance[44] and Z_W is the Warburg impedance. The Gerischer impedance shows a finite low-frequency asymptote

$$\lim_{K\to 0} Z_G = \frac{1}{\sqrt{k_{dim}}} \tag{11.129}$$

due to the contribution of local homogeneous generation of reacting species. The shape of the normalized Gerischer impedance, given in Figure 11.5, is independent of the value of k_{dim}, but the characteristic frequency depends on k_{dim} following

$$K_c = 1.7783k_{dim} \tag{11.130}$$

or, as frequency and $k = k_f + k_b$ are made dimensionless using the same terms,

$$\omega_c = 1.7783k \tag{11.131}$$

The influence of k_{dim} is visible in Figure 11.6.

Thus, for a given value of ω_c or f_c taken from the high-frequency Gerischer loop, the thickness of the reaction layer may be obtained as

$$\delta_r = \sqrt{D_i/k} \tag{11.132}$$

or

$$\delta_r = \sqrt{1.778D_i/\omega_c} \tag{11.133}$$

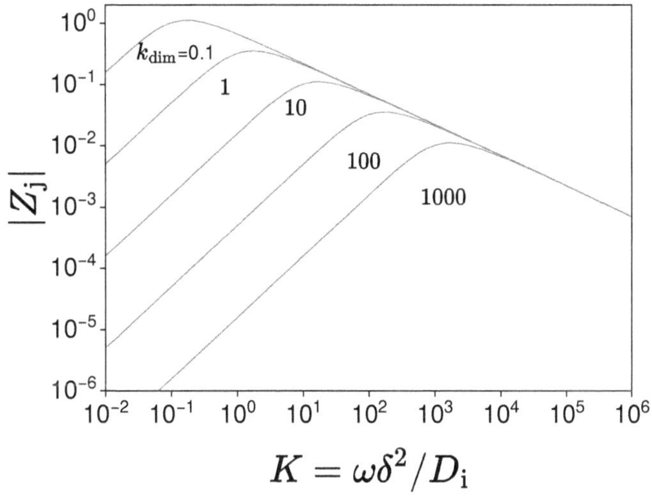

Figure 11.6: Imaginary part of the Gerischer impedance as a function of dimensionless frequency with k_{dim} as a parameter.

This simple approach is only valid for conditions under which the Gerischer impedance applies.

11.12 For an electrode covered by a porous film, the conservation equation

$$\frac{\partial^2 \widetilde{c}_i^{(1)}}{\partial y^2} - \frac{j\omega}{D_{t,i}}\widetilde{c}_i^{(1)} = 0 \tag{11.134}$$

applies for the concentration phasor inside the film. At the fluid–film interface ($y = \delta_f$ or $y_e = 0$), the concentration gradient and concentration phasor of the fluid $\widetilde{c}_i^{(2)}$ are linked by

$$\left.\frac{\partial \widetilde{c}_i^{(2)}}{\partial y_e}\right|_{y_e=0} = \frac{\widetilde{c}_i^{(2)}(0)}{\delta_{N,i}}\theta'(0) \tag{11.135}$$

Boundary conditions at $y = \delta_f$ are invoked to find the solution to equation (11.134). What are these boundary conditions?

Solution 11.12. The boundary conditions coupling the inner film diffusion and outer convective diffusion are evaluated at $y = \delta_f$ or $y_e = 0$. The conditions are

$$c_i^{(1)}(\delta_f) = c_i^{(2)}(0) \tag{11.136}$$

and

$$D_{\mathrm{f,i}}\frac{\partial c_{\mathrm{i}}^{(1)}}{\partial y} = D_{\mathrm{i}}\frac{\partial c_{\mathrm{i}}^{(2)}}{\partial y_{\mathrm{e}}} \tag{11.137}$$

The boundary condition for the solution of the impedance equations can be found by application of

$$c_{\mathrm{i}} = \overline{c}_{\mathrm{i}} + \mathrm{Re}\{\widetilde{c}_{\mathrm{i}}\exp(\mathrm{j}\omega t)\} \tag{11.138}$$

to be

$$\widetilde{c}_{\mathrm{i}}^{(1)}(\delta_{\mathrm{f}}) = \widetilde{c}_{\mathrm{i}}^{(2)}(0) \tag{11.139}$$

and

$$D_{\mathrm{f,i}}\left.\frac{\mathrm{d}\widetilde{c}_{\mathrm{i}}^{(1)}}{\mathrm{d}y}\right|_{y=\delta_{\mathrm{f}}} = D_{\mathrm{i}}\left.\frac{\mathrm{d}\widetilde{c}_{\mathrm{i}}^{(2)}}{\mathrm{d}y_{\mathrm{e}}}\right|_{y_{\mathrm{e}}=0} \tag{11.140}$$

where

$$\left.\frac{\mathrm{d}\widetilde{c}_{\mathrm{i}}^{(2)}}{\mathrm{d}y_{\mathrm{e}}}\right|_{y_{\mathrm{e}}=0} = \frac{\widetilde{c}_{\mathrm{i}}^{(2)}(0)}{\delta_{\mathrm{N,i}}}\theta'(0) \tag{11.141}$$

The complete solution for the impedance and electrohydrodynamic impedance (see Chapter 16) response of a coated rotating disk electrode was presented by Deslouis et al.[45]

11.13 Consider a system for which the faradaic impedance is given by

$$Z_{\mathrm{F}} = R_{\mathrm{t}} + R_{\mathrm{d}}\left(\frac{-1}{\theta'(0)}\right) \tag{11.142}$$

The dimensionless diffusion impedance corresponds to a film of thickness δ, i.e.,

$$\frac{-1}{\theta'(0)} = \frac{\tanh\sqrt{\mathrm{j}K}}{\sqrt{\mathrm{j}K}} \tag{11.143}$$

and the dimensionless frequency is expressed as

$$K = \frac{\omega\delta^2}{D_{\mathrm{i}}} \tag{11.144}$$

(a) Write an expression for the overall impedance associated with this system, including a parallel double-layer capacitance C_{dl} and an ohmic resistance R_e The frequency should be written in terms of the dimensionless frequency given in equation (11.143).

(b) Derive an expression for a normalized impedance $(Z - R_e)/R_d$. In doing so, you should identify a dimensionless parameter that represents a ratio of diffusion and $R_d C_{dl}$ time constants.

(c) Show that the equation you obtained in part 11.13.b represents a single loop in a Nyquist plot under the assumption that the kinetic rate constant is sufficiently large. You need not plot the results to show this.

(d) Use the parameter you derived in part 11.13.b to identify the conditions under which the loop has the appearance of a diffusion impedance. What is the value of the characteristic dimensionless frequency (where $-Z_j$ has a maximum value) of this loop?

(e) Use the parameter you derived in part 11.13.b to identify the conditions under which the loop has the appearance of a semicircle. Is the characteristic dimensionless frequency of this loop the same as you found for part 11.13.d?

Solution 11.13. The solution to this problem was developed by Gao et al.[46]

(a) The impedance may be expressed as

$$Z = R_e + \frac{R_t + R_d\left(\dfrac{-1}{\theta_i'(0)}\right)}{1 + j\omega C_{dl}\left(R_t + R_d\left(\dfrac{-1}{\theta_i'(0)}\right)\right)} \tag{11.145}$$

In terms of the dimensionless frequency given in equation (11.143),

$$Z = R_e + \frac{R_t + R_d\left(\dfrac{-1}{\theta_i'(0)}\right)}{1 + jK\dfrac{D_i C_{dl}}{\delta^2}\left(R_t + R_d\left(\dfrac{-1}{\theta_i'(0)}\right)\right)} \tag{11.146}$$

(b) The corresponding impedance may be expressed as

$$\frac{Z - R_e}{R_d} = \frac{\dfrac{R_t}{R_d} + \left(\dfrac{-1}{\theta_i'(0)}\right)}{1 + jK\xi\left(\dfrac{R_t}{R_d} + \left(\dfrac{-1}{\theta_i'(0)}\right)\right)} \tag{11.147}$$

where the parameter

$$\xi = \frac{D_i R_d C_{dl}}{\delta^2} \tag{11.148}$$

represents the ratio of diffusion and $R_d C_{dl}$ time constants.

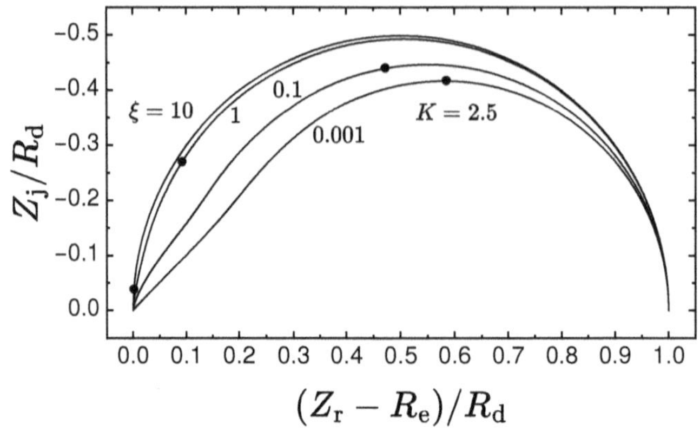

Figure 11.7: Dimensionless impedance response calculated from equation (11.147) with $R_t/R_d \ll 1$ and with ξ as a parameter (see equation (11.148)). The filled circles represent the impedance at a dimensionless frequency of $K = 2.5$, corresponding to the characteristic frequency for diffusion through a stagnant film.

(c) A large value for rate constant corresponds to a small value for the charge transfer resistance. Under the imposed conditions, $R_t/R_d \ll 1$, and

$$\frac{Z - R_e}{R_d} = \frac{\left(\dfrac{-1}{\theta_i'(0)}\right)}{1 + jK\xi\left(\dfrac{-1}{\theta_i'(0)}\right)} \tag{11.149}$$

In the absence of the charge-transfer resistance, equation (11.147) has only a single capacitive loop in a Nyquist representation.

(d) The impedance associated with equation (11.147) is presented in Nyquist format in Figure 11.7 under the assumption that $R_t/R_d \ll 1$ and with ξ as a parameter. The filled circles represent the impedance at a dimensionless frequency of $K = 2.5$, corresponding to the characteristic frequency for diffusion through a stagnant film. For small values of ξ, the impedance approaches

$$\frac{Z - R_e}{R_d} = \left(\frac{-1}{\theta_i'(0)}\right) = \frac{\tanh\sqrt{jK}}{\sqrt{jK}} \tag{11.150}$$

The resulting impedance is a diffusion impedance with characteristic frequency $K_c = 2.5$.

(e) As ξ increases, the value of K associated with the characteristic frequency decreases. For small values of dimensionless frequency K, the dimensionless diffusion impedance approaches unity. Thus, the characteristic frequency has a value $K_c = 1/\xi$. The impedance is transformed into a semicircle of width R_d.

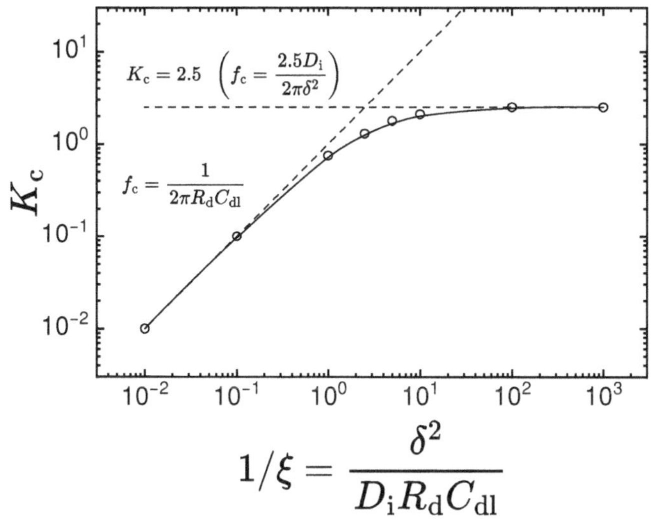

Figure 11.8: Dimensionless characteristic frequency taken from Figure 11.7 as a function of $1/\xi$ (see equation (11.148)). The solid line is given by equation (11.153).

The characteristic frequency is presented in Figure 11.8 as a function of $1/\xi$. For small values of ξ (large values of $1/\xi$), the characteristic dimensionless frequency approaches a value $K_c = 2.5$. Thus, in terms of dimensional frequency,

$$\lim_{\xi \to 0} f_c = \frac{2.5D_i}{2\pi\delta^2} \tag{11.151}$$

For large values of ξ, the characteristic dimensionless frequency approaches a value $K_c = 1/\xi$. Thus, in terms of dimensional frequency,

$$\lim_{\xi \to \infty} f_c = \frac{1}{2\pi R_d C_{dl}} \tag{11.152}$$

An interpolation formula

$$K_c = \frac{1}{1/2.5 + \xi} \tag{11.153}$$

is presented as a solid line in Figure 11.8.

A large value of ξ (or a small value of $1/\xi$) may be achieved for a large value of capacitance, or by a large value of diffusion resistance R_d. A large capacitance may be attributed to an enhanced electrode surface area.[47] A large value of diffusion resistance may be observed on the mass-transfer-limited plateau.

Chapter 12

Impedance of Materials

In many cases, the impedance measurement provides information concerning the properties of materials. At high frequencies, the impedance measurement may be used to explore dielectric relaxation. The problems presented in this chapter provide an overview of the impedance of materials, including dielectric relaxation, geometric capacitance, and Mott–Schottky analysis. Problems 12.1–12.8 are based on problems presented by Orazem and Tribollet.[1]

Problems

12.1 Calculate the impedance response for $L = 2$ cm and for $L = 10$ cm for an electrochemical cell consisting of two electrodes, which fill the cross section of a cylinder, as shown in Figure 12.1, with a cross-sectional area $A = 1\ \text{cm}^2$. The cylindrical design of the electrochemical cell was chosen to avoid the influence of nonuniform current and potential distributions. The electrolyte has a relative permittivity (or dielectric constant) $\varepsilon = 4$ and a resistivity $\rho = 2 \times 10^{10}\ \Omega\text{cm}$. Present the results as a Nyquist plot. What is the corresponding ohmic resistance, geometric capacitance, and characteristic frequency?

Solution 12.1. The ohmic resistance is given by

$$R_e = \frac{\rho L}{A} \tag{12.1}$$

The ohmic resistance for a resistivity $\rho = 2 \times 10^{10}\ \Omega\text{cm}$ and a cross-sectional area $A = 1\ \text{cm}^2$ is $R_e = 4 \times 10^{10}\ \Omega$ for $L = 2$ cm and $R_e = 2 \times 10^{11}\ \Omega$ for $L = 10$ cm. The geometric capacitance, given by

$$C_g = \frac{\varepsilon \varepsilon_0 A}{L} \tag{12.2}$$

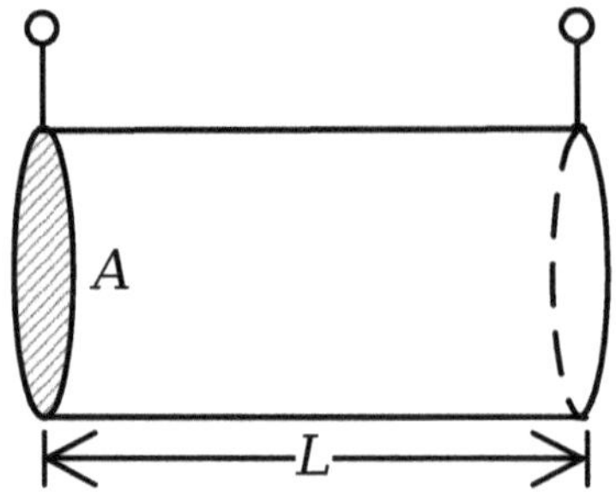

Figure 12.1: Cylindrical cell used to obtain the geometrical impedance as described in Problem 12.1. Taken from Orazem and Tribollet[1] with permission of John Wiley & Sons.

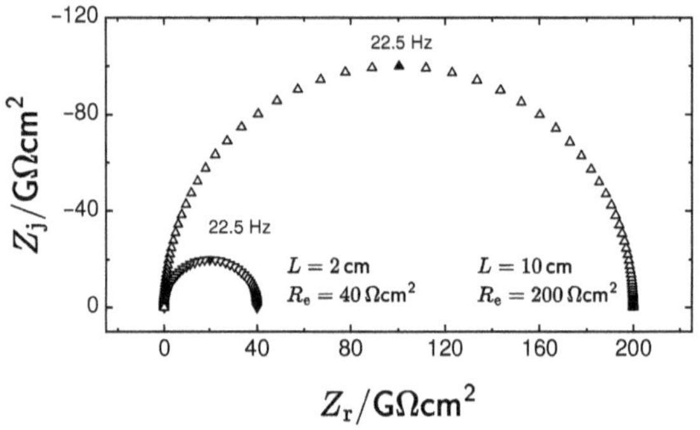

Figure 12.2: The impedance response corresponding to a resistivity $\rho = 2 \times 10^{10}$ Ωcm, a dielectric constant $\varepsilon = 4$, a cross-sectional area $A = 1$ cm^2, and distances between electrodes of $L = 2$ and 10 cm.

yields $C_g = 1.77 \times 10^{-13}$ F for $L = 2$ cm and $C_g = 3.54 \times 10^{-14}$ F for $L = 10$ cm. The characteristic frequency

$$f_c = \frac{1}{2\pi\rho\varepsilon\varepsilon_0} \tag{12.3}$$

is independent of L, yielding $f_c = 2.25$ Hz for both $L = 2$ cm and $L = 10$ cm.

The impedance response is presented in Figure 12.2 for distances between electrodes of $L = 2$ and 10 cm. The high-frequency asymptote is at the origin, which is characteristic of the impedance associated with a geometric capacitance. The low-frequency asymptote is at the ohmic resistance. Characteristic frequencies are marked in Figure 12.2 and, as indicated in equation (12.3), are equal for the two electrode spacings indicated.

12.2 Estimate the frequency at which a geometric capacitance is apparent for:

(a) an electrolyte containing 0.1 M NaCl at a temperature of 25 °C.

(b) an electrolyte containing 1.0 M NaCl at a temperature of 25 °C.

Solution 12.2. The question raised here is whether the geometric capacitance is visible in ordinary electrochemical impedance measurements in aqueous electrolytes.

(a) The electrolyte conductivity is expressed as

$$\kappa = \frac{F^2}{RT}\sum_i z_i^2 D_i c_i \tag{12.4}$$

where values for the diffusivity $D_{)}$ for some species are presented in Table A.13. For a 0.1 M NaCl electrolyte at 25 °C, the diffusion coefficients are $D_{Na^+} = 1.334 \times 10^{-5}$ cm^2/s and $D_{Cl^-} = 2.032 \times 10^{-5}$ cm^2/s. Equation (12.4) yields $\kappa = 0.1265$ S/cm or $\rho = 7.9$ Ωcm. The characteristic frequency is given by

$$f_c = \frac{1}{2\pi\rho\varepsilon\varepsilon_0} \tag{12.5}$$

For a dielectric constant $\varepsilon = 80$, equation (12.5) yields a characteristic frequency $f_c = 2.8$ GHz, a value that is outside the range of usual electrochemical impedance measurements.

(b) For a 1 M NaCl electrolyte at 25 °C, the conductivity is $\kappa = 1.265$ S/cm or $\rho = 0.79$ Ωcm. For a dielectric constant $\varepsilon = 80$, equation (12.5) yields a characteristic frequency $f_c = 28$ GHz, a value that is also outside the range of usual electrochemical impedance measurements.

This problem suggests that the geometric capacitance plays a role only for very resistive electrolytes.

12.3 Calculate the impedance of a film with a resistivity distribution given as

$$\rho(y) = \rho_0 \exp\left(-y/\lambda\right) \tag{12.6}$$

Solution 12.3. Please see the development presented for Problem 13.14.b.

12.4 Develop the relationship needed to convert the mobility given in Table A.16, e.g., μ_n, to diffusivity.

Solution 12.4. The mobility given in Table A.16 has units of cm^2/Vs; whereas the diffusivity has units of cm^2/s. The Nernst–Einstein relationship

$$D_i = RTu_i \tag{12.7}$$

provides a relationship between diffusivity D_i and mobility u_i, but the mobility used in equation (12.7) has units of cm^{2}mol/Js. Given that 1V[=]1J/C,

$$u_i[=]\frac{\text{cm}^2\text{mol}}{\text{Js}} = \frac{1}{\text{F}z_i}\mu_i[=]\frac{\text{equiv}}{\text{C}}\frac{\text{mol}}{\text{equiv}}\frac{\text{cm}^2\text{C}}{\text{Js}} \tag{12.8}$$

Thus, to account for the difference in units, equation (12.7) can be modified as

$$D_i = \frac{RT}{\text{F}z_i}\mu_i \tag{12.9}$$

The mobility of electrons at $T = 300$ K is given in Table A.16 to be $\mu_n = 8,500$ cm^2/Vs. The corresponding diffusivity is calculated to be $D_n = 220$ cm^2/s. The mobility of holes at $T = 300$ K is given in Table A.16 to be $\mu_p = 400$ cm^2/Vs. The corresponding diffusivity is calculated to be $D_p = 10$ cm^2/s.

12.5 Calculate the Debye length in units of μm expected for an n-type GaAs semiconductor with a dopant concentration of 10^{16}. Compare the value you obtain to the Debye length obtained for an electrolytic system with a NaCl concentration of 0.1 M.

Solution 12.5. The Debye length is given by

$$\lambda = \sqrt{\frac{\varepsilon\varepsilon_0 RT}{F^2\sum_i z_i^2 c_i(\infty)}} \tag{12.10}$$

The dominant species in the semiconductor should be the dopant and the oppositely charged species, i.e., electrons or holes. The concentration of dopant is obtained in units of mol/cm^2 by dividing by Avogadro's number $N_A = 6.0221415 \times 10^{23}$ mol^{-1} to be $c_{dopant} = 1.66 \times 10^{-8}$ mol/cm^3. The dielectric constant for GaAs is obtained from Table A.16 to be $\epsilon = 12.9$. Equation (12.10) yields a Debye length of $\lambda = 30.4$ nm (3.04×10^{-2} μm). A concentration of NaCl of 0.1 M corresponds to $c_{NaCl} = 10^{-4}$ mol/cm^2. For $\varepsilon = 80$, equation (12.10) yields a Debye length of $\lambda = 0.97$ nm (9.7×10^{-4} μm).

12.6 The capacity of the space-charge region can be related to the dopant concentration (or fixed charge) in a semiconductor. The space-charge region is essentially equivalent to the diffuse double layer treated in electrolytes with the exception that ionized impurities are present that, at room temperatures, are immobile. For this case, Poisson's equation becomes

$$\nabla^2\Phi = -\frac{\mathrm{F}}{\varepsilon\varepsilon_0}\left\{n - p - (N_\mathrm{d} - N_\mathrm{a})\right\} \tag{12.11}$$

Show that the capacity can be related to doping level $(N_\mathrm{d} - N_\mathrm{a})$ and potential by the Mott–Schottky relationship

$$\frac{1}{C^2} = -\frac{2}{\mathrm{F}\varepsilon\varepsilon_0}\frac{(V + \mathrm{R}T/\mathrm{F})}{(N_\mathrm{d} - N_\mathrm{a})} \tag{12.12}$$

Solution 12.6. Solution of equation (12.11) is facilitated by expressing the concentration of electrons and holes in terms of potential as

$$n = N\exp\left(\mathrm{F}\Phi/\mathrm{R}T\right) \tag{12.13}$$

and

$$n = P\exp\left(-\mathrm{F}\Phi/\mathrm{R}T\right) \tag{12.14}$$

for electrons and holes, respectively. Thus, Poisson's equation can be expressed as

$$\frac{\mathrm{d}^2\Phi}{\mathrm{d}y^2} = -\frac{\mathrm{F}}{\varepsilon\varepsilon_0}\left[P\mathrm{e}^{-\mathrm{F}\Phi/\mathrm{R}T} - N\mathrm{e}^{\mathrm{F}\Phi/\mathrm{R}T} + (N_\mathrm{d} - N_\mathrm{a})\right] \tag{12.15}$$

where Φ is the electrostatic potential, $(N_\mathrm{d} - N_\mathrm{a})$ is the doping level, and P and N are the hole and electron concentrations, respectively, at the flat-band potential.

The charge density used in equation (12.15), i.e.,

$$\rho_\mathrm{sc}(\Phi) = P\mathrm{e}^{-\mathrm{F}\Phi/\mathrm{R}T} - N\mathrm{e}^{\mathrm{F}\Phi/\mathrm{R}T} + (N_\mathrm{d} - N_\mathrm{a}) \tag{12.16}$$

is now an explicit function of potential rather than position. Integration of Poisson's equation is facilitated by posing equation (12.15) in terms of the electric field

$$E = -\frac{\mathrm{d}\Phi}{\mathrm{d}y} \tag{12.17}$$

The second derivation of potential with respect to position can be expressed in terms of a derivative with respect to potential as

$$\frac{\mathrm{d}^2\Phi}{\mathrm{d}y^2} = -\frac{\mathrm{d}E}{\mathrm{d}y} = -\frac{\mathrm{d}E}{\mathrm{d}\Phi}\frac{\mathrm{d}\Phi}{\mathrm{d}y} = E\frac{\mathrm{d}E}{\mathrm{d}\Phi} = \frac{1}{2}\frac{\mathrm{d}E^2}{\mathrm{d}\Phi} \tag{12.18}$$

Thus,

$$\frac{dE^2}{d\Phi} = -\frac{2}{\varepsilon\varepsilon_0}\rho_{sc}(\Phi) \tag{12.19}$$

Integration yields

$$E^2 = -\frac{2}{\varepsilon\varepsilon_0}\int_{\Phi_{fb}}^{\Phi}\rho_{sc}(\Phi)d\Phi \tag{12.20}$$

where the flat-band potential Φ_{fb} is the potential in the electrically neutral region far from the interface.

The charge held within the space-charge region is given by

$$q_{sc} = \int_0^{\infty}\rho_{sc}dy \tag{12.21}$$

Under the assumption that there are no surface states or specific adsorption of charged species, the space-charge q_{sc} in a semiconductor in contact with an electrolyte is balanced by the charge in the diffuse part of the double layer q_d; thus, $q_{sc} = q_d$. Gauss's law can therefore be used to provide a boundary condition for the electric field at the surface of the semiconductor as

$$E(\Phi(0)) = -\left.\frac{d\Phi}{dy}\right|_{y=0} = \frac{q_{sc}}{\varepsilon\varepsilon_0} \tag{12.22}$$

Thus

$$q_{sc} = \varepsilon\varepsilon_0 E(\Phi(0)) \tag{12.23}$$

The space-charge capacitance is given by

$$C_{sc} = -\frac{dq_{sc}}{d\Phi(0)} = \varepsilon\varepsilon_0\frac{dE(\Phi(0))}{d\Phi(0)} \tag{12.24}$$

or

$$C_{sc} = -\frac{\rho_{sc}(\Phi(0))}{E(\Phi(0))} \tag{12.25}$$

Equation (12.25) provides a relationship between the capacitance of the space-charge region of the semiconductor, the electric field at the interface, and the charge density at the interface.

Under the convention that the potential is referenced to the flat-band potential Φ_{fb}, i.e., the potential is equal to zero far from the interface where the electric field is also equal to zero, integration of equation (12.20) yields

$$E^2 = \frac{2RT}{\varepsilon\varepsilon_0}\left[P(e^{-F\Phi/RT} - 1) + N(e^{F\Phi/RT} - 1) - \frac{F\Phi}{RT}(N_d - N_a)\right] \tag{12.26}$$

A general expression for capacity can be found to be

$$\frac{1}{C_{sc}^2} = \frac{2RT}{F^2\varepsilon\varepsilon_0}\frac{P(e^{-F\Phi(0)/RT} - 1) + N(e^{F\Phi(0)/RT} - 1) - \frac{F\Phi(0)}{RT}(N_d - N_a)}{\left[-Pe^{-F\Phi(0)/RT} + Ne^{F\Phi(0)/RT} - (N_d - N_a)\right]^2} \tag{12.27}$$

The values for N and P can be evaluated at the flat-band potential under the assumptions of electroneutrality (i.e., $\rho_{sc} = 0$) and equilibrium, i.e.,

$$np = n_i^2 \tag{12.28}$$

where n_i is the intrinsic concentration. Thus,

$$N = \frac{1}{2}\left[(N_d - N_a) + \sqrt{(N_d - N_a)^2 + 4n_i^2}\right] \tag{12.29}$$

and

$$P = \frac{1}{2}\left[-(N_d - N_a) + \sqrt{(N_d - N_a)^2 + 4n_i^2}\right] \tag{12.30}$$

Equations (12.29) and (12.30) can be inserted into equation (12.27) to provide an expression for capacitance as a function of potential, with doping level as a parameter.

For an n-type semiconductor, $(N_d - N_a)$ is positive and much larger than the intrinsic concentration n_i; thus, $N \to (N_d - N_a)$ and $P \to 0$. In the denominator, the concentration of holes $P\exp(-F\Phi(0)/RT)$ can be neglected. In the numerator, the term $P(\exp(-F\Phi(0)/RT) - 1)$ can also be neglected. The linear portion of the curve lies at negative potentials, referenced to the flat-band potential; thus, $\exp(F\Phi(0)/RT) \to 0$ and

$$N(\exp(-F\Phi(0)/RT) - 1) \approx -(N_d - N_a) \tag{12.31}$$

Equation (12.27) can be expressed as

$$\frac{1}{C_{sc}^2} = \frac{2RT}{F^2\varepsilon\varepsilon_0}\frac{-(N_d - N_a)\left(\dfrac{F\Phi(0)}{RT} + 1\right)}{(N_d - N_a)^2} \tag{12.32}$$

or

$$\frac{1}{C_{sc}^2} = -\frac{2(\Phi(0) - \Phi_{fb} + RT/F)}{\varepsilon\varepsilon_0 F(N_d - N_a)} \tag{12.33}$$

Equation (12.12) is recovered for $V = \Phi(0) - \Phi_{fb}$.

12.7 The capacity of the electrolytic diffuse double layer is often ignored when Mott–Schottky plots are used to characterize semiconductor–electrolyte interfaces. Under what conditions is this assumption justified?

Solution 12.7. As the impedance of the double layer on the electrolyte side and the space-charge region on the semiconductor side are in series,

$$\frac{1}{C_{total}} = \frac{1}{C_{sc}} + \frac{1}{C_{dl}} \tag{12.34}$$

The capacitance of the double layer may be ignored if $C_{dl} \gg C_{sc}$.

12.8 Mott–Schottky plots are often generated by using measurements at a single frequency, often 1 kHz. Explain the limitations of this approach.

Solution 12.8. Mott–Schottky theory applies specifically for cases in which the measured capacitance is independent of frequency. It does not apply for semiconductors influenced by deep-level electronic states, surface states, and impedance responses influenced by distributions of time constants. For such systems, regression of a suitable process model is preferred.

Chapter 13

Time-Constant Dispersion

Time-constant dispersion observed in impedance spectroscopy may be attributed to variation of properties along the electrode surface or to a distribution in a direction normal to the electrode. The problems presented in the present chapter address the impedance response described by transmission-line models, including that associated with porous electrodes, hierarchical porous electrodes comprising macro-, meso-, and micropores, and thin-layer cells. Other systems require solution of appropriate governing equations such as Laplace's equation for geometry-induced current and potential distributions, convective diffusion equations for nonuniform mass transfer, and multicomponent transport for coupled charging and faradaic currents. Other distributed-time-constant behavior can result from exponential distributions of resistivity in a film. Problems 13.1–13.15 are based on problems presented by Orazem and Tribollet.[1]

Problems

13.1 Provide an analytic solution to

$$\frac{d^2\theta_{i,0}}{d\eta^2} + \frac{\eta^2}{3}\frac{d\theta_{i,0}}{d\eta} = 0 \tag{13.1}$$

with the boundary condition $\theta_{i,0}(0) = 1$ and $\theta_{i,0}(\infty) = 0$. Given

$$\frac{d^2\theta_{i,m}}{d\eta^2} + \frac{\eta^2}{3}\frac{d\theta_{i,m}}{d\eta} - \frac{2m\eta}{3}\theta_{i,m} = \theta_{i,m-1} \tag{13.2}$$

write explicitly the equation corresponding to $m = 1$.

Solution 13.1. The present problem arises from the study of the diffusion impedance associated with flow past a small electrode resulting in a nonuniform diffusion boundary layer. By using

$$\lambda = \frac{d\theta_{i,0}}{d\eta} \tag{13.3}$$

equation (13.1) becomes

$$\frac{d\lambda}{d\eta} + \frac{\eta^2}{3}\lambda = 0 \tag{13.4}$$

The solution is

$$\lambda = A_1 \exp\left(\frac{-\eta^3}{9}\right) \tag{13.5}$$

where A_1 is a constant. The boundary conditions are in terms of $\theta_{i,0}$, which can be written as

$$\theta_{i,0} = \int_0^\eta A_1 \exp\left(\frac{-x^3}{9}\right) dx + A_2 \tag{13.6}$$

The boundary condition at $\eta = 0$ yields $A_2 = 1$. The boundary condition at $\eta \to \infty$ yields

$$\int_0^\infty A_1 \exp\left(\frac{-x^3}{9}\right) dx + 1 = 0 \tag{13.7}$$

or

$$A_1 = \frac{-1}{\displaystyle\int_0^\infty \exp\left(\frac{-x^3}{9}\right) dx} \tag{13.8}$$

and

$$\theta_{i,0} = 1 - \frac{\displaystyle\int_0^\eta \exp\left(\frac{-x^3}{9}\right) dx}{\displaystyle\int_0^\infty \exp\left(\frac{-x^3}{9}\right) dx} \tag{13.9}$$

By using the variable substitution $t = x^3/9$,

$$\int_0^\infty \exp\left(\frac{-x^3}{9}\right) dx = 3^{-1/3}\int_0^\infty t^{-2/3}\exp(-t)\, dt = \frac{\Gamma(1/3)}{3^{1/3}} \tag{13.10}$$

as $\Gamma(z+1) = z\Gamma(z)$,

$$\frac{\Gamma(1/3)}{3^{1/3}} = 3^{2/3}\Gamma(4/3) \tag{13.11}$$

Thus, the solution to equation (13.1) is

$$\theta_{i,0} = 1 - \frac{\displaystyle\int_0^\eta \exp\left(\frac{-x^3}{9}\right) dx}{3^{2/3}\Gamma(4/3)} \tag{13.12}$$

For $m = 1$, equation (13.2) may be written as

$$\frac{d^2\theta_{i,1}}{d\eta^2} + \frac{\eta^2}{3}\frac{d\theta_{i,1}}{d\eta} - \frac{2\eta}{3}\theta_{i,1} = 1 - \frac{\displaystyle\int_0^\eta \exp\left(\frac{-x^3}{9}\right) dx}{3^{2/3}\Gamma(4/3)} \tag{13.13}$$

The extension of the present analysis may be found in Deslouis et al.[48]

13.2 Consider a 0.25 cm radius Pt disk in a 0.1 M NaCl solution at room temperature. Estimate the frequency above which geometry-induced time-constant dispersion will influence the impedance response.

Solution 13.2. The disk radius is $r_0 = 0.25$ cm, and the electrode capacitance may be assumed to have a value of $C_0 = 20\ \mu\text{F/cm}^2$. The conductivity is obtained from

$$\kappa = \text{F}^2 \sum_{\text{i}} z_{\text{i}}^2 u_{\text{i}} c_{\text{i}} = \frac{\text{F}^2}{RT} \sum_{\text{i}} z_{\text{i}}^2 D_{\text{i}} c_{\text{i}} \tag{13.14}$$

to be $\kappa = 1.26 \times 10^{-2}$ S/cm, where the required diffusion coefficients were obtained from Table A.13. The ohmic resistance may be obtained from

$$R_{\text{e}} = \frac{\pi r_0}{4\kappa} = 15.5\ \Omega\text{cm}^2 \tag{13.15}$$

The characteristic frequency may be obtained from

$$f_{\text{c,disk}} = \frac{1}{2\pi} \frac{1}{R_{\text{e}} C_0} \tag{13.16}$$

yielding an answer of $f_{\text{c}} = 510$ Hz. The expression

$$f_{\text{c,disk}} = \frac{1}{2\pi} \frac{\kappa}{C_0 \ell_{\text{disk}}} \tag{13.17}$$

yields a value for the characteristic frequency $f_{\text{c}} = 400$ Hz for $\ell_{\text{disk}} = r_0$. The difference between these two values for characteristic frequency is due to the assumption that $\ell_{\text{disk}} = r_0$. Application of $\ell_{\text{disk}} = \pi r_0/4$ yields the answer obtained by equation (13.16).

13.3 Consider a 0.25 cm radius steel disk covered with a native oxide layer. The electrolyte is a 0.1 M NaCl solution at room temperature. Estimate the frequency above which geometry-induced time-constant dispersion will influence the impedance response.

Solution 13.3. The disk radius is $r_0 = 0.25$ cm, and the capacitance may be assumed to have a value of $C_0 = 3\ \mu\text{F/cm}^2$. The conductivity is obtained from equation (13.14) to be $\kappa = 1.26 \times 10^{-2}$ S/cm. Under the assumption that $\ell_{\text{disk}} = \pi r_0/4$, equation (13.17) yields a value for the characteristic frequency $f_{\text{c}} = 3.4$ kHz.

13.4 Consider a 0.25 cm radius steel disk covered with a polymer coating that has a thickness of 100 μm. The electrolyte is a 0.1 M NaCl solution at room temperature. Estimate the frequency above which geometry-induced time-constant dispersion will influence the impedance response.

Solution 13.4. The electrolyte conductivity for this problem is taken from Problem 13.2, and the corresponding ohmic resistance may be taken from equation (13.15). Estimation of capacitance requires assumption of a dielectric constant, which, for a polymeric coating, may be assumed to have a value $\varepsilon = 3$. The capacitance is obtained from

$$C = \frac{\varepsilon\varepsilon_0}{\delta} \tag{13.18}$$

where $\varepsilon_0 = 8.8542 \times 10^{-14}$ F/cm. The capacitance may be calculated to have a value $C = 2.7 \times 10^{-11}$ F/cm^2. The characteristic frequency may be calculated from equation (13.16) to be $f_c = 390$ MHz, much larger than the usual frequency range for electrochemical measurements.

13.5 The dimensionless form of

$$\mathrm{j}K(\widetilde{V} - \widetilde{\Phi}_0) = -r_0 \left.\frac{\partial \widetilde{\Phi}}{\partial \xi}\right|_{\xi=0} \tag{13.19}$$

was used to justify the use of a dimensionless frequency, defined as

$$K = \frac{\omega C_0 r_0}{\kappa} \tag{13.20}$$

in which the characteristic dimension was the radius of the disk r_0. Beginning with

$$K = \omega C_0 R_\mathrm{e} \tag{13.21}$$

and

$$R_\mathrm{e} = \frac{1}{4\kappa r_0}\pi r_0^2 = \frac{\pi r_0}{4\kappa} \tag{13.22}$$

identify an alternative characteristic dimension.

Solution 13.5. The solution to this problem was suggested in Problem 13.2. By setting $K = 1$ in equation (13.20), a characteristic angular frequency may be obtained as

$$\omega_\mathrm{c} = \frac{\kappa}{C_0 r_0} \tag{13.23}$$

Setting $K = 1$ in equation (13.21) yields

$$\omega_\mathrm{c} = \frac{1}{R_\mathrm{e} C_0} \tag{13.24}$$

Introduction of equation (13.22) into equation (13.24) yields

$$\omega_\mathrm{c} = \frac{4\kappa}{\pi C_0 r_0} \tag{13.25}$$

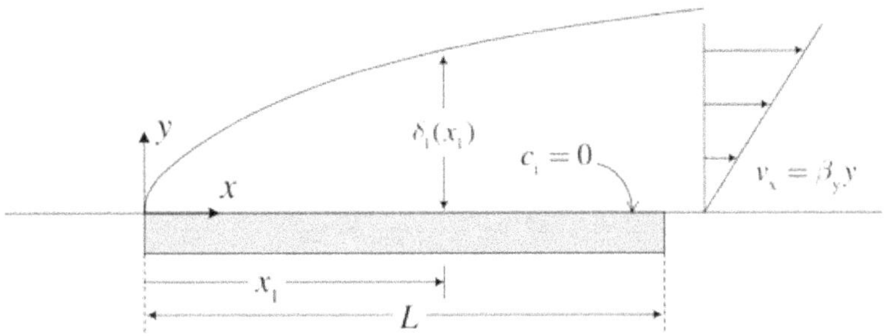

Figure 13.1: Schematic representation of flow past a small electrode with a linear velocity profile and a position-dependent boundary layer thickness. Taken from Orazem and Tribollet[1] with permission of John Wiley & Sons.

If the resistance can be expressed as a characteristic dimension divided by the conductivity

$$R_e = \frac{\ell_c}{\kappa} \tag{13.26}$$

and

$$K = \frac{\omega C_0 \ell_c}{\kappa} \tag{13.27}$$

then the characteristic dimension for a disk electrode may be expressed as $\ell_{\text{disk}} = \pi r_0/4$.

13.6 Find the equations that are necessary to solve the response of a small electrode to a rotation speed modulation.

Solution 13.6. Using a local frame of reference (x, y) attached to the electrode, shown in Figure 13.1, the mass conservation equation governing the concentration distribution c_i of a species transported by convective diffusion can be expressed as

$$\frac{\partial c_i}{\partial t} + \beta_i(t) y \frac{\partial c_i}{\partial x} = D_i \frac{\partial^2 c_i}{\partial y^2} \tag{13.28}$$

The velocity v_x is given by the expression

$$v_x = \overline{\beta}_y y + \text{Re}\left\{\widetilde{\beta}_y \exp(j\omega t)\right\} \tag{13.29}$$

The nonsteady part of equation (13.28) may be written as

$$j\omega\widetilde{c}_i + \overline{\beta}_i(t) y \frac{\partial \widetilde{c}_i}{\partial x} - D_i \frac{\partial^2 \widetilde{c}_i}{\partial y^2} = \widetilde{\beta}_i(t) y \frac{\partial \overline{c}_i}{\partial x} \tag{13.30}$$

with boundary conditions $\widetilde{c}_\text{i} = \widetilde{c}_\text{i}(0)$ for $y = 0$ and $0 < x < L$, $\partial\widetilde{c}_\text{i}/\partial y = 0$ for $y = 0$ and $x < 0$, and $\widetilde{c}_\text{i} = 0$ for $y \to \infty$ and all x.

13.7 A thin-layer cell is composed of an insulated plane at a very short distance ε from a working electrode. By considering the system with a cylindrical symmetry, calculate the impedance of a disk electrode in this configuration.

Solution **13.7.** The model sketched in Figure 13.2 is very similar to that of de Levie[50] for describing impedance response of electrodes located in a cylindrical thin-layer cell. The cylindrical geometry of the thin-layer cell is taken into account by introducing the dependency on the radial coordinate of the ring shape electrode surface elements $\text{d}S_\text{elec}(r)$ and on the lateral surface of the cylindrical electrolyte film confined within the cell $\text{d}S_\text{lat}(r)$, i.e.,

$$\text{d}S_\text{elec}(r) = 2\pi r \text{d}r \tag{13.31}$$

and

$$\text{d}S_\text{lat}(r) = 2\pi r\delta \tag{13.32}$$

where r is the radial coordinate and δ is the film thickness.

The resistance and admittance can be expressed as

$$\text{d}R = R_\text{e}\text{d}r \tag{13.33}$$

and

$$\text{d}Y = \frac{\text{d}r}{Z_0} \tag{13.34}$$

respectively, where $\text{d}R$ is the resistance of a length element of the liquid layer, R_e is the electrolyte resistance, $\text{d}Y$ is the admittance of a ring-shape surface element $\text{d}S_\text{elec}(r)$, and Z_0 is the impedance of the interface for a radial unit length of electrode. Equations (13.33) and (13.34) can be written as

$$\text{d}R = \frac{\rho}{2\pi r\delta}\text{d}r \tag{13.35}$$

and

$$\text{d}Y = \frac{2\pi r\text{d}r}{Z_\text{int}} \tag{13.36}$$

where ρ is the electrolyte resistivity.

The local currents i and $\text{d}i$ flowing in the radial and normal directions of the electrode surface through the electrolyte layer, expressed in frequency domain, are linked by charge conservation, i.e.,

$$\text{d}\widetilde{i}(r) = \widetilde{i}(r + \text{d}r) - \widetilde{i}(r) \tag{13.37}$$

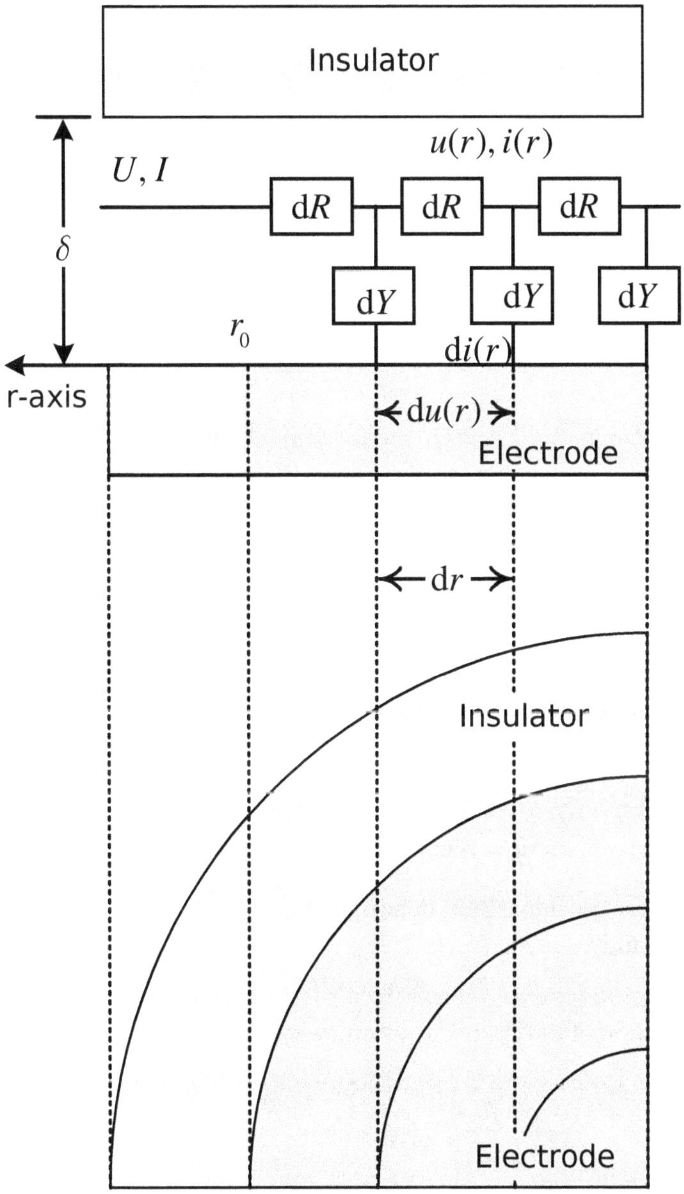

Figure 13.2: Schematic representation of a cylindrical thin-layer cell. Taken from Orazem and Tribollet[1] with permission of John Wiley & Sons.[49]

These currents can also be expressed using Ohm's law according to

$$\widetilde{i} = -\frac{\mathrm{d}\widetilde{u}}{\mathrm{d}R} \tag{13.38}$$

and

$$\mathrm{d}\widetilde{i} = -\widetilde{u}\mathrm{d}Y \tag{13.39}$$

where $\widetilde{u}$ and $\mathrm{d}\widetilde{u}$ are, respectively, the local potential within the thin-layer cell and the local ohmic drop in an length element $\mathrm{d}r$ of the liquid layer.

Using equations (13.35) and (13.36), equations (13.38) and (13.39) may be written as

$$\frac{\mathrm{d}\widetilde{u}}{\mathrm{d}r} = -\frac{\rho}{2\pi r\delta}\widetilde{i} \tag{13.40}$$

and

$$\frac{\mathrm{d}\widetilde{i}}{\mathrm{d}r} = -\frac{2\pi r}{Z_{\text{int}}}\widetilde{u} \tag{13.41}$$

Using equations (13.38) and (13.41), equation (13.40) becomes

$$\frac{\mathrm{d}^2\widetilde{u}}{\mathrm{d}r^2} + \frac{1}{r}\frac{\mathrm{d}\widetilde{u}}{\mathrm{d}r} - \frac{\rho}{\delta Z_{\text{int}}}\widetilde{u} = 0 \tag{13.42}$$

which resembles the Telegrapher equation

$$\frac{\mathrm{d}^2\widetilde{u}(x)}{\mathrm{d}x^2} = \frac{Z_1}{Z_2}\widetilde{u}(x) \tag{13.43}$$

developed in Section 13.1 of Orazem and Tribollet.[1]

The term $(1/r)\mathrm{d}\widetilde{u}/\mathrm{d}r$ may be neglected for a thin-layer cell. The general solution for equation (13.42) is

$$\widetilde{u} = A\exp(r\alpha_{\text{cell}}) + B\exp(-r\alpha_{\text{cell}}) \tag{13.44}$$

where A and B are constants to be determined by application of the appropriate boundary conditions, and

$$\alpha_{\text{cell}} = \sqrt{\rho/\delta Z_{\text{int}}} \tag{13.45}$$

From consideration of the cylindrical symmetry of the cell,

$$\left.\frac{\mathrm{d}\widetilde{u}}{\mathrm{d}r}\right|_{r=0} = 0 \tag{13.46}$$

Thus, $A = B$, and

$$\widetilde{u} = 2A\cosh(r\alpha_{\text{cell}}) \tag{13.47}$$

The global impedance of the transmission line may be expressed in terms of the potential at the periphery of the electrode, $\widetilde{U} = \widetilde{u}(r_0)$, and the global current flowing through the interface $\widetilde{I} = \widetilde{i}(r_0)$ as

$$Z = \frac{\widetilde{U}}{\widetilde{I}}\pi r_0^2 \tag{13.48}$$

The global current may be written as

$$\widetilde{I} = \int_0^{r_0} \mathrm{d}\widetilde{i} = -\int_0^{r_0} \frac{2\pi r}{Z_{\mathrm{int}}} \widetilde{u} \mathrm{d}r \tag{13.49}$$

Combination of equations (13.47) and (13.49) yields

$$\widetilde{I} = \frac{4\pi A}{Z_{\mathrm{int}}} \left(\frac{r_0}{\alpha_{\mathrm{cell}}} \sinh\left(r_0 \alpha_{\mathrm{cell}}\right) + \frac{1}{\alpha_{\mathrm{cell}}^2} \left(1 - \cosh\left(r_0 \alpha_{\mathrm{cell}}\right)\right) \right) \tag{13.50}$$

The impedance of the electrode is given by

$$Z = \frac{r_0^2 Z_{\mathrm{int}}}{2\left(\frac{r_0}{\alpha_{\mathrm{cell}}} \tanh\left(r_0 \alpha_{\mathrm{cell}}\right) + \frac{1}{\alpha_{\mathrm{cell}}^2} \left(\frac{1}{\cosh\left(r_0 \alpha_{\mathrm{cell}}\right)} - 1 \right) \right)} \tag{13.51}$$

When $\alpha_{\mathrm{cell}} \to 0$, i.e., when $\delta \to \infty$ or $\rho \to 0$, $Z \to Z_{\mathrm{int}}$. When $\alpha_{\mathrm{cell}} \to \infty$, i.e., when $\delta \to 0$ or $\rho \to \infty$, the cell may be treated as a one-dimensional system and Z is proportional to $\sqrt{Z_{\mathrm{int}}}$, a result that is similar to de Levie's impedance for a one-dimensional porous electrode.

13.8 Explore the limiting behavior of

$$Z = R_{\mathrm{e}} + \frac{(\rho Z_{\mathrm{eq}})^{1/2}}{\sqrt{2}\pi n r^{3/2}} \coth\left(\ell \sqrt{\frac{2\rho}{r Z_{\mathrm{eq}}}} \right) \tag{13.52}$$

when $\ell\sqrt{2\rho / r Z_{\mathrm{eq}}}$ is very small. What independent parameters or combinations of parameters can be obtained by regression of this model to experimental data? To what geometry does this limit conform?

Solution 13.8. The Taylor series expansion for $\coth(x)$ near $x \to 0$ is given as

$$\coth(x) = \frac{1}{x} + \frac{x}{3} + \ldots \tag{13.53}$$

Thus, using the first term of the expansion, equation (13.52) may be expressed as

$$Z = R_{\mathrm{e}} + \frac{(\rho Z_{\mathrm{eq}})^{1/2}}{\sqrt{2}\pi n r^{3/2}} \left(\frac{1}{\ell} \sqrt{\frac{r Z_{\mathrm{eq}}}{2\rho}} \right) \tag{13.54}$$

or

$$Z = R_{\mathrm{e}} + Z_{\mathrm{eq}} \frac{1}{n 2\pi r \ell} = R_{\mathrm{e}} + \frac{Z_{\mathrm{eq}}}{A} \tag{13.55}$$

The parameters that can be obtained by regression include R_{e} and the parameters from Z_{eq}/A. No parameters associated with the porous structure may be obtained. This corresponds to a planar, nonporous electrode.

13.9 Explore the behavior of equation (13.52) when $\ell\sqrt{2\rho/rZ_{\text{eq}}}$ is neither very small nor very large. What independent parameters or combinations of parameters can be obtained by regression of this model to experimental data?

Solution 13.9. When $\ell\sqrt{2/rZ_{\text{eq}}}$ is neither very small nor very large, no simplification on the hyperbolic cotangent function can be proposed. Equation (13.52) can be written as

$$Z = R_{\text{e}} + \frac{A}{\sqrt{Z_{\text{eq}}}}\coth\left(B\sqrt{Z_{\text{eq}}}\right) \tag{13.56}$$

where

$$A = \frac{\sqrt{\rho}}{\sqrt{2}\pi n r^{3/2}} \tag{13.57}$$

and

$$B = \ell\sqrt{\frac{2\rho}{r}} \tag{13.58}$$

The number of independent parameters depends on the model introduced for Z_{eq}. For example, if

$$Z_{\text{eq}} = \frac{1}{(\text{j}\omega)^{\alpha}Q} \tag{13.59}$$

Equation (13.56) becomes

$$Z = R_{\text{e}} + \frac{A'}{(\text{j}\omega)^{\alpha/2}}\coth\left(B'(\text{j}\omega)^{\alpha/2}\right) \tag{13.60}$$

where $A' = A/\sqrt{Q}$ and $B = B\sqrt{Q}$. The independent parameters are R_{e}, A', B', and α.

13.10 Numerical methods were used to generate the characteristic frequencies shown in Table 13.1 for a rectangular electrode of different dimensions with a counterelectrode located at infinity. The capacitance was assigned a value of 20 μF/cm^2, and the resistivity was 1,000 Ωcm. Show that the characteristic dimension associated with this characteristic frequency can be expressed as $d = d_1 d_2/(d_1 + d_2)$.

Solution 13.10. The characteristic frequency is given by

$$f_{\text{c}} = \frac{1}{2\pi C_0 d\rho} \tag{13.61}$$

The value for d obtained from

$$d = \frac{1}{2\pi C_0 f_{\text{c}}\rho} \tag{13.62}$$

Table 13.1: Calculated characteristic frequencies for rectangular electrodes with dimensions d_1 and d_2.

f_c, kHz	d_1, mm	d_2, mm
21.59	0.04055	0.00405
9.81	0.04055	0.01014
5.89	0.04055	0.02028
3.92	0.04055	0.04055
3.11	0.02819	0.28190
2.94	0.04055	0.08110
2.35	0.04055	0.20275
2.22	0.04055	0.30413
2.16	0.04055	0.40550
2.11	0.04055	0.52715
1.41	0.07048	0.28190
0.85	0.14095	0.28190
0.56	0.28190	0.28190
0.42	0.56380	0.28190
0.34	1.40950	0.28190
0.32	2.11425	0.28190
0.31	2.81900	0.28190
0.30	4.22850	0.28190

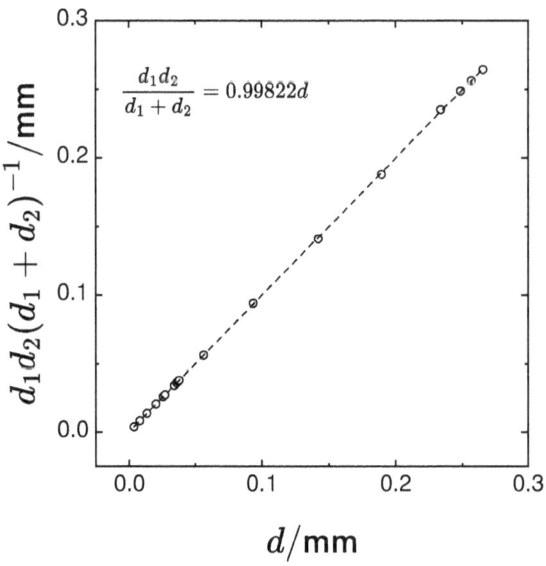

Figure 13.3: Comparison of the dimensions $d_1 d_2/(d_1 + d_2)$ to the value d obtained from equation (13.62) for Problem 13.10.

is compared in Figure 13.3 to the value $d_1 d_2/(d_1 + d_2)$. An excellent agreement is obtained.

13.11 By inspection of Table A.21, identify an expression for the characteristic frequency associated with time-constant dispersion associated with a CPE on a rectangular electrode.

Solution 13.11. As shown in Table A.21, the characteristic frequency for a disk electrode with a CPE is

$$f_c = \frac{1}{2\pi}\left(\frac{\kappa}{Q r_0}\right)^{1/\alpha} \tag{13.63}$$

Thus, for a rectangular electrode, we can expect that

$$f_c = \frac{1}{2\pi}\left(\frac{\kappa(d_1 + d_2)}{C_0 d_1 d_2}\right)^{1/\alpha} \tag{13.64}$$

where d_1 and d_2 are the side lengths of the rectangle.

13.12 Find the smallest frequency for which geometry-induced time-constant dispersion will appear for an electrode for which three electrodes of different dimensions are connected in parallel:

(a) a disk of radius $r_0 = 0.25$ cm, a rectangle with dimensions 0.2 cm and 0.6 cm, and a hemispherical electrode with radius $r_0 = 0.25$ cm.

(b) a disk of radius $r_0 = 0.25$ cm, a disk of radius $r_0 = 0.5$ cm, and a disk of radius $r_0 = 0.75$ cm.

(c) three rectangles with dimensions 0.2×0.2 cm, 0.2×0.6 cm, and 0.2×1.2 cm.

Solution 13.12. To perform these calculations, assume that the capacitance is $C_0 = 20\ \mu\text{F/cm}^2$ and the electrolyte conductivity is $\kappa = 0.01$ S/cm.

(a) The primary and secondary current distributions are uniform for a hemispherical electrode, and no geometry-induced time-constant dispersion will be seen. For a disk electrode

$$f_{c,\text{disk}} = \frac{\kappa}{2\pi C_0 r_0} \tag{13.65}$$

and for a rectangular electrode of dimensions $d_1 \times d_2$

$$f_{c,\text{rect}} = \frac{\kappa(d_1 + d_2)}{2\pi C_0 d_1 d_2} \tag{13.66}$$

Equation (13.65) yields $f_{c,\text{disk}} = 318$ Hz, and equation (13.66) yields $f_{c,\text{disk}} = 530$ Hz. Thus, the smallest frequency for which geometry-induced time-constant dispersion will appear was calculated to be 318 Hz.

(b) The smallest frequency for which geometry-induced time-constant dispersion will appear is calculated for the largest radius to be 106 Hz.

(c) The smallest frequency was calculated for the 0.2 × 1.2 cm electrode to be 464 Hz.

13.13 Find the smallest frequency for which geometry-induced time-constant dispersion will appear for the following:
(a) an electrode consisting of 20 disks of radius $r_0 = 10$ μm.
(b) a single disk electrode of area equal to 20 disks of radius $r_0 = 10$ μm.

Solution 13.13. The area of a single disk electrode equal to the area of 20 disks of radius r_0 can be found as

$$A' = 20\pi r_0^2 = \pi (r_0')^2 \tag{13.67}$$

where r_0' is the radius of the larger electrode, given as

$$r_0' = \sqrt{20} r_0 \tag{13.68}$$

The ratio of characteristic frequencies is

$$\frac{f_\mathrm{c}}{f_\mathrm{c}'} = \frac{\dfrac{\kappa}{2\pi C_0 r_0}}{\dfrac{\kappa}{2\pi C_0 r_0'}} = \frac{r_0'}{r_0} = \sqrt{20} \tag{13.69}$$

To perform these calculations, assume that the capacitance is $C_0 = 20$ μF/cm^2 and the electrolyte conductivity is $\kappa = 0.01$ S/cm.

(a) For an electrode of radius $r_0 = 10$ μm or 0.001 cm, the characteristic frequency is obtained from equation (13.65) to be $f_\mathrm{c} = 7.96 \times 10^4$ Hz.

(b) For an electrode of radius $r_0 = 10\sqrt{20}$ μm or 0.00447 cm, the characteristic frequency is obtained from equation (13.65) to be $f_\mathrm{c}' = 1.78 \times 10^4$ Hz.

Thus, $f_\mathrm{c}/f_\mathrm{c}' = 4.47$.

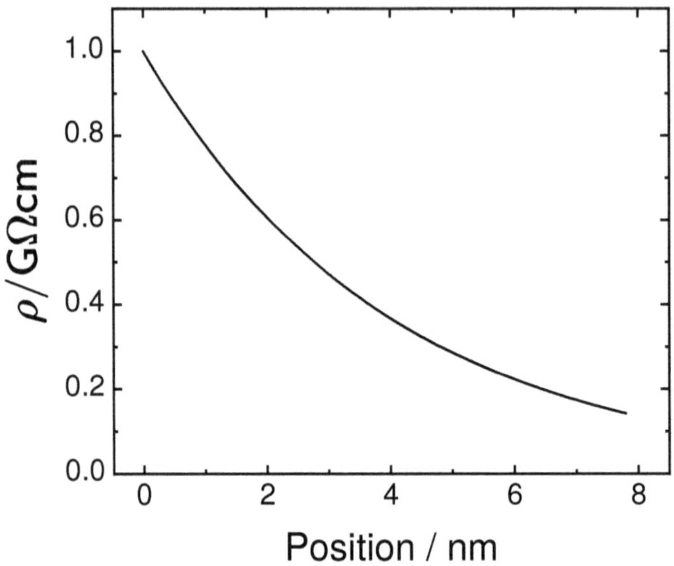

Figure 13.4: The distribution of resistivity corresponding to equation (13.70) with parameters given in part 13.14.a.

13.14 Consider the impedance response of a film in which the resistivity varies with the position perpendicular to the plane of the film according to

$$\rho(y) = \rho_0 \exp(-y/\lambda) \tag{13.70}$$

where λ is a characteristic dimension over which resistivity varies.

(a) Plot the resulting resistivity as a function of position for a film of thickness $\delta = 8$ nm. You may assume that $\lambda = 4$ nm and $\rho_0 = 1 \times 10^9$ Ωcm.

(b) Develop an expression for the impedance response of a film of thickness $\delta = 8$ nm.

(c) Plot the resulting impedance as a Nyquist plot for $\varepsilon = 12$, $\lambda = 4$ nm, $\delta = 8$ nm, and $\rho_0 = 1 \times 10^9$ Ωcm. Does this impedance represent a depressed semicircle? Does this impedance represent a CPE? You may use a plot of the absolute value of the imaginary part of the impedance as a function of frequency on a log-log scale to justify your answers.

Solution 13.14.

(a) The distribution of resistivity given by equation (13.70) for $\delta = 8$ nm, $\lambda = 4$ nm, and $\rho_0 = 1 \times 10^9$ Ωcm is given in Figure 13.4.

(b) The impedance of a layer can be considered to correspond to a large number of Voigt elements as represented in Figure 13.5. The capacity $C(y)$ is the capacity

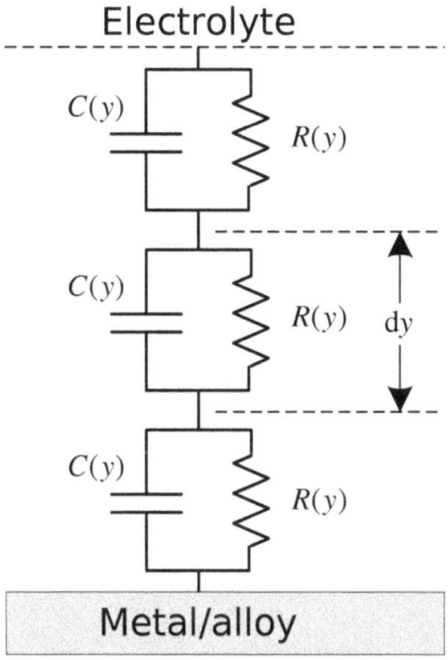

Figure 13.5: Equivalent circuit corresponding to an oxide layer with an axial distribution of dielectric and resistive properties. Taken from Orazem and Tribollet[1] with permission of John Wiley & Sons.

of a dielectric with a thickness $\mathrm{d}y$, and the resistance $R(y)$ corresponds to a layer of same thickness $\mathrm{d}y$ with a resistivity $\rho(y)$. The local impedance is obtained by integration along the y-direction from 0 to the coating thickness δ, i.e.,

$$Z_{\mathrm{Y}} = \int_0^{\delta} \frac{\rho(y)}{1 + \mathrm{j}\omega\varepsilon\varepsilon_0\rho(y)} \mathrm{d}y \tag{13.71}$$

As shown in Table A.14, the dielectric constant ε varies in a narrow range for a metal oxide. Thus, to a first approximation, ε can be considered to be independent of y, and the resistivity ρ may be assumed to be a function of y.

For an oxide layer, Young[51] assumed that the nonstoichiometry of the oxide layer resulted in an exponential variation of the conductivity with respect to the normal distance to the electrode as shown in equation (13.70). The corresponding Young impedance is given by

$$Z_{\mathrm{Y}} = \frac{\lambda}{\mathrm{j}\omega\varepsilon\varepsilon_0} \ln\left(\frac{1 + \mathrm{j}\omega\rho_0\varepsilon\varepsilon_0}{1 + \mathrm{j}\omega\rho_0\varepsilon\varepsilon_0 \exp(-\delta/\lambda)}\right) \tag{13.72}$$

where δ is the film thickness.

(c) The Young impedance for the given parameters is presented in Nyquist format in Figure 13.6(a). The Young impedance has the appearance of a slightly depressed semicircle as compared to the semicircle represented by a solid line. The observation that, as shown in Figure 13.6(b), the derivative $\mathrm{d}\log|Z_{\mathrm{j}}|/\mathrm{d}\log(f)$ approaches

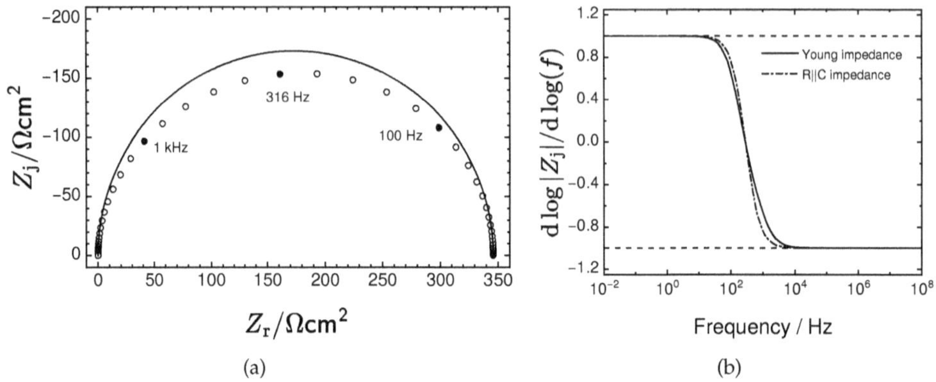

Figure 13.6: The Young impedance, equation (13.72), for $\varepsilon = 12$, $\lambda = 4$ nm, $\delta = 8$ nm, and $\rho_0 = 1 \times 10^9$ Ωcm: (a) Nyquist plot, where the symbols are the Young impedance and the line represents a semicircle, and (b) derivative $d\log|Z_j|/d\log(f)$ as a function of frequency.

limits of $+1$ at low frequency and -1 at high frequency indicates that the Young impedance is not equivalent to a constant-phase element, for which the low- and high-frequency limits would be $+\alpha$ and $-\alpha$, respectively. The approach to the low- and high-frequency limits of $+1$ and -1, respectively, differs from that of the parallel combination of a resistor and a capacitor, shown as a solid line in Figure 13.6(b).

13.15 The impedance for a system has been modeled by

$$Z = R_e + \frac{1}{(j\omega)^\alpha Q} \tag{13.73}$$

where $\alpha = 0.5$. Equation (13.73) is normally associated with a distribution of time constants through a film or along a surface, but other explanations may apply when $\alpha = 0.5$. Suggest two other possible physical explanations that could account for this result.

Solution 13.15. One possible physical impedance corresponds to the Warburg impedance. In this case, a mass-transport limitation must exist and must appear in the physical model. A second physical impedance could correspond to the high-frequency behavior of a porous electrode impedance in a frequency range where the interfacial impedance corresponds to a capacitance.

13.16 Under the assumptions that Z_0 and R_0 are independent of position in cylindrical pores, the impedance of the porous electrode Z can be expressed as equation (13.52).[50]

(a) Develop an expression for the impedance of a porous electrode for an interfacial impedance expressed as a charge-transfer resistance in parallel with a double layer capacitance, i.e.,

$$Z_{\mathrm{eq}} = \frac{R_{\mathrm{t}}}{1 + \mathrm{j}\omega C R_{\mathrm{t}}} \tag{13.74}$$

where Z_{eq} is the interfacial impedance per unit area, with units of $\Omega\mathrm{cm}^2$.

(b) Find the limiting behavior as $\omega \to 0$.

(c) Find the limiting behavior as $\omega \to \infty$.

Solution 13.16.

(a) Equation (13.52) can be written as

$$Z = R_{\mathrm{e}} + \frac{A}{\sqrt{1 + \mathrm{j}\omega C R_{\mathrm{t}}}} \coth\left(B\sqrt{1 + \mathrm{j}\omega C R_{\mathrm{t}}}\right) \tag{13.75}$$

where

$$A = \frac{\sqrt{\rho R_{\mathrm{t}}}}{\sqrt{2}\pi n r^{3/2}} \tag{13.76}$$

and

$$B = \ell\sqrt{\frac{2\rho}{r R_{\mathrm{t}}}} \tag{13.77}$$

The impedance depends on only four parameters, R_{e}, A, B, and CR_{t}. The individual parameters such as R_{t}, C, n, r, ρ, and ℓ cannot be determined separately by regression analysis.

(b) In the limit as $x \to 0$,

$$\lim_{x \to 0} \coth x = \frac{1}{x} \tag{13.78}$$

and in the limit as $x \to \infty$,

$$\lim_{x \to \infty} \coth x = 1 + 2\exp(-2x) \tag{13.79}$$

Thus, the value of equation (13.75), as $\omega \to 0$, depends on the value of B. If $B \gg 1$, the estimation of the coth term follows equation (13.79), and

$$\lim_{\omega \to 0} Z - R_{\mathrm{e}} = A \tag{13.80}$$

Examination of equation (13.77) shows that B represents the ratio of the ohmic resistance over the charge transfer resistance. A large value of B indicates that the ohmic resistance dominates.

If $B \ll 1$, the estimation of the coth term follows equation (13.78), and

$$\lim_{\omega \to 0} Z - R_{\mathrm{e}} = \frac{A}{B} \tag{13.81}$$

Manipulation of equations (13.76) and (13.77) yields

$$\frac{A}{B} = \frac{R_{\mathrm{t}}}{n 2 \pi r \ell} \tag{13.82}$$

where the area term represents the wall area of n pores.

(c) As $\omega \to \infty$, The coth tends to unity. Following equation (13.79),

$$\lim_{\omega \to 0} Z - R_{\mathrm{e}} = \frac{A}{\sqrt{\mathrm{j}\omega R_{\mathrm{t}} C}} = \frac{D}{\sqrt{\mathrm{j}\omega}} \tag{13.83}$$

where

$$D = \sqrt{\frac{\rho}{2C}} \frac{1}{\pi n r^{3/2}} \tag{13.84}$$

which is a function of electrolyte resistivity and capacitance.

13.17 Develop an expression for the impedance of a porous electrode for which an interfacial impedance is expressed as a double layer capacitance, i.e.,

$$Z_{\mathrm{eq}} = \frac{1}{\mathrm{j}\omega C} \tag{13.85}$$

Use the general expression for the impedance of cylindrical pores given in equation (13.52).[50] Find the limiting behavior as $\omega \to 0$ and as $\omega \to \infty$.

Solution 13.17. Equation (13.52) can be expressed as

$$Z = R_{\mathrm{e}} + \frac{A}{\sqrt{\mathrm{j}\omega}} \coth\left(B\sqrt{\mathrm{j}\omega}\right) \tag{13.86}$$

where

$$A = \frac{\sqrt{\rho}}{\sqrt{2C}} \frac{1}{\pi n r^{3/2}} \tag{13.87}$$

and

$$B = \ell \sqrt{\frac{2\rho C}{r}} \tag{13.88}$$

The independent variables are R_e, A, and B. The capacitance does not appear as an independent variable. Following equation (13.78), $\coth B\sqrt{j\omega}$ tends toward $B\sqrt{j\omega}$. Thus,

$$\lim_{\omega \to 0} Z - R_e = \frac{A}{\sqrt{j\omega}} \frac{1}{B\sqrt{j\omega}} = \frac{A}{B} \frac{1}{j\omega} \tag{13.89}$$

The apparent capacitance is

$$\frac{B}{A} = Cn2\pi r\ell \tag{13.90}$$

which represents the capacitance per unit area of the pore wall multiplied by the area of the pore wall. Extraction of C from equation (13.90) requires independent knowledge of the pore-wall area. As $\omega \to \infty$, the impedance $Z - R_e$ tends toward $A/\sqrt{j\omega}$.

13.18 Develop an expression for the impedance of a porous electrode for which an interfacial impedance is expressed as a constant-phase element, i.e.,

$$Z_{eq} = \frac{1}{(j\omega)^{\alpha} Q} \tag{13.91}$$

where α and Q are CPE parameters. Use the general expression for the impedance of cylindrical pores given in equation (13.52).[50] Find the limiting behavior as $\omega \to 0$ and as $\omega \to \infty$.

Solution 13.18. Equation (13.52) can be expressed as

$$Z = R_e + \frac{A}{(j\omega)^{\alpha/2}} \coth\left(B\,(j\omega)^{\alpha/2}\right) \tag{13.92}$$

where

$$A = \frac{\sqrt{\rho}}{\sqrt{2Q}} \frac{1}{\pi n r^{3/2}} \tag{13.93}$$

and

$$B = \ell\sqrt{\frac{2\rho Q}{r}} \tag{13.94}$$

The independent variables are R_e, A, B, and α. The CPE coefficient Q does not appear as an independent variable. As the frequency tends toward zero,

$$\lim_{\omega \to 0} Z - R_e = \frac{A}{(j\omega)^{\alpha/2}} \frac{1}{B\,(j\omega)^{\alpha/2}} = \frac{A}{B} \frac{1}{(j\omega)^{\alpha}} \tag{13.95}$$

The apparent CPE coefficient

$$\frac{B}{A} = Qn2\pi r\ell \tag{13.96}$$

is the value of Q scaled per unit area of the pore wall multiplied by the area of the pore wall. Extraction of Q from equation (13.96) requires independent knowledge of the pore-wall area. As $\omega \to \infty$, the impedance $Z - R_e$ tends toward $A/(j\omega)^{\alpha/2}$.

13.19 Lasia[52] writes the expression for the impedance of a porous electrode as

$$Z_{\text{pore}} = \frac{R_{\text{pore}}}{\sqrt{\Lambda}} \coth \sqrt{\Lambda} \tag{13.97}$$

Show that this form is consistent with

$$Z_{\text{pore}} = \sqrt{R_0 Z_0} \coth\left(\ell \sqrt{R_0/Z_0}\right) \tag{13.98}$$

Solution 13.19. Comparison of the argument for the coth function from equations (13.97) and (13.98) yields

$$\Lambda = \frac{\ell^2 R_0}{Z_0} \tag{13.99}$$

thus,

$$R_{\text{pore}} = \ell R_0 \tag{13.100}$$

As the total impedance is

$$Z = R_{\text{e}} + \frac{Z_{\text{pore}}}{n} \tag{13.101}$$

where n is the number of pores, the effective pore resistance is $R_{\text{p}} = R_{\text{pore}}/n$, and

$$Z = R_{\text{e}} + \frac{R_{\text{p}}}{\sqrt{\Lambda}} \coth \sqrt{\Lambda} \tag{13.102}$$

Equation (13.102) has the advantage that the pore resistance is clearly identified. The value of R_{p} was used, for example, by Lutz et al.[53] as the ohmic resistance term in the Brug formula,[54] e.g.,

$$C_{\text{p}} = Q^{1/\alpha} R_{\text{p}}^{(1-\alpha)/\alpha} \tag{13.103}$$

to convert pore CPE parameters into a capacitance.

Chapter 14

Constant-Phase Elements

Electrical circuits invoking constant-phase elements (CPE) are often used to fit impedance data arising from a broad range of experimental systems. The CPE is often used simply as a way to improve the fit of a model to impedance data, justified by a vague assertion that a distribution of time constants is present in the system under investigation. The problems in the present chapter explore the physical meaning of CPE parameters. The interpretation of CPE parameters in terms of physically meaningful properties such as capacitance requires an understanding of the nature of the time-constant distribution. Problems 14.1–14.10 are based on problems presented by Orazem and Tribollet.[1]

Problems

14.1 Why does the following represent a constant-phase elephant? 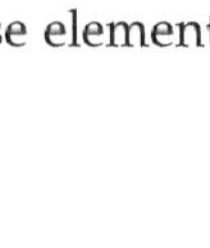

Solution 14.1. The impedance response of a constant-phase element may be expressed as

$$Z_{\mathrm{CPE}} = \frac{1}{(\mathrm{j}\omega)^{\alpha} Q} \tag{14.1}$$

and has the appearance, in a Nyquist plot, of a straight line with an angle referenced to the real axis of $\varphi = \alpha 90°$. As shown in Figure 14.1, the angle of the trunk of the constant-phase elephant with respect to the ground is about $\varphi = 74.1°$. The corresponding value of α is 0.82. A value of $\alpha = 0.82$ may correspond to the frequency dispersion associated with electrode geometry or with the dielectric response of human skin.

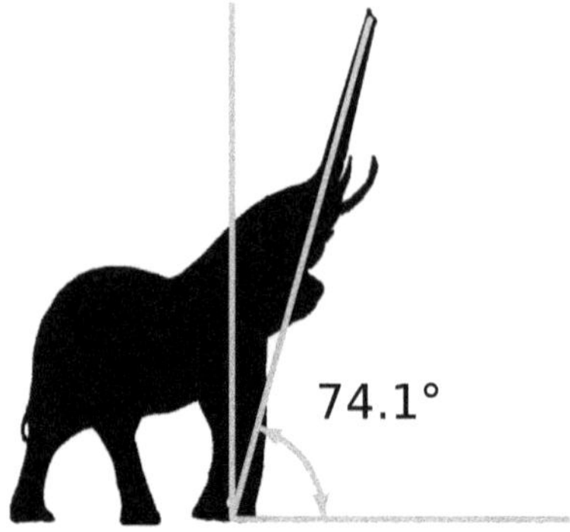

Figure 14.1: Schematic representation of the angle of the trunk of the constant-phase elephant with respect to ground.

The constant-phase elephant was so named because its trunk is extended at an angle relative to the ground similar to the manner in which the impedance of a constant-phase element is seen in a Nyquist plot.

14.2 With the aid of Table A.17 and under the assumption that Q has units of $\mathrm{F/s^{(1-\alpha)}cm^2}$ and resistances have units of $\Omega\mathrm{cm}^2$, identify the units of capacitance given in the following:

(a)

$$C_{\mathrm{eff},Q} = Q \tag{14.2}$$

(b)

$$C_{\mathrm{eff,norm}} = Q^{1/\alpha} R_{||}^{(1-\alpha)/\alpha} \tag{14.3}$$

(c)

$$C_{\mathrm{eff,surf}} = Q^{1/\alpha} \left(\frac{R_{\mathrm{e}} R_{||}}{R_{\mathrm{e}} + R_{||}} \right)^{(1-\alpha)/\alpha} \tag{14.4}$$

(d)

$$C_{\mathrm{eff,PL}} = gQ \left(\rho_\delta \varepsilon \varepsilon_0 \right)^{1-\alpha} \tag{14.5}$$

(e)

$$C_{\mathrm{eff,PL}} = (gQ)^{1/\alpha} \left(\delta \rho_\delta \right)^{(1-\alpha)/\alpha} \tag{14.6}$$

Solution 14.2.

(a) As C and Q have different units, equation (14.2) is not dimensionally consistent. Thus,

$$C_{\mathrm{eff},Q} \; [=] \; \frac{\mathrm{F}}{\mathrm{cm^2 s^{1-\alpha}}} \; [\neq] \; \frac{\mathrm{F}}{\mathrm{cm^2}} \tag{14.7}$$

(b) The key step for analyzing equation (14.8) is to recognize that, from Table A.17, $1\Omega = 1\text{s}/F$. Thus,

$$\begin{aligned} C_{\text{eff,norm}} &[=] \left[\frac{\text{F}}{\text{cm}^2\text{s}^{1-\alpha}}\right]^{1/\alpha} \left[\Omega\text{cm}^2\right]^{(1-\alpha)/\alpha} \\ &[=] \left[\frac{\text{F}}{\text{cm}^2\text{s}^{1-\alpha}}\right]^{1/\alpha} \left[\frac{\text{s}}{F}\right]^{(1-\alpha)/\alpha} \left[\text{cm}^2\right]^{(1-\alpha)/\alpha} \\ &[=] \frac{\text{F}^{1/\alpha-(1-\alpha)/\alpha}}{(\text{cm}^2)^{1/\alpha-(1-\alpha)/\alpha}} [=] \frac{\text{F}}{\text{cm}^2} \end{aligned} \tag{14.8}$$

Equation (14.8) is dimensionally consistent.

(c) The quantity $R_e R_{||}/(R_e + R_{||})$ has units of Ωcm^2. Thus, the answer for part (c) is the same as for part (b). Equation (14.4) is dimensionally consistent.

(d) The quantities g and ε are dimensionless.

$$\begin{aligned} C_{\text{eff,PL}} &[=] \left[\frac{\text{F}}{\text{cm}^2\text{s}^{1-\alpha}}\right] \left[\Omega\text{cm}\left[\frac{\text{F}}{\text{cm}}\right]\right]^{1-\alpha} \\ &[=] \left[\frac{\text{F}}{\text{cm}^2\text{s}^{1-\alpha}}\right] \text{s}^{1-\alpha} [=] \frac{\text{F}}{\text{cm}^2} \end{aligned} \tag{14.9}$$

Equation (14.5) is dimensionally consistent.

(e) The quantity gQ has units of $\text{F}/(\text{cm}^2\text{s}^{1-\alpha})$, and the quantity $\delta\rho_\delta$ has units of Ωcm^2. Thus, the answer for part (e) is the same as for part (b). Equation (14.6) is dimensionally consistent.

14.3 Time-constant distributions leading to CPE behavior may be described as being in the direction normal to the electrode surface or along the electrode surface.

(a) Under what conditions could a system show CPE behavior resulting from a distribution in only the direction normal to the electrode surface?

(b) Under what conditions could a system show CPE behavior resulting from a distribution along the electrode surface?

Solution 14.3. The concept of CPE behavior caused by normal or surface distributions was described by Jorcin et al.[55]

(a) CPE behavior resulting from a distribution in the direction normal to the electrode surface could be caused by a distribution of properties of a film. Hirschorn et al.[56] showed that a power-law distribution of resistivity could result in a CPE response. They assumed that the relative dielectric permittivity had a uniform value.

(b) In the absence of a film, CPE behavior may result from a distribution of time constants along the electrode surface. The origin of this distribution is still under investigation. The formula used to extract capacitance from CPE parameters, i.e., equation (14.4), was developed by Brug et al.[54] under the assumption of a surface distribution of time constants. Electrode heterogeneity, such as roughness, distributed capacitance, and distributed reactivity, has often been invoked as a source of CPE behavior, but Alexander et al.[47,57–59] showed that roughness and capacitance distributions show frequency dispersion at frequencies larger than that caused by the electrode geometry itself.

14.4 Starting from

$$Z = R_\mathrm{e} + \frac{R_{||}}{1 + (\mathrm{j}K_\mathrm{norm})^\alpha} \tag{14.10}$$

verify equation (14.3).

Solution 14.4. Equation (14.10) has, as an origin,

$$Z = R_\mathrm{e} + \frac{R_{||}}{1 + (\mathrm{j}\omega)^\alpha R_{||} Q} \tag{14.11}$$

The quantity K_norm in equation (14.10) is a dimensionless frequency, which can be expressed for an impedance as

$$K_\mathrm{norm} = \omega \left(R_{||} Q\right)^{1/\alpha} \tag{14.12}$$

A characteristic frequency can be found for $K_\mathrm{norm} = 1$ such that

$$f_\mathrm{c,norm} = \frac{1}{2\pi \left(R_{||} Q\right)^{1/\alpha}} \tag{14.13}$$

The essential hypothesis of the approach presented in this section is that the characteristic frequency can also be expressed in terms of an effective capacitance as

$$f_\mathrm{c,norm} = \frac{1}{2\pi R_{||} C_\mathrm{eff,norm}} \tag{14.14}$$

By solving equations (14.13) and (14.14) for $C_\mathrm{eff,norm}$, an expression for the effective capacitance associated with the CPE is found to be equation (14.3).

To illustrate the approach, synthetic data are presented in Figure 14.2 for a system with $R_\mathrm{e} = 10\ \Omega\mathrm{cm}^2$, $R_{||} = 10\ \mathrm{k}\Omega\mathrm{cm}^2$, $Q = 10\ \mu\mathrm{F/cm^2 s^{1-\alpha}}$, and $\alpha = 0.9$. The characteristic frequency has a value of about 2 Hz. Equation (14.14) yields a capacitance with a value of 7.96 $\mu\mathrm{F/cm^2}$, and equation (14.3) yields a capacitance with a value of

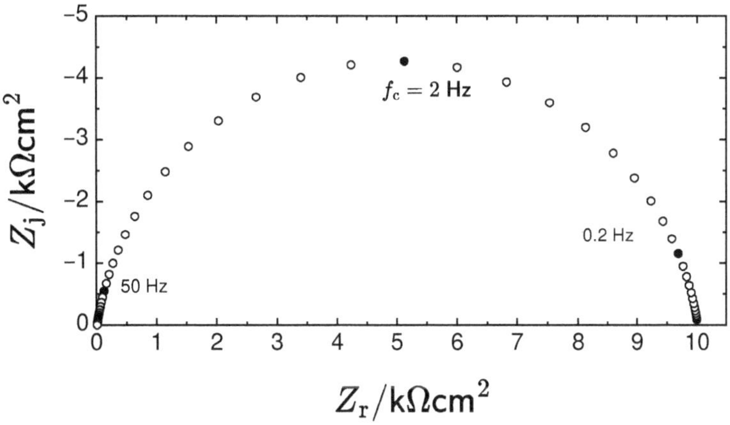

Figure 14.2: Impedance response for a system with $R_e = 10\ \Omega\text{cm}^2$, $R_{||} = 10\ \text{k}\Omega\text{cm}^2$, $Q = 10\ \mu\text{F/cm}^2\text{s}^{1-\alpha}$, and $\alpha = 0.9$.

$7.74\ \mu\text{F/cm}^2$. The difference between these values may be attributed to the approximate evaluation of the characteristic frequency in Figure 14.2. Note that, even for $\alpha = 0.9$, the numerical value of the capacitance obtained by equation (14.3) differs substantially from the numerical value of 10 for Q.

14.5 Starting from

$$Y = \frac{1}{R_e}\left[1 - \frac{R_{||}}{R_e + R_{||}}\left(1 + \frac{R_e R_{||}}{R_e + R_{||}} Q(j\omega)^\alpha\right)^{-1}\right] \tag{14.15}$$

verify equation (14.4).

Solution 14.5. Equation (14.15) can be expressed as

$$Y = \frac{1}{R_e}\left[1 - \frac{R_{||}}{R_e + R_{||}}\left(1 + (jK_{surf})^\alpha\right)^{-1}\right] \tag{14.16}$$

where K_{surf} is a dimensionless frequency, expressed for an admittance as

$$K_{surf} = \left(\frac{R_e R_{||}}{R_e + R_{||}} Q\right)^{1/\alpha} \omega \tag{14.17}$$

A characteristic frequency can be found for $K_{surf} = 1$ such that

$$f_{c,surf} = \frac{1}{2\pi\left(QR_eR_{||}/(R_e + R_{||})\right)^{1/\alpha}} \tag{14.18}$$

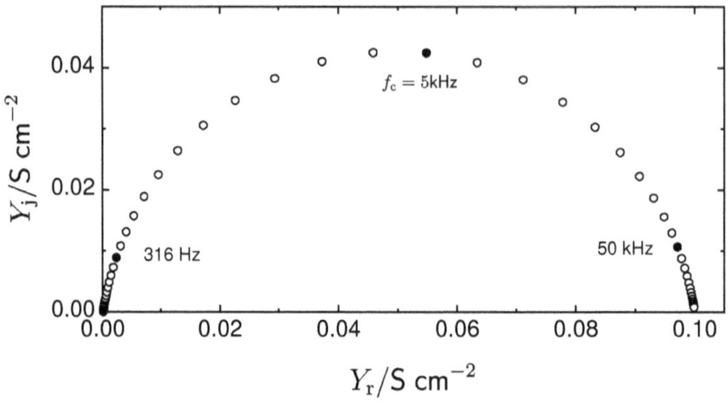

Figure 14.3: Admittance response for the system presented in Figure 14.2 with $R_e = 10\ \Omega\text{cm}^2$, $R_{||} = 10\ \text{k}\Omega\text{cm}^2$, $Q = 10\ \mu\text{F/cm}^2\text{s}^{1-\alpha}$, and $\alpha = 0.9$.

This frequency depends on the ohmic and parallel resistances as well as the CPE parameters. The characteristic frequency can also be expressed in terms of an effective capacitance as

$$f_{c,\text{surf}} = \frac{1}{2\pi\left(\frac{R_e R_{||}}{R_e + R_{||}} C_{\text{eff,surf}}\right)} \tag{14.19}$$

Equations (14.18) and (14.19) yield an expression for the effective capacitance given as in equation (14.4). In the limit that $R_{||}$ becomes much larger than R_e, equation (14.4) becomes

$$C_{\text{eff,surf}} = Q^{1/\alpha} R_e^{(1-\alpha)/\alpha} \tag{14.20}$$

which is equivalent to equation (5) presented by Brug et al.[54] for a blocking electrode.

To illustrate the approach, synthetic data are presented in Figure 14.3 for a system with $R_e = 10\ \Omega\text{cm}^2$, $R_{||} = 10\ \text{k}\Omega\text{cm}^2$, $Q = 10\ \mu\text{F/cm}^2\text{s}^{1-\alpha}$, and $\alpha = 0.9$. The characteristic frequency has a value of about 5 kHz, and the effective resistance $R_{\text{eff}} = R_e R_{||}/(R_e + R_{||})$ has a value of 9.99 Ωcm^2, close to the value of R_e. Equation (14.4) yields a capacitance with a value of 3.59 μF/cm^2, and equation (14.19) yields a capacitance with a value of 3.19 μF/cm^2. The difference between these values may be attributed to the approximate evaluation of the characteristic frequency in Figure 14.2. Note that the value of capacitance obtained is sensitive to the assumption that the CPE is caused by either a surface or a normal distribution of time constants.

14.6 Starting from

$$Z_f(\omega) = g\frac{\delta\rho_\delta^{1/\gamma}}{(\rho_0^{-1} + j\omega\varepsilon\varepsilon_0)^{(\gamma-1)/\gamma}} \tag{14.21}$$

verify equations (14.5) and (14.6).

Solution 14.6. Equation (14.21) was derived from the solution of

$$Z_f(f) = \int_0^\delta \frac{\rho(y)}{1 + j\omega\varepsilon\varepsilon_0\rho(y)}\mathrm{d}y \tag{14.22}$$

for a resistivity distribution given as

$$\frac{\rho}{\rho_\delta} = \left(\frac{\rho_\delta}{\rho_0} + \left(1 - \frac{\rho_\delta}{\rho_0}\right)\xi^\gamma\right)^{-1} \tag{14.23}$$

where ρ_0 and ρ_δ are the boundary values of resistivity at the respective interfaces. Equation (14.21) applies under the conditions that $\rho_0 \gg \rho_\delta$ and $\omega < (\rho_\delta\varepsilon\varepsilon_0)^{-1}$. The variable g is a function of γ.

Equation (14.21) is in the form of a CPE for $\omega > (\rho_0\varepsilon\varepsilon_0)^{-1}$, i.e.,

$$Z_f(\omega) = g\frac{\delta\rho_\delta^{1/\gamma}}{(j\omega\varepsilon\varepsilon_0)^{(\gamma-1)/\gamma}} = \frac{1}{(j\omega)^\alpha Q} \tag{14.24}$$

Thus, for $(\rho_0\varepsilon\varepsilon_0)^{-1} < \omega < (\rho_\delta\varepsilon\varepsilon_0)^{-1}$, equation (14.21) yields the impedance associated with a CPE. Inspection of equation (14.24) suggests that

$$\alpha = \frac{\gamma - 1}{\gamma} \tag{14.25}$$

or $1/\gamma = 1 - \alpha$, where $\gamma \geq 2$ for $0.5 \leq \alpha \leq 1$. Numerical integration was used to develop the interpolation formula

$$g = 1 + 2.88(1 - \alpha)^{2.375} \tag{14.26}$$

A relationship among the CPE parameters Q and α and the dielectric constant ε, resistivity ρ_δ, and film thickness δ can be found to be

$$Q = \frac{(\varepsilon\varepsilon_0)^\alpha}{g\delta\rho_\delta^{1-\alpha}} \tag{14.27}$$

As

$$C = \frac{\varepsilon\varepsilon_0}{\delta} \tag{14.28}$$

equation (14.27) yields an expression for the effective capacitance given as in equation (14.5). Equation (14.6) is obtained by introducing equation (14.28) to eliminate $\varepsilon\varepsilon_0$.

Table 14.1: Regression results for free-machining 18/8 stainless steel in an electrolyte consisting of 22 g/L boric acid with NaOH added to bring the pH to 7.2.[60]

Parameter	As received	After proprietary treatment
R_e, Ωcm^2	15.3	13.3
$R_{\|\|}$, MΩcm^2	2.33	16.8
α	0.91	0.91
Q, μF/s$^{(1-\alpha)}$cm^2	11	30.5
δ by XPS, nm	6.3	2.5

Table 14.2: Calculated results corresponding to application of equation (14.5) to the data presented in Table 14.1.

ρ_δ, Ωcm	ε	g	$C_{\text{eff,PL}}$, μF/cm^2	δ, nm
As received				
500	12	1.0095	1.62	6.5
500	20	1.0095	1.70	10.
10,000	12	1.0095	2.13	5.0
After treatment				
500	12	1.0095	4.50	2.4
500	20	1.0095	4.72	3.8
10,000	12	1.0095	5.90	1.8

Equation (14.5) depends on the CPE parameters Q and α, which would normally be obtained by regression to experimental data, and the dielectric constant ε and the smaller value of the resistivity ρ_δ. Unlike the development for equations (14.3) and (14.4), the characteristic frequency is not invoked, and, in contrast with equation (14.3), the results depend only on the high-frequency data.

14.7 Use the power-law model[60,56] to calculate the film thickness for the data presented in Table 14.1.

(a) Assume that $\varepsilon = 12$ and that $\rho_\delta = 500$ Ωcm.
(b) Assume that $\varepsilon = 20$ and that $\rho_\delta = 500$ Ωcm.
(c) Assume that $\varepsilon = 12$ and that $\rho_\delta = 10,000$ Ωcm.

Solution 14.7. The results presented in Table 14.2 show significant sensitivity

Table 14.3: Calculated results corresponding to application of equation (14.4) to the data presented in Table 14.1.

Condition	ε	$C_{\text{eff,surf}}$, μF/cm^2	δ, nm
As received	12	4.66	2.3
	20	4.66	3.8
After treatment	12	14.09	0.75
	20	14.09	1.3

Table 14.4: Calculated results corresponding to assigning the numerical value for Q to the film capacitance for the data presented in Table 14.1.

Condition	ε	$C_{\text{eff,Q}}$, μF/cm^2	δ, nm	% Error
As received	12	11.00	0.97	85
	20	11.00	1.6	74
After treatment	12	30.50	0.35	86
	20	30.50	0.58	77

to the value of dielectric constant used but much less sensitivity to the value of ρ_δ.

14.8 For the data presented in Table 14.1, use the Brug formula[54] to estimate the film thickness.

Solution 14.8. The solution to this problem requires a value for dielectric constant. The solution presented in Table 14.3 makes use of the values provided in Problem 14.7. The film thicknesses obtained do not agree with the values reported in Table 14.1 obtained by XPS.

14.9 For the data presented in Table 14.1, estimate the error in capacitance obtained if the film capacitance is assigned the numerical value for Q.

Solution 14.9. The solution to this problem requires a value for dielectric constant. The solution presented in Table 14.4 makes use of the values provided in Problem 14.7. The film thicknesses obtained do not agree with the values reported in Table 14.1 obtained by XPS. The average error is on the order of 80 percent.

Table 14.5: Regression results for the data collected on gold and platinum electrodes in deionized water.[61]

Parameter	Au	Pt
R_e, kΩcm^2	5.85$\pm$0.07	4.93$\pm$0.05
Q, μF/s$^{(1-\alpha)}$cm^2	27.6$\pm$0.3	27.6$\pm$0.3
α	0.66$\pm$0.01	0.69$\pm$0.01
$R_{\parallel}$, kΩcm^2	790$\pm$210	750$\pm$170

Table 14.6: Solution to Problem 14.10 providing capacitance from equation (14.4) and characteristic frequency for the data presented in Table 14.5.

Electrode	$C_{\text{eff,surf}}$, μF/cm^2	f_c, Hz
Au	10.75	2.5
Pt	11.23	2.9

14.10 Open-circuit electrochemical impedance measurements were performed on 0.25 mm diameter gold and platinum disk electrodes in deionized water. The frequency range was 54.5 Hz – 46.5 mHz. The resulting regression parameters are provided in Table 14.5.

(a) Use the Brug formula to estimate the capacitance.

(b) Determine the characteristic frequency at which time-constant dispersion may be expected. Comment on the relevance of the value of characteristic frequency.

Solution 14.10. Equation (14.4) is used to calculate the capacitance from equation (14.4). The characteristic frequency at which geometry-induced frequency dispersion may be expected to affect the impedance response can be obtained from

$$f_c = \frac{1}{2\pi C_{\text{eff,surf}} R_e} \tag{14.29}$$

The results are presented in Table 14.6. The results show a capacitance that is of the order expected for a metal–electrolyte interface, and a characteristic frequency that is on the order of 3 Hz.

14.11 A microelectrode with a nominal diameter of 50 μm is considered for use as a sensor. The microelectrode is encased in an insulating sheath such that only the circular face of the wire is exposed to the electrolyte.

(a) Calculate the frequency at which the geometry of the electrode may be expected to influence the impedance response. You may assume that the electrolyte is phosphate buffered saline solution (with a conductivity of 19 mS/cm) and that the capacitance is typical of a metal-electrolyte interface.

(b) Impedance measurements yield CPE behavior with $R_e = 1.33 \times 10^3\ \Omega\text{cm}^2$, $\alpha = 0.91$, and $Q = 7.7 \times 10^{-5}\ \text{F/s}^{1-\alpha}\text{cm}^2$. Estimate the corresponding capacitance under assumption that the electrode is not covered by a film. Is the resulting capacitance consistent with the assumption of a bare metal in contact with an electrolyte?

Solution 14.11.

(a) The frequency at which geometry-induced frequency dispersion is seen for a disk electrode may be estimated from

$$f_c = \frac{\kappa}{2\pi C_0 r_0} \tag{14.30}$$

For a metal–electrolyte interface, as shown in Table A.14, the capacitance value may range from 10 to 50 μF/cm^2. Under assumption that the capacitance is $C_0 = 2 \times 10^{-5}$ F/cm^2, the conductivity is $\kappa = 1.9 \times 10^{-2}$ S/cm, and the disk radius is $r_0 = 2.5 \times 10^{-3}$ cm, the characteristic frequency is obtained from equation (14.30) to be $f_c = 60$ kHz.

(b) In the absence of a film on the electrode surface, the capacitance may be obtained from the formula presented by Brug et al.,[54] i.e., equation (14.4). As no value for a parallel resistance $R_{||}$ was given, we may assume that $R_{||}$ is much larger than R_e. Thus, equation (14.4) becomes equation (14.20). The effective capacitance, calculated from equation (14.20), is $C_{\text{eff,surf}} = 61$ μF/cm^2. This value is somewhat larger than expected for a metal–electrolyte interface, but it is much larger than expected for a film. Thus, the assumption that the electrode is not covered by a film is justified.

Chapter 15

Generalized Transfer Functions

Electrical techniques are extremely efficient for disentangling the coupling between mass-transport and chemical and electrochemical reactions or for performing a test, because they allow in situ study of the electrochemical system. The techniques placed at the electrochemist's disposal are founded on an application of signal processing to electrochemistry. By using a small-amplitude sine-wave perturbation, electrochemical systems can be considered to be linear, and they can be investigated on the basis of a frequency analysis of a transfer function involving at least one electrical quantity (current or potential). Most significant results have been obtained by measurements of the electrochemical impedance, which leads to kinetic characterization of the phenomena in terms of process rates for mass transport, electrochemical reactions, or chemical reactions. Use of nonelectrical quantities has been introduced in impedance spectroscopy, which complements those obtained by measuring the electrochemical impedance. The problems presented in this chapter explore the framework developed by Gabrielli and Tribollet[62] for generalized transfer-function measurements for electrochemical systems. Problems 15.1–15.4 are based on problems presented by Orazem and Tribollet.[1]

Problems

15.1 Derive the steady-state concentration gradient corresponding to

$$j\omega\widetilde{c}_{\mathrm{i}} + \overline{v}_{\mathrm{x}}\frac{\partial \widetilde{c}_{\mathrm{i}}}{\partial x} + \overline{v}_{\mathrm{y}}\frac{\partial \widetilde{c}_{\mathrm{i}}}{\partial y} - D_{\mathrm{i}}\frac{\partial^2 \widetilde{c}_{\mathrm{i}}}{\partial y^2} = -\widetilde{v}_{\mathrm{x}}\frac{\mathrm{d}\overline{c}_{\mathrm{i}}}{\mathrm{d}x} - \widetilde{v}_{\mathrm{y}}\frac{\mathrm{d}\overline{c}_{\mathrm{i}}}{\mathrm{d}y} \qquad (15.1)$$

representing the transient material balance for the small electrode shown in Figure 13.1.

Solution 15.1. Equation (15.1) was obtained by a transient material balance for a two-dimensional electrode. Under assumption of a steady state,

$$\overline{v}_{\mathrm{x}}\frac{\partial\overline{c}_{\mathrm{i}}}{\partial x}+\overline{v}_{\mathrm{y}}\frac{\partial\overline{c}_{\mathrm{i}}}{\partial y}=D_{\mathrm{i}}\frac{\partial^2\overline{c}_{\mathrm{i}}}{\partial y^2} \tag{15.2}$$

The length of the electrode in the direction of mean flow is sufficiently small that the diffusion layer thickness δ_{i} is very small, minimizing the effect of the normal velocity component. The magnitude of the normal velocity component is proportional to y^2; whereas the magnitude of the longitudinal component is proportional to y, where y is the coordinate normal to the wall. Thus, $\overline{v}_{\mathrm{x}}=y\beta_{\mathrm{y}}$, $\overline{v}_{\mathrm{y}}\approx 0$, and

$$y\beta_{\mathrm{y}}\frac{\partial\overline{c}_{\mathrm{i}}}{\partial x}=D_{\mathrm{i}}\frac{\partial^2\overline{c}_{\mathrm{i}}}{\partial y^2} \tag{15.3}$$

The normal distance to the wall can be defined to be

$$\eta=y/\delta_{\mathrm{i}}=y\left(\frac{\beta_{\mathrm{y}}}{D_{\mathrm{i}}x}\right)^{1/3} \tag{15.4}$$

Equation (15.4) is a similarity transformation variable. Upon introduction of equation (15.4), equation (15.3) becomes

$$\frac{\eta^2}{3}\frac{\mathrm{d}\overline{c}_{\mathrm{i}}}{\mathrm{d}\eta}+\frac{\mathrm{d}^2\overline{c}_{\mathrm{i}}}{\mathrm{d}\eta^2}=0 \tag{15.5}$$

which can be solved subject to boundary conditions that $\overline{c}_{\mathrm{i}}=\overline{c}_{\mathrm{i}}(0)$ at $\eta=0$ and $\overline{c}_{\mathrm{i}}=\overline{c}_{\mathrm{i}}(\infty)$ at $\eta=\infty$ (corresponding to $x=0$ or $y=\infty$). Introduction of a new function $\lambda_{\mathrm{i}}=\mathrm{d}\overline{c}_{\mathrm{i}}/\mathrm{d}\eta$ yields a first-order homogeneous equation written as

$$\frac{\eta^2}{3}\lambda_{\mathrm{i}}+\frac{\mathrm{d}\lambda_{\mathrm{i}}}{\mathrm{d}\eta}=0 \tag{15.6}$$

The solution is

$$\lambda_{\mathrm{i}}=\lambda_{\mathrm{i}}(0)\exp\left(-\frac{\eta^3}{9}\right) \tag{15.7}$$

Thus,

$$\frac{\mathrm{d}\overline{c}_{\mathrm{i}}}{\mathrm{d}y}=\frac{\lambda_{\mathrm{i}}(0)}{\delta_{\mathrm{i}}}\exp\left(-\frac{(y/\delta_{\mathrm{i}})^3}{9}\right) \tag{15.8}$$

and

$$\overline{c}_{\mathrm{i}}=\frac{\lambda_{\mathrm{i}}(0)}{\delta_{\mathrm{i}}}\int_0^y\exp\left(-\frac{(y/\delta_{\mathrm{i}})^3}{9}\right)\mathrm{d}y+\overline{c}_{\mathrm{i}}(0) \tag{15.9}$$

Evaluation of equation (15.9) at $y=\infty$ is simplified by observation that

$$\int_0^\infty\exp\left(-\left(\frac{y}{3^{2/3}\delta_{\mathrm{i}}}\right)^3\right)\mathrm{d}\left(\frac{y}{3^{2/3}\delta_{\mathrm{i}}}\right)=\Gamma(4/3) \tag{15.10}$$

Thus,

$$\bar{c}_{\mathrm{i}}(\infty)=\frac{\lambda_{\mathrm{i}}(0)}{\delta_{\mathrm{i}}}3^{2/3}\Gamma(4/3)\delta_{\mathrm{i}}+\bar{c}_{\mathrm{i}}(0) \tag{15.11}$$

Introduction of $\lambda_{\mathrm{i}}=\mathrm{d}\bar{c}_{\mathrm{i}}/\mathrm{d}\eta$ yields

$$\left.\frac{\mathrm{d}\bar{c}_{\mathrm{i}}}{\mathrm{d}y}\right|_0=\frac{\bar{c}_{\mathrm{i}}(\infty)-\bar{c}_{\mathrm{i}}(0)}{\delta_{\mathrm{i}}3^{2/3}\Gamma(4/3)} \tag{15.12}$$

Equation (15.1) was developed to account for the modulated velocity near a rectangular electrode. Equation (15.12) is used in the solution for the associated transfer function.

15.2 Derive the high-frequency solution for a thermoelectrochemical transfer function corresponding to

$$\tilde{i}_{\mathrm{rect}}=L\int_0^{\ell}n\mathrm{F}D_{\mathrm{i}}\left.\frac{\partial\tilde{c}_{\mathrm{i}}}{\partial y}\right|_0\mathrm{d}x \tag{15.13}$$

and, in particular, show that the phase shift is constant and equal to $-135°$.

Solution 15.2. The ac current on a rectangular electrode is given by equation (15.13). The general expression for the material balance is given by

$$\mathrm{j}\mathrm{K_x}\tilde{c}_{\mathrm{i}}+\frac{4\xi\mathrm{K_x}}{9}\frac{\partial\tilde{c}_{\mathrm{i}}}{\partial\mathrm{K_x}}-\frac{\xi^2}{3}\frac{\partial\tilde{c}_{\mathrm{i}}}{\partial\xi}-\frac{\partial^2\tilde{c}_{\mathrm{i}}}{\partial\xi^2}=\frac{\xi^2}{3}\frac{\mathrm{d}\bar{c}_{\mathrm{i}}}{\mathrm{d}\xi}\frac{\tilde{\mathcal{A}}}{\bar{\bar{\mathcal{A}}}} \tag{15.14}$$

where $\mathcal{A}$ is a temperature-dependent coefficient used in the development of thermoelectrochemical impedance spectroscopy. In the high-frequency range, the temperature perturbation is rapidly damped close to the wall; thus, the convective term can be disregarded, and equation (15.14) reduces to

$$\mathrm{j}\mathrm{K_x}\tilde{c}_{\mathrm{i}}-\frac{\partial^2\tilde{c}_{\mathrm{i}}}{\partial\xi^2}=\frac{\xi^2}{3}\frac{\mathrm{d}\bar{c}_{\mathrm{i}}}{\mathrm{d}\xi}\frac{\tilde{\mathcal{A}}}{\bar{\bar{\mathcal{A}}}} \tag{15.15}$$

The solution of the homogeneous equation is

$$\theta=\exp\left(-\sqrt{jK_x}\xi\right) \tag{15.16}$$

As the frequency is large, the distance over which a temperature perturbation has an influence is small, and the gradient $\mathrm{d}\bar{c}_{\mathrm{i}}/\mathrm{d}\xi$ can be considered to be equal to the gradient at the wall at $\xi=0$. Thus, the heterogeneous equation can be written as

$$\left(-\frac{\mathrm{d}^2\Phi}{\mathrm{d}\xi^2}+2\frac{\mathrm{d}\Phi}{\mathrm{d}\xi}(\mathrm{j}K_x)^{1/2}\right)\theta=\frac{\xi^2}{3}\left.\frac{\mathrm{d}\bar{c}_i}{\mathrm{d}\xi}\right|_0\frac{\tilde{\mathcal{A}}}{\bar{\bar{\mathcal{A}}}} \tag{15.17}$$

where

$$\widetilde{c}_i = \theta\Phi \tag{15.18}$$

With $\lambda = d\Phi/d\xi$, the solution of the homogeneous equation corresponding to equation (15.17) is

$$\lambda = \exp\left(2(jK_x)^{1/2}\xi\right) = \theta^{-2} \tag{15.19}$$

With $\lambda = f\theta^{-2}$, the heterogeneous equation becomes

$$\frac{df}{d\xi}\theta^{-2} = \frac{\xi^2}{3}\left.\frac{d\bar{c}_i}{d\xi}\right|_0 \left(\frac{\widetilde{A}}{\overline{\overline{A}}}\right)\theta^{-1} \tag{15.20}$$

with the solution

$$f = \int_0^{\xi'} \frac{\xi''^2}{3}\left.\frac{d\bar{c}_i}{d\xi}\right|_0 \left(\frac{\widetilde{A}}{\overline{\overline{A}}}\right)\theta d\xi'' + K_1 \tag{15.21}$$

where ξ' and ξ'' are integration variables. Following equation (15.18), the concentration phasor is given by

$$\widetilde{c}_i = \theta\left[\int_0^{\xi}\theta^{-2}\left(\int_0^{\xi'}\frac{\xi''^2}{3}\left.\frac{d\bar{c}_i}{d\xi}\right|_0\left(\frac{\widetilde{A}}{\overline{\overline{A}}}\right)\theta d\xi'' + K_1\right)d\xi'\right] + \widetilde{c}_i(0) \tag{15.22}$$

The boundary condition for the concentration phasor is that $\widetilde{c}_i \to 0$ for $\xi \to \infty$. Thus,

$$K_1 = \frac{1}{3}\left.\frac{d\bar{c}_i}{d\xi}\right|_0\left(\frac{\widetilde{A}}{\overline{\overline{A}}}\right)\int_0^{\infty}\xi^2\theta(\xi)d\xi \tag{15.23}$$

or

$$K_1 = \frac{1}{3}\left.\frac{d\bar{c}_i}{d\xi}\right|_0\left(\frac{\widetilde{A}}{\overline{\overline{A}}}\right)\frac{1}{(jK_x)^{3/2}}\int_0^{\infty}X^2\exp(-X)dX \tag{15.24}$$

where $X = (jK_x)^{1/2}\xi$.

$$\left.\frac{d\widetilde{c}_i}{d\xi}\right|_0 = K_1 \tag{15.25}$$

As shown in equation (15.24), the transfer function

$$\frac{\left.\frac{d\widetilde{c}_i}{d\xi}\right|_0}{\widetilde{A}} \propto j^{-3/2} \tag{15.26}$$

where $j^{-3/2}$ corresponds to a phase angle of $-135°$.

15.3 Derive the relationship among the thermoelectrochemical transfer function measured under potentiostatic regulation, the thermoelectrochemical transfer function measured under galvanostatic regulation, and the electrochemical impedance.

Solution 15.3. The thermoelectrochemical transfer function measured under potentiostatic regulation is defined as

$$Z_{\mathrm{I,T}} = \frac{\widetilde{I}}{\widetilde{T}} \tag{15.27}$$

The thermoelectrochemical transfer function measured under galvanostatic regulation is defined as

$$Z_{\mathrm{U,T}} = \frac{\widetilde{U}}{\widetilde{T}} \tag{15.28}$$

The current phasor can be written as

$$\widetilde{I} = Z^{-1}\widetilde{U} + Z_{\mathrm{I,T}}\widetilde{T} \tag{15.29}$$

where Z is the electrochemical impedance. The potential phasor can be written as

$$\widetilde{U} = Z\widetilde{I} + Z_{\mathrm{U,T}}\widetilde{T} \tag{15.30}$$

As equations (15.29) and (15.30) are both satisfied,

$$\widetilde{I} = Z^{-1}\left(Z\widetilde{I} + Z_{\mathrm{U,T}}\widetilde{T}\right) + Z_{\mathrm{I,T}}\widetilde{T} \tag{15.31}$$

is obtained by replacing $\widetilde{U}$ in equation (15.29) with equation (15.30). Equation (15.31) may be written as

$$(Z^{-1}Z_{\mathrm{U,T}} + Z_{\mathrm{I,T}})\widetilde{T} = 0 \tag{15.32}$$

Thus,

$$Z^{-1}Z_{\mathrm{U,T}} + Z_{\mathrm{I,T}} = 0 \tag{15.33}$$

provides the relationship among the thermoelectrochemical transfer function measured under potentiostatic regulation $Z_{\mathrm{U,T}}$, the thermoelectrochemical transfer function measured under galvanostatic regulation $Z_{\mathrm{I,T}}$, and the electrochemical impedance Z.

15.4 Derive the equation relating the electrogravimetric transfer function measured under potentiostatic regulation, the electrogravimetric transfer function measured under galvanostatic regulation, and the electrochemical impedance.

Solution 15.4. Under galvanostatic regulation, the phasor associated with the mass response is given by

$$\widetilde{M} = Z_{\mathrm{M,I}}\widetilde{I} \tag{15.34}$$

where $Z_{M,I}$ is the electrogravimetric transfer function measured under galvanostatic regulation. Under potentiostatic regulation, the phasor associated with the mass response is given by

$$\widetilde{M} = Z_{M,U}\widetilde{U} \tag{15.35}$$

where $Z_{M,U}$ is the electrogravimetric transfer function measured under potentiostatic regulation. As

$$Z = \frac{\widetilde{U}}{\widetilde{I}} = \frac{Z_{M,I}}{Z_{M,U}} \tag{15.36}$$

then

$$Z^{-1}Z_{M,I} - Z_{M,U} = 0 \tag{15.37}$$

provides the relationship among the electrogravimetric transfer function measured under potentiostatic regulation $Z_{M,U}$, the electrogravimetric transfer function measured under galvanostatic regulation $Z_{M,I}$, and the electrochemical impedance Z.

Chapter 16

Electrohydrodynamic Impedance

Electrohydrodynamic (EHD) impedance provides a practical example of a generalized transfer function involving nonelectrical quantities. In this chapter, a rotating disk electrode is considered. The current is a function of the rotation speed, which means that the current is totally or partially limited by mass transport. The problems presented in this chapter explore applications of electrohydrodynamic impedance. Problems 16.1–16.3 are based on problems presented by Orazem and Tribollet.[1]

Problems

16.1 From

$$\left.\frac{\mathrm{d}\widetilde{c}_{\mathrm{i}}}{\mathrm{d}y}\right|_{y=0} = \frac{\widetilde{c}_{\mathrm{i}}(0)}{\delta_{\mathrm{i}}}\theta_{\mathrm{i}}'(0) + \frac{\Delta\Omega}{\overline{\Omega}}\left.\frac{\mathrm{d}\overline{c}_{\mathrm{i}}}{\mathrm{d}y}\right|_{y=0} W_{\mathrm{i}} \tag{16.1}$$

deduce the expressions

$$Z_{U,\mathrm{m}}\, Z^{-1} + Z_{I,\mathrm{m}} = 0 \tag{16.2}$$

and

$$\frac{\widetilde{U}}{\widetilde{W}_{\mathrm{m}}} = -\frac{\widetilde{U}}{\widetilde{I}}\frac{\widetilde{I}}{\widetilde{W}_{\mathrm{m}}} \tag{16.3}$$

Solution 16.1. The relationship between the concentration phasor and the concentration phasor derivative, both evaluated at the electrode surface, is expressed in terms of the dimensional distance y by equation (16.1), where $\theta_{\mathrm{i}} = \widetilde{c}_{\mathrm{i}}/\widetilde{c}_{\mathrm{i}}(0)$, $-1/\theta_{\mathrm{i}}'(0)$ is the dimensionless convective diffusion impedance, $\Delta\Omega$ is the perturbation amplitude for the modulated rotating disk electrode, $\overline{\Omega}$ is the steady rotation speed for the disk electrode, and W_{i} is a complex function that is derived from the convective diffusion equations for a modulated rotation speed. For electrohydrodynamic impedance under potentiostatic regulation, i.e., at a fixed potential, the concentration at the wall of the electrode is constant and $\tilde{c}_i(0) = 0$. Equation (16.1) becomes

$$\left.\frac{d\tilde{c}_i}{dy}\right|_{y=0} = \frac{\Delta\Omega}{\overline{\Omega}}\left.\frac{d\bar{c}_i}{dy}\right|_{y=0} W_i \tag{16.4}$$

or

$$\frac{\left.\frac{d\tilde{c}_i}{dy}\right|_{y=0}}{\Delta\Omega} = \frac{1}{\overline{\Omega}}\left.\frac{d\bar{c}_i}{dy}\right|_{y=0} W_i = Z_{c,\Omega} \tag{16.5}$$

For electrohydrodynamic impedance under galvanostatic regulation, i.e., at a fixed current, the concentration gradient at the wall is constant and $d\tilde{c}_i/dy|_0 = 0$. Equation (16.1) becomes

$$\tilde{c}_i(0) = \frac{\Delta\Omega}{\overline{\Omega}}\left.\frac{d\bar{c}_i}{dy}\right|_0 W_i \frac{1}{\delta_i}\left(\frac{-1}{\theta'(0)}\right) \tag{16.6}$$

or

$$\frac{\tilde{c}_i(0)}{\Delta\Omega} = \frac{1}{\overline{\Omega}}\left.\frac{d\bar{c}_i}{dy}\right|_0 W_i \frac{1}{\delta_i}\left(\frac{-1}{\theta'(0)}\right) = Z_{dc/dy,\Omega} \tag{16.7}$$

Combination of equations (16.5) and (16.7) yields

$$Z_{dc/dy,\Omega} = Z_{c,\Omega}\frac{1}{\delta_i}\left(\frac{-1}{\theta'(0)}\right) \tag{16.8}$$

or

$$\frac{\tilde{c}_i(0)}{\Delta\Omega} = \frac{\left.\frac{d\tilde{c}_i}{dy}\right|_{y=0}}{\Delta\Omega}\frac{1}{\delta_i}\left(\frac{-1}{\theta'(0)}\right) \tag{16.9}$$

16.2 Derive the EHD impedance for the anodic dissolution of copper in a chloride medium, where the reactions proceed according to

$$Cu + Cl^- \rightleftarrows CuCl_{ads} + e^- \tag{16.10}$$

where $CuCl_{ads}$ is an adsorbed intermediate that reacts to form $CuCl_2^-$ by

$$CuCl_{ads} + Cl^- \rightleftarrows CuCl_2^- \tag{16.11}$$

The mass-transfer limitation is due only to $CuCl_2^-$.

Solution 16.2. The faradaic current may be written as

$$\frac{i_{F,Cu}}{F} = K_1^* c_{Cl^-}\Gamma(1-\gamma)\exp(bV) - K_{-1}\Gamma\gamma\exp(-bV) \tag{16.12}$$

where γ is the fractional coverage of CuCl, Γ is the maximum surface coverage by CuCl in units of moles per unit area, $b = \alpha F/RT$, and $V = \Phi_m - \Phi_0$ is the local potential driving force for the reaction. Application of a conservation equation for the adsorbed intermediate yields

$$\Gamma\frac{d\gamma}{dt} = K_1 c_{Cl^-}(1-\gamma)\exp(bV) - K_{-1}\Gamma\gamma\exp(-bV) - k_f\Gamma\gamma c_{Cl^-} + k_b(1-\gamma)c_{CuCl_2^-}(0) \tag{16.13}$$

Under the assumption that $\gamma \ll 1$, equations (16.12) and (16.13) can be written

$$\frac{i_{F,Cu}}{F} = K_1 c_{Cl^-}\exp(bV) - K_{-1}\Gamma\gamma\exp(-bV) \tag{16.14}$$

and

$$\Gamma\frac{d\gamma}{dt} = K_1 c_{Cl^-}\exp(bV) - K_{-1}\Gamma\gamma\exp(-bV) - k_f\Gamma\gamma c_{Cl^-} + k_b c_{CuCl_2^-}(0) \tag{16.15}$$

respectively. In equation (16.14), $K_1 = K_1^*\Gamma$.

Consider EHD impedance under potentiostatic regulation, where $\widetilde{V} = 0$. Under a sinusoidal perturbation of the rotation speed, equation (16.14) yields

$$\frac{\widetilde{i}_{F,Cu}}{F} = -K_{-1}\Gamma\exp(-b\overline{V})\widetilde{\gamma} \tag{16.16}$$

and equation (16.15) yields

$$j\omega\Gamma\widetilde{\gamma} = \left(-K_{-1}\Gamma\exp(-b\overline{V}) - k_f\Gamma c_{Cl^-}\right)\widetilde{\gamma} + k_b\widetilde{c}_{CuCl_2^-}(0) \tag{16.17}$$

The reaction stoichiometry yields an expression for the flux of $CuCl_2^-$

$$D_{CuCl_2^-}\left.\frac{d\widetilde{c}_{CuCl_2^-}}{dy}\right|_0 = k_f c_{Cl^-}\Gamma\widetilde{\gamma} - k_b\widetilde{c}_{CuCl_2^-}(0) \tag{16.18}$$

The general relationship between the concentration phasor and the concentration phasor derivative, equation (16.1), is expressed as

$$\left.\frac{d\widetilde{c}_{CuCl_2^-}}{dy}\right|_{y=0} = \frac{\widetilde{c}_{CuCl_2^-}(0)}{\delta_{CuCl_2^-}}\theta'_{CuCl_2^-}(0) + \frac{\Delta\Omega}{\overline{\Omega}}\left.\frac{d\overline{c}_{CuCl_2^-}}{dy}\right|_{y=0} W_{CuCl_2^-} \tag{16.19}$$

From equation (16.17),

$$\widetilde{\gamma} = \frac{k_b}{j\omega\Gamma + K_{-1}\Gamma\exp(-b\overline{V}) + k_f\Gamma c_{Cl^-}}\widetilde{c}_{CuCl_2^-}(0) \tag{16.20}$$

From equations (16.18) and (16.19),

$$\frac{k_f c_{Cl^-}\Gamma\widetilde{\gamma} - k_b\widetilde{c}_{CuCl_2^-}(0)}{D_{CuCl_2^-}} = \frac{\widetilde{c}_{CuCl_2^-}(0)}{\delta_{CuCl_2^-}}\theta'_{CuCl_2^-}(0) + \frac{\Delta\Omega}{\overline{\Omega}}\left.\frac{d\overline{c}_{CuCl_2^-}}{dy}\right|_{y=0} W_{CuCl_2^-} \tag{16.21}$$

or

$$k_f c_{Cl^-} \Gamma \widetilde{\gamma} = \left(k_b + D_{CuCl_2^-} \frac{\theta'_{CuCl_2^-}(0)}{\delta_{CuCl_2^-}} \right) \widetilde{c}_{CuCl_2^-}(0) + D_{CuCl_2^-} \frac{\Delta\Omega}{\overline{\Omega}} \left. \frac{d\overline{c}_{CuCl_2^-}}{dy} \right|_{y=0} W_{CuCl_2^-} \tag{16.22}$$

Introduction of $\widetilde{c}_{CuCl_2^-}(0)$ from equation (16.20) yields

$$k_f c_{Cl^-} \Gamma \widetilde{\gamma} = \left(k_b + D_{CuCl_2^-} \frac{\theta'_{CuCl_2^-}(0)}{\delta_{CuCl_2^-}} \right) \frac{j\omega\Gamma + K_{-1}\Gamma \exp\left(-b\overline{V}\right) + k_f \Gamma c_{Cl^-}}{k_b} \widetilde{\gamma} + D_{CuCl_2^-} \frac{\Delta\Omega}{\overline{\Omega}} \left. \frac{d\overline{c}_{CuCl_2^-}}{dy} \right|_{y=0} W_{CuCl_2^-} \tag{16.23}$$

and

$$\widetilde{\gamma} = \frac{D_{CuCl_2^-} \frac{\Delta\Omega}{\overline{\Omega}} \left. \frac{d\overline{c}_{CuCl_2^-}}{dy} \right|_0 W_{CuCl_2^-}}{k_f c_{Cl^-} \Gamma - \left(k_b + D_{CuCl_2^-} \frac{\theta'_{CuCl_2^-}}{\delta_{CuCl_2^-}} \right) \frac{j\omega\Gamma + K_{-1}\Gamma \exp\left(-b\overline{V}\right) + k_f \Gamma c_{Cl^-}}{k_b}} \tag{16.24}$$

Introduction of equation (16.24) into equation (16.16) yields

$$\frac{\widetilde{i}_{F,Cu}}{F} = \frac{-K_{-1}\Gamma \exp\left(-b\overline{V}\right) D_{CuCl_2^-} \frac{\Delta\Omega}{\overline{\Omega}} \left. \frac{d\overline{c}_{CuCl_2^-}}{dy} \right|_0 W_{CuCl_2^-}}{k_f c_{Cl^-} \Gamma - \left(k_b + D_{CuCl_2^-} \frac{\theta'_{CuCl_2^-}}{\delta_{CuCl_2^-}} \right) \frac{j\omega\Gamma + K_{-1}\Gamma \exp\left(-b\overline{V}\right) + k_f \Gamma c_{Cl^-}}{k_b}} \tag{16.25}$$

and the electrohydrodynamic impedance may be expressed as

$$\frac{\widetilde{i}_{F,Cu}}{\Delta\Omega} = \frac{-FK_{-1}\Gamma \exp\left(-b\overline{V}\right) D_{CuCl_2^-} \frac{1}{\overline{\Omega}} \left. \frac{d\overline{c}_{CuCl_2^-}}{dy} \right|_0 W_{CuCl_2^-}}{k_f c_{Cl^-} \Gamma - \left(k_b + D_{CuCl_2^-} \frac{\theta'_{CuCl_2^-}}{\delta_{CuCl_2^-}} \right) \frac{j\omega\Gamma + K_{-1}\Gamma \exp\left(-b\overline{V}\right) + k_f \Gamma c_{Cl^-}}{k_b}} \tag{16.26}$$

The EHD impedance generally corresponds to a low-frequency range. If

$$j\omega\Gamma \ll K_{-1}\Gamma \exp\left(-b\overline{V}\right) + k_f \Gamma c_{Cl^-} \tag{16.27}$$

then $\widetilde{i}_{F,Cu}/\Delta\Omega$ is proportional to $W_{CuCl_2^-}$, which is the EHD impedance for $CuCl_2^-$. This result is in agreement with the experimental observations of Barcia et al.[34]

16.3 Beginning with

$$\widetilde{U} = Z\widetilde{I} + \frac{\Delta\Omega}{\overline{\Omega}} \frac{1}{1 + j\omega C_D(R_t + Z_D)} \frac{\overline{I}R_t}{nF} \sum_i \left.\frac{\partial f}{\partial c_i(0)}\right|_{E,c_j(0),j\neq i} \frac{W_i \delta_i}{\theta_i'(0)} \frac{s_i}{D_i} \tag{16.28}$$

write an expression for the electrohydrodynamic impedance under potentiostatic regulation $Z_{EHD,p}$, the electrohydrodynamic impedance under galvanostatic regulation $Z_{EHD,g}$, and the relationship between $Z_{EHD,p}$, $Z_{EHD,g}$, and the electrochemical impedance.

Solution 16.3. The general relationship between potential $\widetilde{U}$, current $\widetilde{I}$, and rotation speed is given by equation (16.28). For electrohydrodynamic impedance under potentiostatic regulation, the potential is held constant and $\widetilde{U} = 0$. Equation (16.28) becomes

$$\begin{aligned} Z_{EHD,p} &= \frac{\widetilde{I}}{\Delta\Omega} \\ &= \frac{1}{Z} \frac{1}{\overline{\Omega}} \frac{1}{1 + j\omega C_D(R_t + Z_D)} \frac{\overline{I}R_t}{nF} \sum_i \left.\frac{\partial f}{\partial c_i(0)}\right|_{E,c_j(0),j\neq i} \frac{W_i \delta_i s_i}{D_i} \left(\frac{-1}{\theta_i'(0)}\right) \end{aligned} \tag{16.29}$$

Under galvanostatic regulation, the current is maintained at a constant value, and $\widetilde{I} = 0$. Equation (16.28) becomes

$$\begin{aligned} Z_{EHD,g} &= \frac{\widetilde{U}}{\Delta\Omega} \\ &= \frac{1}{\overline{\Omega}} \frac{1}{1 + j\omega C_D(R_t + Z_D)} \frac{\overline{I}R_t}{nF} \sum_i \left.\frac{\partial f}{\partial c_i(0)}\right|_{E,c_j(0),j\neq i} \frac{W_i \delta_i}{\theta_i'(0)} \frac{s_i}{D_i} \end{aligned} \tag{16.30}$$

Note that the units for the electrohydrodynamic impedance depend on the mode of regulation. Under galvanostatic regulation, the units of $Z_{EHD,g}$ may be expressed as V/rpm; whereas under potentiostatic regulation, the units of $Z_{EHD,p}$ may be expressed as A/rpm.

Chapter 17

Methods for Representing Impedance

Impedance data are presented in different formats to emphasize specific classes of behavior. The impedance format emphasizes the values at low frequency, which typically are of greatest importance for electrochemical systems that are influenced by mass transfer and reaction kinetics. The admittance format, which emphasizes the capacitive behavior at high frequencies, is often employed for solid-state systems. The complex capacity format is used for dielectric systems in which the capacity is often the feature of greatest interest. The method of representing impedance data has great impact on the use of graphical methods to visualize and interpret data. The problems presented in this chapter address different ways to represent impedance data. Problems 17.1–17.5 are based on problems presented by Orazem and Tribollet.[1]

Problems

17.1 Consider the electrical circuit given in Figure 4.1(a). Use a spreadsheet program to plot the results for $R_0 = 10\ \Omega$, $R_1 = 50\ \Omega$, $C_1 = 20\ \mu\text{F}$, $R_2 = 500\ \Omega$, and $C_2 = 10\ \mu\text{F}$ for the following representations:

(a) impedance in Nyquist, Bode, and ohmic-resistance-corrected Bode plots.

(b) admittance in Nyquist, real, and imaginary plots.

(c) complex capacitance in a Cole–Cole plot.

(d) effective capacitance as a function of frequency.

Solution 17.1.

(a) The impedance for the electrical circuit given in Figure 4.1(a) may be expressed

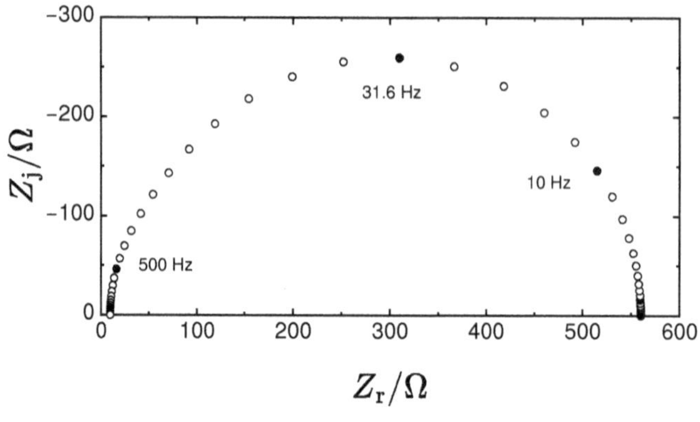

Figure 17.1: Nyquist plot for the impedance of the electrical circuit given in Figure 4.1(a).

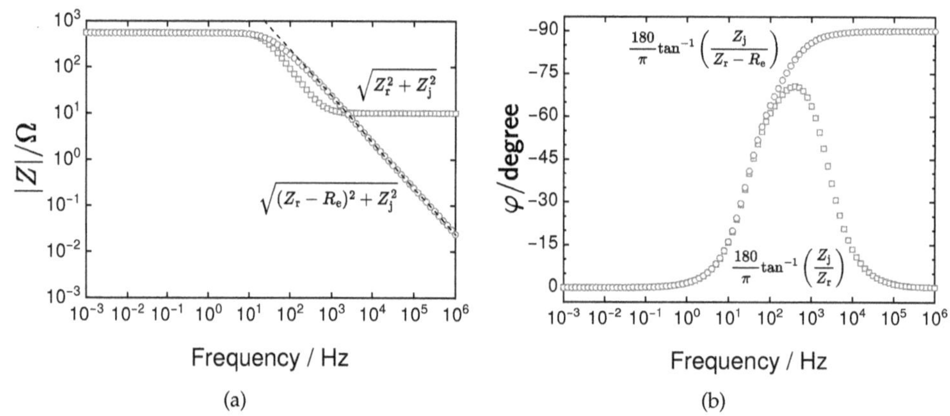

Figure 17.2: Impedance in Bode format of the electrical circuit given in Figure 4.1(a): (a) magnitude and (b) phase. The dashed line in (a) has a slope of −1.

as

$$Z = R_0 + \frac{R_1}{1 + j\omega R_1 C_1} + \frac{R_2}{1 + j\omega R_2 C_2} \tag{17.1}$$

The corresponding Nyquist plot is given in Figure 17.1 Even though the impedance corresponds to two RC elements, the Nyquist plot does not show discrete semicircles. The characteristic frequencies $f_c = 1/(2\pi RC)$ are 318 and 15.9 Hz. These are sufficiently separated to broaden the capacitive loop, giving the appearance of a slightly depressed semicircle.

The impedance is shown in both ohmic-resistance-corrected and non-corrected Bode format in Figure 17.2. Without correction, the magnitude shown in Figure 17.2(a) approaches the ohmic resistance $R_0 = 10\ \Omega$; whereas correction for the ohmic resistance yields a line with a well-defined slope equal, in this case, to −1, as shown by the dashed line. The uncorrected phase angle shown in Fig-

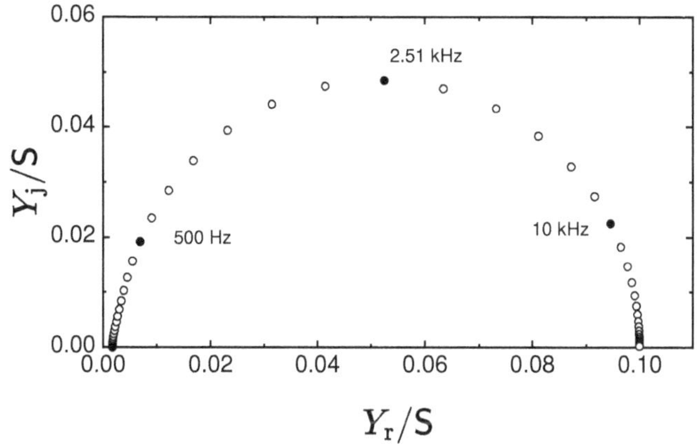

Figure 17.3: Nyquist plot for the admittance of the electrical circuit given in Figure 4.1(a).

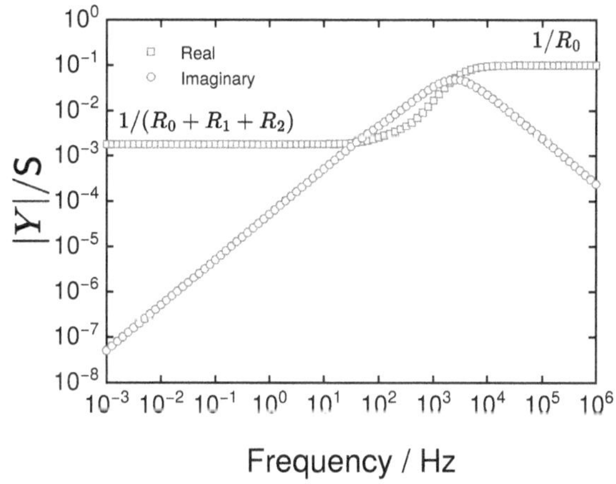

Figure 17.4: The admittance of the electrical circuit given in Figure 4.1(a).

ure 17.2(b) approaches zero at high frequency because the interfacial impedance tends toward zero, and the overall impedance is dominated by the ohmic resistance. Correction for the ohmic resistance yields a phase angle equal to $-90°$ at high frequency, as would be expected for the capacitive behavior of the circuit.

(b) The Nyquist plot of the admittance $Y = 1/Z$, given in Figure 17.3, appears as a loop with lower frequencies at the origin and larger frequencies at larger values of admittance. The unit of admittance is the Siemens (S). The admittance corresponding to Figure 17.3 is given in Figure 17.4. The real part has asymptotic values $1/R_0$ at high frequency and $1/(R_0 + R_1 + R_2)$ at low frequency.

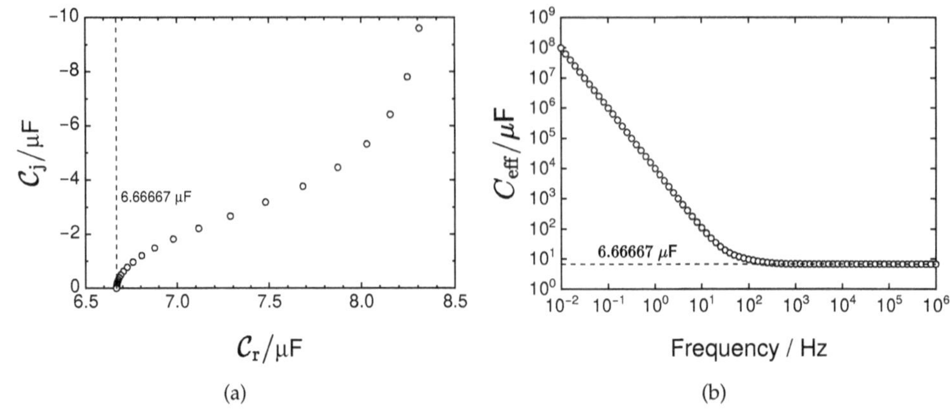

Figure 17.5: Calculated capacitance for the electrical circuit given in Figure 4.1(a): (a) Cole–Cole plot for the complex capacitance and b effective capacitance calculated from equation (17.4).

(c) The complex capacitance, given as

$$\mathcal{C} = \frac{1}{j\omega(Z - R_e)} \tag{17.2}$$

is presented in Figure 17.5(a). The high-frequency part of the complex capacitance, highlighted in Figure 17.5(a), approaches an asymptotic value

$$\lim_{f\to\infty} \mathcal{C}_r = \frac{C_1 C_2}{C_1 + C_2} \tag{17.3}$$

which, for the given values of capacitance, has a value of 6.66667 μF.

(d) The effective capacitance, calculated as

$$C_{eff} = \frac{1}{j\omega Z_j} \tag{17.4}$$

is presented in Figure 17.5(b) as a function of frequency. At low frequencies, equation (17.4) yields an incorrect value for capacitance, but the capacitance at high frequencies approaches the value obtained from equation (17.3).

17.2 Use a spreadsheet program to plot the results of Problem 10.7 for the following representations:

(a) impedance in Nyquist, Bode, and ohmic-resistance-corrected Bode plots.

(b) admittance in Nyquist, real, and imaginary plots.

(c) complex capacitance in a Cole–Cole plot.

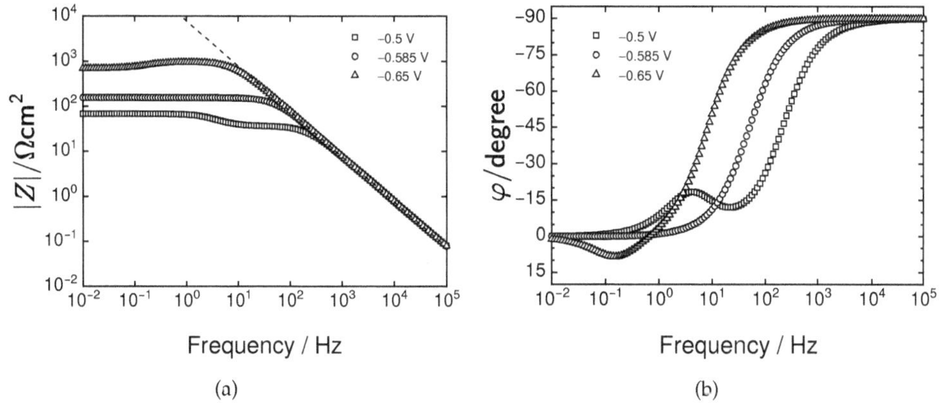

Figure 17.6: Impedance in Bode format for the results of Problem 10.7: (a) magnitude and (b) phase. The dashed line in (a) has a slope of -1.

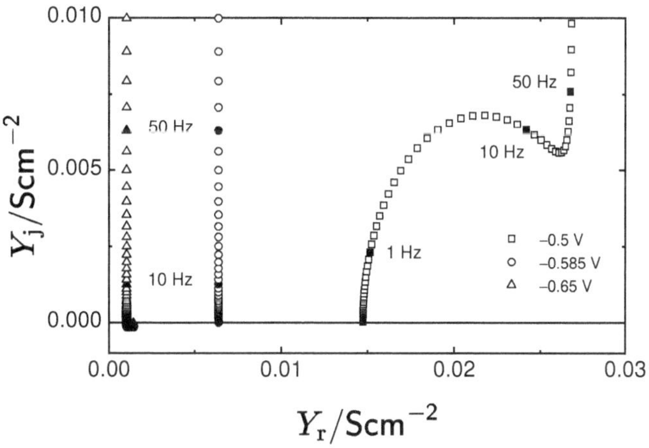

Figure 17.7: The admittance in Nyquist format for the results of Problem 10.7.

Solution 17.2.

(a) The Nyquist plots corresponding to Problem 10.7 are given in Figure 10.4. The Bode plots are presented in Figure 17.6. As no ohmic resistance was used in the calculations for Problem 10.7, the ohmic-resistance-corrected Bode plots are identical to that shown in Figure 17.6.

(b) The admittance $Y = 1/Z$ is presented in Figure 17.7. While the Nyquist plots for impedance are shown in a $1:1$ aspect ratio in Figure 10.4. The aspect ratio is not $1:1$ in Figure 17.7 to enable presentation of the admittance for all three potentials on the same plot. The real and imaginary parts of the admittance are presented

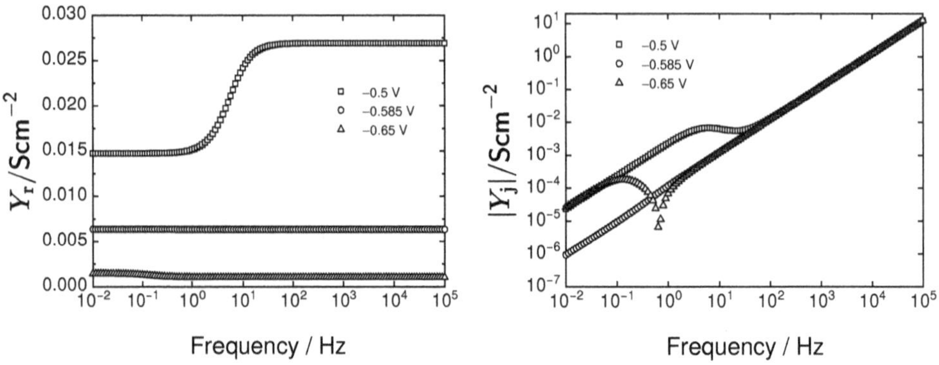

(a) (b)

Figure 17.8: Admittance as a function of frequency for the results of Problem 10.7: (a) real and (b) imaginary.

in Figure 17.8 as functions of frequency. The absolute value of the imaginary part is shown in log format in Figure 17.8(b) to emphasize differences that are not apparent in a linear format.

(c) The complex capacitance, given in equation (17.2), is presented in Figure 17.9. The high-frequency part of the plot is emphasized to show the asymptotic approach to a common real part of the capacitance, with a value of 20 μF/cm^2.

17.3 Consider the equation used for a so-called constant-phase element, e.g.,

$$Z(\omega) = R_e + \frac{R_t}{1 + (j\omega)^{\alpha} R_t Q} \tag{17.5}$$

Plot the model results in Nyquist, Bode, impedance, and admittance formats, letting $R_e = 10\ \Omega\text{cm}^2$, $R_t = 100\ \Omega\text{cm}^2$, and $Q = 20\ \mu\text{Fs}^{\alpha-1}/\text{cm}^2$. Allow α to be a parameter such that $1 > \alpha > 0.5$.

Solution 17.3. The impedance for $\alpha = 1$ and 0.9 was calculated over a range from 10 mHz to 10 kHz with 10 frequencies per decade. The impedance for $\alpha = 0.7$ and 0.5 was calculated over a range from 10 mHz to 10 MHz with 5 frequencies per decade. The resulting Nyquist plots are shown in Figure 17.10 with α as a parameter. Asymptotic values are given as R_e at high frequency and $(R_e + R_t)$ at low frequency. The characteristic frequency changes with α.

The real and imaginary parts of the impedance are presented in Figures 17.11(a) and 17.11(b), respectively. The high- and low-frequency limits for the real part of the impedance are given as dashed lines in Figure 17.11(a). The characteristic frequency

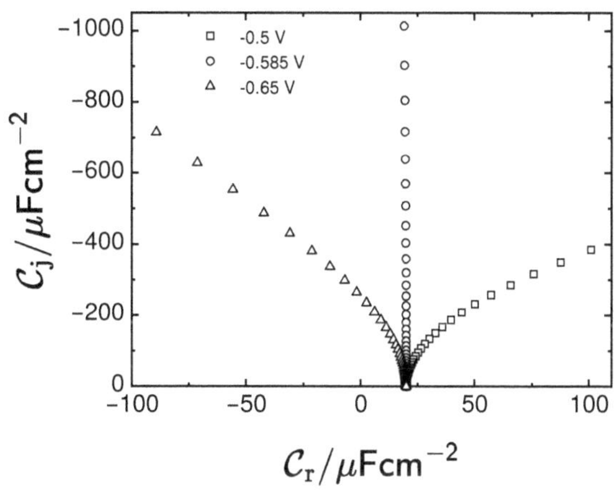

Figure 17.9: Cole–Cole plot for the complex capacitance for the results of Problem 10.7. The high-frequency part of the plot is emphasized to show the asymptotic approach to a common real part of the capacitance, with a value of 20 μF/cm^2.

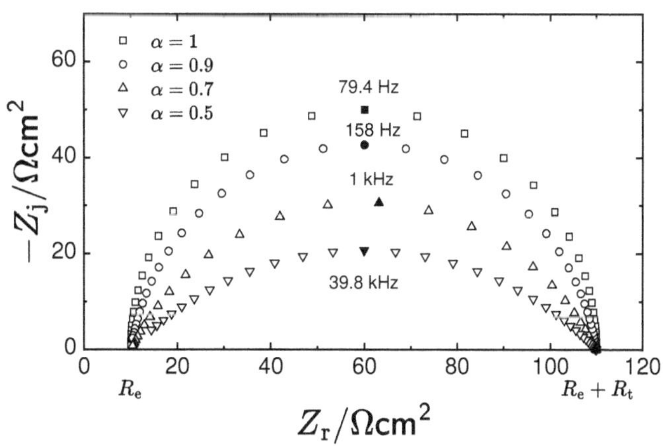

Figure 17.10: Nyquist plot for the complex impedance for the results of Problem 17.3 with α as a parameter. The characteristic frequency changes with α.

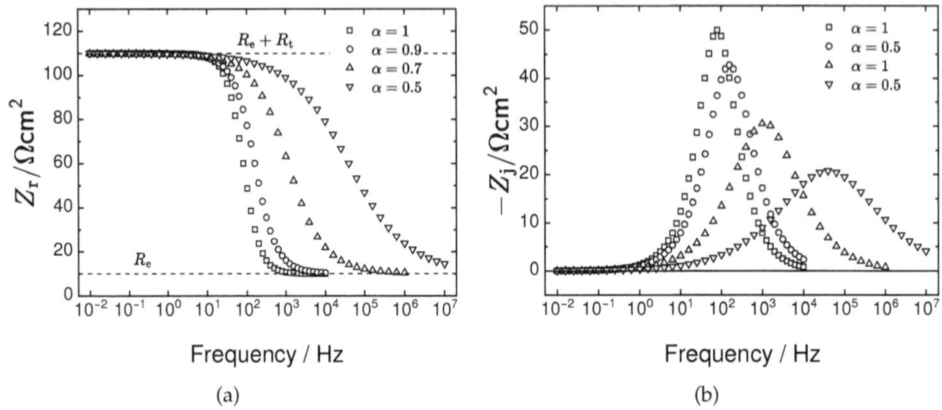

Figure 17.11: Impedance as a function of frequency for the results of Problem 17.3 with α as a parameter: (a) real part of the impedance and (b) imaginary part of the impedance.

corresponding to the peak in Figure 17.11(b) changes with frequency, in accordance with the behavior shown in Figure 17.10.

The impedance is presented in Bode format in Figure 17.12. in two representations. The magnitude

$$|Z| = \sqrt{Z_r^2 + Z_j^2} \tag{17.6}$$

is also represented as an ohmic-resistance-corrected magnitude

$$|Z|_{adj} = \sqrt{(Z_r - R_e)^2 + Z_j^2} \tag{17.7}$$

The phase angle

$$\varphi = \tan^{-1}\left(\frac{Z_j}{Z_r}\right) \tag{17.8}$$

is represented as an ohmic-resistance-corrected phase angle as

$$\varphi_{adj} = \tan^{-1}\left(\frac{Z_j}{Z_r - R_e}\right) \tag{17.9}$$

Open symbols in Figure 17.12 represent the ohmic-resistance-corrected magnitude and phase. The slope of the ohmic-resistance-corrected magnitude at high frequency in Figure 17.12(a) is equal to $-\alpha$, and the high-frequency limit of the ohmic-resistance-corrected phase angle in Figure 17.12(b) is $-\alpha 90°$.

The admittance, given by $Y = 1/Z$, is presented in Nyquist format in Figure 17.13. As with the impedance shown in Figure 17.10, the characteristic frequency of the admittance changes with α. Asymptotic values are given as $1/R_e$ at high frequency and $1/(R_e + R_t)$ at low frequency.

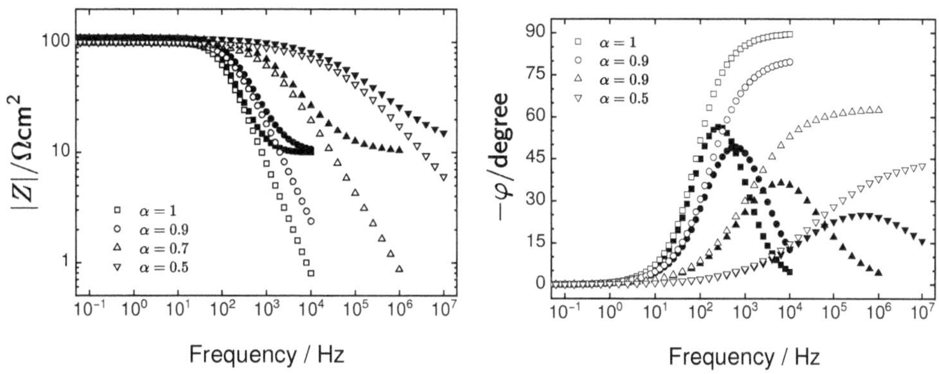

(a) (b)

Figure 17.12: Impedance in a Bode representation as a function of frequency for the results of Problem 17.3 with α as a parameter: (a) magnitude of the impedance and (b) phase. Open symbols represent the ohmic-resistance-corrected magnitude and phase.

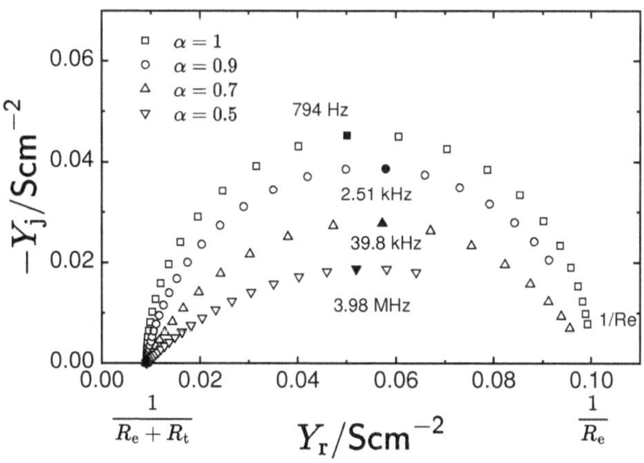

Figure 17.13: Nyquist plot for the complex admittance for the results of Problem 17.3 with α as a parameter. As with Figure 17.10, the characteristic frequency changes with α.

17.4 We are interested in the meaning of the effective capacitance obtained using

$$C_{\text{eff}} = -\frac{1}{\omega Z_{\text{j}}} \tag{17.10}$$

for the convective-diffusion impedance expressed as

$$Z(\omega) = R_{\text{e}} + \frac{R_{\text{t}} + Z_{\text{D}}(\omega)}{1 + \text{j}\omega C_{\text{dl}}\left(R_{\text{t}} + Z_{\text{D}}(\omega)\right)} \tag{17.11}$$

where $Z_{\text{D}}(\omega)$ is a complex function of frequency that tends toward zero as frequency tends toward ∞. Show that

$$C_{\text{eff}} = \lim_{\omega \to \infty} -\frac{1}{\omega Z_{\text{j}}} = C_{\text{dl}} \tag{17.12}$$

Solution 17.4. As frequency becomes large, equation (17.11) becomes

$$\lim_{\omega \to \infty} Z = R_{\text{e}} + \frac{1}{\text{j}\omega C_{\text{dl}}} \tag{17.13}$$

or

$$\lim_{\omega \to \infty} Z_{\text{j}} = \frac{1}{\text{j}\omega C_{\text{dl}}} \tag{17.14}$$

Thus, as frequency becomes sufficiently large, equation (17.12) is satisfied.

17.5 Electrochemical impedance spectroscopy has become a powerful tool to assess the response of human skin to external stimuli. Two approaches have been taken to analyze the skin impedance data. The skin properties have been assumed to be independent of spatial position but dependent on frequency, and, alternatively, skin properties have been assumed to be independent of frequency but dependent on spatial position.

(a) Under the assumption that the impedance response of human skin can be described by

$$Z_{\epsilon} = \frac{R}{1 + \text{j}\omega RC} \tag{17.15}$$

derive expressions for the frequency-dependent complex relative permittivity

$$\varepsilon(\omega) = \delta\frac{C(\omega)}{\epsilon_0} \tag{17.16}$$

(b) Express your results in terms of real frequency-dependent quantities: the real part of the relative permittivity and the resistivity.

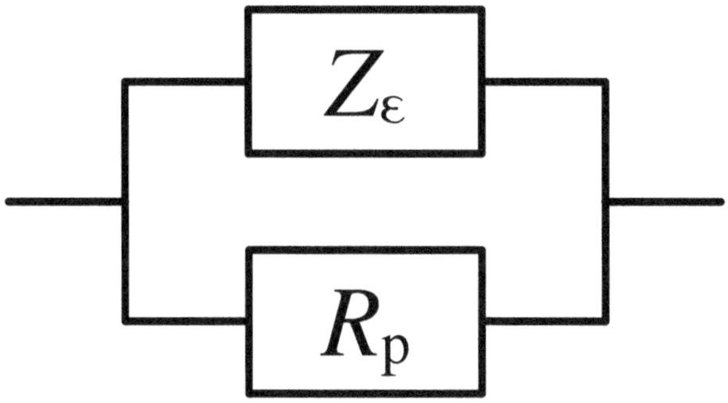

Figure 17.14: An electrical circuit illustrating the method used to extract the impedance associated with the dielectric properties of skin.

Solution 17.5. In this analysis, the frequency-independent dc contribution to the impedance is accounted for separately, as is represented schematically in Figure 17.14, where R_p is the resistance of parallel pathways and Z_ϵ is the impedance associated with the dielectric response of the skin. The complex capacitance can be expressed as

$$C = \frac{1}{j\omega Z_\epsilon(\omega)} = \frac{\epsilon(\omega)\epsilon_0}{\delta} \tag{17.17}$$

where δ is the skin thickness, $\epsilon(\omega)$ is the frequency-dependent relative permittivity, ϵ_0 is the permittivity of vacuum with value 8.8542×10^{-14}, and the impedance Z_ϵ has units of Ωcm^2.

(a) If the skin response can be described as a capacitor,

$$Z_\epsilon = -j\frac{1}{\omega C} \tag{17.18}$$

and

$$\epsilon = \frac{\delta C}{\epsilon_0} \tag{17.19}$$

is a real number and is independent of frequency. Under the assumption that the skin response can be described by equation (17.15), the dielectric response is

$$\epsilon(\omega) = \frac{\delta C}{\epsilon_0} - j\frac{\delta}{\epsilon_0 \omega R} \tag{17.20}$$

(b) Since $\rho = R/\delta$,

$$\epsilon(\omega) = \frac{\delta C}{\epsilon_0} - j\frac{1}{\epsilon_0 \omega \rho} \tag{17.21}$$

where the real part of ϵ is the frequency-independent relative permittivity and imaginary part is a function of frequency.

Chapter 18

Graphical Methods

Graphical methods provide a first step toward interpretation and evaluation of impedance data. The problems presented in this chapter explore general graphical techniques that do not depend on any specific model. The approaches, therefore, can provide a qualitative interpretation. Surprisingly, even in the absence of specific models, values of such meaningful parameters as the double-layer capacitance can be obtained from high- and low-frequency asymptotes. Problems 18.1–18.6 are based on problems presented by Orazem and Tribollet.[1]

Problems

18.1 Choose appropriate parameters for the circuit presented in Figure 18.1 that will yield time constants in a measurable frequency range. Prepare Nyquist plots, Bode and ohmic-resistance-corrected Bode plots, plots of the logarithm of the absolute value of the imaginary impedance as a function of frequency, plots of the derivative of the logarithm of the absolute value of the imaginary impedance as a function of frequency, and plots of the effective capacitance C_{eff} or CPE coefficient Q_{eff} as a function of frequency.

Solution 18.1. There are, of course, many answers to this question. The parameters for the low-frequency loop can correspond to a charge-transfer resistance in parallel to a double-layer capacitance, for example, $R_{\text{low}} = 1{,}000\ \Omega\text{cm}^2$ and $C_{\text{low}} = 40\ \mu\text{F/cm}^2$. For the high-frequency loop, the CPE parameters can correspond to a coating, e.g., $Q_{\text{high}} = 0.01\ \mu\text{F/s}^{1-\alpha}\text{cm}^2$ and $\alpha_{\text{high}} = 0.9$. The resistance R_{high} can be also equal to $1{,}000\ \Omega\text{cm}^2$. The electrolyte resistance is $200\ \Omega\text{cm}^2$. With the above parameters, the Nyquist plot is shown as Figure 18.2(a). The Bode plots are presented in uncorrected and ohmic-resistance-corrected formats in Figure 18.2(b) for modulus and Figure 18.2(c) for phase. At high frequencies, the slope of the ohmic-resistance-corrected mod-

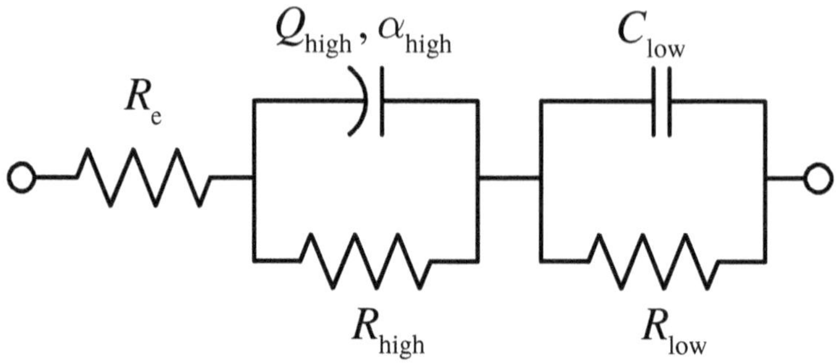

Figure 18.1: Equivalent circuit for Problem 18.1. Taken from Orazem and Tribollet[1] with permission of John Wiley & Sons.

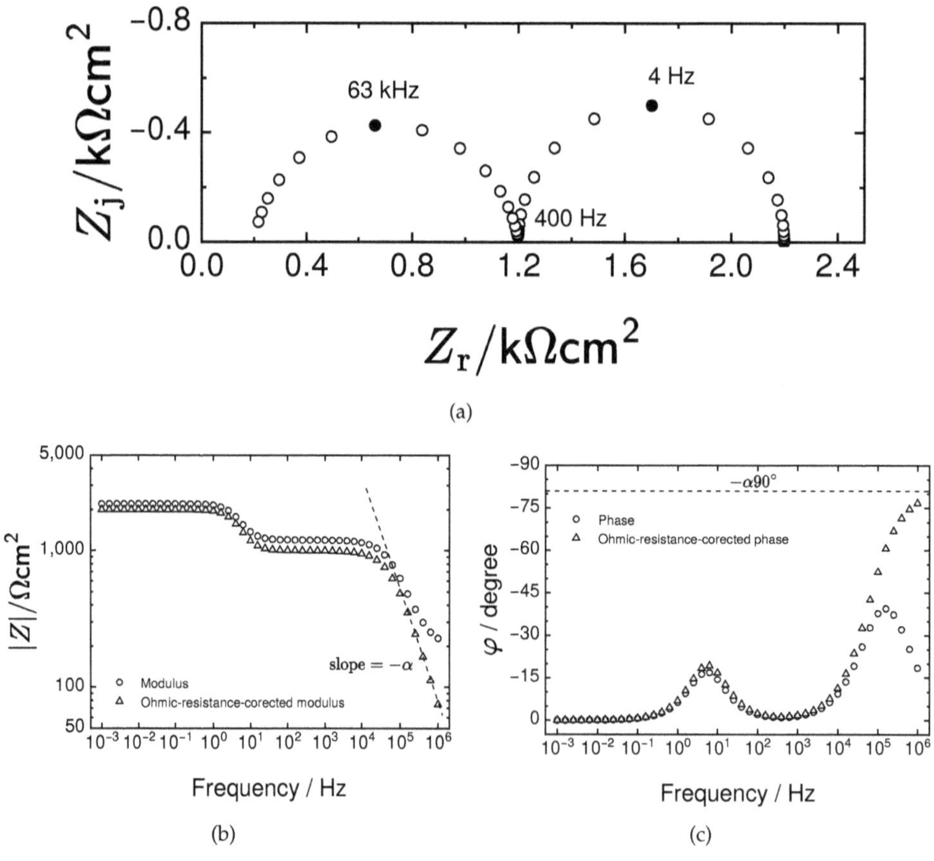

Figure 18.2: Impedance in Bode format of the electrical circuit given as Figure 18.1: (a) Nyquist, (b) magnitude, and (c) phase. Parameter values are $R_{\mathrm{low}} = 1{,}000\ \Omega\mathrm{cm}^2$, $C_{\mathrm{low}} = 40\ \mu\mathrm{F/cm}^2$, $Q_{\mathrm{high}} = 0.01\ \ \mu\mathrm{F/s}^{1-\alpha}\mathrm{cm}^2$, $\alpha_{\mathrm{high}} = 0.9$, $R_{\mathrm{high}} = 1{,}000\ \Omega\mathrm{cm}^2$, and $R_{\mathrm{e}} = 200\ \Omega\mathrm{cm}^2$.

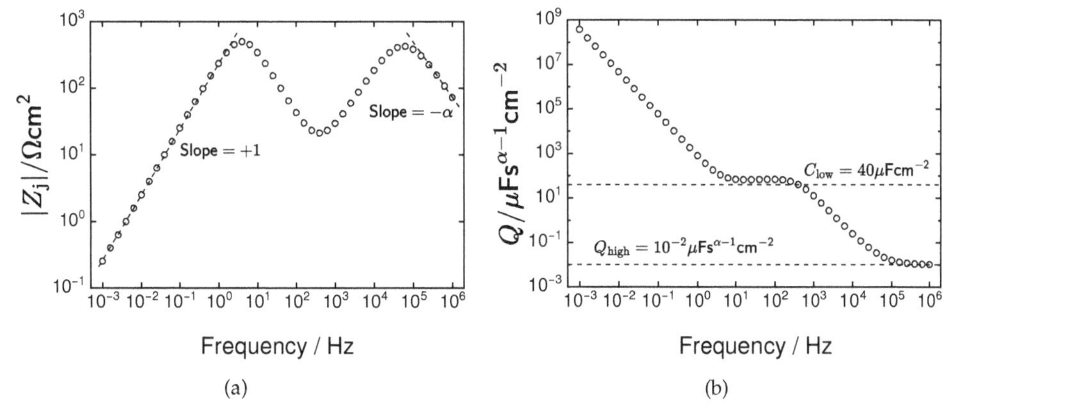

Figure 18.3: Impedance as described in Figure 18.2 as a function of frequency: (a) magnitude of the imaginary part of the impedance on a logarithmic scale and (b) effective CPE parameter Q.

ulus approaches $-\alpha$, and the ohmic-resistance-corrected phase approaches $-\alpha 90 = 81^\circ$.

The magnitude of the imaginary part of the impedance is presented in Figure 18.3(a) on a logarithmic scale as a function of frequency, also on a logarithmic scale. The low-frequency slope has a value of $+1$, consistent with a capacitor; whereas, the high-frequency slope has a value of $-\alpha$. The effective value of Q

$$Q = -\sin\left(\frac{\alpha\pi}{2}\right)\frac{1}{(2\pi f)^{\alpha} Z_j} \tag{18.1}$$

is given in Figure 18.3(b). The value at high frequency is in good agreement with Q_{high}, and the value in mid-frequency ranges approximates the capacitance C_{low}.

18.2 Consider the circuit presented in Figure 9.5 for an electrode covered by a film. Choose appropriate parameters that will yield time constants in a measurable frequency range. Prepare Nyquist, Bode, and ohmic-resistance-corrected Bode plots, plots of the logarithm of the absolute value of the imaginary impedance as a function of frequency, plots of the derivative of the logarithm of the absolute value of the imaginary impedance as a function of frequency, and plots of the effective capacitance C_{eff} or CPE coefficient Q_{eff} as a function of frequency.

Solution 18.2. The parameters selected for Figure 9.5 are as follows: the capacitance for a film is given by

$$C_\ell = \frac{\varepsilon\varepsilon_0}{\delta} \tag{18.2}$$

where $\varepsilon_0 = 8.8542 \times 10^{-14}$ F/cm is the permittivity of vacuum, ε is the dielectric constant, and δ is the film thickness. For a thick film, e.g., $\delta = 10$ μm, and $\varepsilon = 5.4$, a value

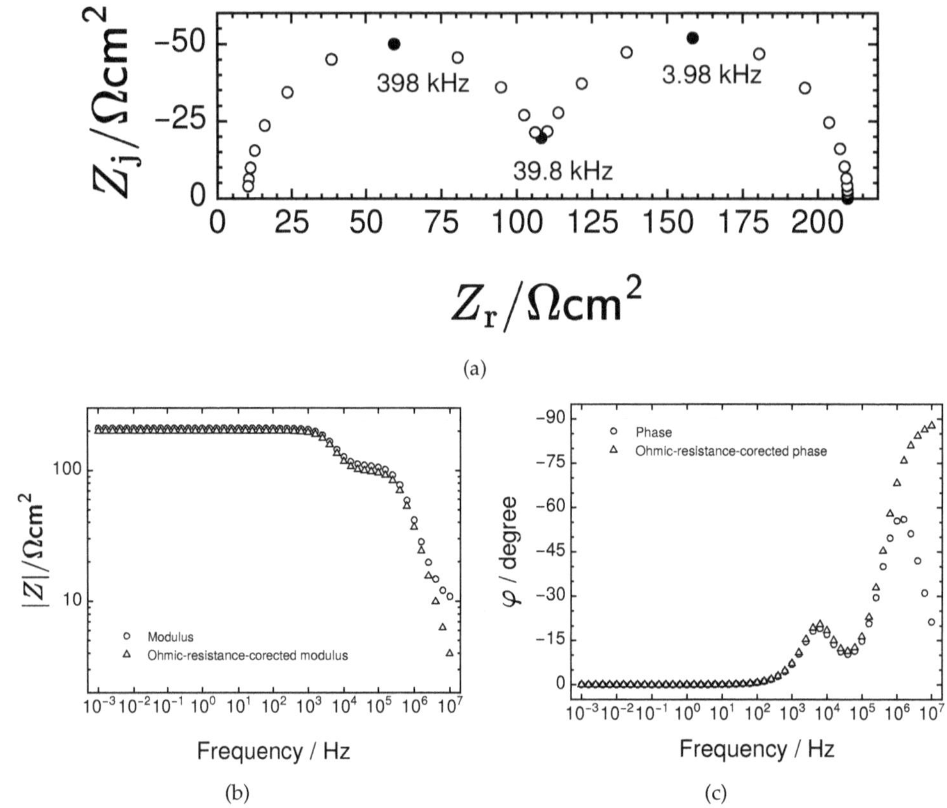

Figure 18.4: Impedance in Bode format of the electrical circuit given as Figure 9.5: (a) Nyquist, (b) magnitude, and (c) phase. Parameter values are $C_\ell = 4 \times 10^{-9}$ F/cm^2, $C_{\text{dl}} = 0.4$ µF/cm^2, $R_{\text{t}} = 100\ \Omega\text{cm}^2$, $R_\ell = 100\ \Omega\text{cm}^2$, and $R_{\text{e}} = 10\ \Omega\text{cm}^2$.

corresponding to a polymeric film, $C_\ell = 4 \times 10^{-9}$ F/cm^2. If the film covers 99% of the electrode and the capacitance of the bare electrode has a value of 40 µF/cm^2, the capacitance scaled by the entire electrode area is $C_{\text{dl}} = 0.4$ µF/cm^2. If the charge-transfer resistance for the bare electrode is 1 Ωcm^2, the value scaled by the entire electrode area is $R_{\text{t}} = 100\ \Omega\text{cm}^2$. Let the pore resistance be $R_\ell = 100\ \Omega\text{cm}^2$ and the ohmic resistance be $R_{\text{e}} = 10\ \Omega\text{cm}^2$. The resulting impedance is presented in Nyquist format in Figure 18.2(a). Two loops are seen with a high-frequency asymptote corresponding to the ohmic resistance, and the low-frequency limit corresponds to the sum of R_{e}, R_{t}, and R_ℓ. The Bode plots are presented in uncorrected and ohmic-resistance-corrected formats in Figure 18.4(b) for modulus and Figure 18.4(c) for phase. At high frequencies, the slope of the ohmic-resistance-corrected modulus approaches -1, and the ohmic-resistance-corrected phase approaches $-90°$.

The magnitude of the imaginary part of the impedance is presented in Figure 18.5(a) on a logarithmic scale as a function of frequency, also on a logarithmic scale. As shown

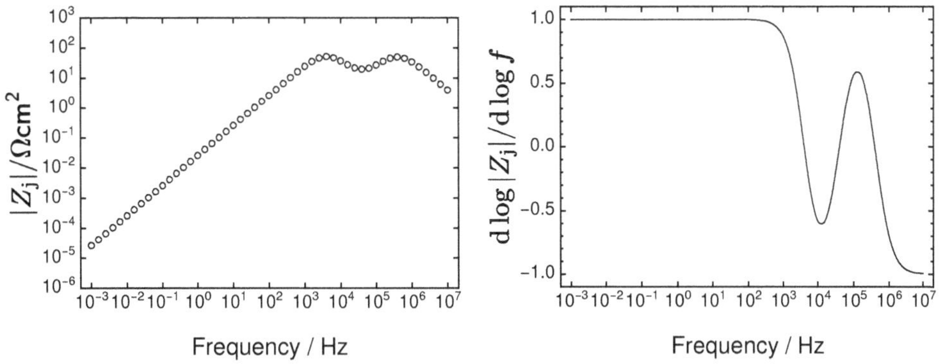

Figure 18.5: Impedance as described in Figure 18.4 as a function of frequency: (a) magnitude of the imaginary part of the impedance on a logarithmic scale and (b) derivative of the plot shown in panel (a).

in Figure 18.5(b), the low-frequency derivative has a value of $+1$, consistent with a capacitor; whereas, the high-frequency slope has a value of -1, also consistent with a capacitor.

The effective capacitance, obtained from

$$Q = -\frac{1}{2\pi f Z_{\mathrm{j}}} \tag{18.3}$$

is presented in Figure 18.6 as a function of frequency. At high frequencies, the capacitance approaches the input value for C_ℓ. An inflection is seen near $f = 10^4$ Hz, and the capacitance at this frequency is close to C_{dl}.

18.3 The value of the Schmidt number may be used to determine which species participate in an electrochemical reaction. Estimate the value for the Schmidt number that could be expected to be determined by impedance measurements at 25 °C for the system 0.01 M $K_3Fe(CN)_6$, 0.01 M $K_4Fe(CN)_6$, and 1 M KCl:

(a) At the open circuit potential.

(b) At the cathodic mass-transfer-limited current.

(c) At the anodic mass-transfer-limited current.

Solution 18.3. The diffusivity for ferricyanide at 25 °C and infinite dilution is given in Table A.13 as $D_{\mathrm{Fe(CN)}_6^{3-}} = 0.896 \times 10^{-5}$ cm^2/s, and that for ferrocyanide is $D_{\mathrm{Fe(CN)}_6^{4-}} = 0.739 \times 10^{-5}$ cm^2/s. For aqueous systems at 25 °C, the kinematic viscosity of water is approximately $\nu = 0.01$ cm^2/s. The Schmidt number for species i is given as $\mathrm{Sc} = \nu / D_{\mathrm{i}}$.

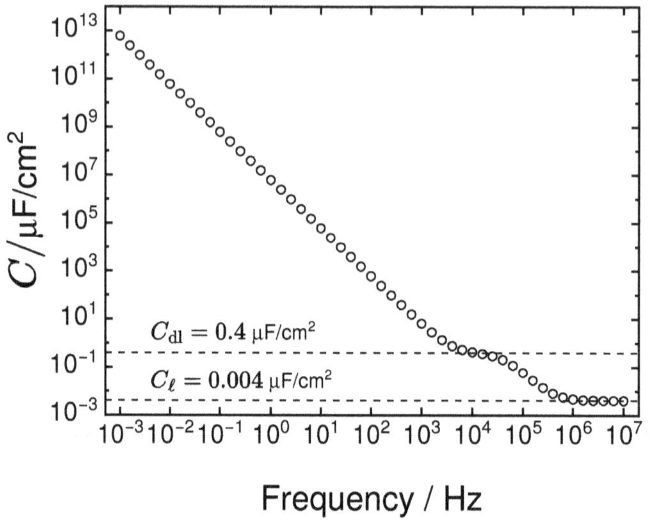

Figure 18.6: Effective capacitance for the impedance described in Figure 18.4 as a function of frequency.

(a) The Schmidt number for ferricyanide is $Sc_{Fe(CN)_6^{3-}} = 1116$, and that for ferrocyanide is $Sc_{Fe(CN)_6^{3-}} = 1353$. At open circuit, the reduction of ferricyanide is balanced by the oxidation of ferrocyanide, and the effective diffusion coefficient should be between that of the two species. An effective Schmidt number of $Sc = 1250$ has been reported.

(b) At the cathodic mass-transfer limit, reduction of ferricyanide should dominate, yielding $Sc = 1116$.

(c) At the anodic mass-transfer limit, oxidation of ferrocyanide should dominate, yielding $Sc = 1353$.

18.4 The experimental diffusion impedance data presented in Table 18.1 were obtained at a rotation speed of 600 rpm. Find the numerical value of the Schmidt number by using

$$1.2261{Sc_i}^{2/3} + (0.84 - s)\,{Sc_i}^{1/3} + 0.63 = 0 \tag{18.4}$$

developed by Tribollet et al.[63]

Solution 18.4. The data from Table 18.1 are presented in Nyquist format in Figure 18.7. The data are quite noisy, and the frequency range is not sufficient to see the low-frequency data reaching the real axis. The scaled real impedance $Z_r/Z_{r,max}$ is

Table 18.1: Experimental diffusion impedance data obtained at a rotation speed of 600 rpm.

f, Hz	Z_r, Ωcm^2	Z_j, Ωcm^2	f, Hz	Z_r, Ωcm^2	Z_j, Ωcm^2
0.07942	997.9	−65.1	3.16	309	−273
0.1	992.2	−85.2	3.98	270.6	−253.7
0.1258	991	−97.8	5.02	239	−259.2
0.1584	973	−118	6.3	208.6	−197.9
0.1996	967	−149	7.94	188.8	−179.1
0.2512	940	−190	10	165.2	−159.1
0.316	920	−231	12.58	147.3	−141.6
0.398	882	−263	15.84	131.9	−126.4
0.502	831	−320	19.94	116.3	−112.8
0.63	764	−324.6	25.12	103.7	−100.4
0.794	692	−347	31.624	91.9	−89.04
1	613	−354	39.812	82.3	−79.43
1.258	544	−342	50.12	73.14	−70.85
1.584	470	−348	79.43	59.7	−57.6
1.994	409	−328	100	52.11	−50.11
2.52	356	−303			

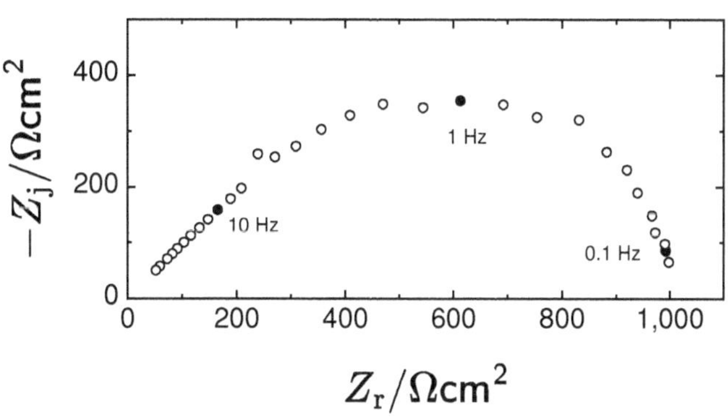

Figure 18.7: Nyquist representation of the data shown in Table 18.1.

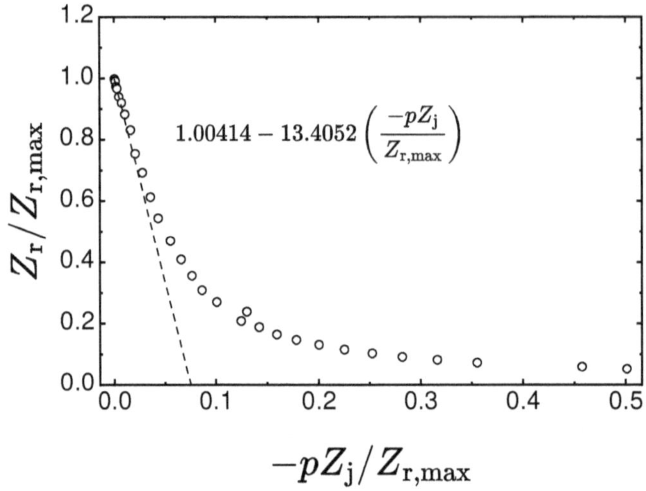

Figure 18.8: Graphical analysis for the data that are shown in Table 18.1, where $Z_r/Z_{r,max}$ is presented as a function of $-pZ_j/Z_{r,max}$. The line yields the asymptotic slope as frequency tends toward zero.

presented in Figure 18.8 as a function of $-pZ_j/Z_{r,max}$, where $p = \omega/\Omega = f/10$. The regressed line has an asymptote of 1.0041 ± 0.0026 and a slope of -13.4052 ± 1.0. The asymptote may be expected to be slightly larger than unity because $Z_{r,max}$ represents an approximation to $Z_r(0)$.

For $s = 13.4052$, equation (18.4) yields Sc $= 1060$. The uncertainty in the value of s is large. For $s = 14.4052$, equation (18.4) yields Sc $= 1337$; for $s = 12.4052$, equation (18.4) yields Sc $= 824$. A Monte Carlo simulation yields the histogram provided in Figure 18.9. The graphical method will provide more reliable results for impedance data that has less noise and extends to lower frequencies.

18.5 For simple reactions, the charge-transfer resistance may be given as a function of potential by

$$R_{t,M} = \frac{1}{K_M \exp\left(b_M\overline{V}\right) b_M} = \frac{1}{\overline{i}_M b_M} \tag{18.5}$$

Develop a means of superposing impedance data taken at different potentials for such a system where the impedance response is given by

$$Z_M(\omega) = R_e + \frac{R_{t,M}}{1 + j\omega R_{t,M} C_{dl}} \tag{18.6}$$

Solution 18.5. The synthetic data shown in Figure 18.10 were generated from

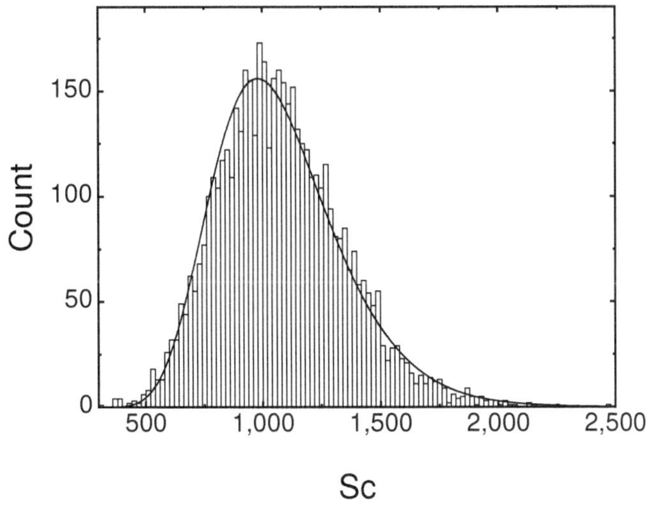

Figure 18.9: Histogram showing the uncertainty in the value of Schmidt number obtained by the method shown in Figure 18.8.

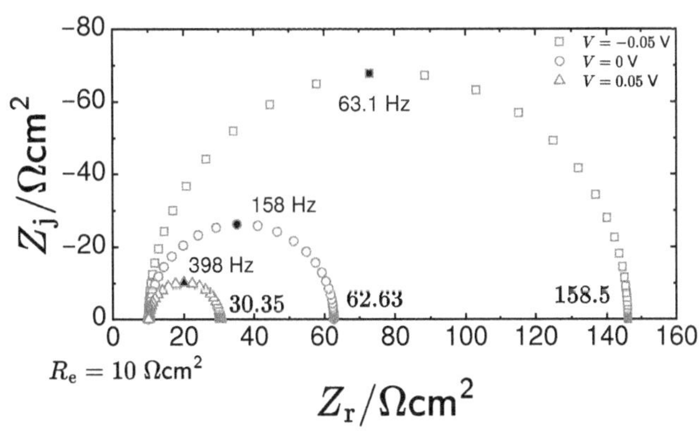

Figure 18.10: Nyquist plot for synthetic data generated from equations (18.5) and (18.6) for $K_M = 10^{-3}$ A/cm^2, $b_M = 19$ V^{-1}, $R_e = 10\ \Omega$cm^2, and $C_{dl} = 2 \times 10^{-5}$ F/cm^2.

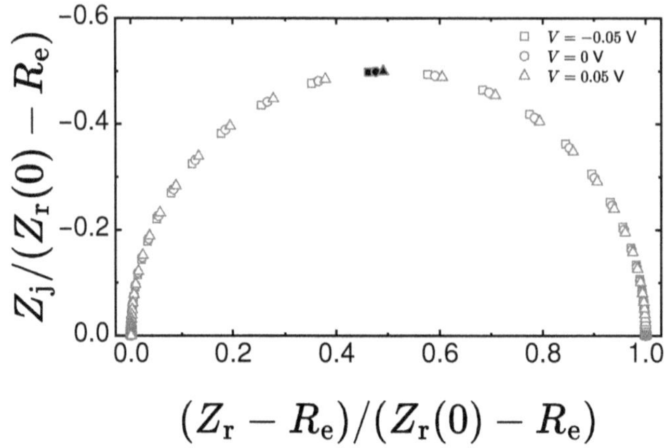

Figure 18.11: Scaled Nyquist plot for synthetic data that are shown in Figure 18.10, where $(Z_r - R_e)/(Z_r(0) - R_e)$ is presented as a function of $Z_j/(Z_r(0) - R_e)$.

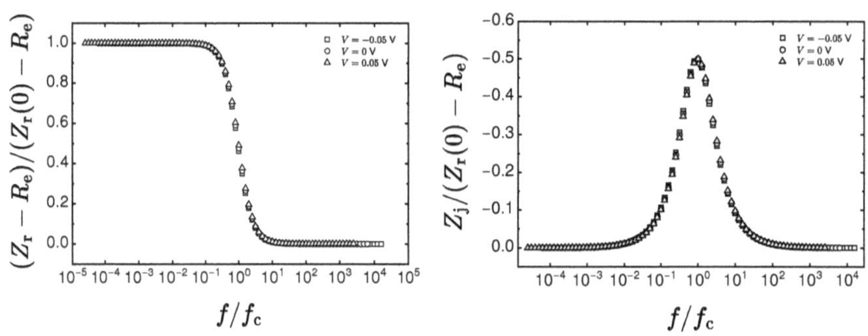

(a) (b)

Figure 18.12: Scaled impedance for the synthetic data that are shown in Figure 18.10 as a function of scaled frequency: (a) real part of the impedance and (b) imaginary part of the impedance.

equations (18.5) and (18.6) for $K_M = 10^{-3}$ A/cm^2, $b_M = 19$ V^{-1}, $R_e = 10$ Ωcm^2, and $C_{dl} = 2 \times 10^{-5}$ F/cm^2. Parameters may be extracted from Figure 18.10 to allow calculation of a dimensionless frequency f/f_c, a dimensionless real part of the impedance $(Z_r - R_e)/(Z_r(0) - R_e)$, and a dimensionless imaginary part of the impedance $Z_j/(Z_r(0) - R_e)$. The scaled Nyquist plot is shown in Figure 18.11.

The scaled real and imaginary impedances are presented in Figure 18.12 as a function of scaled frequency. The information obtained from Figure 18.10 enables superposition of the impedance in both Nyquist format and as a function of scaled frequency. The approach shown here is general, though, for more complex systems, one must choose how to select the maximum real impedance and the characteristic frequency.

18.6 Show how a linear dependence of capacitance on potential would influence the superposition developed in Problem 18.5.

Solution 18.6. The method presented in the solution to Problem 18.5 is general. A potential dependent capacitance should not affect the ability to superpose impedance data as shown in Figures 18.11 and 18.12.

18.7 Nyquist and Bode plots are commonly used to represent impedance data. For typical impedance measurements in electrochemical systems, the phase angle approaches zero at high frequency, even though we anticipate that the electrode–electrolyte interface experiences a charging current, for which the phase angle should approach -90°. Explain (in one or two sentences) why the phase angle approaches zero at high frequency.

Solution 18.7. The interfacial impedance tends toward zero at high frequency; thus, the impedance approaches the ohmic resistance, for which the phase angle is equal to zero.

18.8 Impedance is measured under three different conditions. An attempt is made to demonstrate superposition of scaled Nyquist plots, i.e., $Z_j/(Z_r(0) - R_e)$ as a function of $(Z_r - R_e)/(Z_r(0) - R_e)$. In one or two sentences, what do you infer if superposition is not possible?

Solution 18.8. The inability to superpose Nyquist plots suggests that the fundamental chemistry and/or physics are not the same for the individual spectra. For example, Erol et al.[64] demonstrate that impedance spectra collected before and after overcharge of a $LiCoO_2|C$ battery could not be superposed, as shown in Figure 18.13(a); whereas, impedance spectra collected before and after over-discharge could be superposed, shown in Figure 18.13(b). The failure to superpose was attributed to a change in battery chemistry after overcharging.

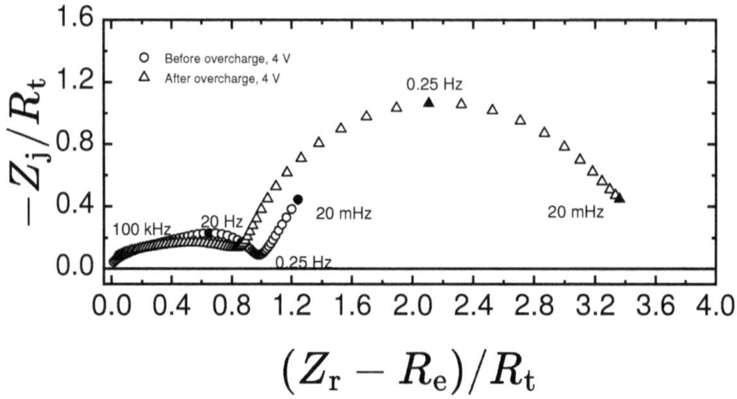

(a)

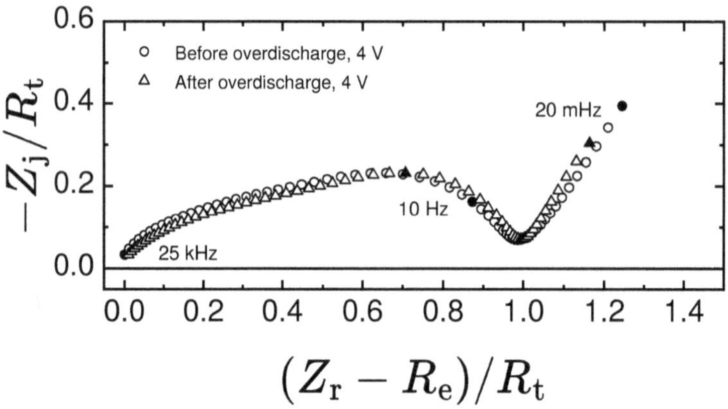

(b)

Figure 18.13: Scaled impedance for a $LiCoO_2|C$ cell at 4 V; (a) before and after the cell was overcharged and (b) before and after the cell was over-discharged. Data taken from Erol et al.[64]

Chapter 19

Complex Nonlinear Regression

The problems presented in this chapter address application of complex nonlinear least-squares (CNLS) regression techniques to impedance measurements. Problems 19.1–19.4 are based on problems presented by Orazem and Tribollet.[1] The solutions presented here makes use of the measurement model software published by Watson and Orazem[16] under GNU General Public License (GPL) version 3.0.

Problems

19.1 Create a synthetic dataset for a system with a constant-phase element as given in

$$Z = R_e + \frac{R_{||}}{1 + (j\omega)^{\alpha} R_{||} Q} \tag{19.1}$$

with parameters $R_e = 10\ \Omega\text{cm}^2$; $R_{||} = 10^3\ \Omega\text{cm}^2$; $Q = 100\ \text{s}^{\alpha}/\text{M}\Omega\text{cm}^2$; and $\alpha = 0.7$ with frequencies ranging from 10^{-2} to 10^4 Hz spaced logarithmically with 10 measurements per decade. Use a commercial regression software or create your own to regress a Voigt model, i.e.,

$$Z = R_e + \sum_{k}^{K} \frac{R_k}{1 + j\omega\tau_k} \tag{19.2}$$

to the data using modulus weighting. Find the maximum number of Voigt elements that can be added without including zero in the 2σ (95.45%) confidence interval.

Solution 19.1. The measurement model software[16] was used to regress equation (19.2) to the synthetic data described in Problem 19.1. The results shown in Figure 19.1 suggest a very good fit was obtained for 21 Voigt elements. The regressed values are presented in Table 19.1. The software employed a Levenberg–Marquardt algo-

Table 19.1: Regressed parameters obtained for regression of equation (19.2) to the synthetic data generated for Problem 19.1. The regression yielded K = 21 and $\chi^2/\nu = 4.12 \times 10^{-15}$.

k	τ_k, s	R_k, Ωcm^2
0	–	30.71 ± 0.30
1	$(2.979 \pm 0.054) \times 10^{-6}$	70.93 ± 0.91
2	$(1.000 \pm 0.016) \times 10^{-5}$	98.88 ± 1.2
3	$(2.106 \pm 0.030) \times 10^{-5}$	135.3 ± 1.5
4	$(3.842 \pm 0.054) \times 10^{-5}$	168.4 ± 1.4
5	$(6.693 \pm 0.098) \times 10^{-5}$	165.5 ± 1.3
6	$(1.168 \pm 0.018) \times 10^{-4}$	126.0 ± 1.5
7	$(2.084 \pm 0.035) \times 10^{-4}$	82.3 ± 1.2
8	$(3.816 \pm 0.067) \times 10^{-4}$	51.01 ± 0.82
9	$(7.16 \pm 0.13) \times 10^{-4}$	31.35 ± 0.52
10	$(1.378 \pm 0.026) \times 10^{-3}$	19.34 ± 0.32
11	$(2.715 \pm 0.054) \times 10^{-3}$	11.96 ± 0.20
12	$(5.49 \pm 0.11) \times 10^{-3}$	7.38 ± 0.13
13	0.01141 ± 0.00025	4.520 ± 0.081
14	0.02443 ± 0.00055	2.736 ± 0.051
15	0.0541 ± 0.0013	1.629 ± 0.031
16	0.1247 ± 0.0031	0.951 ± 0.019
17	0.3006 ± 0.0078	0.542 ± 0.011
18	0.764 ± 0.021	0.2998 ± 0.067
19	2.072 ± 0.060	0.1606 ± 0.0038
20	6.19 ± 0.19	0.0842 ± 0.0021
21	24.67 ± 0.87	0.0488 ± 0.0011

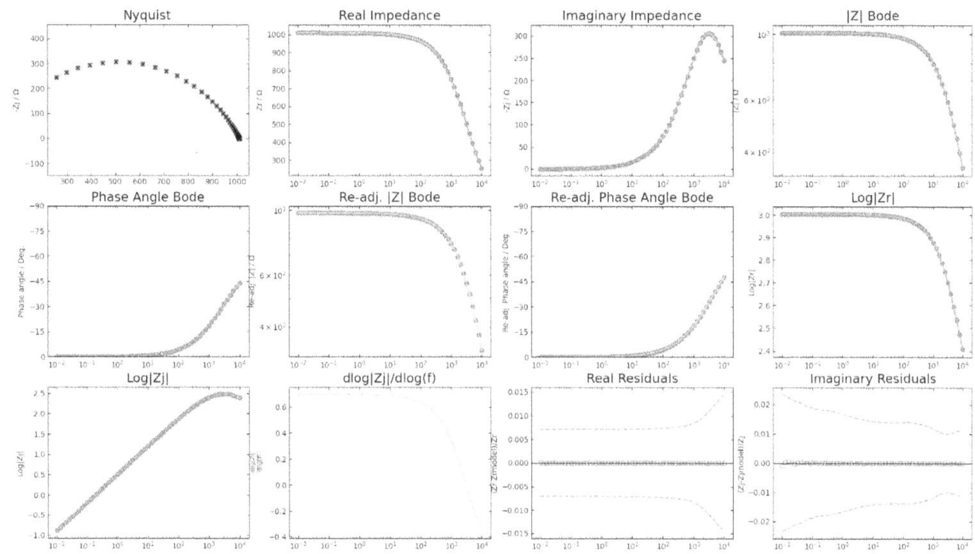

Figure 19.1: Plots generated by the measurement model program for regression of the measurement model equation (19.2) to the synthetic data generated for Problem 19.1. Symbols represent the synthetic data, and the solid lines represent the fitted model. Dashed lines represent the 95.4% confidence interval for the model. The labels for the abscissa were omitted in the measurement model program to increase the size of the plots.

rithm, which is sensitive to initial guesses. The "Magic Finger" feature of the program was used to select initial guesses corresponding to the low frequency region, where regression errors were proportionally largest.

The χ^2/ν statistic had a value of 4.12×10^{-15}. In the absence of a formal error structure for the regression, this value does not have meaning other than the number is small. The plots shown in Figure 19.1 show that the normalized fitting errors are smaller than 0.1%.

19.2 Generate the synthetic data described in Problem 19.1, adding independent normally distributed random numbers $N(0, \sigma)$ to both real and imaginary parts of the impedance, where $\sigma = a|Z|$ and the value of a is given below:

(a) Generate the synthetic dataset for $a = 0.01$. Perform the regression using modulus weighting for which $\sigma = 0.01|Z|$. Identify the value of the resulting χ^2 statistic and the number of parameters that can be obtained without including zero in the 2σ (95.45%) confidence interval.

(b) Generate the synthetic dataset for $a = 0.05$. Perform the regression using modulus weighting for which $\sigma = 0.01|Z|$. Identify the value of the resulting χ^2 statistic and the number of parameters that can be obtained without including zero in the 2σ (95.45%) confidence interval.

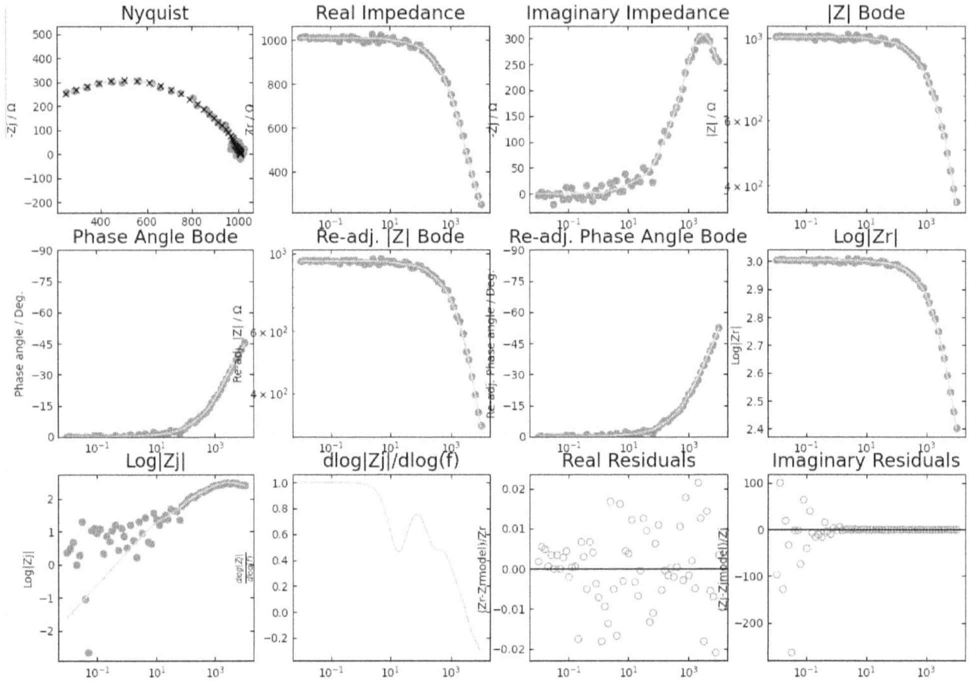

Figure 19.2: Plots generated by the measurement model program for regression of the measurement model equation (19.2) to the synthetic data generated for Problem 19.2.a. Symbols represent the synthetic data and the solid lines represent that fitted model. The labels for the abscissa were omitted in the measurement model program to increase the size of the plots.

Solution 19.2. Synthetic data were generated from equation (19.1) with the parameters specified in Problem 19.1. Normally distributed noise was added to the synthetic data, i.e.,

$$Z_r = Z_{r,model} + N_1(\mu, \sigma^2) \tag{19.3}$$

and

$$Z_j = Z_{j,model} + N_2(\mu, \sigma^2) \tag{19.4}$$

where the means of independent normal distributions N_1 and N_2 had value $\mu = 0$. The standard deviations for the independent normal distributions were $\sigma = 0.01|Z|$ for part (a) and $\sigma = 0.05|Z|$ for part (b).

(a) The fit of the measurement model to synthetic data with $\sigma = 0.01|Z|$ is shown in Figure 19.2. The regressed results are presented in Table 19.2. The data shown in Figure 19.2 are very noisy, especially at low frequencies where the magnitude of the impedance is large. Impedance data may often be collected with much smaller levels of noise. The confidence intervals for the regressed parameters are

Table 19.2: Regressed parameters obtained for regression of equation (19.2) to the synthetic data generated for Problem 19.2.a. The regression was weighted using $\sigma = 0.01|Z|$ and yielded $K = 5$ and $\chi^2/\nu = 0.968$.

k	τ_k, s	R_k, Ωcm^2
0		57 ± 23
1	$(1.18 \pm 0.38) \times 10^{-5}$	242 ± 54
2	$(4.6 \pm 1.4) \times 10^{-5}$	347 ± 60
3	$(1.44 \pm 0.49) \times 10^{-4}$	236 ± 82
4	$(7.2 \pm 1.9) \times 10^{-4}$	104 ± 27
5	0.0011 ± 0.0049	22.9 ± 5.9

Table 19.3: Regressed parameters obtained for regression of equation (19.2) to the synthetic data generated for Problem 19.2.b. The regression was weighted using $\sigma = 0.01|Z|$ and yielded $K = 2$ and $\chi^2/\nu = 28$.

k	τ_k, s	R_k, Ωcm^2
0		115 ± 19
1	$(3.18 \pm 0.29) \times 10^{-5}$	552 ± 27
2	$(2.89 \pm 0.44) \times 10^{-4}$	325 ± 32

large, but still yield parameters that do not include zero in the $\pm 2\sigma$ (95.4%) confidence interval. The χ^2/ν statistic had a value of 0.968. Interestingly, a regression with four Voigt elements yielded a fit with smaller confidence intervals and a χ^2/ν statistic of 1.03. A good fit will yield a weighted χ^2/ν statistic on the order of unity, i.e.,

$$\chi^2/\nu = 1 + \sqrt{2/\nu} \tag{19.5}$$

where ν is the degree of freedom for the regression.

(b) The regressed results are presented in Table 19.3. The increased level of noise in the synthetic data served to reduce the number of parameters that could be extracted. he regression was weighted using $\sigma = 0.01|Z|$ and yielded $\chi^2/\nu = 28$.

19.3 Generate the synthetic data described in Problem 19.1, adding independent normally distributed random numbers $N(0, \sigma)$ to both real and imaginary parts of the impedance where $\sigma = 0.01|Z|$.

(a) Truncate the synthetic dataset by removing the data below a frequency of 1 Hz. Perform the regression using modulus weighting for which $\sigma = 0.01|Z|$. Identify the value of the resulting χ^2 statistic and the number of parameters that can be obtained without including zero in the 2σ (95.45%) confidence interval.

Table 19.4: Regressed parameters obtained for regression of equation (19.2) to the synthetic data generated for Problem 19.2.a. The data were truncated such that the lowest frequency was 1 Hz. The regression was weighted using $\sigma = 0.01|Z|$ and yielded $K = 4$ and $\chi^2/\nu = 1.4$.

k	τ_k, s	R_k, Ωcm^2
0		57 ± 23
1	$(1.18 \pm 0.38) \times 10^{-5}$	242 ± 54
2	$(4.6 \pm 1.4) \times 10^{-5}$	347 ± 60
3	$(1.44 \pm 0.49) \times 10^{-4}$	236 ± 82
4	$(7.2 \pm 1.9) \times 10^{-4}$	104 ± 27

(b) Truncate the synthetic dataset by removing the data below a frequency of 10 Hz. Perform the regression using modulus weighting for which $\sigma = 0.01|Z|$. Identify the value of the resulting χ^2 statistic and the number of parameters that can be obtained without including zero in the 2σ (95.45%) confidence interval.

Solution 19.3. Synthetic data were generated from equation (19.1) with the parameters specified in Problem 19.1. Normally distributed noise was added to the synthetic data, i.e.,

$$Z_r = Z_{r,model} + N_1(\mu, \sigma^2) \tag{19.6}$$

and

$$Z_j = Z_{j,model} + N_2(\mu, \sigma^2) \tag{19.7}$$

where the means of independent normal distributions N_1 and N_2 had value $\mu = 0$ and $\sigma = 0.01|Z|$.

(a) The fit of the measurement model to synthetic data with $\sigma = 0.01|Z|$ is shown in Figure 19.3 where the data were truncated such that the lowest frequency was 1 Hz. The regressed results are presented in Table 19.4.

(b) The fit of the measurement model to synthetic data with $\sigma = 0.05|Z|$ is shown in Figure 19.4. where the data were truncated such that the lowest frequency was 10 Hz. The regressed results are presented in Table 19.5.

19.4 Perform the regression for the data presented in Problem 19.2(a) using a weighting based on $\sigma = 0.05|Z|$. Compare the resulting value of the χ^2 statistic and the number of parameters that can be obtained without including zero in the 2σ (95.45%) confidence interval to that obtained for Problem 19.2(a).

Solution 19.4. Regression weighted by $\sigma = 0.05|Z|$ yielded the same parameters as shown in Table 19.3, but the χ^2/ν statistic had a value of 1.1, showing a good

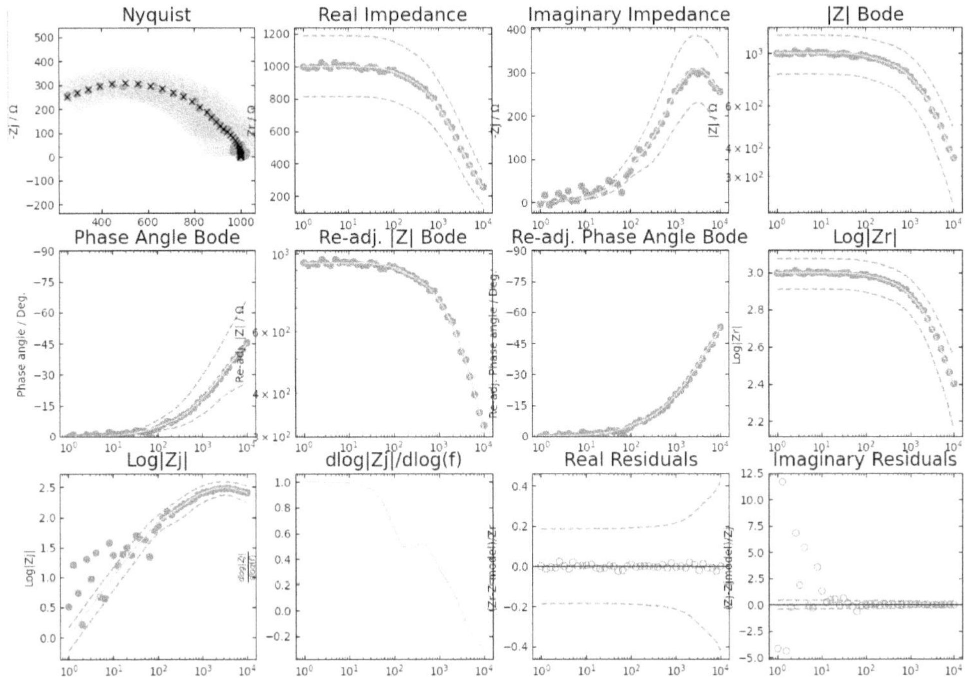

Figure 19.3: Plots generated by the measurement model program for regression of the measurement model equation (19.2) to the synthetic data generated for Problem 19.2.a. The data were truncated such that the lowest frequency was 1 Hz. Symbols represent the synthetic data, and the solid lines represent the fitted model. Dashed lines represent the 95.4% confidence interval for the model. The labels for the abscissa were omitted in the measurement model program to increase the size of the plots.

Table 19.5: Regressed parameters obtained for regression of equation (19.2) to the synthetic data generated for Problem 19.2.b. The data were truncated such that the lowest frequency was 10 Hz. The regression was weighted using $\sigma = 0.01|Z|$ and yielded $\mathrm{K} = 4$ and $\chi^2/\nu = 1.2$.

k	τ_k, s	R_k, $\Omega\mathrm{cm}^2$
0		57 ± 24
1	$(1.20 \pm 0.38) \times 10^{-5}$	246 ± 50
2	$(4.8 \pm 1.3) \times 10^{-5}$	361 ± 49
3	$(1.60 \pm 0.46) \times 10^{-4}$	234 ± 69
4	$(9.5 \pm 2.0) \times 10^{-4}$	97 ± 21

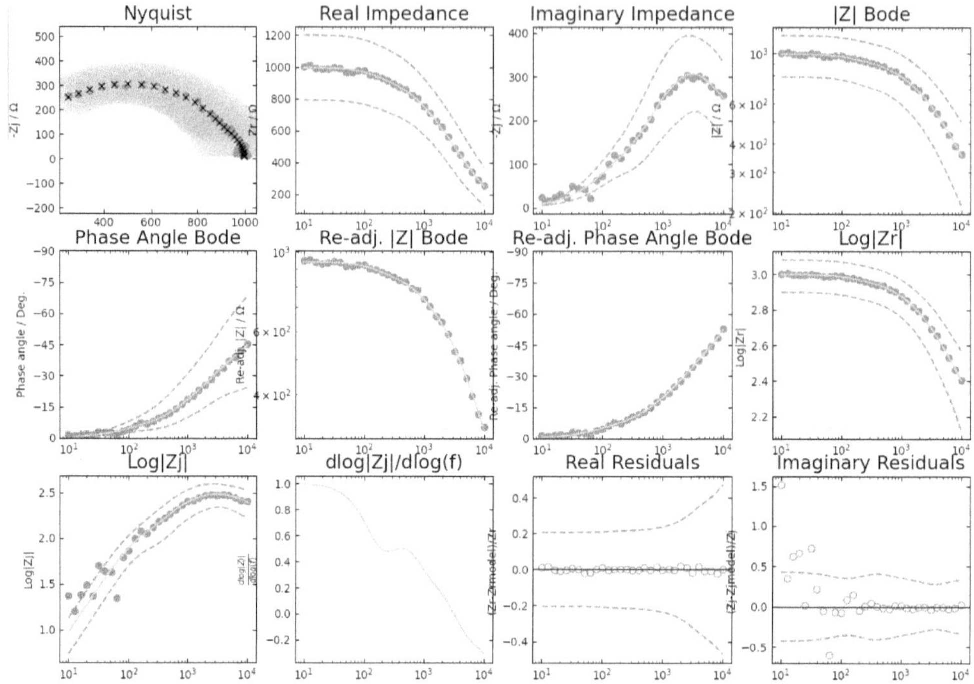

Figure 19.4: Plots generated by the measurement model program for regression of the measurement model equation (19.2) to the synthetic data generated for Problem 19.2.a. The data were truncated such that the lowest frequency was 10 Hz. Symbols represent the synthetic data, and the solid lines represent the fitted model. Dashed lines represent the 95.4% confidence interval for the model. The labels for the abscissa were omitted in the measurement model program to increase the size of the plots.

fit. This exercise shows that the meaning of the χ^2/ν statistic relies on the use of a weighting strategy that is the inverse of the variance of the data.

Chapter 20

Assessing Regression Quality

The problems presented in this chapter explore a systematic approach to determine whether the model provides a statistically adequate description of the data. Problems 20.1–20.6 are based on problems presented by Orazem and Tribollet.[1]

Problems

20.1 Compare the plots described in Table A.26 for the regression mentioned in Problem 19.2.

Solution 20.1. A series of plots are presented in Figure 19.2. The Nyquist plot shows the general behavior of the data, and the large noise at low frequency is evident. The plots of real and imaginary impedance as functions of frequency show that the model provides a good fit, as do the Bode plots and ohmic-resistance corrected Bode plots. The lack of fit is shown most clearly in the plot of $|Z_{\rm j}|$ as a function of frequency on a logarithmic scale. This plot emphasizes the role of added noise. The normalized residual error plots also give insight into the quality of the fit; however, the normalized errors for the imaginary part of the impedance seem excessively large at low frequency, where the imaginary part of the impedance tends toward zero.

20.2 Plot χ^2/ν as a function of the number of Voigt elements for the regressions mentioned in Problem 19.2.

Solution 20.2. The χ^2/ν statistic is shown in Figure 20.1 as a function of the number of Voigt elements for regression of the measurement model equation (19.2) to the synthetic data generated for Problem 19.2.a.

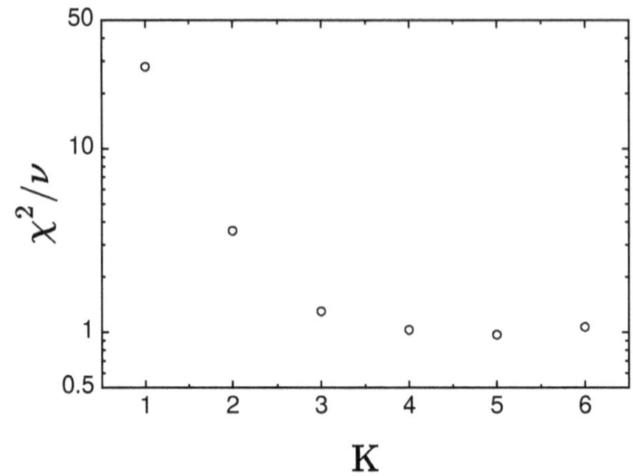

Figure 20.1: The χ^2/ν statistic as a function of the number of Voigt elements for regression of the measurement model equation (19.2) to the synthetic data generated for Problem 19.2.a.

The maximum number of statistically significant parameters was obtained for five Voigt elements. That number corresponds to the minimum value $\chi^2/\nu = 0.968$. A value $\chi^2/\nu = 1.07$ is obtained for six Voigt elements.

20.3 Statistics textbooks show different ways to visualize residual errors. A useful alternative representation involves scaling the residual errors by the standard deviation. Create such plots for the regression mentioned in Problem 19.2.

Solution 20.3. The residual errors for Problem 19.2(a), scaled by the standard deviation, are presented in Figure 20.2 as functions of frequency. Scaling by the standard deviation allows assessment of the extent to which the residual errors fall within the confidence interval for the data. The dashed lines show that almost all residual errors fall within the $\pm 2\sigma$ (95.4%) confidence interval.

20.4 Monte Carlo simulation can be used to explore the influence of random noise on the parameter estimation. Consider the following protocol:

(a) Use a measurement model to identify error structure.

(b) Use a measurement model on Kramers–Kronig-consistent data set to get a model for data.

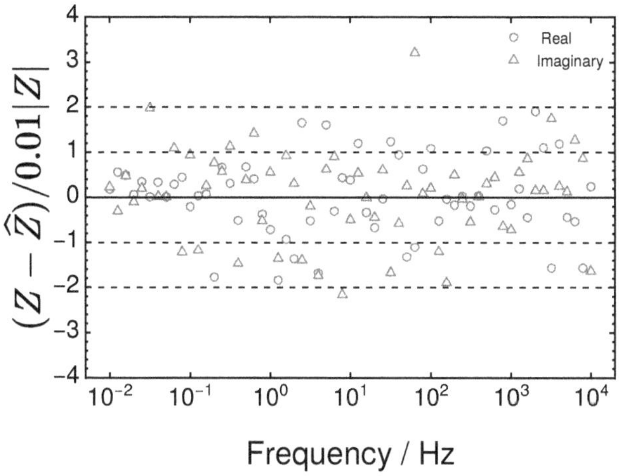

Figure 20.2: The real and imaginary residual errors for the solution of Problem 19.2(a), scaled by the standard deviation, as functions of frequency.

(c) Add to the model values obtained in the previous step a normally distributed random error with an experimentally determined standard deviation.

(d) Fit the resulting model to each synthetic data set, and obtain a distribution of parameter values.

Contrast this approach to direct use of Monte Carlo simulations for estimation of the confidence interval for the diffusivity expressed as $D_i = \nu/\mathrm{Sc}_i$ for which experimentally determined standard deviations are available for ν and Sc_i.

Solution 20.4. The solution for this problem makes use of the data presented in Chapter 30. The error structure is given by equation (30.2), and the regression measurement model results for the data collected after an elapsed time of 12 hours are given in Table 20.1. The measurement model was used to calculate a theoretical impedance response, to which noise was added as given by

$$Z_{r,1} = \widehat{Z}_r + N_1(\mu, \sigma) \tag{20.1}$$

and

$$Z_{j,1} = \widehat{Z}_j + N_2(\mu, \sigma) \tag{20.2}$$

where N_1 and N_2 are independent normal distributions with $\mu = 0$. A similar approach was used to generate datasets 2 and 3 such that all normal distributions were independent. Equation (30.6) was regressed to the synthetic data, and the results are presented in Table 20.2.

Table 20.1: Parameter estimates for regression of the measurement model to data presented in Chapter 30 after an elapsed time of 12 h.

k	τ_k, s	σ_{τ_k}, s	R_k, Ωcm^2	σ_{R_k}, Ωcm^2
1	4.84×10^{-6}	0.85×10^{-6}	1.90	0.34
2	2.75×10^{-5}	0.99×10^{-5}	2.92	0.73
3	2.30×10^{-4}	0.66×10^{-4}	19.3	6.2
4	1.72×10^{-3}	0.42×10^{-3}	185	54
5	0.0112	0.0019	1720	300
6	0.0762	0.0080	1.86×10^4	0.23×10^4
7	0.444	0.032	1.60×10^5	0.13×10^5
8	2.73	0.17	1.57×10^6	0.16×10^6
9	10.75	0.74	8.11×10^6	0.62×10^6
10	63.7	6.6	3.53×10^7	0.15×10^7

Table 20.2: Parameter estimates for regression of equation (30.6) to synthetic data generated by adding stochastic errors to the measurement model fit to data presented in Chapter 30 after an elapsed time of 12 h.

Parameter	Set 1	Set 2	Set 3
R_t, MΩcm^2	62.3 ± 2.2	61.2 ± 2.0	63.1 ± 2.3
Q, μF/cm^{2}s$^{1-\alpha_p}$	0.1364 ± 0.0020	0.1370 ± 0.0020	0.1356 ± 0.0020
α_p	0.5172 ± 0.0077	0.5221 ± 0.0074	0.5152 ± 0.0077
Q_f, μF/cm^{2}s$^{1-\alpha_f}$	0.4077 ± 0.0022	0.4044 ± 0.0021	0.4083 ± 0.0022
α_f	0.98392 ± 0.00061	0.98425 ± 0.00059	0.98373 ± 0.00060
R_e, Ωcm^2	69.66 ± 0.24	69.87 ± 0.23	69.84 ± 0.23
χ^2/ν	25.4	23.2	25.1

Table 20.3: Statistical results for regression results shown in Table 20.2.

Parameter	μ_{1-3}	σ_{1-3}	$\langle\sigma\rangle$	$\langle\sigma\rangle/\sigma_{1-3}$
R_t, MΩcm^2	62.2	1	2.2	2.2
Q, μF/cm^{2}s$^{\beta_p}$	0.1363	0.00067	0.002	3.0
α_p	0.5182	0.0035	0.0076	2.2
Q_f, μF/cm^{2}s$^{\beta_f}$	0.4075	0.00088	0.0022	2.5
α_f	0.98397	0.00026	0.0006	2.3
R_e, Ωcm^2	69.79	0.11	0.23	2.1

The mean value for regression results are presented in Table 20.3 as μ_{1-3}, and the corresponding standard deviation is presented as σ_{1-3}. The standard deviation of regressed parameters from fits to three independent noise-added spectra is compared in Table 20.3 to $\langle\sigma\rangle$, the mean standard deviation provided by the Levenberg–Marquardt regression. As shown in Table 20.3, $\langle\sigma\rangle$ is larger than the sample standard deviation σ_{1-3} by more than a factor of 2.

The standard deviation provided by the regression algorithm represents the confidence in the parameter; whereas the standard deviation of parameters obtained from regression to independent replicate spectra represents the repeatability of the regression. The standard deviation provided by the regression algorithm should be used for calculation of error propagation and for assessing the statistical significance of regressed parameters.

Chapter 21

Error Structure of Impedance Measurements

The problems presented in this section are used to explore an assessment of measurement errors. Problems 21.1–21.5 are based on problems presented by Orazem and Tribollet.[1]

Problems

21.1 Consider a system with a single electrochemical reaction with $R_e = 10\ \Omega$, $R_t = 100\ \Omega$, and $R_tC_{dl} = 1$ s. The noise level of the measurement is often assumed to follow

$$\sigma_{Z_r}(\omega) = \sigma_{Z_j}(\omega) = \alpha_M |Z(\omega)| \tag{21.1}$$

with $\sigma = 0.03\,|Z|$.

(a) Plot the standard deviation of the real and imaginary parts of the impedance measurement as functions of frequency.

(b) Plot the normalized standard deviation of the real and imaginary parts of the impedance measurement, scaled to the respective impedance component, as functions of frequency.

(c) Plot the weighting factor $w = 1/\sigma^2$ for the real and imaginary parts of the impedance measurement as functions of frequency.

Solution 21.1. This problem emphasizes the weighting strategy used for regression of models to impedance data. Modulus weighting, based on equation (21.1), is appropriate for potentiostatic measurements made with a constant amplitude for the input potential. Frequency-independent weighting, based on

$$\sigma_{Z_r}(\omega) = \sigma_{Z_j}(\omega) = \alpha \tag{21.2}$$

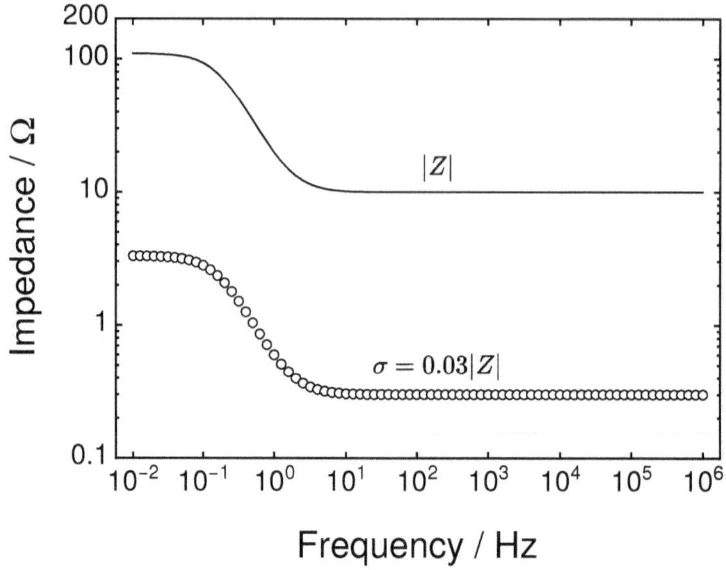

Figure 21.1: Magnitude and standard deviations for the parameters described in Problem 21.1.

applies for galvanostatic measurements made with a constant current-perturbation amplitude. Both equations (21.1) and (21.2) assume that the standard deviation for the real part of the impedance is equal to that of the imaginary part of the impedance. Proportional weighting, in which

$$\sigma_{Z_r}(\omega) = \alpha \left| Z_r(\omega) \right| \tag{21.3}$$

and

$$\sigma_{Z_j}(\omega) = \alpha \left| Z_j(\omega) \right| \tag{21.4}$$

is best suited for regression to synthetic data for which the noise level is controlled by machine precision.

(a) The magnitude of the impedance and the corresponding standard deviation given by equation (21.1) with $\alpha = 0.03$ is given in Figure 21.1. In well-performed experiments, values of $\alpha = 0.002$ are commonly achieved. The error structure shown in Figure 21.1 represents a good first approximation for experiments performed under potentiostatic modulation.

(b) Equation (21.1) is in accordance with the observation that the standard deviation for the real and imaginary parts of the impedance are equal. The scaling shown in Figure 21.2 suggests that the weight applied to real and imaginary parts of the impedance will be very different and the magnitude of the real and imaginary parts of the impedance differ.

(c) The weight applied for regressions under the assumption that the standard deviation of the impedance is as described in Problem 21.1 is shown in Figure 21.3.

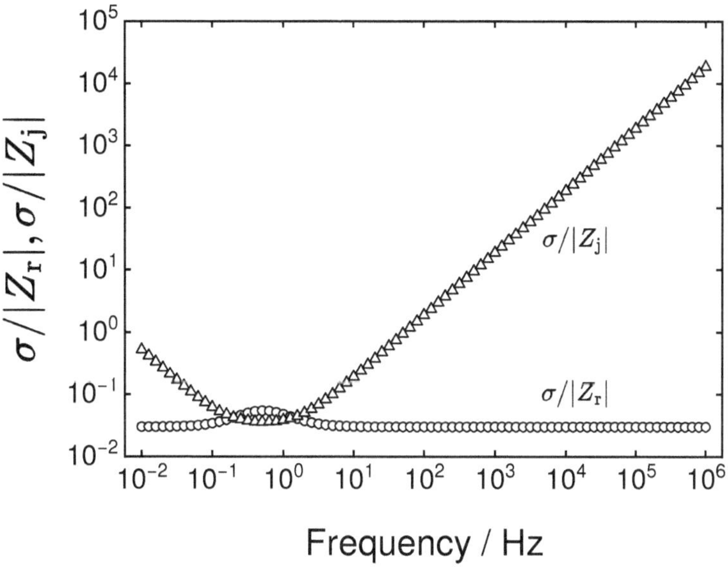

Figure 21.2: Standard deviations scaled by the real and imaginary parts of the impedance for the parameters described in Problem 21.1.

The weight is larger at high frequency and lower at low frequency to account for the observation that the impedance increases at low frequency and approaches the ohmic resistance at high frequency.

21.2 Use Figure 21.4 to estimate the time required for an impedance measurement from 100 kHz to 0.1 Hz with eight measurements per decade. How does this value change if the measurement requires 10 measurements per decade?

Solution 21.2. The elapsed time per frequency may be expressed as the maximum of $3/f$ s and 5 s. Frequencies larger than 0.6 Hz will require 5 s per measurement. Frequencies smaller than 0.6 Hz will require three cycles to make a good measurement. For a frequency range of 100 kHz to 0.1 Hz with eight measurements per decade, the elapsed time is 314 s (5.23 min). The elapsed time increases for 10 points per decade to 388 s (6.46 min).

21.3 Use Figure 21.4 to estimate the time required for an impedance measurement from 100 kHz to 0.1 mHz with eight measurements per decade. How does this value change if the measurement requires three measurements per decade?

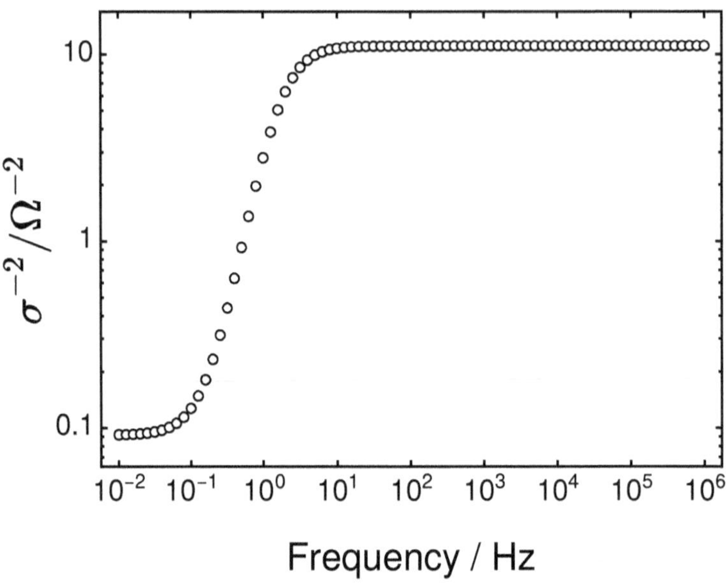

Figure 21.3: The weight applied for regressions under the assumption that the standard deviation of the impedance is as described in Problem 21.1.

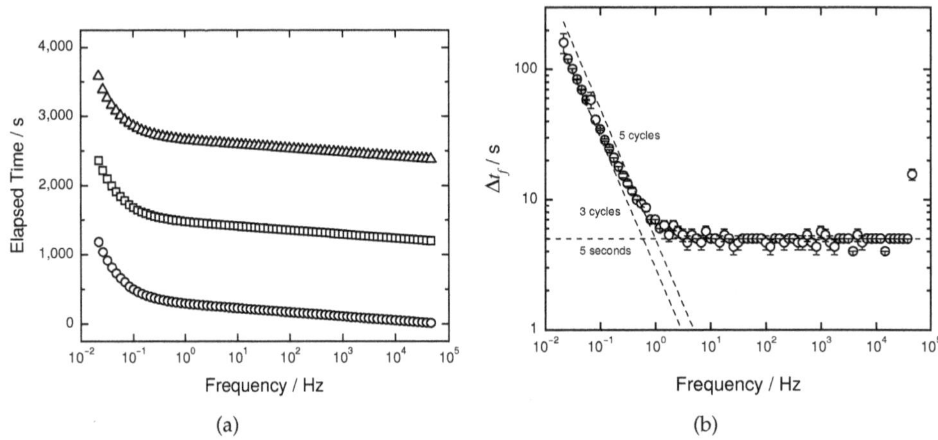

Figure 21.4: Elapsed time for a series of three sequential impedance measurements collected for reduction of ferricyanide on a rotating Pt disk electrode: (a) elapsed time and (b) time required for measurement at each frequency. The standard deviation reflects variability among the three impedance measurements. Taken from Orazem and Tribollet[1] with permission of John Wiley & Sons.

Solution 21.3. The elapsed time per frequency may be expressed as the maximum of $3/f$ s and 5 s. Frequencies larger than 0.6 Hz will require 5 s per measurement. Frequencies smaller than 0.6 Hz will require three cycles to make a good measurement. For a frequency range of 100 kHz to 0.1 mHz with eight measurements per decade, the elapsed time is 1.2×10^5 s (33.4 h). The elapsed time decreases for three points per decade to 5.61×10^4 s (15.6 h).

21.4 Use a spreadsheet program such as Microsoft Excel® or a computational programming environment such as MATLAB® to reproduce the results presented in Figures 21.5 and 21.6.

Solution 21.4. Stochastic noise can be added to both the potential and the current, but the noise in the current is propagated through the impedance of the cell. The percent noise used in the input is defined to be a percent of the amplitude, i.e.,

$$\sigma_{\mathrm{V}} = \frac{P_{\mathrm{V}} \Delta V}{100} \tag{21.5}$$

where σ_{V} is the standard deviation for the potential signal and P_{V} is the percent noise input for the potential signal. The standard deviation for current includes propagation through the electrochemical cell as

$$\sigma_{\mathrm{I}} = \frac{P_{\mathrm{I}} \Delta V}{100|Z|} = \frac{P_{\mathrm{I}} \Delta V}{100} \left(\frac{\sqrt{1 + \omega^2 R_{\mathrm{t}}^2 C_{\mathrm{dl}}^2}}{R_{\mathrm{t}}} \right) \tag{21.6}$$

where σ_{I} is the standard deviation for the current signal, P_{V} is the percent noise input for the current signal, and $\omega = 2\pi f$.

The potential was expressed as

$$V(t) = \overline{V} + \Delta V \cos(\omega t) + N_1(\mu_{\mathrm{V}}, \sigma_{\mathrm{V}}^2) \tag{21.7}$$

where N_1 is a normally distributed random number with mean $\mu_{\mathrm{V}} = 0$ and variance σ_{V}^2. The current was expressed as

$$I(t) = I + N_2(\mu_{\mathrm{I}}, \sigma_{\mathrm{I}}^2) \tag{21.8}$$

where N_2 is an independent normally distributed random number with mean $\mu_{\mathrm{I}} = 0$ and variance σ_{I}^2, and I was given as

$$I(t) = i_0 \left(\exp(b_{\mathrm{a}} V) + \exp(-b_{\mathrm{c}} V) \right) - \omega C_{\mathrm{dl}} \Delta V \sin(\omega t) \tag{21.9}$$

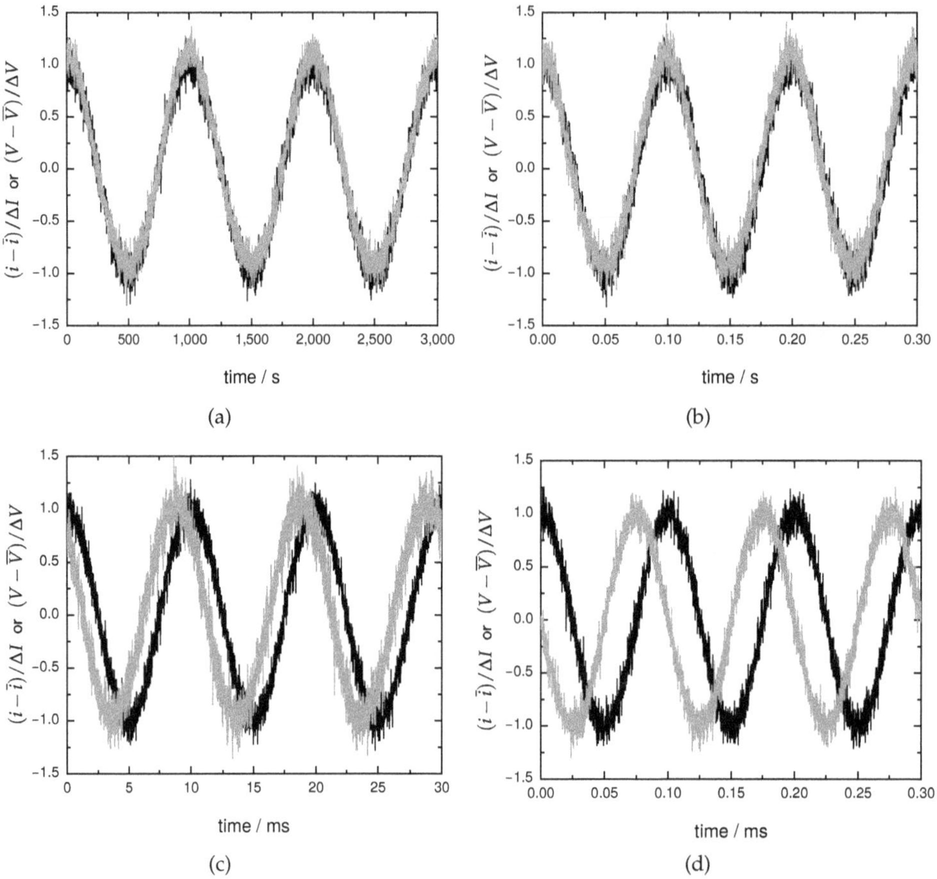

Figure 21.5: The current density response to a sinusoidal potential input for the system with parameters $C_{\text{dl}} = 31\ \mu\text{F/cm}^2$, $i_0 = 0.14\ \text{mA/cm}^2$, $b_{\text{a}} = 19.5\ \text{V}^{-1}$, $b_{\text{c}} = 19.5\ \text{V}^{-1}$, and $\overline{V} = 0.1$ V: (a) 1 mHz; (b) 10 Hz; (c) 100 Hz; and (d) 10 kHz. The signals include normally distributed additive errors with a magnitude of 10 percent of the magnitude of the respective signal. The gray line represents the potential input, and the black line represents the resulting current density, scaled by $\Delta I = \Delta V/|Z|$. Taken from Orazem and Tribollet[1] with permission of John Wiley & Sons.

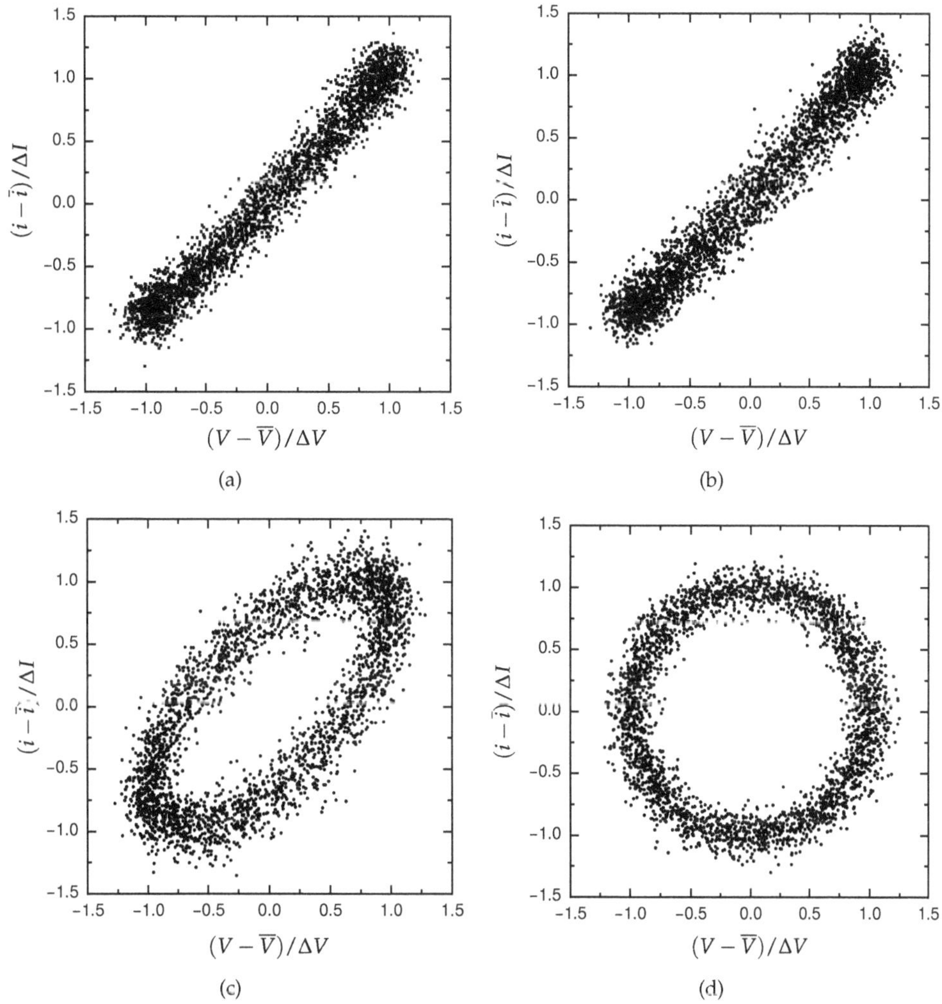

Figure 21.6: Lissajous plots for the system presented in Figure 21.5: (a) 1 mHz; (b) 10 Hz; (c) 100 Hz; and (d) 10 kHz. The potential and current signals were scaled by ΔV and $\Delta I = \Delta V/|Z|$, respectively. Taken from Orazem and Tribollet[1] with permission of John Wiley & Sons.

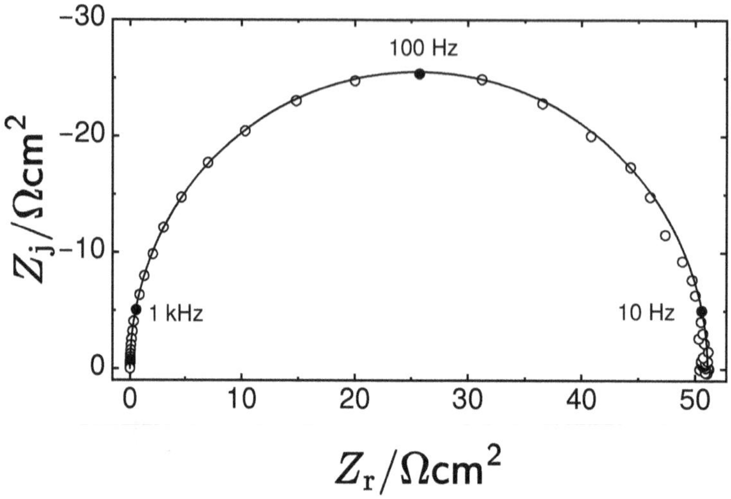

Figure 21.7: Nyquist plot generated using the parameters presented in Figure 21.5. The line represents the theoretical value for linear impedance.

where exchange current density i_0, capacitance C_{dl}, and parameters b_a and b_c are inputs to the program. The code used to generate Figures 21.5 and 21.6 is found in Section B.3.

The calculated impedance, presented in Nyquist format in Figure 21.7, shows the influence of noise at low frequencies. The impedance presented on a logarithmic scale in Figure 21.8 shows the influence of noise in time domain signals when the real part of the impedance approaches zero at high frequencies in Figure 21.8(a) and when the imaginary part of the impedance approaches zero at low frequencies in Figure 21.8(b).

21.5 Add normally distributed stochastic errors to the time-domain potential and current signals for a system with capacity $C_{dl} = 31\ \mu F/cm^2$, charge-transfer resistance $R_t = 51.34\ \Omega cm^2$, and a potential perturbation $\Delta V = 0.01$ V. Then apply the Fourier analysis to calculate the impedance response at the characteristic frequency. Repeat this process, refreshing the random numbers used, so as to calculate the standard deviation of the resulting impedance. How does this result depend on the number of cycles used for the integration?

Solution 21.5. The integration code was replaced with a summation, i.e.,

$$V_r = (V(t) - V)\cos(\omega t) \tag{21.10}$$

was used to calculate the real part of the potential, and

$$V_j = (V(t) - V)\sin(\omega t) \tag{21.11}$$

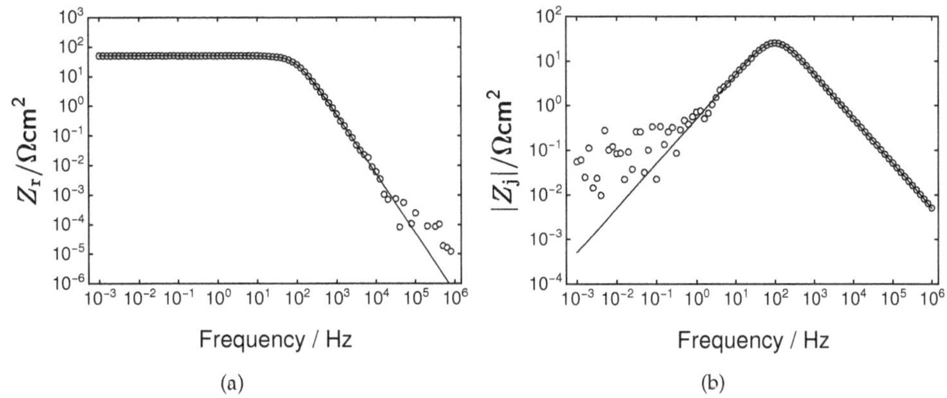

Figure 21.8: Impedance as a function of frequency calculated using the parameters presented in Figure 21.5: (a) real part of the impedance and (b) absolute value of the imaginary part of the impedance. The line represents the theoretical value for linear impedance.

was used to calculate the imaginary part of the potential. Similarly,

$$I_r = (I(t) - \mathrm{I})\cos(\omega t) \tag{21.12}$$

was used to calculate the real part of the current, and

$$I_j = (I(t) - \mathrm{I})\sin(\omega t) \tag{21.13}$$

was used to calculate the imaginary part of the current. The impedance was calculated from

$$Z = \frac{V_r + jV_j}{I_r + jI_j} \tag{21.14}$$

As the integration was a simple summation, the results are sensitive to the number of time intervals. For the data shown here, the 3,000 time intervals comprised one cycle. The code is presented in Section B.4.

The standard deviations for the impedance calculated at a frequency of 100 Hz are presented in Figure 21.9 as a function of number of cycles. The standard deviations, presented for five or more cycles, decrease with increasing number of cycles. The values obtained show standard deviations on the order of 0.3% of the modulus. The standard deviations for the real and imaginary parts of the impedance are equal, as is expected for a Fourier analysis of time-domain signals.

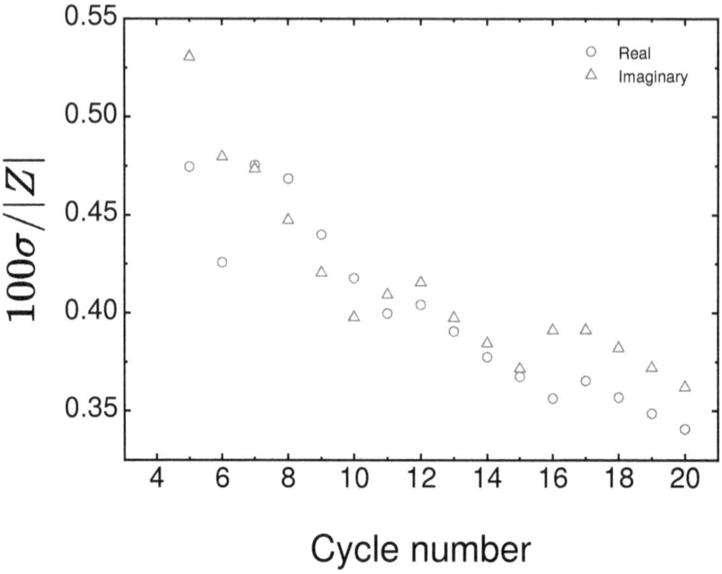

Figure 21.9: Standard deviation as a function of number of cycles for impedance calculated at a frequency of 100 Hz for the parameters described in Figure 21.5. The standard deviations were scaled by the magnitude of the impedance and multiplied by 100 to give units of percentage.

21.6 We are interested in the factors that may influence the error structure of impedance measurements. Please indicate in Table 21.1, by use of a check mark, the corresponding characteristics of impedance data.

Solution 21.6. The answers are given below:

- Potentiostats use different current ranges, e.g., 1, 10, and 100 mA, to measure current. A current of 0.1 mA is measured using the 10 mA range rather than the more appropriate 1 mA range.

  ✓**(A.)** Use of an suboptimal current range will result in increased stochastic error as the resolution of current measurement will be inadequate.

- A long cable is used to measure impedance.

  ✓**(D.)** Cables have an associated resistance, inductance, and capacitance. The result is a Kramers–Kronig-consistent artifact.

Table 21.1: Examples of experimental situations that may cause changes to the impedance measurement error structure.

	Impedance Measurement Characteristic			
	A.	B.	C.	D.
Situation for Impedance Measurement	Increased stochastic error	Failure of causality	Failure of linearity	Kramers-Kronig-consistent artifact
Potentiostats use different current ranges, e.g., 1 mA, 10 mA, and 100 mA, to measure current. A current of 0.1 mA is measured using the 10 mA range rather than the more appropriate 1 mA range.				
A long cable is used to measure impedance.				
A 100 mV perturbation amplitude is used to measure the impedance for a system dominated by electrochemical reactions.				
Impedance is measured for a system that is changing during the course of the measurement.				
Impedance is measured using a setting for which one cycle is used for each measured frequency.				
System temperature is not constant during the course of an impedance measurement.				
Bubbles are generated during the course of the measurement.				

- A 100 mV perturbation amplitude is used to measure the impedance for a system dominated by electrochemical reactions.

  ✓**(C.)** and **(D.)** The 100 mV perturbation amplitude is very large for electrochemical systems, for which a 3 to 10 mV perturbation amplitude is more common. The resulting spectrum will be influenced by failure of linearity and, as the value of the impedance is affected, a Kramers–Kronig -consistent artifact.

- Impedance is measured for a system that is changing during the course of the measurement.

  ✓**(B.)** Nonstationary behavior is associated with failure of causality.

- Impedance is measured using a setting for which one cycle is used for each measured frequency.

  ✓**(A.)** Use of Fourier integration over only once cycle per frequency results in excessive noise in the impedance measurement. Thus, the spectrum will be characterized by increased stochastic error.

- System temperature is not constant during the course of an impedance measurement.

  ✓ **(B.)** System parameters such as rate constants and diffusion coefficients are strongly dependent on temperature. The change in temperature results in non-stationary behavior that itself is associated with failure of causality.

- Bubbles are generated during the course of the measurement. ✓ **(A.)** The presence of bubbles will influence the ohmic resistance. The formation and release of gas bubbles result in an increase of stochastic errors at low frequency.

Chapter 22

The Kramers–Kronig Relations

The problems presented in this chapter explore the Kramers–Kronig relations, integral equations that constrain the real and imaginary components of complex quantities for systems that satisfy conditions of linearity, causality, and stability. These relationships, derived independently by Kronig[65,66] and Kramers,[67,68] were initially developed from the constitutive relations associated with the Maxwell equations for description of an electromagnetic field at interior points in matter.[69] Problems 22.1–22.6 are based on problems presented by Orazem and Tribollet.[1]

Problems

22.1 Use a circuit with $R_e = 10\ \Omega\text{cm}^2$, $R_t = 100\ \Omega\text{cm}^2$, and $C_{dl} = 20\ \mu\text{F/cm}^2$ to verify the equations given in Table A.25. This problem requires use of a spreadsheet program such as Microsoft Excel® or a computational programming environment such as MATLAB®.

(a) Equation (A.84)
(b) Equation (A.85)
(c) Equation (A.86)
(d) Equation (A.87)
(e) Equation (A.88)
(f) Equation (A.89)
(g) Equation (A.90)
(h) Equation (A.91)
(i) Equation (A.92)
(j) Equation (A.93)
(k) Equation (A.94)
(l) Equation (A.95)

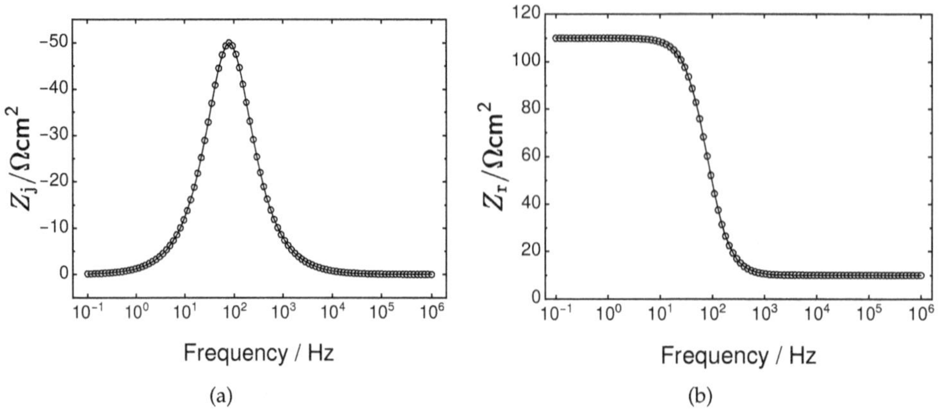

Figure 22.1: Validation of the Kramers–Kronig relations: (a) the imaginary part of the impedance as a function of frequency, where symbols are synthetic data obtained from equation (22.1) and the lines are obtained by integration of equation (A.84) and (b) the real part of the impedance as a function of frequency, where symbols are synthetic data obtained from equation (22.1) and the lines are obtained by integration of equation (A.85).

Solution 22.1. The impedance is given as

$$Z = R_e + \frac{R_t}{1 + j\omega R_t C} \tag{22.1}$$

This problem could be solved using a spreadsheet or by writing a script in a programming language such as MATLAB.® The first step is to use equation (22.1) to create paired values for real and imaginary impedance at discrete frequencies. The second step is to create a continuous function for either the real or imaginary parts of the impedance to be used in the integrand. The most accurate interpolation formula will be obtaining by using equation (22.1). In the following, the symbols represent the synthetic data and the lines represent the transformation obtained by the respective Kramers–Kronig expression.

(a) The program listing is given in Section B.6.1. The results are presented in Figure 22.1(a) as a function of frequency. The superposition of symbols and the results of the integral provide a verification of equation (A.84). The integration is sensitive to the value used as a surrogate for $\pm\infty$ and the value of ϵ used to avoid the singularity at $x = \omega$.

(b) The program listing is given in Section B.6.2. The results are presented in Figure 22.1(b) as a function of frequency. The superposition of symbols and the results of the integral provide a verification of equation (A.85). The integration is sensitive to the value used as a surrogate for $\pm\infty$ and the value of ϵ used to avoid the singularity at $x = \omega$.

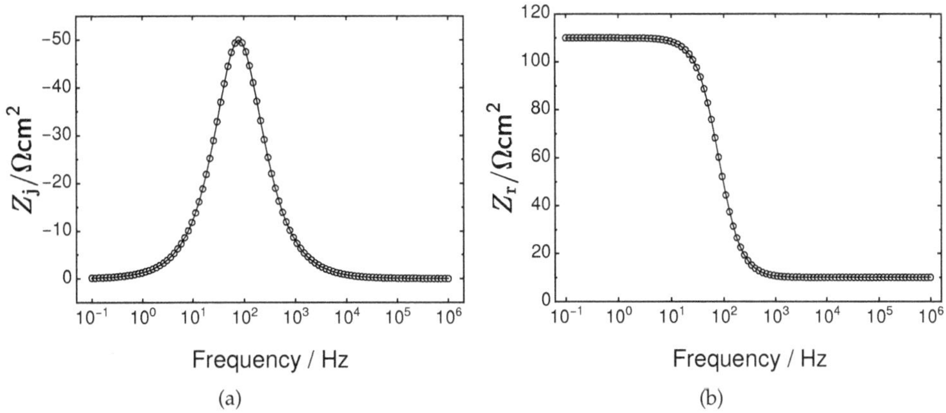

Figure 22.2: Validation of the Kramers–Kronig relations: (a) the imaginary part of the impedance as a function of frequency, where symbols are synthetic data obtained from equation (22.1) and the lines are obtained by integration of equation (A.86) and (b) the real part of the impedance as a function of frequency, where symbols are synthetic data obtained from equation (22.1) and the lines are obtained by integration of equation (A.87).

(c) The program listing is given in Section B.6.3. The results are presented in Figure 22.2(a). The superposition of symbols and the results of the integral provide a verification of equation (A.86). The integration is sensitive to the value used as a surrogate for ∞ and the value of ϵ used to avoid $x = \omega$.

(d) The program listing is given in Section B.6.4. The results are presented in Figure 22.2(b) as a function of frequency. The superposition of symbols and the results of the integral provide a verification of equation (A.87). The integration is sensitive to the value used as a surrogate for ∞ and the value of ϵ used to avoid $x = \omega$.

(e) The program listing is given in Section B.6.5. The results are presented in Figure 22.3(a) as a function of frequency. The superposition of symbols and the results of the integral provide a verification of equation (A.88) of the textbook. The integration is sensitive to the value used as a surrogate for ∞ and the value of ϵ used to avoid $x = \omega$.

(f) The program listing is given in Section B.6.6. The results are presented in Figure 22.3(b) as a function of frequency. The superposition of symbols and the results of the integral provide a verification of equation (A.89) of the textbook. The integration is sensitive to the value used as a surrogate for ∞ and the value of ϵ used to avoid $x = \omega$.

(g) The program listing is given in Section B.6.7. The results are presented in Figure 22.4 as a function of frequency. To get a good agreement at high frequency, the approximation for infinity needed to be increased. Note that the sign in equation (A.90) differs from that published as equation (22.7) in Orazem and Tribollet.[1]

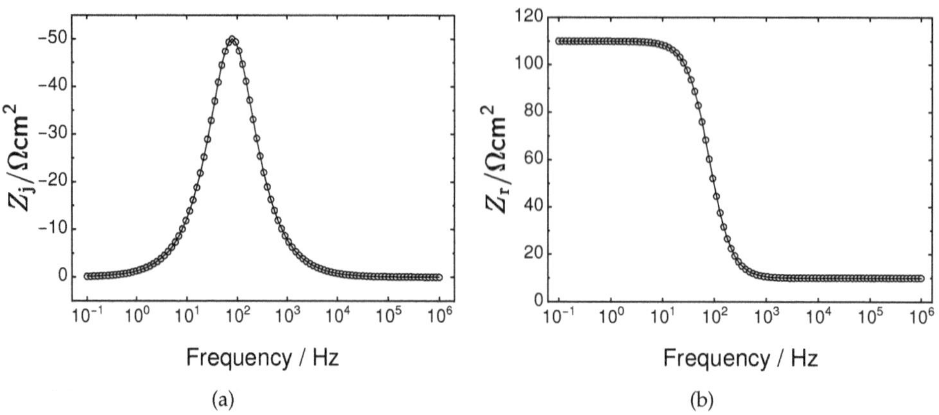

Figure 22.3: Validation of the Kramers–Kronig relations: (a) the imaginary part of the impedance as a function of frequency, where symbols are synthetic data obtained from equation (22.1) and the lines are obtained by integration of equation (A.88) and (b) the real part of the impedance as a function of frequency, where symbols are synthetic data obtained from equation (22.1) and the lines are obtained by integration of equation (A.89).

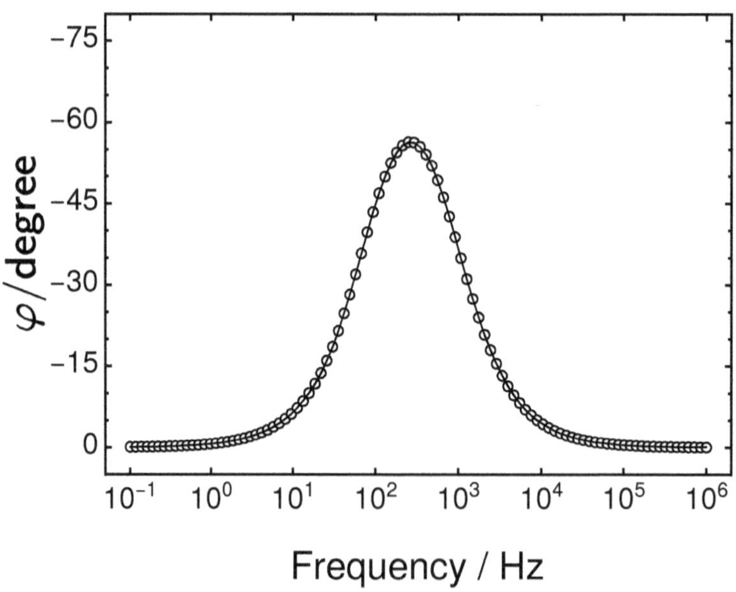

Figure 22.4: Validation of the Kramers–Kronig relations: the phase of the impedance as a function of frequency, where symbols are synthetic data obtained from equation (22.1) and the lines are obtained by integration of equations (A.88), (A.91), and (A.92). The lines are all superposed.

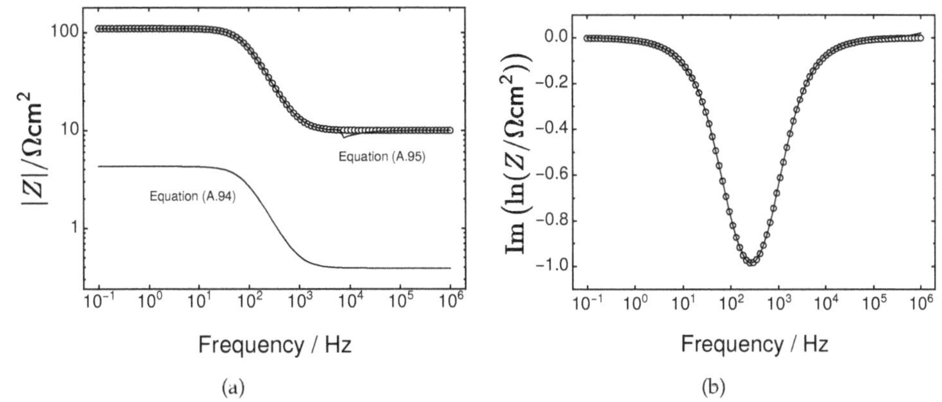

Figure 22.5: Validation of the Kramers–Kronig relations: (a) the magnitude of the impedance as a function of frequency, where symbols are synthetic data obtained from equation (22.1) and the lines are obtained by integration of equations (A.93) and (A.94) and (b) the imaginary part of $\ln(Z)$ as a function of frequency, where symbols are synthetic data obtained from equation (22.1) and the lines are obtained by integration of equation (A.95).

(h) The program listing is given in Section B.6.8. The results presented in Figure 22.4 show good agreement with the synthetic phase calculated from equation (22.1). Note that the sign in equation (A.91) differs from that published as equation (22.8) in Orazem and Tribollet.[1]

(i) The program listing is given in Section B.6.9. The results presented in Figure 22.4 show good agreement with the synthetic phase calculated from equation (22.1). Note that equation (A.92) differs by a factor of 2 from that published as equation (22.9) in Orazem and Tribollet.[1]

(j) The program listing is given in Section B.6.10. The results presented in Figure 22.5(a) show poor agreement with the expected results. Integration of equation (A.93) yields a magnitude that is shifted by a constant amount for all frequencies. The problem may be attributed to the periodic character of the phase. A similar problem was observed for the numerical integration of

$$\ln|Z(\omega)| = \frac{2}{\pi}\int_0^\infty \frac{x\varphi(\omega)}{x^2-\omega^2}\mathrm{d}x \tag{22.2}$$

derived by Kozak et al.[70]

(k) The program listing is given in Section B.6.11. Integration of equation (A.94) yields a magnitude in good agreement with the expected value, with the exception of a discontinuity seen between 6 and 12 kHz.

(l) The program listing is given in Section B.6.12. The results presented in Figure 22.5(b) agrees well with the value obtained from equation (22.1). The deviation at high frequency is very sensitive to the value used to approximate infinity in the integration. Note that the sign in equation (A.95) differs from that published as equation (22.12) in Orazem and Tribollet.[1]

22.2 Verify that the impedance response that includes a CPE, expressed as

$$Z = R_{\mathrm{e}} + \frac{R_{||}}{1 + (\mathrm{j}\omega)^{\alpha} R_{||} Q} \tag{22.3}$$

satisfies the Kramers–Kronig relations.

Solution 22.2. Any form of the Kramers–Kronig relations could be used to verify that the impedance given in equation (22.3) satisfies the Kramers–Kronig relations. This problem may be solved using a spreadsheet or by writing a script in a programming language such as MATLAB.® The first step is to use equation (22.3) to create paired values for real and imaginary impedance at discrete frequencies. The second step is to create a continuous function for either the real or imaginary parts of the impedance to be used in the integrand. The most accurate interpolation formula will be obtaining by using equation (22.3). In the following, the symbols represent the synthetic data and the lines represent the transformation obtained by the respective Kramers–Kronig expression.

The program listings are given in Section B.6.13. The confirmation that equation (22.3) is consistent with the Kramers–Kronig relations is shown in Figure 22.6 using integration of equations (A.86) and (A.87). The parameters used for the simulation were $R_{\mathrm{e}} = 10\ \Omega\mathrm{cm}^2$, $R_{\mathrm{t}} = 100\ \Omega\mathrm{cm}^2$, $Q = 20\ \mu\mathrm{F/cm}^2\mathrm{s}^{1-\alpha}$, and $\alpha = 0.9$. A similar demonstration may be obtained by the regression of the measurement model discussed in Chapter 25.

22.3 Using the results of Problem 8.6, explore whether the Kramers–Kronig relations can be used to detect a nonlinearity in the impedance response. Discuss the implication of your result on experimental design.

Solution 22.3. In general, the Kramers–Kronig relations do not serve to detect nonlinearity so long as the measurements do not encompass a transition frequency. Hirschorn and Orazem[71] showed that, for a faradaic reaction in parallel with a double-layer capacitance, the transition frequency can be expressed as

$$f_{\mathrm{t}} = \frac{1}{2\pi R_{\mathrm{t,obs}} C_{\mathrm{dl}}} \left(1 + \frac{R_{\mathrm{t,obs}}}{R_{\mathrm{e}}}\right) \tag{22.4}$$

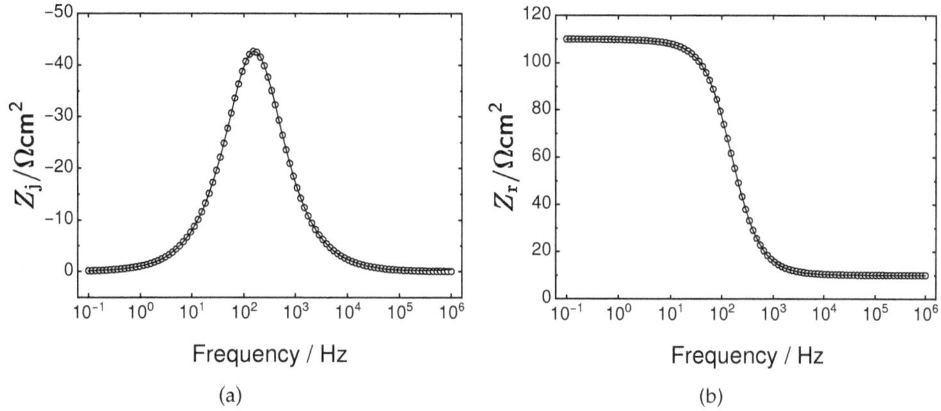

Figure 22.6: Validation of the Kramers–Kronig relations for a CPE given in equation (22.3) for $R_e = 10\ \Omega cm^2$, $R_t = 100\ \Omega cm^2$, $Q = 20\ \mu F/cm^2 s^{1-\alpha}$, and $\alpha = 0.9$: (a) the imaginary part of the impedance as a function of frequency, where symbols are synthetic data obtained from equation (22.3) and the lines are obtained by integration of equation (A.86) and (b) the real part of the impedance as a function of frequency, where symbols are synthetic data obtained from equation (22.3) and the lines are obtained by integration of equation (A.87).

where $R_{t,obs}$ is the observed charge-transfer resistance at the given perturbation amplitude. They provided a simple plot to facilitate determining whether the transition frequency was within the measured frequency range. A version is presented in Figure 22.7. For $R_t = 100\ \Omega cm^2$, $C_{dl} = 20\ \mu F/cm^2$, and $R_e = 1\ \Omega cm^2$, the parameters used in Figure 22.7 may be calculated to be $R_t C_{dl} = 2 \times 10^{-3}$ s and $R_e/R_t = 0.01$. The intersection of dashed lines in Figure 22.7 shows that the transition frequency is 8 kHz. This is almost equal to the characteristic frequency $f_c = 1/(2\pi R_e C_{dl})$ at which nonuniform current and potential distributions associated with the electrode geometry will distort the impedance response.

In short, the Kramers–Kronig relations are not a reliable tool to assess nonlinearity caused by large perturbation amplitudes. Some better approaches include evaluation of distortions in Lissajous curves and total harmonic distortion.[72]

22.4 Why are the assumptions of linearity, causality, and stability necessary for the derivation of the Kramers–Kronig relations?

Solution 22.4. Linearity holds that the output is a linear function of the input. Stability holds that the response to an input impulse cannot increase indefinitely with time. Primitive causality holds that the output response to a perturbation cannot precede the input perturbation. The derivation of the Kramers–Kronig relations involves manipulation of Cauchy's theorem, which states that the integral of an analytic function over a closed contour is equal to zero. In Section 22.2 of Orazem and Tribollet,[1] the

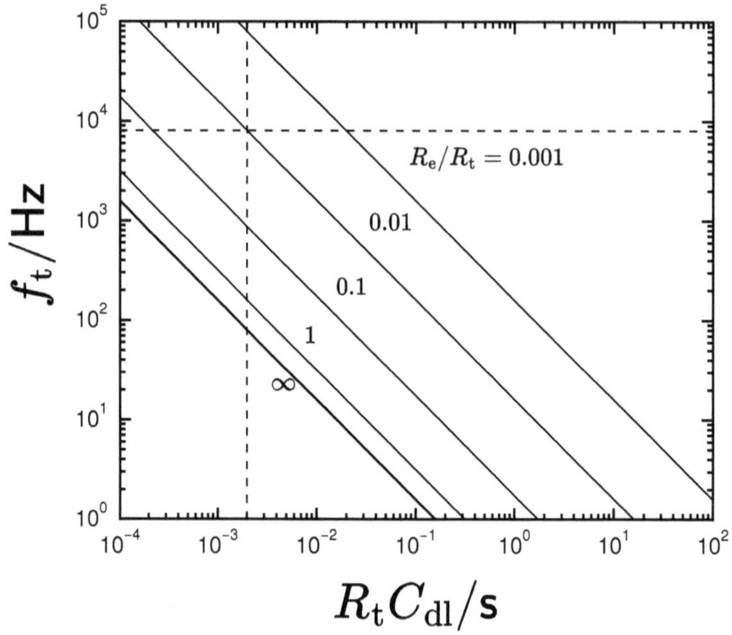

Figure 22.7: The transition frequency given in equation (22.4) as a function of R_tC_{dl} with R_e/R_t as a parameter. Adapted from Hirschorn and Orazem.[71]

concepts of linearity and causality were applied in time domain to establish a causal relationship between input and output. A Fourier integral transform was used to convert the time-domain functions into frequency-domain transfer functions, resulting in the transfer function

$$G(\omega) = \int_0^{+\infty} g(\tau) \exp(-j\omega\tau) d\tau \tag{22.5}$$

which means that G is analytic. Equation (22.5) is not sufficient for derivation of the Kramers–Kronig relations. To be a causal function, the value of $G(\omega)$ must tend toward zero as $\omega \to \infty$, a statement that does not apply for impedance spectroscopy. A weaker restriction of stability can be applied, which can be expressed as requiring that the total output energy cannot exceed the total input energy. Mathematically,

$$|G(\omega)|^2 \leq A \tag{22.6}$$

where A is a constant. The impedance Z is not a causal transform because the imaginary part of Z does not contain all the information needed to obtain the real part. An additive constant cannot be determined by the transformation. By subtracting an additive constant, a modified function $Z - Z_r(\infty)$ is obtained that is a causal transform.

22.5 What is a causal transform? Why is the impedance Z not a causal transform?

Solution 22.5. As discussed in the solution to Problem 22.4, in order for the impedance $Z(\omega)$ to represent a causal transform, Z must satisfy

$$\lim_{\omega\to\infty} Z(\omega) = 0 \tag{22.7}$$

While the imaginary part of Z tends toward zero as $1/\omega$ for $\omega \to \infty$, the real part of the impedance tends toward a finite value. Thus, $Z(\omega)$ is not a causal transform, but $Z(\omega) - Z_r(\infty)$ is a causal transform.

22.6 It is sometimes said that a finite impedance is needed in order for application of the Kramers–Kronig relations to an electrochemical system. Yet, blocking electrodes do not have a finite impedance. Do the Kramers–Kronig relations apply for blocking electrodes? If so, how can they be applied?

Solution 22.6. The Kramers–Kronig relations do apply for electrodes with impedance that tends toward infinity as frequency tends toward zero. The integral equations such as are presented in Table A.25 of the textbook may be applied to a transformed variable, such as admittance $Y = 1/Z$, that has finite values at both high- and low-frequency limits.

22.7 Derive the expression given in equation (A.87).

Solution 22.7. The transfer function $Z(x) - Z_r(\infty)$ tends toward zero with increasing frequency. As $Z(x)$ is analytic, Cauchy's integral theorem can be written as

$$\oint (Z(x) - Z_r(\infty))\, dx = 0 \tag{22.8}$$

where x is an independent and continuous variable that represents the complex frequency.

The first step in the analysis is to combine Z with some other function in such a way that the result vanishes as rapidly as $1/\omega^2$ as ω tends to infinity. In this way, the contribution of the large semicircular path to the contour integral (see, for example, Figure 22.8) can be neglected.

In order to evaluate the real or imaginary part of impedance at a particular frequency, a pole is created at ω, as shown in Figure 22.8. From considerations of symmetry, a complementary pole is created at $-\omega$. If $Z_r(\omega)$, the value assumed by Z_r at

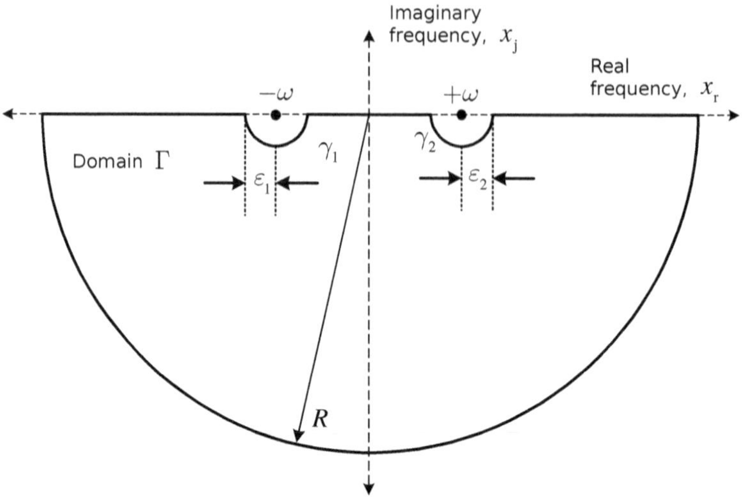

Figure 22.8: Domain of integration for application of Cauchy's integral formula. Poles are placed at frequencies $\pm\omega$ on the real frequency axis. Taken from Orazem and Tribollet[1] with permission of John Wiley & Sons.

$x = \omega$, is subtracted from Z, the resulting contour integral can be written as:

$$\oint \left[\frac{Z\,(x) - Z_r(\omega)}{x - \omega} - \frac{Z(x) - Z_r(\omega)}{x + \omega} \right] \mathrm{d}x = 0 \tag{22.9}$$

Note that the constant $Z_r(\infty)$ used in equation (22.8) cancels in the integrand such that

$$\lim_{x \to \infty} \frac{(Z\,(x) - Z_r(\omega))}{x} \to 0 \tag{22.10}$$

according to $1/x^3$ rather than $1/x$. Equation (22.9) is equal to zero because, following the condition of causality, the integrand is analytic in the domain considered and the path of integration (as shown in Figure 22.8) is a closed loop. Under the assumptions that $\varepsilon_1 \to 0$ and $\varepsilon_2 \to 0$ at γ_1 and γ_2, respectively, the contributions to equation (22.9) are given as

$$\oint \left[\frac{Z(x)-Z_r(\omega)}{x-\omega} - \frac{Z(x)-Z_r(\omega)}{x+\omega}\right]dx =$$

$$\begin{aligned}
&\fint_{-\infty}^{+\infty}\left(\frac{Z(x)-Z_r(\omega)}{x-\omega} - \frac{Z(x)-Z_r(\omega)}{x+\omega}\right)dx \\
&+\int_{\gamma_1}\left(\frac{Z(x)-Z_r(\omega)}{x-\omega} - \frac{Z(x)-Z_r(\omega)}{x+\omega}\right)dx \\
&+\int_{\gamma_2}\left(\frac{Z(x)-Z_r(\omega)}{x-\omega} - \frac{Z(x)-Z_r(\omega)}{x+\omega}\right)dx \\
&+\int_{\Gamma}\left(\frac{Z(x)-Z_r(\omega)}{x-\omega} - \frac{Z(x)-Z_r(\omega)}{x+\omega}\right)dx = 0
\end{aligned} \tag{22.11}$$

where $\fint_{-\infty}^{+\infty} f(x)dx$ represents the Cauchy principal value of the integral. As shown in Figure 22.8, integration at γ_1 and γ_2 refers to semicircular paths centered at $-\omega$ and $+\omega$, respectively, and integration at Γ refers to integration along the large semicircle of radius R.

Under the assumption that $Z(x) - Z(\omega)$ approaches a finite value as $x \to \infty$,

$$\lim_{R\to\infty}\int_{\Gamma}\left(\frac{Z(x)-Z_r(\omega)}{x-\omega} - \frac{Z(x)-Z_r(\omega)}{x+\omega}\right)dx = 0 \tag{22.12}$$

where R is the radius of the semicircular path shown in Figure 22.8. Therefore, the non-zero contributions to the integral in equation (22.9) result from the path of integration along the real frequency axis and at the two small semicircular indentations.

If the radii ε_1 and ε_2 of the semicircular paths γ_1 and γ_2 approach zero, the term $1/(x+\omega)$ dominates along path γ_1, and $1/(x-\omega)$ is the dominant term along path γ_2. Thus,

$$\begin{aligned}
&\fint_{-\infty}^{+\infty}\left(\frac{Z(x)-Z_r(\omega)}{x-\omega} - \frac{Z(x)-Z_r(\omega)}{x+\omega}\right)dx \\
&\quad -\int_{\gamma_1}\left(\frac{Z(x)-Z_r(\omega)}{x+\omega}\right)dx + \int_{\gamma_2}\left(\frac{Z(x)-Z_r(\omega)}{x-\omega}\right)dx = 0
\end{aligned} \tag{22.13}$$

From an application of Cauchy's Integral Formula to a half-circle,

$$-\int_{\gamma_1}\frac{Z(x)-Z_r(\omega)}{x+\omega}dx = -j\pi\left(Z(-\omega)-Z_r(\omega)\right) = -\pi Z_j(\omega) \tag{22.14}$$

and

$$\int_{\gamma_2}\frac{Z(x)-Z_r(\omega)}{x-\omega}dx = j\pi\left(Z(\omega)-Z_r(\omega)\right) = -\pi Z_j(\omega) \tag{22.15}$$

Thus, equation (22.11) takes the form

$$\fint_{-\infty}^{+\infty}\left(\frac{(Z_r(x)+jZ_j(x))-Z_r(\omega)}{x-\omega}-\frac{(Z_r(x)+jZ_j(x))-Z_r(\omega)}{x+\omega}\right)dx = 2\pi Z_j(\omega) \quad (22.16)$$

Equation (22.16) can be expressed as

$$2\omega\fint_{-\infty}^{+\infty}\left(\frac{(Z_r(x)+jZ_j(x))-Z_r(\omega)}{x^2-\omega^2}\right)dx = 2\pi Z_j(\omega) \quad (22.17)$$

The integral from $-\infty$ to $+\infty$ can be separated as

$$\begin{aligned}\fint_{-\infty}^{+\infty}\left(\frac{(Z_r(x)+jZ_j(x))-Z_r(\omega)}{x^2-\omega^2}\right)dx = \\ \fint_{-\infty}^{0}\left(\frac{(Z_r(x)+jZ_j(x))-Z_r(\omega)}{x^2-\omega^2}\right)dx + \fint_{0}^{+\infty}\left(\frac{(Z_r(x)+jZ_j(x))-Z_r(\omega)}{x^2-\omega^2}\right)dx\end{aligned} \quad (22.18)$$

where the odd and even properties of the imaginary and real components of $Z(x)$ are again used to yield

$$\fint_{-\infty}^{0}\left(\frac{(Z_r(x)+jZ_j(x))-Z_r(\omega)}{x^2-\omega^2}\right)dx = \fint_{0}^{+\infty}\left(\frac{(Z_r(x)-jZ_j(x))-Z_r(\omega)}{x^2-\omega^2}\right)dx \quad (22.19)$$

Thus,

$$Z_j(\omega) = \frac{2\omega}{\pi}\fint_{0}^{\infty}\frac{Z_r(x)-Z_r(\omega)}{x^2-\omega^2}dx \quad (22.20)$$

Evaluation of the integral at $x \to \omega$ poses no particular problem because, by l'Hôpital's rule,

$$\lim_{x\to\omega}\frac{Z_r(x)-Z_r(\omega)}{x^2-\omega^2} = \frac{1}{2x}\frac{dZ_r(x)}{dx} = \frac{1}{2}\frac{dZ_r(x)}{d\ln(x)} \quad (22.21)$$

A similar development can be used to obtain the real part of the impedance as a function of the imaginary part.

Chapter 23

An Integrated Approach to Impedance Spectroscopy

The problems presented in this chapter explore a philosophy for understanding impedance spectroscopy that integrates experimental observation, model development, and error analysis. Problems 23.1–23.4 are based on problems presented by Orazem and Tribollet.[1]

Problems

23.1 Suggest experiments that could be used to support or reject the model described in:

(a) Corrosion of iron in anaerobic electrolytes for which the proposed reactions are

$$\mathrm{Fe} \rightarrow \mathrm{Fe}^{2+} + 2\mathrm{e}^- \tag{23.1}$$

and

$$\mathrm{H_2O} + \mathrm{e}^- \rightarrow \frac{1}{2}\mathrm{H_2} + \mathrm{OH}^- \tag{23.2}$$

(b) Corrosion of iron in an aerated electrolyte for which the proposed reactions are

$$\mathrm{Fe} \rightarrow \mathrm{Fe}^{2+} + 2\mathrm{e}^- \tag{23.3}$$

and

$$\mathrm{O_2} + 2\mathrm{H_2O} + 4\mathrm{e}^- \rightarrow 4\mathrm{OH}^- \tag{23.4}$$

(c) Problem 10.4
(d) Problem 10.5
(e) Problem 10.6

Solution 23.1. While not mentioned explicitly, the emphasis here is on validating the proposed model for the impedance response at open circuit potential.

(a) The impedance at open circuit can be augmented by impedance measurements at potentials slightly larger and smaller than the open-circuit potential, for example, ± 10 mV, as described in Section A.7. The meaning of different outcomes is described in Table 9.1. As always, it is also necessary to plot the polarization curve. A Tafel analysis is possible. With the model corresponding to the electrochemical reactions (23.1) and (23.2), the impedance must be a semicircle or a depressed semicircle, which can be analyzed with a CPE in parallel to a charge-transfer resistance.

(b) Once again, the impedance at open circuit can be augmented by impedance measurements at potentials slightly larger and smaller than the open-circuit potential, for example, ± 10 mV. The cathodic part of the steady-state polarization curve must show the mass-transport limitation. To control the mass transport, it could be necessary to use an iron rotating disk electrode or an impinging jet cell. The cathodic part of the faradaic impedance must be a charge-transfer resistance in series with a convective diffusion impedance. The anodic faradaic impedance is a charge-transfer resistance.

(c) The model considers a mass-transport limitation; thus, in this case, it is also necessary to control the mass transport by using a rotating disk electrode or an impinging jet cell. If the study is performed at the open-circuit potential, as previously, the impedance at open circuit can be augmented by impedance measurements at potentials slightly larger and smaller than the open-circuit potential, for example, ± 10 mV. For measurements near open circuit, the cathodic reaction must be taken into account.

If the study is considered in the anodic potential range, it is also necessary to perform some impedance measurements around this anodic potential. Some measurements at different concentrations of the species could also help to verify the model.

(d) The impedance measurements at potentials slightly larger and smaller than the open-circuit potential are always necessary. Due to the mass-transport limitation, the use of a rotating disk electrode or an impinging jet cell to provide a controlled convection is also necessary. Some measurements at different Cl^- concentrations could be useful.

(e) The use of a rotating disk electrode of magnesium could show the effect of the mass-transport limitation due to the convection when the layer of $Mg(OH)_2$ is thin.

23.2 Suggest experiments that could be used to support or reject the model for the impedance response of an *n*-GaAs Schottky diode at different temperatures. One model uses a CPE that accounts for implicit distributed relaxation processes.[73] This model fits the data very well with a minimal number of parameters. A second model accounts for discrete energy levels and provides a fit of equivalent regression quality at each temperature.[74,75]

Solution 23.2. Deep-Level Transient Spectroscopy (DLTS) is commonly used to explore the existence of discrete electronic states within the band gap.[76] Orazem et al.[75] used DLTS to justify a model that accounted for discrete deep-level states.

23.3 Several prominent leaders in impedance research have proposed creating a catalog of impedance spectra with unique interpretations associated with different shapes of the impedance diagrams. Discuss the potential and limitations of such a catalog.

Solution 23.3. Models for impedance are not unique, meaning that models based on different physical descriptions can yield identical fits. Thus, a catalog of shapes, for example, of Nyquist plots, may provide a useful guidance for model development but cannot be definitive.

23.4 Explain the role of error analysis in:

(a) Experimental design

(b) Regression analysis

(c) Model identification

Solution 23.4. The contributions to the error structure include stochastic contributions, which have a mean value equal to zero, bias contributions, which cause inconsistency with the Kramers–Kronig relations, and lack-of-fit.

(a) Excessive stochastic noise may suggest changes in the experimental parameters, including ensuring use of the optimal current measuring range, increasing the integration time, increasing the amplitude of modulated signal, introducing a delay time to avoid undesired error caused by the transient caused by changing from one frequency to another, and avoiding the line frequency and its first harmonic. Low-frequency bias errors are associated with nonstationary behavior. The solution is to reduce the time required for the measurement by reducing the time allowed for integration at each frequency, thereby increasing the magnitude of stochastic errors in the measurement, reducing the frequency range, or reducing the number of frequencies measured per decade.

(b) The standard deviation of the measurements is used to weight regressions, and the assessment of consistency with the Kramers–Kronig relations determines the frequency range used for the analysis.

(c) The error structure affects model identification because the number of parameters that can be extracted is influenced by the noise level of the measurements and the useable frequency range.

23.5 Why is impedance spectroscopy not considered to be a stand-alone technique? Are other electrochemical techniques such as cyclic voltammetry considered to be stand-alone? Explain your answers.

Solution 23.5. Impedance spectroscopy applies a frequency-domain transfer-function measurement to a complex system that cannot be easily visualized. The quantities measured, for example, current and potential for electrochemical or electronic systems and stress and strain for mechanical systems, are macroscopic values that represent the spatial average of individual events. These quantities are influenced by the desired physical properties, such as diffusivity, rate constants, and viscosity, but do not provide a direct measure of them. This means that system-specific models must be developed to interpret the impedance data in terms of physical quantities. The same constraint applies to all electrochemical measurements, including cyclic voltammetry. As the models developed are not unique, these techniques require other measurements to enhance confidence in the model.

Chapter 24

Complex Integrals

Chapter 1 provides a framework for the analysis of complex variables. The problems presented in this chapter explore complex integration in greater detail in the context of the derivation of the Kramers–Kronig relations, see Chapter 22. Problems 24.1–24.3 are based on problems taken from Orazem and Tribollet.[1]

Problems

24.1 Show that e^z is an analytic function in any domain.

Solution 24.1. The derivative of the function

$$f(z) = \exp(z) \tag{24.1}$$

is given as

$$\frac{df(z)}{dz} = \exp(z)\frac{dz}{dz} \tag{24.2}$$

or

$$\frac{df(z)}{dz} = \exp(z) \tag{24.3}$$

The function e^z has a derivative for all finite values of z. Thus, it is an analytic function in any domain.

24.2 Show that the exponential of the complex conjugate $e^{\overline{z}}$ is not an analytic function in any domain.

Solution 24.2. The derivative of the function

$$f(z) = \exp(\overline{z}) \tag{24.4}$$

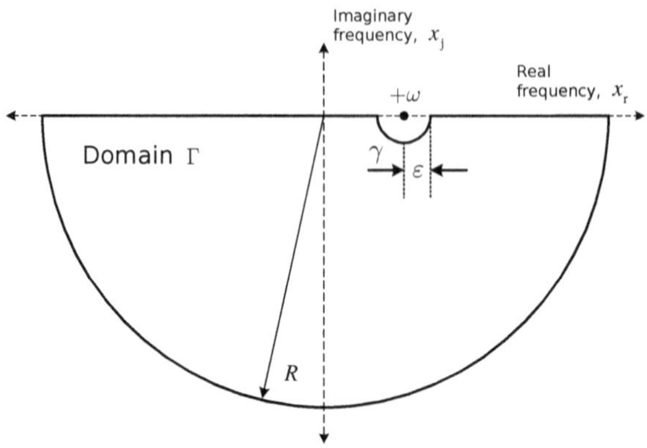

Figure 24.1: Domain of integration for development application of Cauchy's integral formula. A pole is placed at frequency $+\omega$ on the real frequency axis. Taken from Orazem and Tribollet[1] with permission of John Wiley & Sons.

is given as

$$\frac{\mathrm{d}f(z)}{\mathrm{d}z} = \exp(z)\frac{\mathrm{d}\bar{z}}{\mathrm{d}z} \tag{24.5}$$

The complex conjugate is not analytic as it does not satisfy the Cauchy–Riemann conditions, which state that the complex function

$$f(z) = u(x,y) + \mathrm{j}v(x,y) \tag{24.6}$$

is analytic in a region G if and only if $\partial u/\partial x$, $\partial u/\partial y$, $\partial v/\partial x$, and $\partial v/\partial y$ are continuous and satisfy

$$\frac{\partial u}{\partial x} = \frac{\partial v}{\partial y} \quad \text{and} \quad \frac{\partial u}{\partial y} = -\frac{\partial v}{\partial x} \tag{24.7}$$

in the region G. For a function (24.6) that satisfies the Cauchy–Riemann conditions, the complex conjugate, given as

$$f(z) = u(x,y) - \mathrm{j}v(x,y) \tag{24.8}$$

does not satisfy the Cauchy–Riemann conditions.

24.3 Beginning with

$$\oint_{\Gamma} \frac{G(x)}{x-\omega}\mathrm{d}x = 0 \tag{24.9}$$

derive equations (A.84) and (A.85). The domain Γ is shown in terms of complex frequency x in Figure 24.1.

Solution 24.3. Let $G(x)$ be $Z(x) - Z_r(\infty)$, which approaches zero as $\omega \to \infty$. Then,

$$\oint_{\Gamma} \frac{Z(x) - Z_r(\infty)}{x - \omega} dx = \int_{\Gamma} \frac{Z(x) - Z_r(\infty)}{x - \omega} dx + \int_{-R}^{\omega-\varepsilon} \frac{Z(x) - Z_r(\infty)}{x - \omega} dx \quad (24.10)$$

$$+ \int_{\gamma} \frac{Z(x)}{x - \omega} dx + \int_{\omega+\varepsilon}^{R} \frac{Z(x) - Z_r(\infty)}{x - \omega} dx = 0$$

In the limits that $R \to \infty$ and $\epsilon \to 0$,

$$\int_{\Gamma} \frac{Z(x) - Z_r(\infty)}{x - \omega} dx = 0 \quad (24.11)$$

and

$$\int_{-\infty}^{+\infty} \frac{Z(x) - Z_r(\infty)}{x - \omega} dx = - \int_{\gamma} \frac{Z(x) - Z_r(\infty)}{x - \omega} dx + \quad (24.12)$$

If the radius ε of the semicircular path γ approaches zero, the term $1/(x - \omega)$ is the dominant term along path γ. From an application of Cauchy's Integral Formula to a half-circle,

$$\int_{\gamma} \frac{Z(x) - Z_r(\infty)}{x - \omega} dx = -j\pi \left(Z(\omega) - Z_r(\infty) \right) \quad (24.13)$$

Expression of Z into real and imaginary parts yields

$$Z_r(\omega) - Z_r(\infty) + jZ_j(\omega) = \frac{j}{\pi} \int_{-\infty}^{+\infty} \frac{Z_r(x) - Z_r(\infty) + jZ_j(x)}{x - \omega} dx \quad (24.14)$$

Collection of real quantities yields

$$Z_r(\omega) = Z_r(\infty) - \frac{1}{\pi} \int_{-\infty}^{+\infty} \frac{Z_j(x)}{x - \omega} dx \quad (24.15)$$

which is the same as equation (A.85). Collection of imaginary quantities yields

$$Z_j(\omega) = \frac{1}{\pi} \int_{-\infty}^{+\infty} \frac{Z_r(x) - Z_r(\infty)}{x - \omega} dx \quad (24.16)$$

As

$$\int_{-\infty}^{+\infty} \frac{Z_r(\infty)}{x - \omega} dx = 0 \quad (24.17)$$

$$Z_j(\omega) = \frac{1}{\pi} \int_{-\infty}^{+\infty} \frac{Z_r(x)}{x - \omega} dx \tag{24.18}$$

which is the same as equation (A.84).

Part II

Application to Experimental Data

Chapter 25

Measurement Model for Regression of Impedance Data

The measurement model described in this chapter is based on a program developed in our group in the 1990s. Work on the original measurement model began in 1989 and culminated in a graphical user interface written primarily in MATLAB® that interfaced with Fortran executables. Our original measurement model program was not made generally available because each new version of MATLAB® broke a part of the program, and a part of the FORTRAN code was not ours to share. The new code was written from the ground up in Python, although some portions of the back-end libraries, such as `numpy`, use C.

25.1 Introduction

The measurement model was developed in our group as a means to quantify the error structure of electrochemical impedance spectroscopy measurements. The work represented an extension of an approach developed by García-Rubio[77–79] for optical characterizations to spectra associated with electrochemical impedance measurements. In collaboration with García-Rubio, our group demonstrated that a Voigt series can provide an adequate fit to impedance spectra and could therefore serve as a generalized measurement model.[2] By filtering lack of replication of similar impedance spectra, the measurement model was used to identify the standard deviation representing the stochastic error of impedance measurements.[3] The measurement model was also used to identify the frequency range that was consistent with the Kramers–Kronig relations.[80,4]

Significant work was expended in the mid 1990s to the mid 2000s to verify the stochastic error structure identified by the measurement model. Durbha et al.,[81] for example, showed that the statistical properties of the error structure were consistent with the propagation of errors through the Kramers–Kronig relations. Carson et al.[82,83]

showed, by numerical emulation of impedance instruments, that time-domain noise propagated into the frequency domain had properties similar to those obtained by the measurement model analysis. Shukla et al.[84] used both rational[85,86] and Voigt measurement models to show that the stochastic error structure identified was independent of the form of the measurement model used.

Instrument vendors often provide a means of assessing consistency of impedance data to the Kramers–Kronig relations that are based on the linear regression approach pioneered by Boukamp.[87] You et al.[88] showed that the Voigt measurement model presented in this chapter was more sensitive to failures of causality associated with nonstationary behavior than was the linear regression approach.

The measurement model may be used for preliminary characterization of impedance data. Orazem et al.[89] showed that the measurement model could yield estimates for the polarization and ohmic resistances. This approach was applied, for example, to estimate an upper bound for the rate of corrosion of copper in anaerobic water.[61] Orazem et al.[90] used regression of the Voigt measurement model as a means of deconvolution to identify underlying distributions in impedance spectra, and Chen et al.[91] explored the use of the Voigt measurement model to identify resistivity distributions in oxide films. The measurement model was the foundation for the observation that a power-law distribution of resistivity in a film gives rise to constant-phase-element (CPE) behavior.[56,92] More recently, Liao et al.[93] and You et al.[94] demonstrated that, under many conditions, the measurement model may be used to extract the capacitance of electrochemical systems. The ability to extract capacitance by the use of the measurement model is important, because most impedance systems show distributed-time-constant behavior that obscures the value of the capacitance.

This chapter provides an introductory guide to the measurement model tools used in the Orazem group for the past 30 years. The program allows identification of the stochastic error structure of impedance measurements. The stochastic error structure is used to weight further regressions. The program may be used to determine what part of the impedance measurement is inconsistent with the Kramers–Kronig relations. The program provides estimates for capacitance and ohmic resistance, from which the characteristic frequency may be identified above which the geometry of the electrode may cause frequency dispersion. The program also allows fitting of custom models to data. Some of the concepts which underlie the program are described in Orazem and Tribollet.[1] The objective is to take full advantage of the information that can be extracted from electrochemical impedance spectroscopy measurements.[95–97] A comprehensive reference manual is provided with the installation file for the measurement model program that may be accessed from the *Help and About* tab.

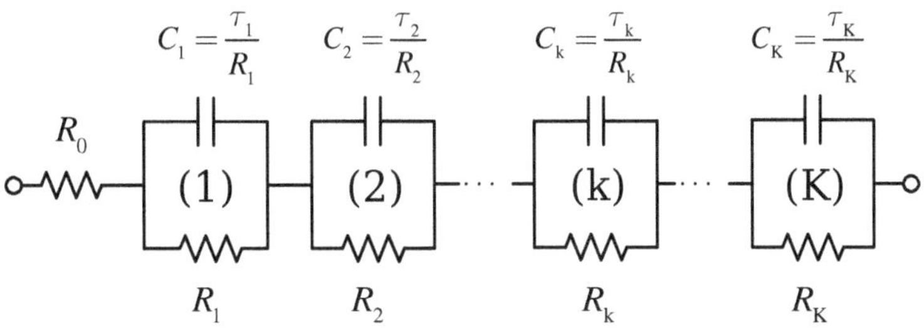

Figure 25.1: A schematic representation of a Voigt circuit used by Agarwal et al.[2–4] as a measurement model. Taken from Liao et al.[93]

25.2 Measurement Model

The measurement model proposed by Agarwal et al.[2] comprises Voigt elements in series with a solution resistance, shown in Figure 25.1. The impedance of the Voigt circuit can be expressed as

$$Z = R_0 + \sum_{k=1}^{K} \frac{R_k}{1 + j\omega\tau_k} \tag{25.1}$$

where the resistance R_k has units of $[\Omega]$ or $[\Omega cm^2]$, τ_k has units of [s], and $j = \sqrt{-1}$ is the imaginary number. Agarwal et al.[2] showed that, by including a sufficient number of terms, a general measurement model based on equation (25.1) can fit impedance data for typical stationary electrochemical systems. The number of Voigt elements K is increased sequentially under the constraint that the 95.4% ($\pm 2\sigma$) confidence interval for each regressed parameter does not include zero. The same value for K can be obtained by minimization of the Akaike information criterion, which assigns a penalty for each additional parameter.[98]

A capacitance for each element k can be expressed as

$$C_k = \frac{\tau_k}{R_k} \tag{25.2}$$

where capacitance C_k has units of [F] or $[F/cm^2]$. In the limit that frequency tends toward infinity,

$$\lim_{\omega \to \infty} Z = R_0 + \sum_{k=1}^{K} \frac{R_k}{j\omega\tau_k} = R_0 - j\sum_{k=1}^{K} \frac{1}{\omega C_k} = R_0 - j\frac{1}{\omega C_{eff}} \tag{25.3}$$

where the effective capacitance can be obtained as

$$\frac{1}{C_{eff}} = \sum_k \frac{1}{C_k} \tag{25.4}$$

In the limit that frequency tends toward zero, the polarization resistance may be obtained as

$$\lim_{\omega \to 0} Z = R_p = R_0 + \sum_{k=1}^{K} R_k \tag{25.5}$$

The confidence intervals for the effective capacitance and polarization resistance can be obtained from the estimated standard errors of the regressed parameters using a linear propagation of error analysis (see Section 3.2 of Orazem and Tribollet[1]).

The ohmic resistance is estimated from the value of R_0. Thus, the measurement model provides estimates for capacitance, polarization resistance, and the ohmic resistance.

25.3 Brief Guide/Instructions

This section provides an overview of the measurement model program and helpful hints. The reference manual, provided with the installation package for the measurement model, provides more detail for the hints mentioned below. The manual may be accessed using the first line in the Help and About tab.

1. The main purposes of the measurement model are listed below.

    (a) Assess the stochastic error structure using repeated measurements (see Section 25.7).

    (b) Determine the frequency range that satisfies the Kramers–Kronig relations, derived under the assumption that the measurement is stable, causal, and linear (see Section 25.8).

    (c) Determine the frequencies that are influenced by nonuniform current and potential distributions (see, for example, Section 26.3).

    (d) Facilitate regression of user-specified models to data (see Section 25.9 and Chapters 26–30).

2. The first impedance data point at high frequency should be removed from the regression because it is affected by the transient change from the steady-state condition to the sinusoidal steady state.

3. Measurement model regression may make use of different strategies.

    (a) The initial regression is usually successful as a good algorithm is used to make an initial guess for parameters. If the initial regression does not succeed, open the edit model parameters dialog and adjust parameters. If the ohmic resistance is very small, fix the value to zero.

    (b) The multi-start option is often successful, but if it is not, consider the use of the Magic Finger.

    (c) Parameters are added sequentially for a measurement model regression until the 95.4% confidence interval for regressed parameters includes zero.

4. Assessment of the frequency-dependent stochastic error structure requires that the number of measurement model parameters and the frequencies analyzed be the same for replicate spectra. For this analysis, the spectra need not be identical, but the shapes of the spectra in, for example, a Nyquist plot, must be similar.

5. It is strongly recommended that error model weighting is used for regressions to assess consistency with the Kramers–Kronig relations and for regression of process models.

6. Use of this program to assess consistency with the Kramers–Kronig relations may require different strategies. A successful complex fit to the data indicates that the data satisfy the Kramers–Kronig relations. It is often more sensitive to fit to the imaginary part of the measurement and use the regressed parameters to predict the real part. The program allows for manual adjustment of the ohmic resistance, which cannot be obtained from regression to the imaginary part of the measurement.

7. Regression weighted by the experimentally determined stochastic error structure yields a χ^2/ν statistic that has physical meaning. For a good fit, $\chi^2/\nu = 1 + \sqrt{2/\nu}$, where ν is the degree of freedom for the regression.

8. For data obtained by Fourier analysis, the test of consistency with the Kramers–Kronig relations provides a means for truncating the frequency range to be used for subsequent regressions. This determination may be iterative.

9. Data collected by Fast Fourier Transform techniques represent an averaged impedance that satisfies the Kramers–Kronig relations.[88]

10. The maximum frequency of the measurement may be affected by the capacitance and inductance associated with the cables connecting the instrument to the system under study.

11. The impedance collected at frequencies larger than $f_c = 1/2\pi R_e C_0$ may be affected by a complex frequency-dependent ohmic impedance.[99] The measurement model provides estimates for R_e and C_0, which can guide assessment of frequencies affected by ohmic impedance.[93]

12. Regression of custom models may be helped by the following hints.

    (a) The step-by-step simplex option may be used to improve the initial guess values and serves to confirm that the model was coded correctly.

    (b) Implementation of the multi-Start option increases the time required to perform a regression but is helpful to adjust for poorly chosen initial guesses. To minimize the increase in required time for regression, use this option for parameters that are most uncertain.

(c) In cases where confidence intervals are not obtained for regressed parameters, ensure that at least one parameter uses the +/− option. The option to perform Monte Carlo simulation for the calculation of the standard deviations of the regressed parameters is not very useful. It is better to refine the regression until confidence intervals are obtained for regressed parameters.

13. A poor experimental design can affect the analysis of impedance data. For example, for repeatable results, make sure the system reaches a steady state before starting the impedance measurement.

25.4 Experimental Methods

Experimental data were collected for an electrode consisting of a 25 μm diameter platinum wire mounted in a glass sleeve such that only the disk-shaped end of the wire was exposed. A platinum mesh counter electrode and a silver/silverchloride (Ag|AgCl) reference were used to form a three-electrode cell configuration. The electrolyte was aerated phosphate-buffered saline containing 0.01 M phosphate buffer, 0.0027 M KCl, and 0.137 M NaCl. The temperature was controlled to 37 ± 0.2 °C by a VWR Scientific temperature control unit. The cell was contained inside a grounded aluminum Faraday cage.

The potentiostat was a Gamry Reference 3000, calibrated following the manufacturer's instructions just before measurements were performed. The frequency range was 100 kHz to 0.1 Hz. A 25 mV rms sinusoidal perturbation was applied at potentials referenced to open circuit of 0 V, 10 mV, and −10 mV. The large perturbation amplitude improves signal-to-noise and is justified by the observation that the zero-frequency limit is not approached at the lowest measured frequency. Thus, the response was linear, as confirmed by observation of well-behaved elliptical low-frequency Lissajous curves.[100] The data were analyzed by the use of the measurement model version 1.8.[16]

25.5 Data Conversion

The first step towards analyzing any impedance spectrum is to convert the file into a format the program will understand. The output of this process is a file with a **.mmfile* extension. These files are ASCII text files with three columns: frequency, real impedance, and imaginary impedance. If the automatic detection of the number of comments is enabled, the program will attempt to parse common file types. The method to detect the different file types was taken from Murbach et al.[101]

While modern potentiostats yield data with very small levels of stochastic noise, impedance measurements taken near the frequency of the power source can nevertheless be noisy. Even one or two data points with excess noise can influence the regression of the impedance spectrum. To avoid this problem, data taken at frequencies close to

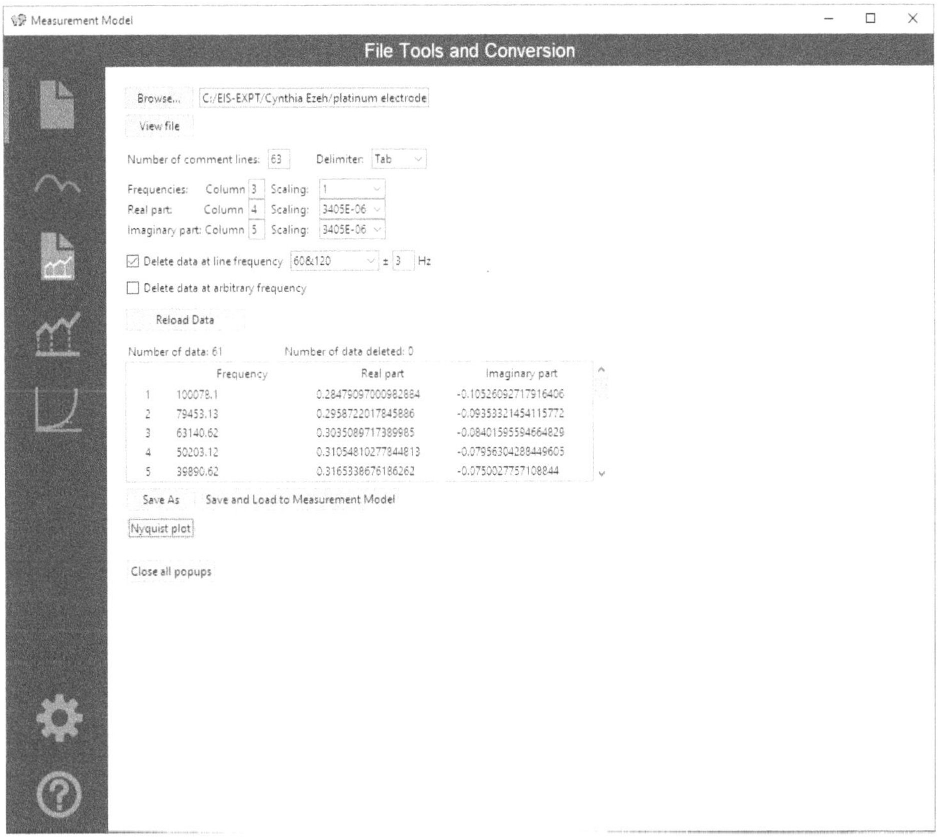

Figure 25.2: Image showing the conversion of a Gamry data file to the *.*mmfile* format used by the measurement model program. Data taken near the line frequency of 60 Hz and its first harmonic will be deleted. The notation above the text box indicates that no data were removed.

the first and second harmonics of the power source frequency are removed from the analysis. For example, as the line frequency in the United States and Canada is 60 Hz, impedance measurements taken near 60 ± 3 Hz and 120 ± 3 Hz should be deleted from the analysis of data collected.

The screenshot showing the conversion of a data file to the format used by the program is presented in Figure 25.2. While the procedure to delete data collected near 60 ± 3 and 120 ± 3 Hz was enabled, the notation above the text box indicates that no data were removed. The data were scaled by the area of the electrode, calculated to be $A = 4.909 \times 10^{-6}$ cm^2.

25.6 Measurement Model Analysis

The following describes a step-by-step procedure for the application of the measurement model to impedance data. The procedure includes the removal of potentially

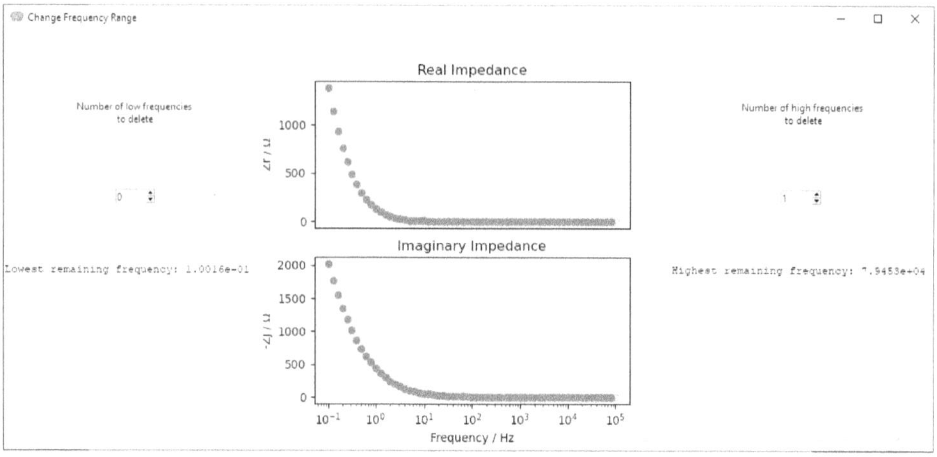

Figure 25.3: When performing regression analysis, it is useful to remove the first frequency measured. This measurement is generally corrupted by the transient response associated with the transition from steady state to an imposed frequency.

corrupted data, regression of the measurement model, estimation of stochastic errors, fitting a model to the stochastic errors, use of the measurement model to assess consistency with the Kramers–Kronig relations, and evaluation of parameters that can be extracted by the use of the measurement model.

As mentioned in the discussion of Figure 25.2, data with excess noise can confound regression analysis. A transient is observed when changing from one frequency to another,[102,103] and instrument vendors mitigate the effects of this transient by skipping the first few cycles before performing the Fourier analysis. These mitigation efforts generally are insufficient to remove the effects of the transition from steady state to the oscillating state. Thus, the first measured frequency should be removed before performing regression analysis. The corresponding dialog box is shown in Figure 25.3.

The regression of the measurement model to data is performed sequentially. The corresponding dialog is shown in Figure 25.4. Generally, the algorithm can make good initial guesses for the first Voigt element. If that algorithm fails, one can manually adjust parameters using the "Edit Model Parameters" dialog. One can use the built-in algorithm to make initial guesses for the next line shape, use an auto-fit feature to iteratively add parameters, or use the Magic Finger to graphically add parameters.

A modulus weighting strategy was selected for the initial measurement model analysis. Modulus weighting provides a good estimate for the error structure of impedance data obtained under potentiostatic modulation or under variable-amplitude galvanostatic modulation. The constant weighting strategy (no weighting) is better for impedance data obtained under fixed-amplitude galvanostatic modulation. Once an experimental model for the error structure has been determined, it should be used for

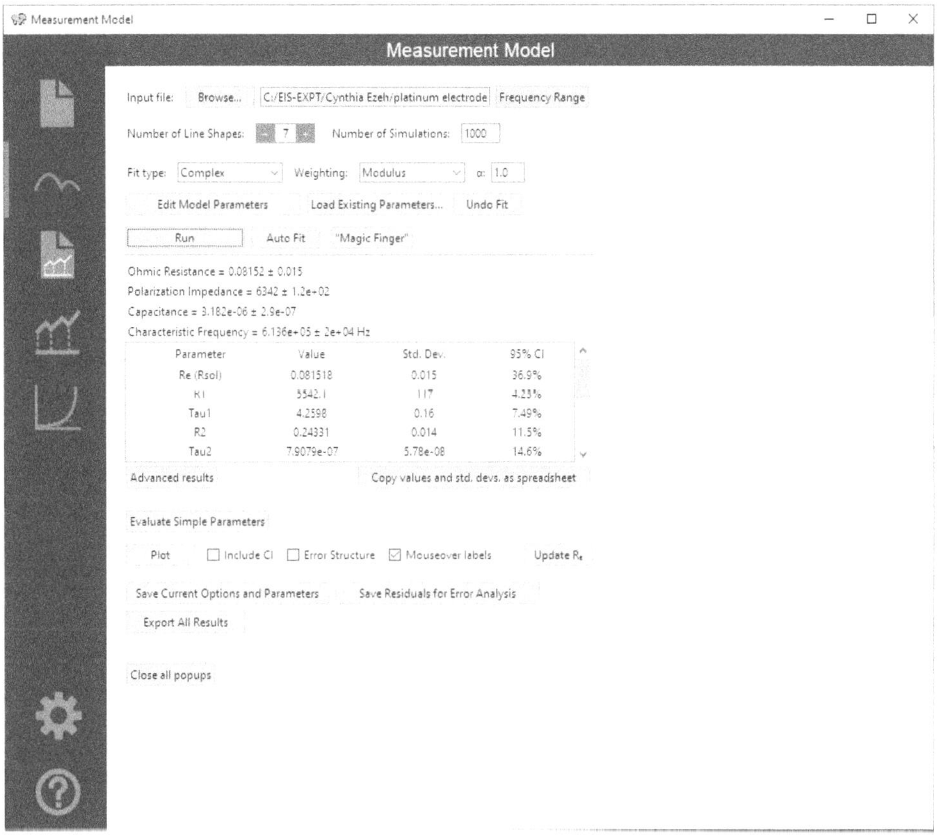

Figure 25.4: The regression of the measurement model to data is performed sequentially. One can use the built-in algorithm to make initial guesses for the next line shape, use an auto-fit feature to iteratively add parameters, or use the Magic Finger to graphically add parameters.

subsequent regression analysis.

The measurement concept is predicated on the requirement that all parameters have statistical significance. For the purpose of the present regression, the 95.4% confidence interval ($\pm 2\sigma$) for a parameter should not include zero. Thus, a value for ohmic resistance of 0.0080 Ωcm^2 with a standard deviation of 0.024 Ωcm^2 will be considered to be statistically insignificant. This value is highlighted in Figure 25.5, and a red warning signal ⚠ is displayed. The "Undo fit" button may be used to recover the previous parameters.

The regression results, presented in Figure 25.6, include Nyquist, real and imaginary impedance as functions of frequency, Bode plots, ohmic-resistance-corrected Bode plots, real and imaginary impedance in a logarithmic scale as functions of frequency, the derivative of the plot of the logarithm of imaginary impedance with respect to the logarithm of frequency, and normalized residual errors as functions of frequency. The red dashed lines are the 95.4% confidence intervals for the regression, calculated by

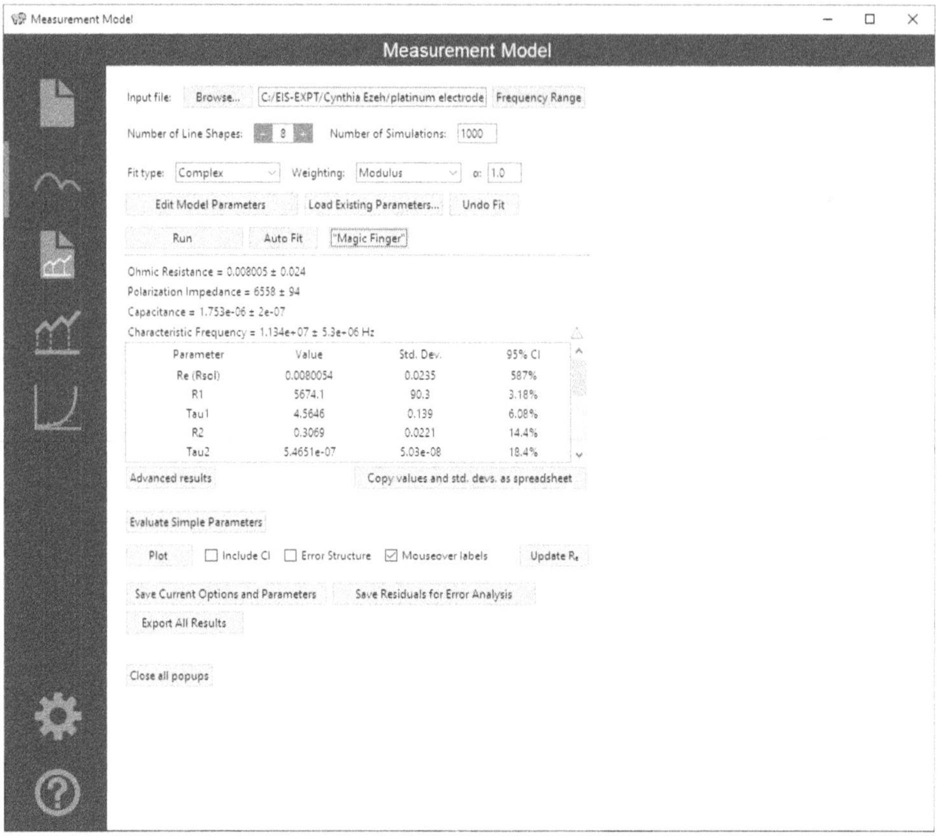

Figure 25.5: While the regression of the measurement model to data is performed sequentially, each regressed parameter must have statistical significance. Over-fitting is indicated by a triangular warning symbol, and the offending parameters are highlighted.

Monte-Carlo simulation using the parameter values and standard deviations obtained from the Levenberg–Marquardt regression. All residual errors fall within the dashed lines, indicating that the data satisfy the Kramers–Kronig relations.

A zoomed-in image of the Nyquist plot is presented in Figure 25.7. The line represents the model fit, and the shaded ellipses indicate the confidence interval for the regression. The beginning of a high-frequency capacitive loop is evident in Figure 25.7, which may be due to the capacitance of the cables connecting the potentiostat to the electrochemical cell. The capacitance calculated by the measurement model, shown in Figure 25.4, was $3.18 \pm 0.29\ \mu\text{F/cm}^2$. This value is small for a double-layer capacitance that is expected to have values between 10 and 50 $\mu\text{F/cm}^2$. The small value could be attributed to an ohmic impedance associated with nonuniform current and potential, but the characteristic frequency for the influence of ohmic impedance is calculated to be 614 kHz, which is much larger than the maximum measured frequency of 100 kHz. The hypothesis that the loop is caused by ohmic impedance may therefore be discarded.

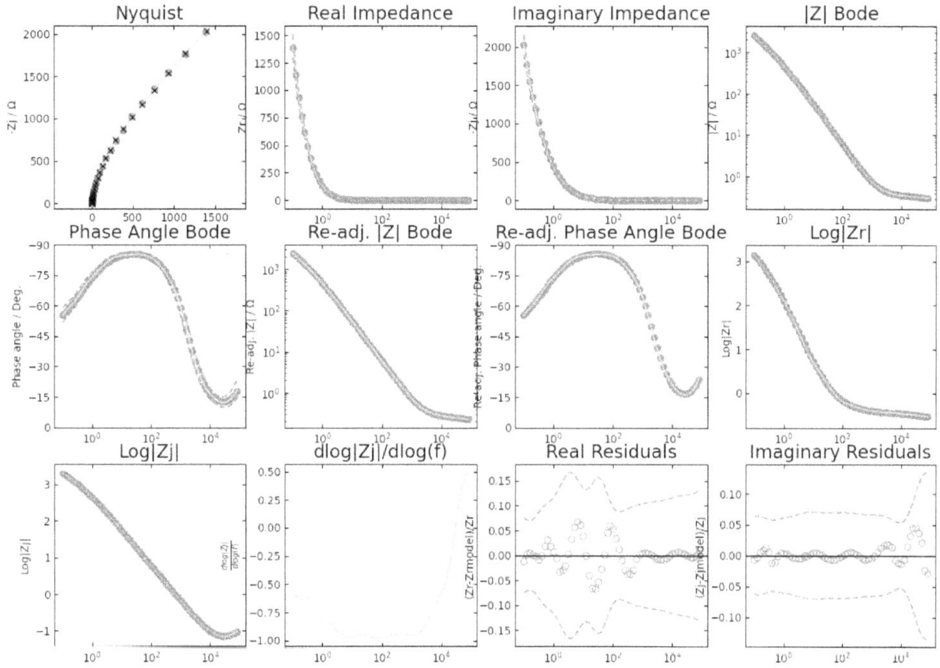

Figure 25.6: The results of the regression of Figure 25.4 are shown in a sequence of plots, including Nyquist, real and imaginary impedance as functions of frequency, Bode plots, ohmic-resistance-corrected Bode plots, real and imaginary impedance in a logarithmic scale as functions of frequency, the derivative of the plot of the logarithm of imaginary impedance with respect to the logarithm of frequency, and normalized residual errors as functions of frequency.

The electrode area is small, and the capacitance in units of farads is 1.6×10^{-11} F, which is on the order of what may be expected for the capacitance of leads.

25.7 Stochastic Error Structure

The regression shown in Figure 25.4 was applied to each of three replicated measurements, and the output was saved using the "Save Residuals for Error Analysis" button. The corresponding error structure is shown in Figure 25.8. The standard deviations for real and imaginary parts of the impedance overlap, as is expected for data that are consistent with the Kramers–Kronig relations. The standard deviations are strong functions of frequency for impedance collected under potentiostatic modulation. This functionality of frequency as the independent variable is converted to impedance following

$$\sigma = \alpha|Z_j| + \beta|Z_r| + \gamma|Z|^2 + \delta \tag{25.6}$$

where α, β, γ, and δ are constants. In equation (25.6), α and β are dimensionless, and the units of γ and δ depend on the units of the impedance under analysis. Equation

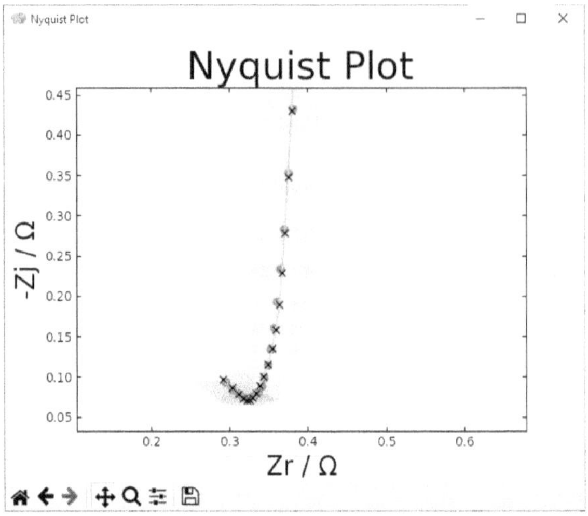

Figure 25.7: The zoomed image of the Nyquist plot, showing the possible influence of cable capacitance on the impedance response. The line represents the model fit, and the shaded ellipses indicate the confidence interval for the regression.

(25.6) was obtained empirically[75] and is often controlled by the potentiostat used with a given set of experimental parameters. The fit of equation (25.6) to the error structure shown in Figure 25.8 is given in Figure 25.9. The error structure can be saved as a file using "Save values and std. devs."

It is useful to compare the standard deviations shown in Figure 25.8 to the modulus of the impedance. The real and imaginary standard deviations, normalized by the averaged modulus, are presented in Figure 25.10 as functions of frequency as a percentage. The standard deviations are on the order of 0.2% at low frequency and 0.03% at high frequency. These values are on the order of what can be expected for modern instrumentation. Selection of settings that increase the speed of impedance measurement can result in larger noise levels.

25.8 Kramers–Kronig Relations

The Kramers–Kronig relations were derived under the assumptions that the system under study is stable, in the sense that perturbations to the system do not grow, that the system responds linearly to a perturbation, and that the system be causal in the sense that a response to a perturbation cannot precede the perturbation. The Kramers–Kronig relationships were found to be general with application to all frequency-domain measurements that could satisfy the above constraints. Bode extended the concept to electrical impedance and tabulated various extremely useful forms of the Kramers–Kronig relations.[104]

There are many forms of the Kramers–Kronig equations. The imaginary part of the

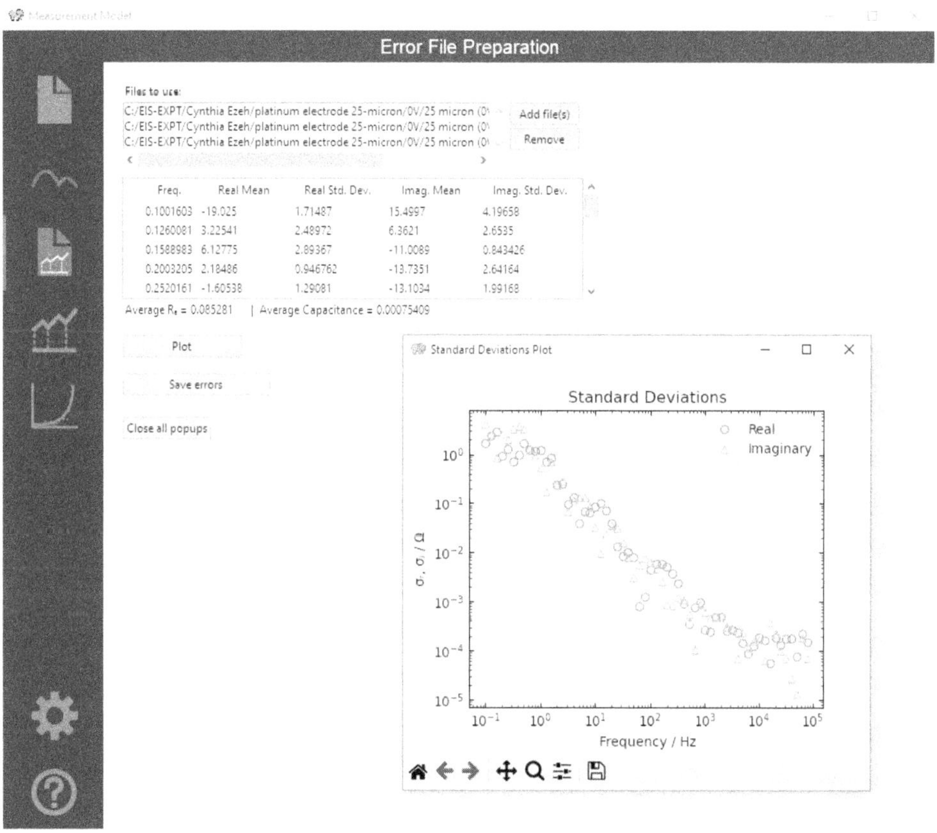

Figure 25.8: The stochastic error structure for three replicated measurements at the open-circuit potential. The standard deviations are strong functions of frequency, and the standard deviations of real ○ and imaginary parts △ of the impedance are equal.

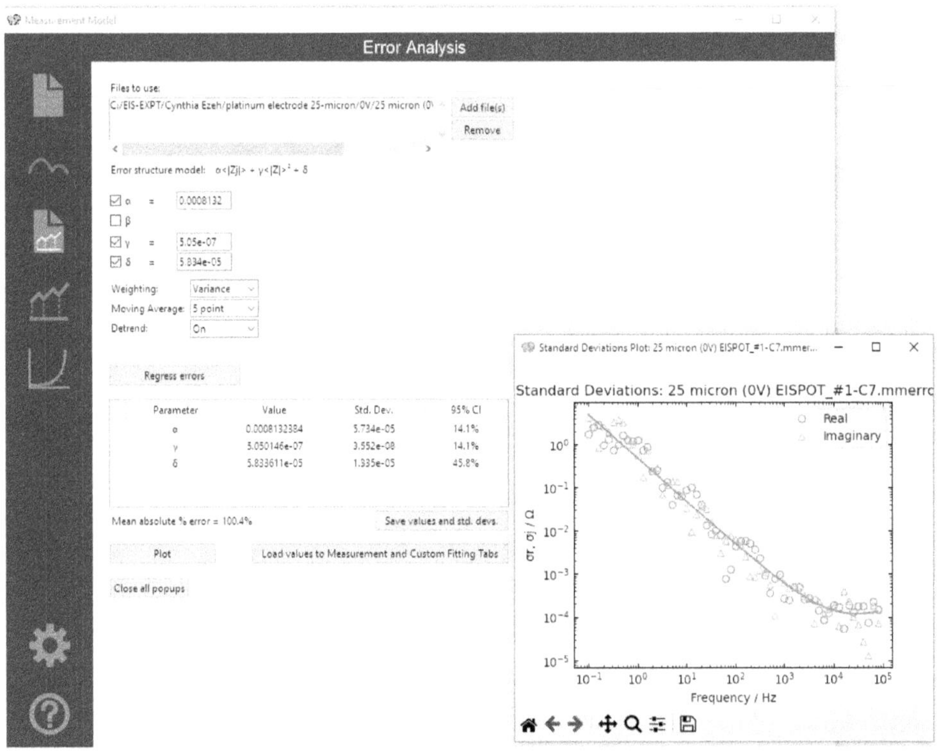

Figure 25.9: Regression of equation (25.6) to the stochastic error structure shown in Figure 25.8. Statistically significant values could not be obtained for β.

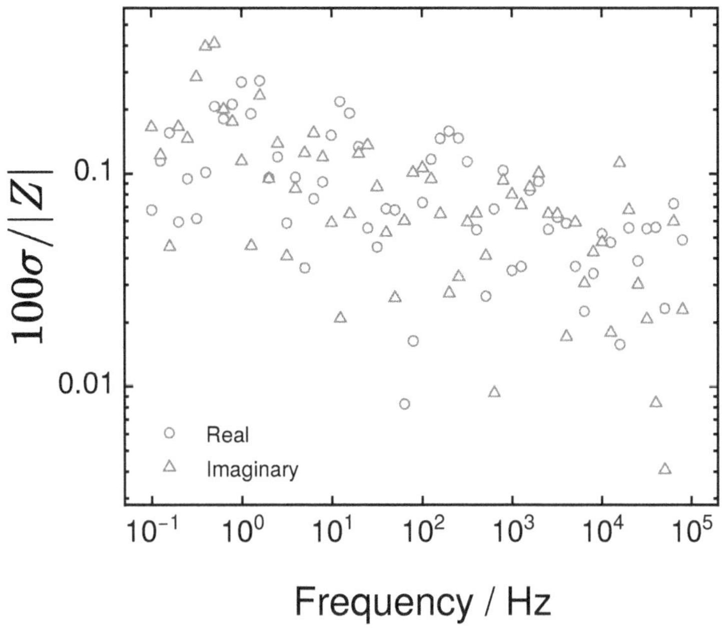

Figure 25.10: The real and imaginary standard deviations from Figure 25.8, normalized by the averaged modulus, as functions of frequency. The values are multiplied by 100 to yield a percentage.

impedance may be obtained at frequency ω from the real part using

$$Z_j(\omega) = -\frac{2\omega}{\pi} \int_0^{\infty} \frac{Z_r(x) - Z_r(\omega)}{x^2 - \omega^2} dx \tag{25.7}$$

and the real at frequency ω may be obtained from the imaginary using

$$Z_r(\omega) = Z_{r,\infty} - \frac{2}{\pi} \int_0^{\infty} \frac{xZ_j(x) - \omega Z_j(\omega)}{x^2 - \omega^2} dx \tag{25.8}$$

where $Z_r(x)$ and $Z_j(x)$ are continuous functions representing the interpolation of the available corresponding data, and the integration yields the Cauchy principal value of the argument. The application of equations (25.7) and (25.8) to electrochemical impedance data was considered controversial due to the observation that data may not be measurable to the limits of integration. The data shown in Figure 25.6, for example, do not extend to the zero-frequency limit. The objection was resolved by the regression of models such as equation (25.1), which satisfy the Kramers–Kronig relations. As shown in Figure 25.6, the measurement model provides a fit for which the residual errors fall within the 95.4% confidence interval of the regression. Thus, these data satisfy the Kramers–Kronig relations.

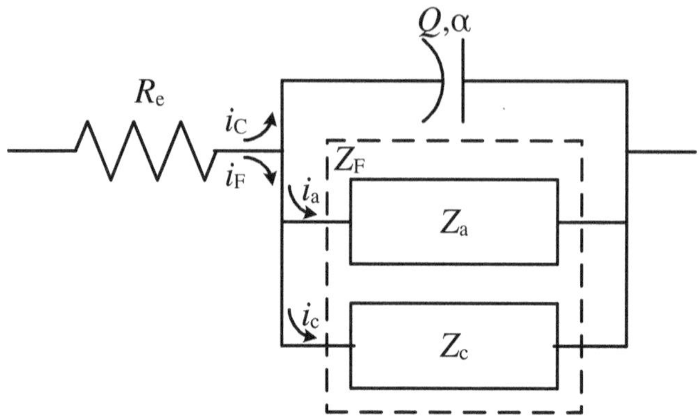

Figure 25.11: Electrical circuit for a measurement conducted at open circuit with a parallel combination of a constant-phase element, anodic impedance, and cathodic impedance.

25.9 Interpretation Model Development

The system under study was a polished Pt electrode in a phosphate-buffered saline electrolyte. The electrolyte was exposed to air; thus, the reversible reduction of oxygen may be expected to be a faradaic reaction. For measurements taken at open circuit, both anodic and cathodic faradaic reactions will take place, yielding a net zero current under steady conditions. Thus, the corresponding electrical circuit may be expressed as in Figure 25.11. If one of the faradaic impedances is large as compared to the other, then the model may be simplified. Thus, if $Z_a \gg Z_c$, Z_a may be eliminated from the model, which would include the parallel combination of the CPE and Z_c.

The method used to test whether all parts of Figure 25.11 are visible in the impedance response is to conduct impedance measurements at potentials slightly larger and slightly smaller than the open-circuit potential. The results of these measurements are compared to the open-circuit results in Figure 25.12. One may expect that at more positive potentials, the anodic reaction will be favored and the corresponding impedance should be reduced and the cathodic reaction will be disfavored and the corresponding impedance should be increased. At more negative potentials, the cathodic reaction will be favored and the corresponding impedance should be reduced and the anodic reaction will be disfavored and the corresponding impedance should be increased. As shown in Figure 25.12, the impedance is smaller at potentials both larger and smaller than open circuit, suggesting that the impedance reflects both anodic and cathodic reactions.

The data from Figure 25.12 are presented in Figure 25.13 in Bode format. As shown in Figure 25.13(a), the magnitude at the potentials OC $\pm$ 10 mV are smaller than that measured at open circuit for frequencies less than 1 kHz. The phase-angle, shown in Figure 25.13(b), tends toward zero at high frequencies and then the magnitude in-

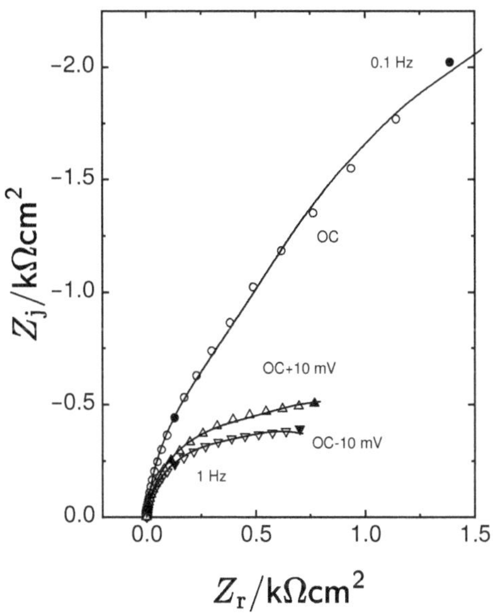

Figure 25.12: Impedance measurements in Nyquist format conducted at open circuit, a potential 10 mV larger than open circuit, and a potential 10 mV lower than open circuit. The impedance is smaller at potentials both larger and smaller than open circuit, suggesting that the impedance reflects both anodic and cathodic reactions.

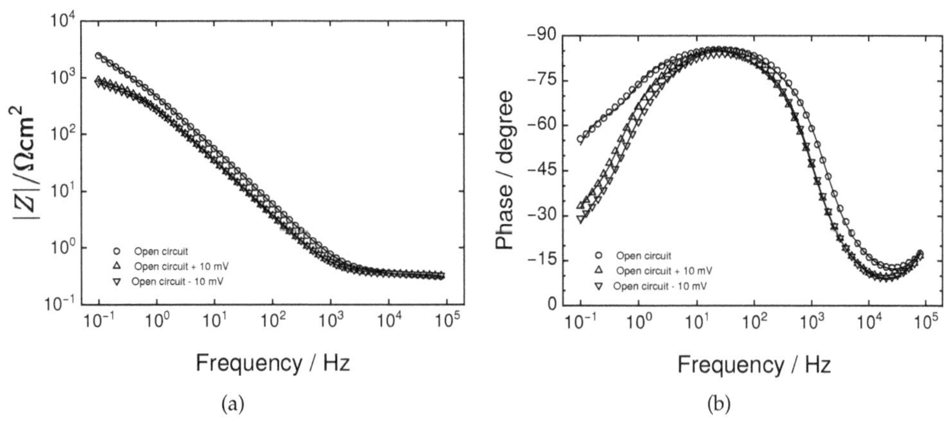

Figure 25.13: Data from Figure 25.12 presented in Bode format: (a) magnitude and (b) phase.

creases due to the influence of parasitic capacitance.

As the electrolyte was open to air, the cathodic reaction was assumed to be oxygen reduction, and the anodic reaction was assumed to be oxidation of water. The reaction mechanism may be given as

$$O_2 + 2H_2O + 4e^- \rightleftarrows 4OH^- \tag{25.9}$$

The current density for the cathodic reaction is given by

$$i_{F,c} = -K_c c_{O_2}(0)\exp\left(-b_c V\right) \tag{25.10}$$

where $K_c c_{O_2}(0)$ has the units of a current density and K_c includes the contribution of the equilibrium potential for the reaction. The current density is a function of both concentration of species O_2 at the electrode surface and the potential difference between the electrode and the solution adjacent to the electrode.

In the present work, the response to an imposed sinusoidal current or potential is expressed in terms of phasors.[1,97] Thus, the faradaic current can be expressed as

$$i_{F,c} = \bar{i}_{F,c} + \mathrm{Re}\{\tilde{i}_{F,c}\exp(j\omega t)\} \tag{25.11}$$

where the phasor $\tilde{i}_{F,c}$ is a complex number and $\bar{i}_{F,c}$ is the steady value of current density. Similarly,

$$c_{O_2} = \bar{c}_{O_2} + \mathrm{Re}\{\tilde{c}_{O_2}\exp(j\omega t)\} \tag{25.12}$$

and

$$V = \overline{V} + \mathrm{Re}\{\widetilde{V}\exp(j\omega t)\} \tag{25.13}$$

The faradaic impedance is given by

$$Z_{F,c} = \frac{\widetilde{V}}{\tilde{i}_{F,c}} \tag{25.14}$$

The current density phasor is obtained from a Taylor series expansion, i.e.,

$$\tilde{i}_{F,c} = \left.\frac{\partial i_{F,c}}{\partial V}\right|_{c_{O_2}(0)}\widetilde{V} + \left.\frac{\partial i_{F,c}}{\partial \bar{c}_{O_2}(0)}\right|_{V}\tilde{c}_{O_2}(0) \tag{25.15}$$

which can be written as

$$\tilde{i}_{F,c} = \frac{1}{R_{t,c}}\widetilde{V} - K_c\exp\left(-b_c\overline{V}\right)\tilde{c}_{O_2}(0) \tag{25.16}$$

or

$$1 = \frac{1}{R_{t,c}}\frac{\widetilde{V}}{\tilde{i}_{F,c}} - K_c\exp\left(-b_c\overline{V}\right)\frac{\tilde{c}_{O_2}(0)}{\tilde{i}_{F,c}} \tag{25.17}$$

where

$$R_{t,c} = \frac{1}{K_c b_c \bar{c}_{O_2}(0)\exp\left(-b_c\overline{V}\right)} \tag{25.18}$$

Equations (25.14) and (25.17) yield an expression for the faradaic impedance in terms of $\theta = \widetilde{c}_{O_2}/\widetilde{c}_{O_2}(0)$ as[1,97]

$$Z_{F,c} = R_{t,c} + R_{d,c}\left(\frac{-1}{\theta'(0)}\right) \tag{25.19}$$

where $R_{d,c}$ is the diffusion resistance, given as,

$$R_{d,c} = \frac{\delta}{b_c\bar{c}_{O_2}(0)nFD_{O_2}} \tag{25.20}$$

and $-1/\theta'(0)$ is the dimensionless diffusion impedance. As shown in equation (25.20), the diffusion resistance is a function of the diffusion layer thickness δ, the kinetic parameter b_c, the concentration of oxygen at the electrode surface, the number of electrons transferred n, and the diffusivity D_{O_2}. Thus, the diffusivity cannot be extracted directly from the diffusion resistance. As demonstrated in the development of equation (25.19), the diffusion impedance is an integral part of the faradaic impedance.

The dimensionless diffusion impedance can be derived for specified geometries. For example, the dimensionless diffusion impedance for a stationary film is expressed as

$$\frac{-1}{\theta_i'(0)} = \frac{\tanh\left(\sqrt{jK_i}\right)}{\sqrt{jK_i}} \tag{25.21}$$

where

$$K_i = \omega\delta_f^2/D_i = \omega\tau_i \tag{25.22}$$

where δ_f is the film thickness, and τ_i is the mass transfer time constant that appears as a parameter in regression models. At high frequency, the film impedance given in equation (11.1) approaches the Warburg impedance for a semi-infinite domain, i.e.,

$$\lim_{\omega\to\infty}\frac{-1}{\theta_i'(0)} = \frac{1}{\sqrt{jK_i}} = \frac{1}{\sqrt{j\omega\delta_f^2/D_i}} \tag{25.23}$$

The Warburg diffusion impedance may be expressed as

$$R_{d,c}\left(\frac{-1}{\theta'(0)}\right) = \frac{1}{b_c c_{O_2}(0)nFD_{O_2}^{1/2}}\frac{1}{\sqrt{j\omega}} = R_{d,c}^*\frac{1}{\sqrt{j\omega}} \tag{25.24}$$

The diffusion layer thickness is eliminated in the development of equation (25.24), and, as in equation (25.20), diffusivity appears in the diffusion resistance as part of a lumped parameter.

The anodic reaction may be developed in a similar way as the cathodic reaction. Under the assumption that the concentration of hydroxide ions is sufficient, the faradaic impedance may be expressed as

$$Z_{F,a} = R_{t,a} \tag{25.25}$$

where

$$R_{t,a} = \frac{1}{K_a b_a \exp\left(b_a\overline{V}\right)} \tag{25.26}$$

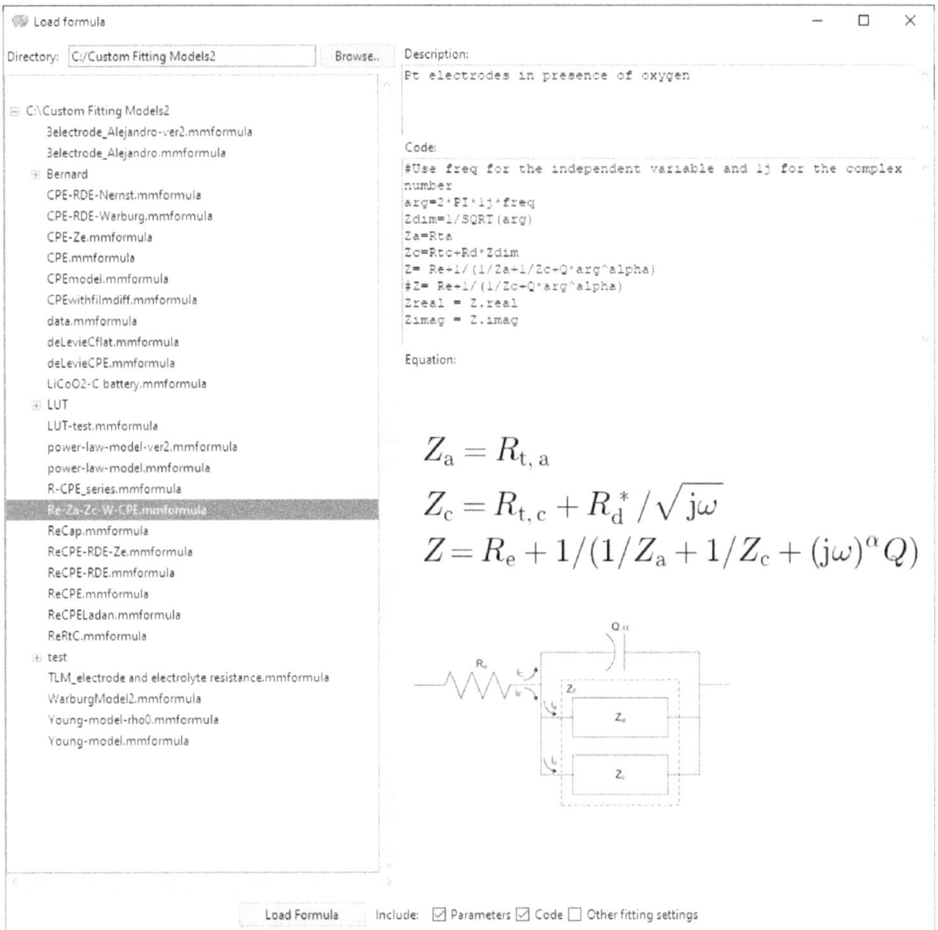

Figure 25.14: The entry for the process model for Pt in saline solution includes a description, the associated Python code, the equations used, and an image of the electrical circuit.

The anodic reaction was assumed to be unaffected by mass transfer.

A model was constructed to account for mass-transfer-influenced reduction of oxygen, the kinetically controlled oxidation of water, and a CPE. The lowest frequencies measured were not sufficient to capture the finite diffusion layer described as equation (11.1); thus, the Warburg impedance given in equations (25.23) or (25.24) for a semi-infinite film was employed. The dialog box used to load the model is shown in Figure 25.14. The anodic impedance is given by a charge-transfer resistance, i.e.,

$$Z_a = R_{t,a} \tag{25.27}$$

The cathodic impedance includes the influence of a charge-transfer resistance and a Warbug impedance, i.e.,

$$Z_c = R_{t,c} + R_d^* / \sqrt{j\omega} \tag{25.28}$$

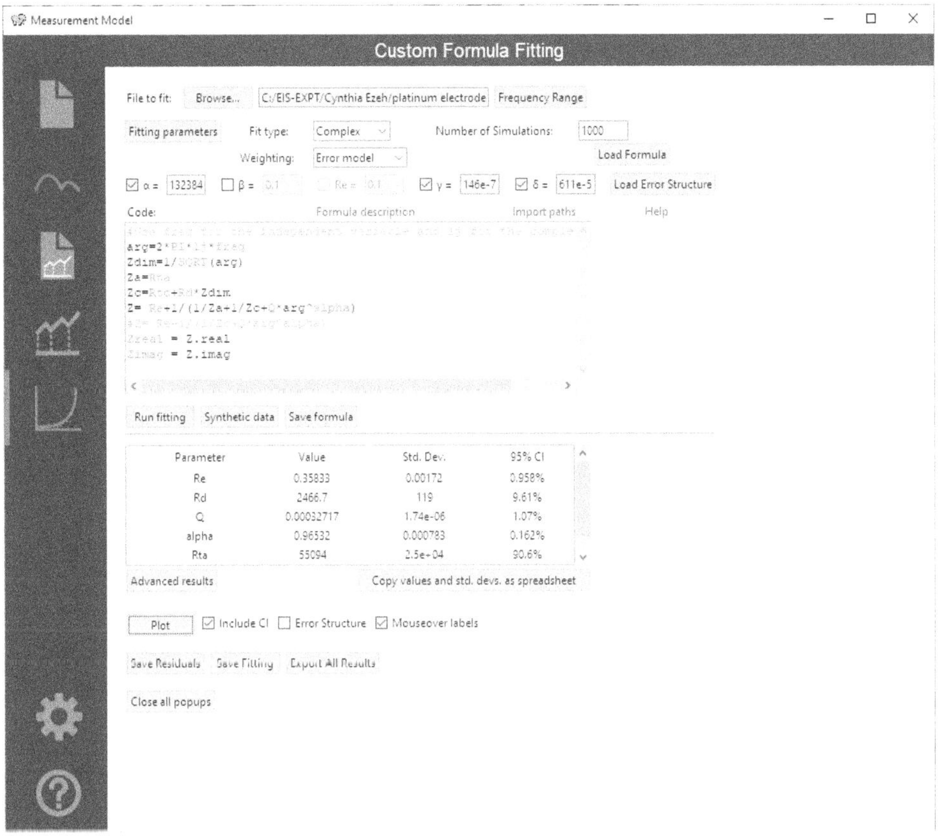

Figure 25.15: Regression of the process model shown in Figure 25.14 to the open-circuit data shown in Figure 25.12. The regression was weighted by the error-structure model shown in Figure 25.9. The maximum frequency used in the regression was 2 kHz to eliminate the influence of parasitic capacitance.

where the diffusion resistance R_{d}^{*} has units of $\Omega\mathrm{cm}^2/\mathrm{s}^{1/2}$. The resulting impedance is given as

$$Z = R_{\mathrm{e}} + 1/(1/Z_{\mathrm{a}} + 1/Z_{\mathrm{c}} + (\mathrm{j}\omega)^{\alpha} Q) \tag{25.29}$$

where R_{e} is the ohmic resistance, and Q and α are parameters associated with the CPE.

The regression of equation (25.29) to the open-circuit data shown in Figure 25.12 was performed for a maximum frequency of 2 kHz to eliminate the influence of parasitic capacitance. The regression was weighted by the error-structure model shown in Figure 25.9. The corresponding dialog is shown in Figure 25.15, and the results are plotted in Figure 25.16. The fit appears to be good. The regressed values are presented in Table 25.1. The weighted χ^2/ν, where ν is the degree of freedom for the regression, was 199. For an excellent regression, the weighted χ^2/ν has a value near unity. The parameters shown in Table 25.1 are statistically significant in the sense that the 95.4%

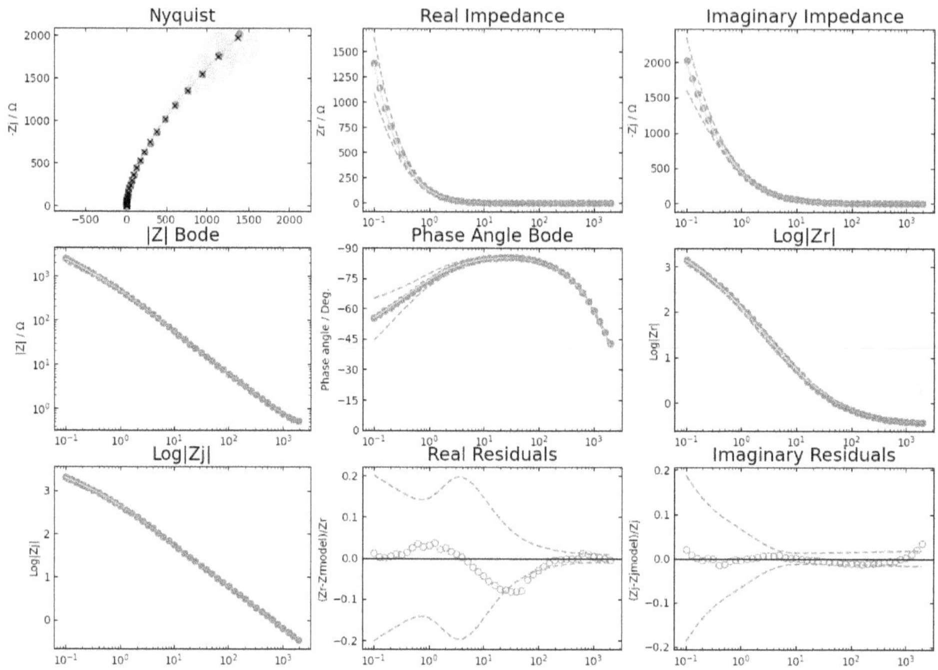

Figure 25.16: The results of the regression of Figure 25.15 are shown in a sequence of plots, including Nyquist, real and imaginary impedance as functions of frequency, Bode plots, real and imaginary impedance in a logarithmic scale as functions of frequency, and normalized residual errors as functions of frequency.

Table 25.1: Regression results corresponding to Figures 25.15 and 25.16.

Parameter	Regressed Value	Units
R_{e}	0.3583 ± 0.0017	$\Omega\mathrm{cm}^2$
R_{d}	2470 ± 120	$\Omega\mathrm{cm}^2$
Q	327.2 ± 1.7	$\mu\mathrm{F/s}^{1-\alpha}\mathrm{cm}^2$
α	0.96532 ± 0.00078	–
$R_{\mathrm{t,a}}$	55 ± 25	$\mathrm{k}\Omega\mathrm{cm}^2$
$R_{\mathrm{t,c}}$	1224 ± 48	$\Omega\mathrm{cm}^2$
χ^2/ν	199	

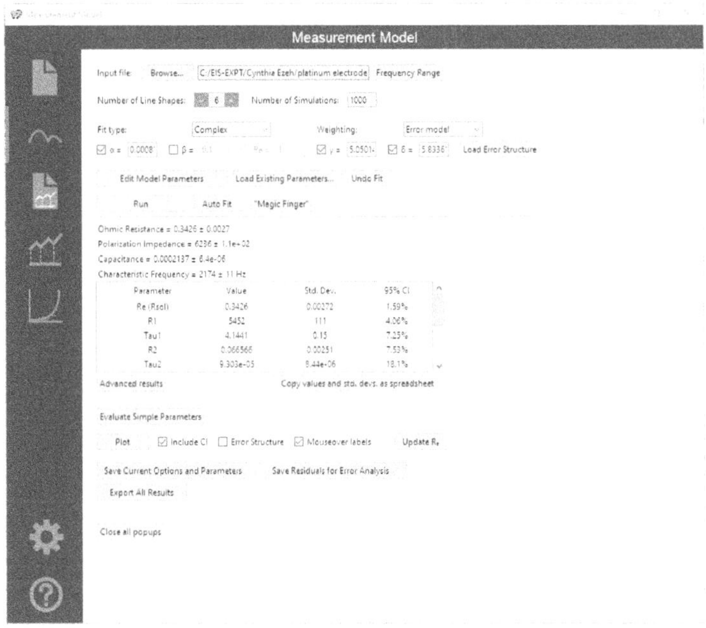

Figure 25.17: Regression of the measurement model to the open-circuit data shown in Figure 25.12. The regression was weighted by the error-structure model shown in Figure 25.9. The maximum frequency used in the regression was 2 kHz to eliminate the influence of parasitic capacitance.

confidence intervals ($\pm 2\sigma$) do not include zero.

25.10 Discussion

For electrodes that are not covered by a film, the effective capacitance may be extracted by the use of an expression derived by Brug et al.[54] for a surface distribution of time constants, i.e.,

$$C_{\text{eff}} = Q^{1/\alpha}\left(\frac{R_\text{e}R_\text{t}}{R_\text{e}+R_\text{t}}\right)^{(1-\alpha)/\alpha} \tag{25.30}$$

In the limit that R_t becomes much larger than R_e, equation (25.30) becomes

$$C_{\text{eff}} = Q^{1/\alpha}R_\text{e}^{(1-\alpha)/\alpha} \tag{25.31}$$

Application to the parameters presented in Table 25.1 yields an effective capacitance $C_{\text{eff}} = 236\ \mu\text{F/cm}^2$, a value that is much larger than expected for a Pt electrode.

Regression of the measurement model to the same frequency range, shown in Figure 25.17, yielded an effective capacitance of 214 $\mu\text{F/cm}^2$, in good agreement with the results of equation (25.31). The capacitance calculated by the measurement model increases as the maximum frequency is reduced. For a maximum frequency of 315 Hz, the effective capacitance is calculated to be 247 $\mu\text{F/cm}^2$. The large value for electrode capacitance is consistent with those obtained for ultramicroelectrodes.

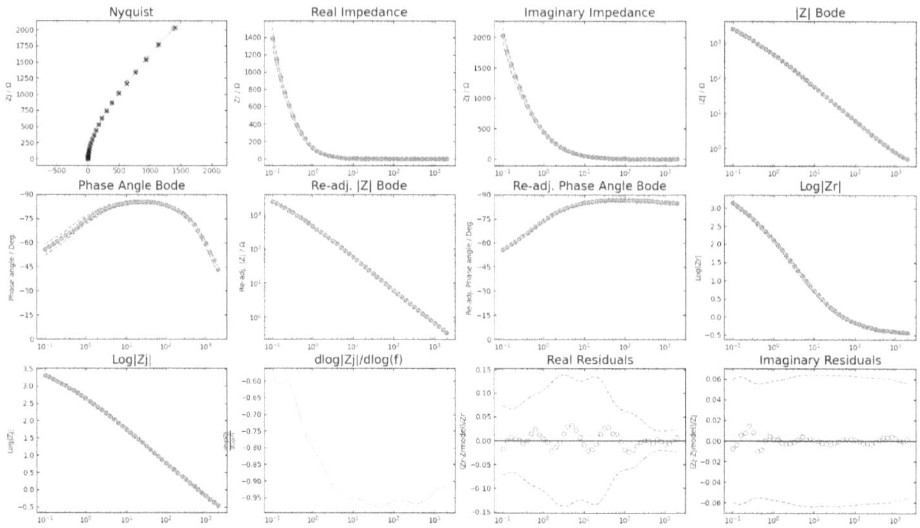

Figure 25.18: The results of the regression of Figure 25.17 are shown in a sequence of plots. The seventh and tenth figures show that the impedance approaches a constant-phase at high frequency with $\alpha \approx 0.95$.

The regression results shown in Figure 25.18 show an excellent fit to the data. The ohmic-resistance-corrected phase angle suggests a high-frequency phase angle close to $-90°$,and the derivative $\mathrm{d}\log|Z_\mathrm{j}|/\mathrm{d}\log f$ yields a high-frequency value close to -0.96, in agreement with the value of α reported in Table 25.1.

The procedure described in this chapter has been used to assess the capacitance of larger Pt electrodes in saline electrolytes. Larger electrodes have a smaller impedance, making them less susceptible to the confounding influence of lead capacitance. For these larger Pt electrodes, the capacitance obtained by the method described here, i.e., the measurement model or equation (25.31), yielded values in good agreement with expected values for double-layer capacitance. For example, the capacitance obtained by the use of equation (25.31) for a 5 mm diameter Pt electrode under different conditions was 15–22 $\mu F/cm^2$, in agreement with expected values for double-layer capacitance.[96]

The measurement model is a powerful tool for analysis of impedance data. It provides the capability to assess the stochastic error structure of impedance data and to assess consistency with the Kramers–Kronig relations. Some parameters such as ohmic resistance, capacitance, and the time constant associated with ohmic impedance may be estimated by the use of the measurement model. The current program also allows regression of custom models to impedance data. The present analysis demonstrates the confounding influence of the capacitance of leads for high-impedance electrodes. The methods described in this chapter yielded capacitances in good agreement with expected values for larger electrodes.

Chapter 26

Physical Properties of Aluminum Oxide

The data presented in this chapter demonstrate the need for a thorough assessment of the consistency with the Kramers–Kronig relations, the need to understand the influence of electrode geometry on impedance response, and the utility of an initial graphical analysis. In the present case, only a single data set is available, which limits the ability to assess the error structure. Nevertheless, the measurement model analysis will provide useful information that complements the process model development. The process model accounts for the influence of an oxide film in which a distribution of resistivity is invoked to account for frequency dispersion. The useable frequency range is shown to be constrained by geometry-induced frequency dispersion.

The data in this chapter involves the impedance response of an AA2024 aluminum alloy. AA2024 aluminum alloy and 2024 aluminum alloy refer to the same material, with AA2024 being a specific designation used in the Aluminum Association system. 2024 aluminum alloy has copper as the primary alloying element. 2024 aluminum alloy is widely used in aerospace applications due to its high strength-to-weight ratio and fatigue resistance. 2024 aluminum alloy is commonly found in aircraft structures, such as wings and fuselages, and is also used in automotive and military applications. This alloy is subject to corrosion, and the tendency for corrosion motivates the present work.

The impedance data presented in Table 26.1 were obtained for an AA2024 aluminum alloy after one hour of immersion in a 0.1 M Na_2SO_4 electrolyte. The impedance spectrum was collected at the open-circuit or corrosion potential. In this solution, with a pH close to 7, the electrode may be expected to be covered completely or partially by an oxide layer, i.e., Al_2O_3. The measurements were performed over a frequency ranging from 35 kHz to 3 mHz.

Table 26.1: Impedance data obtained for an AA2024 aluminum alloy after one hour of immersion in a 0.1 M Na_2SO_4 electrolyte.

f, Hz	Z_r, Ωcm^2	Z_j, Ωcm^2	f, Hz	Z_r, Ωcm^2	Z_j, Ωcm^2
35000	19.152	−0.61628	7.6895	201.89	−864.44
26169	19.171	−0.86972	5.7445	257.73	−1,117.5
19570	19.220	−1.2000	4.2926	336.47	−1,440.9
14638	19.299	−1.6093	3.2150	430.12	−1,846.9
10940	19.447	−2.0983	2.4039	565.72	−2,369.8
8180.0	19.652	−2.7019	1.7952	751.33	−3,049.0
6118.8	19.915	−3.4347	1.3373	1,001.0	−3,923.1
4577.3	20.275	−4.3195	1.0001	1,351.5	−5,027.8
3420.5	20.731	−5.4069	0.7502	1,837.8	−6,409.5
2559.1	21.277	−6.7124	0.5592	2,551.6	−8,220.9
1913.1	21.924	−8.3437	0.4190	3,556.5	−10,406.
1432.3	22.678	−10.359	0.3132	5,053.2	−13,073.
1071.7	23.568	−12.896	0.2350	7,188.1	−16,191.
801.28	24.609	−16.108	0.1756	10,305.6	−19,753.
600.96	25.756	−20.131	0.1313	14,696.6	−23,536.
447.28	27.154	−25.457	0.0983	20,556.9	−26,809.
335.54	28.719	−32.173	0.0734	27,842.5	−29,163.
250.40	30.613	−41.087	0.0549	36,005.7	−29,584.
186.73	32.929	−52.739	0.0410	43,956.9	−28,035.
140.22	35.770	−67.544	0.0307	50,555.1	−24,947.
104.37	39.512	−87.396	0.0229	55,008.4	−20,978.
78.321	44.212	−112.463	0.0172	58,120.9	−16,978.
58.741	50.387	−145.008	0.0128	59,841.3	−13,711.
43.727	58.660	−188.183	0.0096	59,961.5	−11,697.
32.786	69.416	−242.627	0.0072	60,169.9	−10,718.
24.475	83.622	−313.941	0.0054	59,596.7	−11,124.
18.339	102.40	−404.495	0.0040	59,855.4	−12,731.
13.658	127.60	−523.724	0.003	59,897.9	−15,627.
10.253	159.64	−672.937			

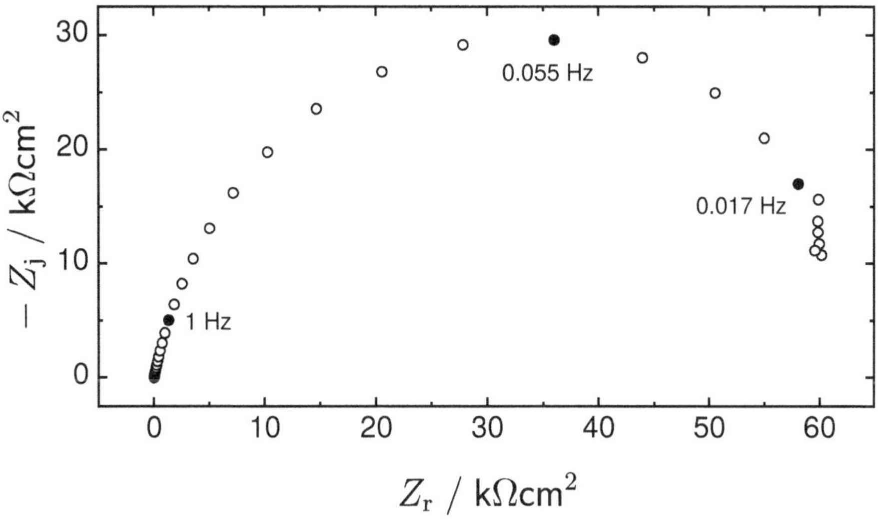

Figure 26.1: Impedance data, presented in Table 26.1, obtained for an AA2024 aluminum alloy after one hour of immersion in a 0.1 M Na_2SO_4 electrolyte in Nyquist format.

26.1 Initial Inspection

As presented in Chapters 27–30, the initial inspection comprises the use of graphical methods and regression of measurement models. The graphical analysis is used to guide model development and extract parameters. The measurement model analysis is used to identify the error structure of the measurement. The stochastic error structure may be identified from replicated measurements. For the system presented in Table 26.1, only one data set is available. Nevertheless, the measurement model may be used to determine the frequency range that is consistent with the Kramers–Kronig relations. In addition, some physical properties may be obtained.

26.1.1 Graphical Analysis

Graphical methods provide a first step toward interpretation and evaluation of impedance data. The graphical techniques presented in this chapter do not depend on any specific model, yet can provide a qualitative if limited interpretation. Even in the absence of specific models, values of such meaningful parameters as the system ohmic resistance and capacitance can be obtained.

The impedance data are presented in Figure 26.1 in a Nyquist format. The characteristic frequency is near 0.055 Hz, and the high-frequency limit is close to zero. The presentation in Bode format is given as Figure 26.2. The ohmic resistance may be obtained from the high-frequency limit of the modulus (Figure 26.2(a)) to be 19.2 Ωcm^2. Other plots may be used to extract the ohmic resistance. The real part of the impedance is shown in Figure 26.3(a), and a zoomed representation of the Nyquist plots is

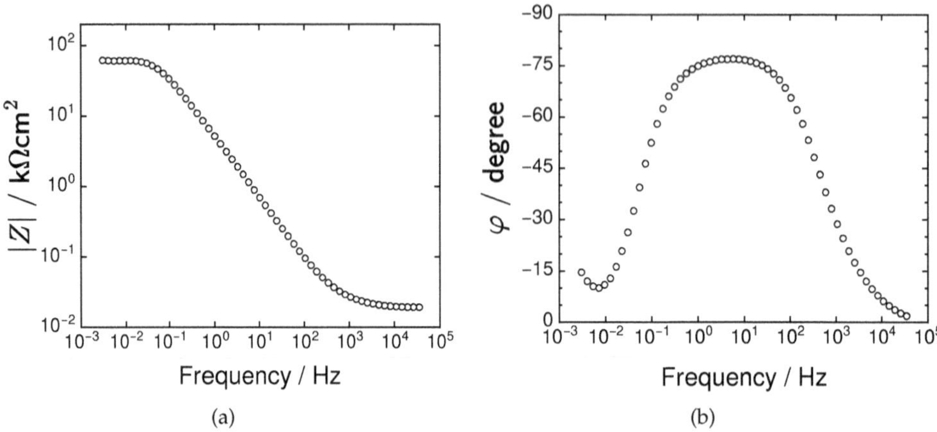

Figure 26.2: Impedance data, presented in Table 26.1, obtained for an AA2024 aluminum alloy after one hour of immersion in a 0.1 M Na_2SO_4 electrolyte as a function of frequency: (a) modulus and (b) phase angle.

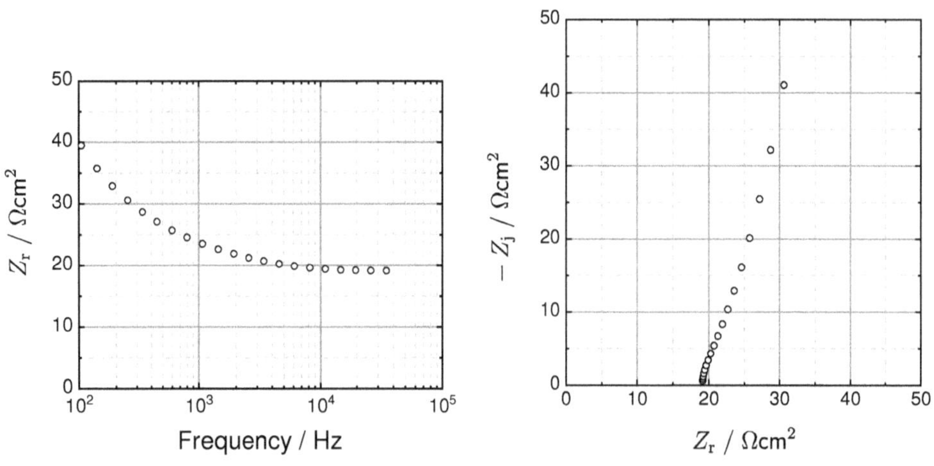

Figure 26.3: Impedance data, presented in Table 26.1, obtained for an AA2024 aluminum alloy after one hour of immersion in a 0.1 M Na_2SO_4 electrolyte: (a) real part of the impedance as a function of frequency at high frequencies and (b) Nyquist plot in the vicinity of the high-frequency asymptote.

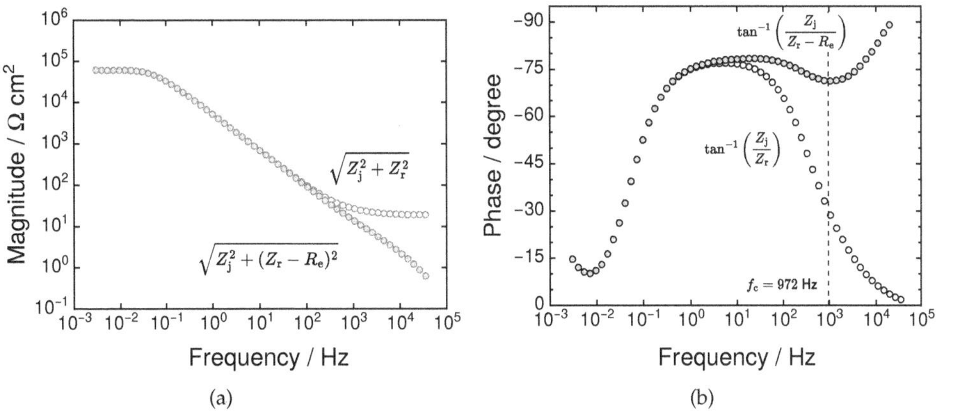

Figure 26.4: Impedance data, presented in Table 26.1, obtained for an AA2024 aluminum alloy after one hour of immersion in a 0.1 M Na_2SO_4 electrolyte as a function of frequency: a) modulus and ohmic-resistance-corrected modulus and b) phase angle and ohmic-resistance-corrected phase angle.

shown in Figure 26.3(b). The ohmic resistance obtained from the high-frequency limit of Figures 26.3(a) and 26.3(b) is also 19.2 Ωcm^2.

The phase angle given in Figure 26.2(b) reveals an increase at low frequencies that suggests the presence of an artifact. At high frequencies, the phase angle tends toward zero due to the influence of the ohmic resistance. The modulus and phase angle presented in Figure 26.2 are compared to the ohmic-resistance-corrected modulus in Figure 26.4(a) and the ohmic-resistance-corrected phase angle in Figure 26.4(b). The ohmic-resistance-corrected phase angle suggests that, when corrected for the influence of the ohmic resistance, the phase angle is approximately independent of frequency over several decades of frequency. At higher frequencies, the impedance may be influenced by the geometry of the disk electrode. The frequency at which the geometry begins to influence the impedance has been shown to be given by[105]

$$f_c = \frac{\kappa}{2\pi C_0 r_0} \tag{26.1}$$

where κ is the electrolyte conductivity, C_0 is the capacitance, and r_0 is the radius of the disk electrode. Equation (26.1) may be expressed in terms of the ohmic resistance R_e as

$$f_c = \frac{1}{8R_eC_0} \tag{26.2}$$

To estimate the characteristic frequency, an estimate for the capacitance is needed.

The complex capacitance may be calculated from the impedance as

$$C = C_r + jC_j = \frac{1}{2\pi f\,(Z - R_e)} \tag{26.3}$$

The complex capacitance is given in a complex-plane representation in Figure 26.5. The extrapolation to high frequency yields a value of 6.7 $\mu F/cm^2$. With an ohmic resis-

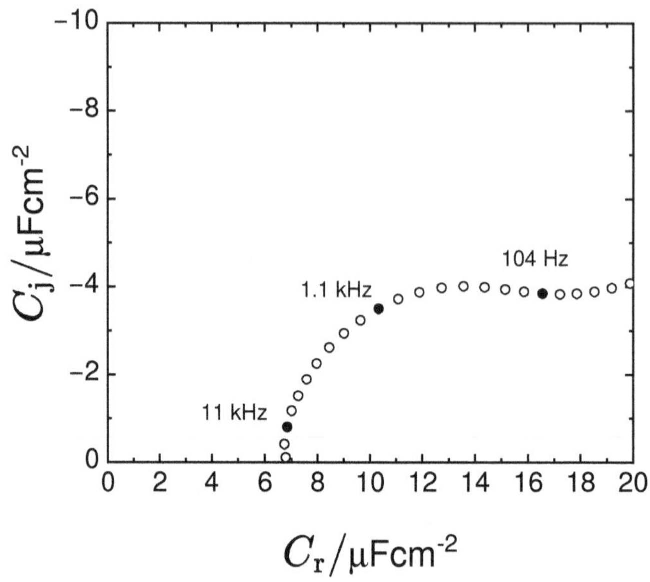

Figure 26.5: The imaginary part of the complex capacitance as a function of the real part of the complex capacitance for the impedance data presented in Table 26.1.

tance of 19.2 Ωcm^2, equation (26.2) yields a characteristic frequency of 972 Hz. Thus, frequency dispersion associated with the disk geometry affects the impedance at frequencies greater than 972 Hz. As shown in Figure 26.4(b), the effect is seen for a decade or so below the characteristic frequency.

26.1.2 Measurement Model Analysis

The measurement model method for distinguishing between bias and stochastic errors is based on using a generalized model as a filter for non-replication of impedance data.[106] The measurement model is composed of a superposition of line-shapes, which can be arbitrarily chosen subject to the constraint that the model satisfies the Kramers–Kronig relations. The model used in the present work is composed of Voigt elements in series with a solution resistance, shown in Figure 25.1 and described mathematically by equation (25.1). Equation (25.1) has been shown to be a useful measurement model. With a sufficient number of parameters, the Voigt model was able to provide a statistically significant fit to a broad variety of impedance spectra.[2]

If replicated measurements are available, the measurement model may be used to extract a model for the stochastic error structure. In the present case, replicates are not available, and modulus weighting was assumed for regression of the measurement model to determine consistency with the Kramers–Kronig relations.[4,80] If the data can be represented by a model that is itself consistent with the Kramers–Kronig relations, the data can be considered to be consistent. The use of an experimental determination

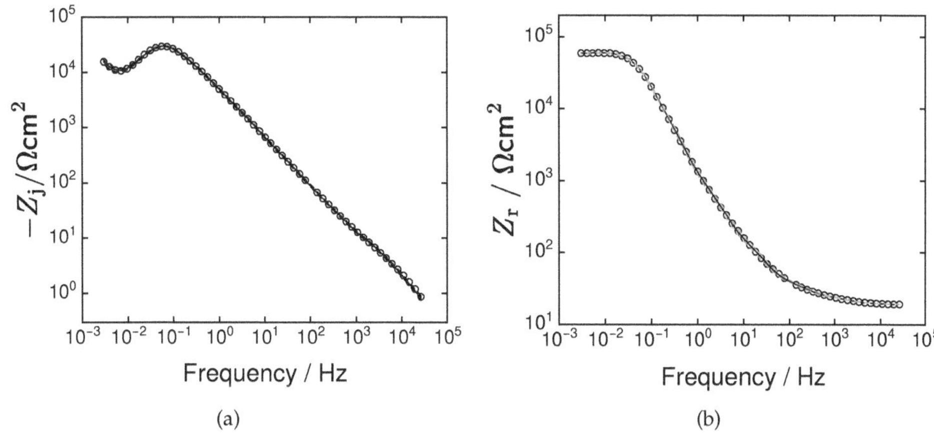

Figure 26.6: Fit of the measurement model to the complete measured spectrum presented in Table 26.1 for an AA2024 aluminum alloy after one hour of immersion in a 0.1 M Na_2SO_4 electrolyte as a function of frequency: (a) imaginary and (b) real parts of the impedance.

of the stochastic error structure allows formal quantification of the extent of agreement with the Kramers–Kronig relations.

The regression of the measurement model to the data presented in Table 26.1 yielded the fit presented in Figures 26.6 and 26.7. To obtain the fit presented in Figures 26.6 and 26.7, the first measured frequency and data taken for frequencies of 50 ± 3 Hz and 100 ± 3 Hz were removed from the data set analyzed. The first measured frequency is generally corrupted by the transient from a steady state to a sinusoidal condition. Data collected near the line frequency and its first harmonic are generally influenced by an inability to filter all the noise in the electrical power lines. Regression of the measurement model to the data is strongly affected by outliers, and the data removed from the data set analyzed often appear as outliers.

The residual errors for the regression are presented in Figure 26.8. The dashed lines reflect the 95.4% (2σ) confidence interval for the regression, obtained by a Monte Carlo simulation. The quality of the measurement model fit is excellent, even at low frequencies at which an inductive feature is apparent.

The fit of the measurement model demonstrates that the entire impedance spectrum satisfies the Kramers–Kronig relations. The use of the measurement model has an additional benefit. The ohmic resistance is estimated to be $19.241 \pm 0.045\ \Omega cm^2$. The capacitance can be estimated for each RC element as

$$C_k = \tau_k / R_k \tag{26.4}$$

and the system capacitance, given by

$$C = \frac{1}{\sum_k 1/C_k} \tag{26.5}$$

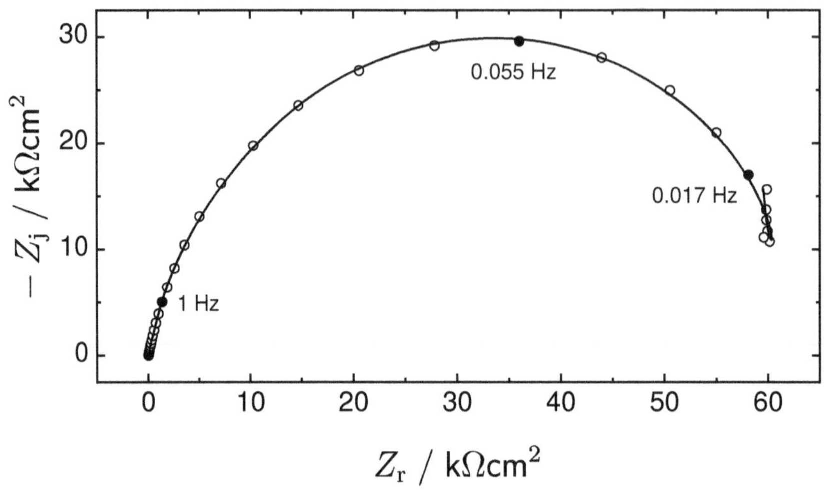

Figure 26.7: Fit of the measurement model to the complete measured spectrum presented in Table 26.1 for an AA2024 aluminum alloy after one hour of immersion in a 0.1 M Na_2SO_4 electrolyte in Nyquist format.

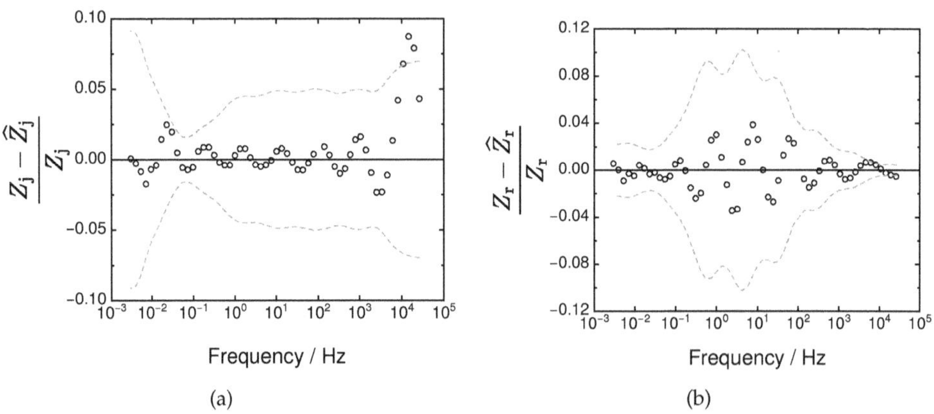

Figure 26.8: Residual errors for a fit of the measurement model to the complete measured spectrum obtained for an AA2024 aluminum alloy after one hour of immersion in a 0.1 M Na_2SO_4 electrolyte as a function of frequency: (a) imaginary and (b) real parts of the impedance.

Table 26.2: Results of the initial graphical and measurement model analysis of the data presented in Table 26.1.

Parameter	Symbol	Graphical	Measurement Model
Ohmic resistance	R_e	19.2 Ωcm^2	19.241 ± 0.045 Ωcm^2
Capacitance	C_0	6.7 μF/cm^2	7.27 ± 0.25 μF/cm^2
Characteristic frequency	f_c	972 Hz	893 Hz

has a value of 7.27 ± 0.25 μF/cm^2.

The quantitative results of the initial analysis of the data are shown in Table 26.2. The ohmic resistance obtained from the measurement model is in good agreement with the result obtained by inspection of Figures 26.2(a) and 26.3. The advantage of the measurement model result is that it provides a confidence interval for the estimation. The ohmic resistance obtained from graphical methods falls within the 95.4% (2σ) confidence interval for the ohmic resistance obtained by regression of the measurement model. The capacitance obtained from extrapolation of Figure 26.5 also falls within the 95.4% (2σ) confidence interval for the capacitance obtained by regression of the measurement model.

From a qualitative perspective, the preliminary analysis revealed that the impedance displayed an asymmetric depressed Nyquist plot with a low-frequency inductive feature. The entire spectrum was found to be consistent with the Kramers–Kronig relations. In addition, the disk geometry affected the impedance response at frequencies higher than roughly 900 Hz.

26.2 Interpretation Model Development

The development of an interpretation model is sequential. An electrical circuit is devised, based on the principles that phenomena that result in the addition of potential drops yield a series summation of respective impedance and phenomena that result in addition of currents yield a summation of respective impedances in parallel. The circuit provides a framework for the models that account for the reaction kinetics, transport and other physical phenomena.

26.2.1 Electrical Circuit Framework

At the corrosion potential, the sum of the anodic and cathodic faradaic currents is equal to zero, i.e., $i_a + i_c = 0$. The faradaic impedance must, therefore, be a parallel combination of Z_a and Z_c. The contribution of the capacitance is added in parallel. As the electrode is expected to be covered by an oxide film, the impedance of the oxide film must also play a role. The corresponding circuit framework is presented in Figure 26.9(a) in which Z_{oxide}, defined to be associated with the passive current, corresponds to the anodic branch. If a cathodic reaction is taken into account, this reaction occurs at

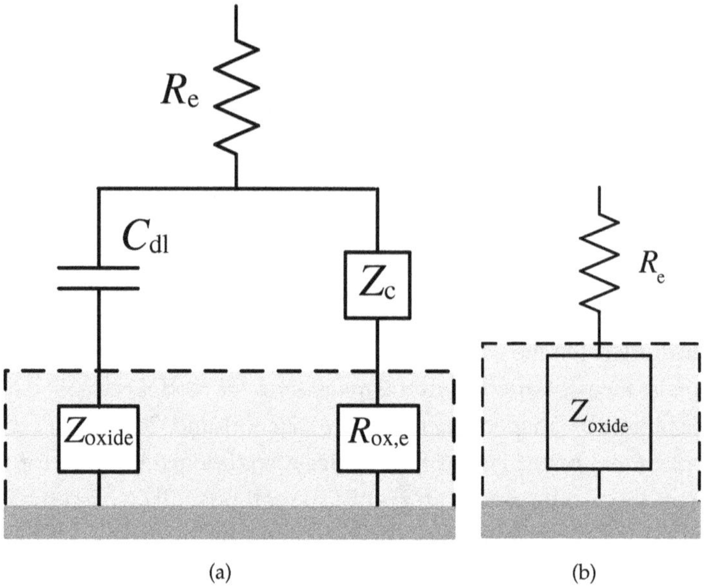

Figure 26.9: Electrical circuit for an oxide film: (a) general circuit taken from Tribollet et al.[107] showing the parallel contributions of anodic and cathodic reactions and (b) simplified circuit including the influence of the dielectric response of the oxide film.

the interface between the electrolyte and the oxide layer, resulting in an impedance that is parallel to the anodic branch, as shown in Figure 26.9(a). The electrons are generated at the interface between the electrolyte and the oxide layer and are transferred to the metal through an electronic resistance.[107]

The ohmic-resistance-corrected phase angle shown in Figure 26.4(b) suggests that the impedance response includes a constant-phase element. The estimated values of the capacitance in Table 26.2 are too small to represent a double-layer. The constant-phase-element behavior can therefore be attributed to the oxide film. Hirschorn et al.[56] showed that a constant-phase-element behavior in a film may be attributed to a power-law distribution of resistivity within the limits of ρ_0 and ρ_δ, representing the resistivity of the oxide at the metal–oxide interface and the oxide–electrolyte interface, respectively. Orazem et al.[60] showed that an asymmetric Nyquist plot may be expected if the resistance associated with the faradaic reactions is larger than the zero-frequency limit of the resistance associated with the oxide film. Thus, the simplified model shown as Figure 26.9(b) may be assumed to apply.

Two major assumptions were made in this analysis. The first is that the double-layer capacitance shown in Figure 26.9(a) may be neglected. The second is neglect of the low-frequency inductive features that were shown to be consistent with the Kramers–Kronig relations, limiting the frequency range of the data used for the regression analysis to frequencies above that where the inductive feature is evident.

26.2.2 Mathematical Model

The ohmic-resistance corrected phase angle suggests the presence of high-frequency CPE behavior. The asymmetric shape of the Nyquist plot is consistent with the power-law model developed for the impedance response of films.[56] The resistivity follows the distribution given as

$$\frac{\rho(y)}{\rho_\delta} = \left(\frac{\rho_\delta}{\rho_0} + \left(1 - \frac{\rho_\delta}{\rho_0}\right)\left(\frac{y}{\delta}\right)^{1/(1-\alpha)}\right)^{-1} \tag{26.6}$$

where δ is the film thickness and ρ_0 and ρ_δ are the boundary values of resistivity at the respective interfaces. The impedance of the film can be written for an arbitrary resistivity distribution $\rho(y)$ as

$$Z_f = \int_0^\delta \frac{\rho(y)}{1 + j\omega\varepsilon\varepsilon_0\rho(y)} dy \tag{26.7}$$

Under the conditions that $\rho_0 \gg \rho_\delta$ and $f < (2\pi\rho_\delta\varepsilon\varepsilon_0)^{-1}$, the impedance may be expressed as

$$Z = R_e + g\frac{\delta\rho_\delta^{1-\alpha}}{(\rho_0^{-1} + j\omega\varepsilon\varepsilon_0)^\alpha} \tag{26.8}$$

where

$$g = 1 + 2.88(1-\alpha)^{2.375} \tag{26.9}$$

Equation (26.8) may be expressed as

$$Z = R_e + \frac{g\delta\rho_\delta^{1-\alpha}\rho_0^\alpha}{(1 + j\omega\varepsilon\varepsilon_0\rho_0)^\alpha} \tag{26.10}$$

or

$$Z = R_e + \frac{K_1}{(1 + j\omega K_2)^\alpha} \tag{26.11}$$

where

$$K_1 = g\delta\rho_\delta^{1-\alpha}\rho_0^\alpha \tag{26.12}$$

and

$$K_2 = \varepsilon\varepsilon_0\rho_0 \tag{26.13}$$

The fitted parameters include R_e, K_1, K_2, and α.

26.3 Regression Analysis

Equation (26.11) was regressed to the portion of the data presented in Table 26.1 ranging from 0.0229 to 78 Hz. Data taken below 0.0229 Hz was assumed to be influenced by the inductive phenomena for which a model has not yet been found. Data taken above 78 Hz was found to be influenced by frequency dispersion associated with the

Table 26.3: Regression results for the fitting of equation (26.11) to the data presented in Table 26.1.

Regressed Parameters	
R_e	18.83 ± 0.80 Ωcm^2
K_1 (see equation (26.12))	63.39 ± 0.31 kΩcm^2
K_2 (see equation (26.13))	18.02 ± 0.14 s
α	0.86764 ± 0.00093
Inferred Parameters	
ρ_0	3.240 ± 0.025 TΩcm
ρ_δ	2.53 MΩcm
δ	1.3 nm

geometry of the electrode. A Levenberg–Marquardt analysis was performed using a weighting based on an assumed error structure in which $\sigma = 0.01|Z|$. The resulting parameters are presented in Table 26.3. The weighted χ^2/ν statistic, based on the assumed error structure, had a value of 0.86.

Regression results for the fit of equation (26.10) to the data presented in Table 26.1 are presented in Figure 26.10 in a logarithmic scale to emphasize the quality of the fit at high frequencies. The data not used in the regression are also shown as gray symbols. As shown in Figure 26.10, equation (26.10) provides an excellent representation of the data.

A more sensitive assessment of the quality of the fit is obtained by examination of the corresponding normalized residual errors, shown in Figure 26.11. The normalized residual errors show trending, suggesting that the model could be improved. The average magnitude of the normalized residual errors was on the order of 0.017 (1.7%) for both the real and imaginary parts of the impedance. The residual errors were larger for the imaginary part of the impedance at low frequencies.

The comparison of the model fit to the data shown in Figure 26.12 in Nyquist format emphasizes the quality of the fit at low frequencies. The discrepancy may be attributed to the influence of the low-frequency inductive feature.

26.3.1 Evaluation of Regression Results

While the regressed values are lumped parameters containing combinations of film properties, a detailed analysis of the film parameters is possible under the assumption of a value for the dielectric constant and by further analysis of the impedance results.

The capacitance may be obtained from the high-frequency asymptote for the complex capacitance, as shown in Figure 26.5. The estimation of capacitance from the

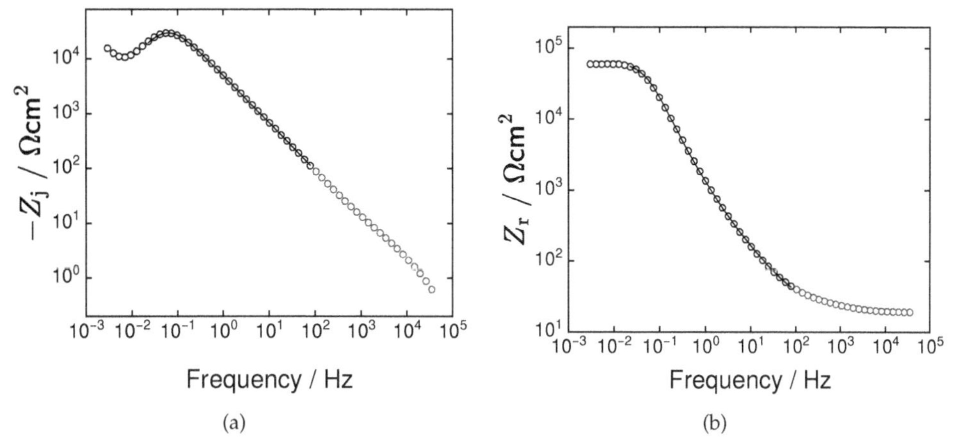

Figure 26.10: Regression results for the fit of equation (26.10) to the data presented in Table 26.1: a) imaginary part of the impedance and b) real part of the impedance. The line represents the fit.

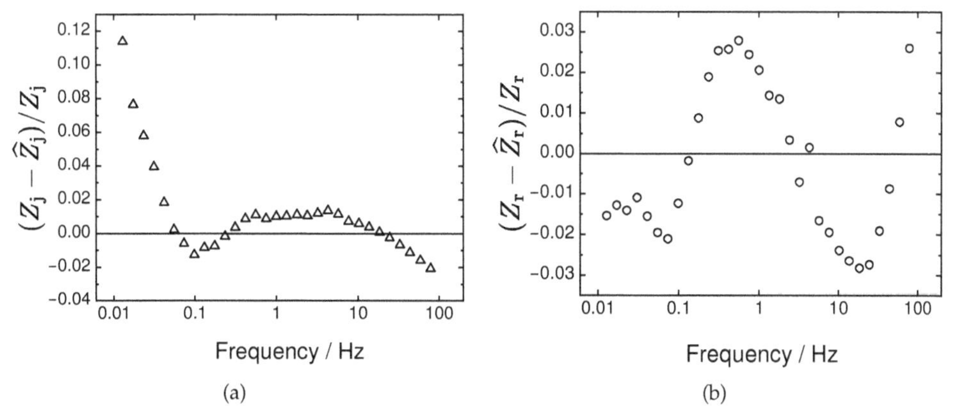

Figure 26.11: Normalized residual errors for the fit of equation (26.10) to the data presented in Table 26.1: (a) imaginary part of the impedance and (b) real part of the impedance.

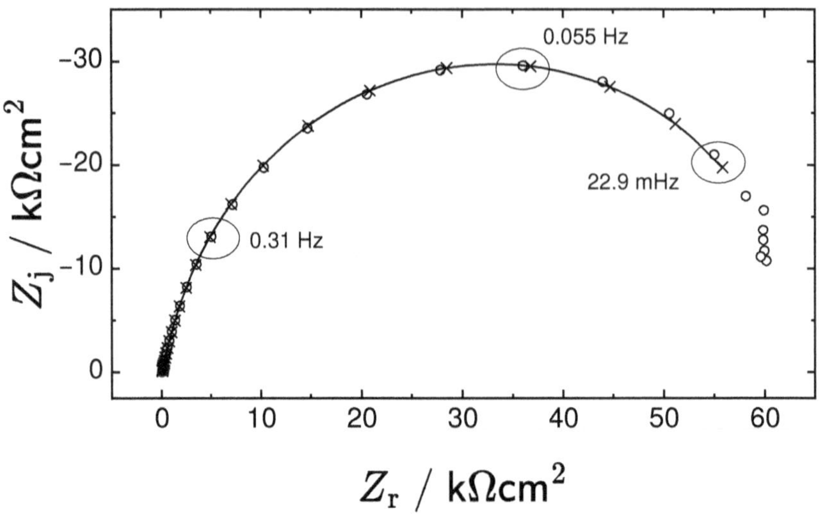

Figure 26.12: Regression results for the fit of equation (26.10) to the data presented in Table 26.1 in Nyquist format. Data are represented by ○, and the fit is represented by ×.

complex-capacitance plot is sensitive to the estimated value for ohmic resistance. As indicated in Table 26.3, the ohmic resistance was estimated by regression to be $R_e = 18.83 \pm 0.80$ Ωcm^2. The corresponding complex capacitance obtained from equation (26.2) is presented in Figure 26.13, where the ohmic resistance was assumed to be $R_e = 18.83 - 0.80$, 18.83, or $18.83 + 0.80$ Ωcm^2. As seen in Figure 26.13, the high-frequency extrapolation of complex capacitance is problematic for the estimated value of ohmic resistance.

As shown in Figure 26.13, modest uncertainty in ohmic resistance causes a large uncertainty in high-frequency limit of complex capacitance. The uncertainty in the evaluation of ohmic resistance was caused by the finite range of frequency used for the regression, which itself was due to the inability of the model to account for geometry-induced frequency dispersion. Fortunately, the value of the primary ohmic resistance is unaffected by geometry-induced frequency dispersion; thus, the general measurement model, fit to all Kramers-Kronig-consistent data, can be used to estimate the ohmic resistance. The complex capacitance evaluated from equation (26.3) with the ohmic resistance obtained by use of the measurement model (see Table 26.2) is presented in Figure 26.14. The tight confidence interval for the ohmic resistance obtained by the measurement model yields a estimates for complex capacitance that show excellent overlap.

The real part of the capacitance shown in Figure 26.14 is presented in Figure 26.15 as a function of frequency. The value of 7.27 μF/cm^2 obtained by use of the measurement model and the value of 6.7 μF/cm^2 corresponding to the apparent plateau are shown as a horizontal dashed lines.

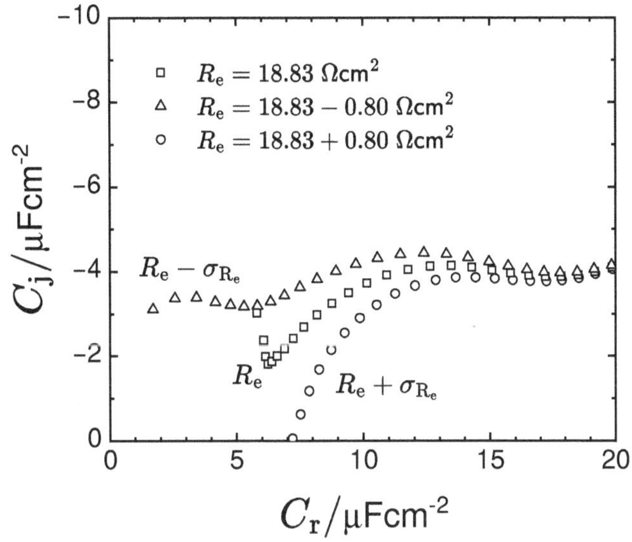

Figure 26.13: The imaginary part of the complex capacitance as a function of the real part of the complex capacitance. The complex capacitance was evaluated by the use of equation (26.3) with the ohmic resistance presented in Table 26.3.

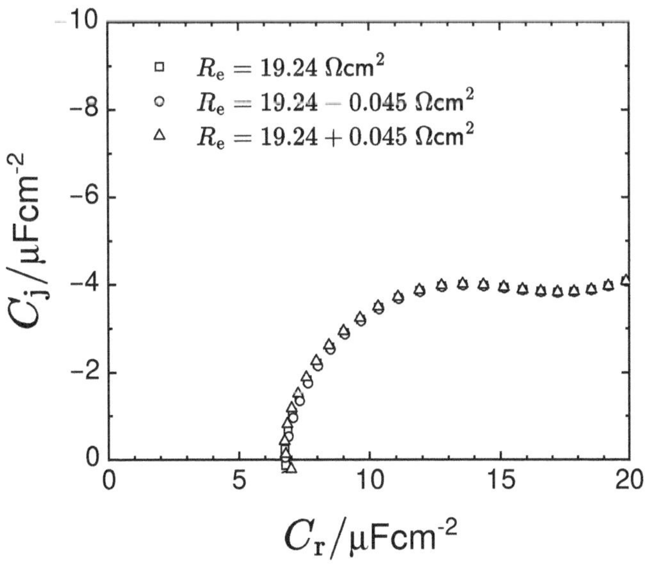

Figure 26.14: The imaginary part of the complex capacitance as a function of the real part of the complex capacitance. The complex capacitance was evaluated from equation (26.3) with the ohmic resistance obtained by the use of the measurement model and presented in Table 26.2.

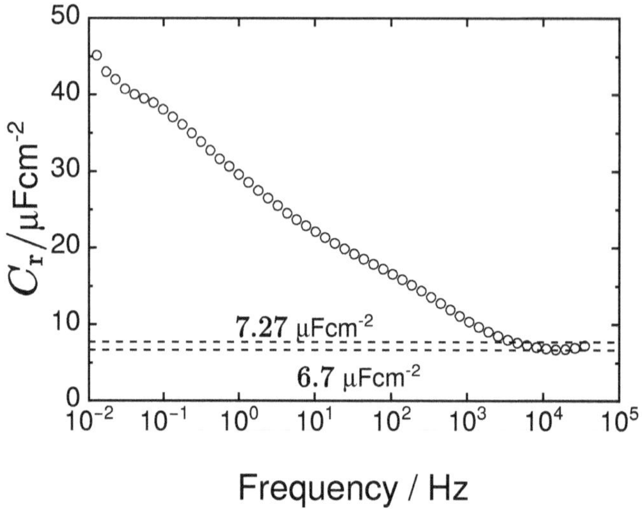

Figure 26.15: The real part of the complex capacitance as a function of frequency. The complex capacitance was evaluated from equation (26.3) with the ohmic resistance obtained by the use of the measurement model and presented in Table 26.2.

The capacitance is related to film properties according to

$$C = \frac{\varepsilon\varepsilon_0}{\delta} \tag{26.14}$$

where ε is the dielectric constant, $\varepsilon_0 = 8.8542 \times 10^{-14}$ F/cm is the permittivity of vacuum, and δ is the film thickness. Under the assumptions that the capacitance is 6.7 μF/cm^2 and the dielectric constant for Al_2O_3 is $\varepsilon = 10$,[108] the film thickness can be calculated to be 1.3 nm. From equation (26.13), the resistivity at the metal–oxide interface can be calculated to be $\rho_0 = 3.240 \times 10^{12}$ Ωcm or 3.240 TΩcm. From equation (26.12), the resistivity at the oxide–electrolyte interface can be obtained as $\rho_\delta = 2.53 \times 10^6$ Ωcm or 2.53 MΩcm. These values are reported in Table 26.3.

The resistivity distribution associated with the power-law model is given by equation (26.6). The resistivity distribution calculated with the values obtained by regression is presented in Figure 26.16. The log–log representation shows a power-law distribution extending from $\rho_\delta = 2.53 \times 10^6$ Ωcm at a position of 1.3 nm to a position of about 0.2 nm, where the resistivity approaches a value of $\rho_0 = 3.240 \times 10^{12}$ Ωcm. Thus, the interpretation model, coupled with identification of the capacitance from Figure 26.15, allows a comprehensive evaluation of the resistivity distribution in the oxide.

26.3.2 Recommendations Resulting from the Regression

The regression presented in this chapter was constrained by the need to truncate data at both high and low frequencies. At low frequencies, data were truncated due to the presence of an inductive feature. Such features were reported by Barcia et al.[34] for the

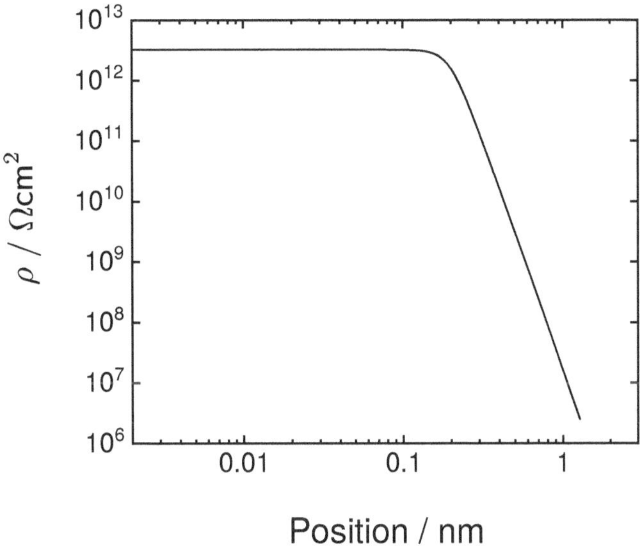

Figure 26.16: The oxide film resistivity as a function of position. Values were inferred from the regression parameters shown in Tables 26.2 and 26.3.

impedance response of copper in a 1 M hydrochloric acid solution. A model for this system was presented in detail by Orazem and Tribollet (Section 11.8 of reference 1), where the low-frequency inductive feature was attributed to the dynamic response of a CuCl salt film thickness to the modulated potential. A similar model has not been developed for metal oxides. To explore this feature, measurements are needed at lower frequencies.

The influence of electrode geometry on impedance was first reported in 1970.[109] Recently, Gharbi et al.[110] showed that the ohmic impedance or frequency dispersion associated with a disk electrode geometry can be fitted by the Havriliak–Negami equation given as[111]

$$Z_e = R_{e,HF} + \frac{R_{e,LF} - R_{e,HF}}{\left(1 + (j\omega\tau)^{1-\nu}\right)^{1-\beta}} \tag{26.15}$$

where τ is the characteristic time constant of the ohmic impedance. The exponents ν and β dictate the limiting phase angles of the ohmic impedance at high and low frequencies. Equation (26.15) shows the presence of two ohmic resistances, one that is seen at the high-frequency limit $R_{e,HF}$ and one that is seen at the low-frequency limit $R_{e,LF}$. The high-frequency ohmic resistance is equal to the ohmic resistance calculated from a steady-state primary current distribution.

Gharbi et al.[110] extended the frequency range by including all high frequency data.

Their model was expressed as

$$Z = R_{e,HF} + \frac{R_{e,LF} - R_{e,HF}}{\left(1 + (j\omega\tau)^{1-\nu}\right)^{1-\beta}} + \int_0^\delta \frac{\rho(y)}{1 + j\omega\varepsilon\varepsilon_0\rho(y)} dy \tag{26.16}$$

and the regression was performed by an in-house simplex routine. The disadvantage of the simplex regression is that, because confidence intervals are not provided for parameter estimates, simplex regressions may over-fit data. A similar Levenberg–Marquardt regression yielded parameters that were not statistically significant. More work is needed to explore the regression of impedance data that include the influence of ohmic impedance.

Impedance is not a stand-alone technique, and models that fit the data are not unique. The model developed in this chapter suggests the use of surface analysis techniques such as high-angle annular dark-field scanning transmission electron microscopy (HAADF-STEM) to verify the estimated film thickness. HAADF-STEM was used by Chen et al.[112] to verify the film thickness of oxides on steel estimated by the use of the power-law model. It would be useful to design experiments that would allow analysis of systems with thicker oxide films.

26.4 Regression of Integral Formula

The measurement model program allows regression of integral equations such as equation (26.7). For the present work, the resistivity was given in equation (26.6). Application of equation (26.7) is under the assumption that the low-frequency resistance evident in the impedance data can be attributed to the properties of the oxide film. The regressed parameters are ρ_δ, ρ_0, δ, ε, R_e, and γ, where

$$\gamma = \frac{1}{1 - \alpha} \tag{26.17}$$

Regression with all parameters yielded a very nice fit with most parameter estimates including zero in the $\pm 2\sigma$ confidence interval. To obtain a fit with statistically significant parameter estimates, some of the variables were fixed to values obtained is Section 26.3. The value $\alpha = 0.86764$ yielded, from equation (26.17), $\gamma = 7.555$. The dielectric constant was assumed to be $\varepsilon = 10$, and the resistivity at the oxide–electrolyte interface was assumed to be $\rho_\delta = 2.53 \times 10^6$ Ωcm^2. The frequency range regressed was from 0.023 to 78.3 Hz, and, as replicate measurements were not available for estimation of the error structure, modulus weighting was employed.

The results are presented in Table 26.4. The values of ρ_0 and δ are in excellent agreement with the values calculated from the results of the regression of equation (26.11) to the data presented in Table 26.1. A comparison of the fit to the data is presented in Nyquist format in Figure 26.17. The corresponding comparison in Bode format is presented in Figure 26.18.

Table 26.4: Results for regression under modulus weighting of equation (26.7) with resistivity given in equation (26.6) to the data presented in Table 26.1. The frequency range regressed was from 0.023 to 78.3 Hz.

Parameter	Analytic	Source	Integral Formula	Source
ρ_δ, MΩcm	2.53	Calculated	2.53	Fixed
ρ_0, TΩcm	3.240±0.025	Calculated	3.240±0.019	Regressed
δ, nm	1.3	Calculated	1.2226±0.0023	Regressed
α	0.86764±0.00093	Regressed	–	
K_1, kΩcm^2	63.39±0.31	Regressed	–	
K_2, s	18.02±0.14	Regressed	–	
γ	–		7.555	Fixed
ϵ	10	Fixed	10	Fixed
R_e, Ωcm^2	18.83±0.80	Regressed	18.87±0.70	Regressed

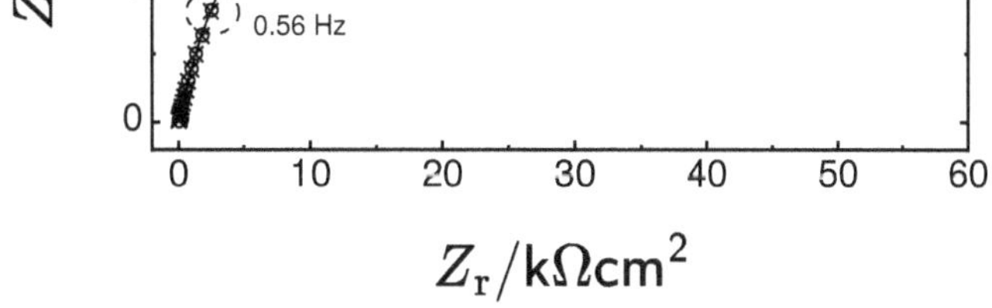

Figure 26.17: Results in Nyquist format for regression under modulus weighting of equation (26.7) with resistivity given as equation (26.6) to the data presented in Table 26.1. The resulting parameters are presented in Table 26.4.

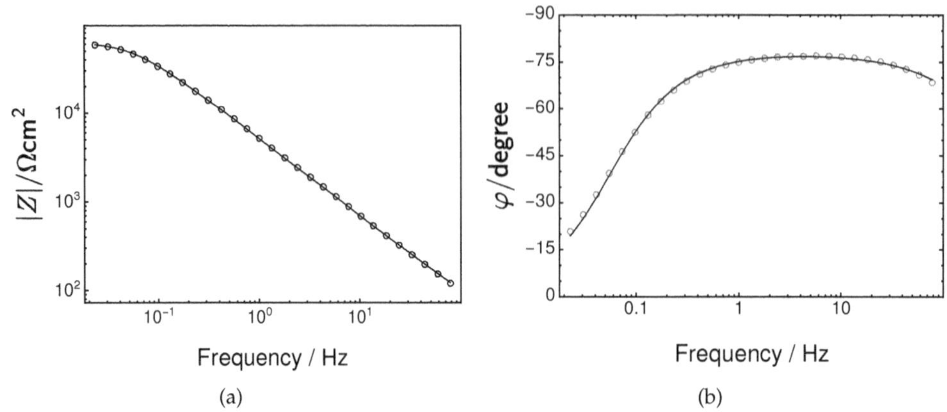

Figure 26.18: Results in Nyquist format for regression under modulus weighting of equation (26.7) with resistivity given as equation (26.6) to the data presented in Table 26.1: (a) modulus and (b) phase. The resulting parameters are presented in Table 26.4.

26.5 Observations

The analysis presented in this chapter shows the value of an initial graphical analysis to extract both physical parameters and insight needed for selecting appropriate models. While the data were shown to be influenced by geometry-induced frequency dispersion and by a strange low-frequency inductive feature, the measurement model analysis showed that the entire spectrum satisfied the Kramers–Kronig relations.

Model selection would have been further enhanced by making impedance measurements at potentials slightly above and slightly below the corrosion potential, as discussed in Problem 9.9. Such an analysis could confirm that the impedance results did not reflect anodic or cathodic reactions, supporting the decision to treat the impedance as being due to the properties of the oxide film. The error structure analysis could be enhanced by replicating the impedance measurements, thus enabling determination of the stochastic contribution to the error structure (see Chapter 27).

The present example demonstrates that regression of a model to data serves as a guide to future experiments. There is a big advantage to analyzing data as experiments are performed and there is still time to design and perform new experiments.

Chapter 27

Mass Transfer in Lithium Cobalt Oxide Coin Cells

The data presented in this chapter show how the measurement model analysis can yield the stochastic error structure from nearly replicate impedance spectra. As is typical for low-impedance systems, the frequency range is constrained at the high-frequency limit by instrumental and wiring artifacts. The model development introduces the concept of anomalous diffusion to account for the low-frequency impedance behavior. The confidence interval for regressed parameters is shown to restrict the number of parameters that can be extracted from the impedance data.

Erol et al.[64,15] report the results of electrochemical experiments performed on commercial $LiCoO_2|C$ coin cells. Impedance spectroscopy was used to monitor changes associated with different states-of-charge and temperatures. The impedance data obtained under normal operation conditions, for example, at 3.6 V, revealed a depressed capacitive loop at high and intermediate frequencies and a straight line at low frequencies, with an angle with respect to the real axis greater than 45°, as shown in Figure 27.1. The impedance data could not be described simply by a Randles circuit design because the high-frequency loop is not a perfect semicircle or depressed semicircle (CPE behavior with one time constant) in a Nyquist plot, and the angle with respect to the real axis of the low-frequency impedance was near 58°, which is much larger than the 45° associated with the Warburg impedance.

The normal operating potential range of the cells reported by the suppliers[113] is between 3.0 V (0% state of charge) and 4.2 V (100% state of charge). In one experiment, a coin cell was initially discharged under constant 2 mA current to 3.00 V. The battery was then subjected to sequential 80 mV steps in cell voltage down to 2.20 V. After reaching a steady state, the battery was allowed to relax for two days at the open-circuit condition. When held at open circuit, the over-discharged battery slowly reached a cell potential within the normal operating range. The results are presented in Figure 27.2 in Nyquist format, with elapsed time after over-discharge as a parameter. These data

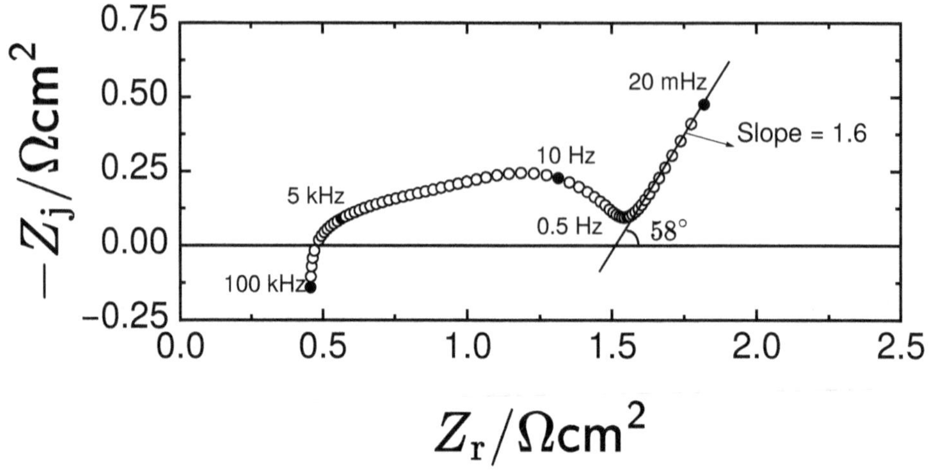

Figure 27.1: Typical impedance response of a $LiCoO_2|C$ coin cell at a potential of 3.6 V, obtained from Erol and Orazem.[15]

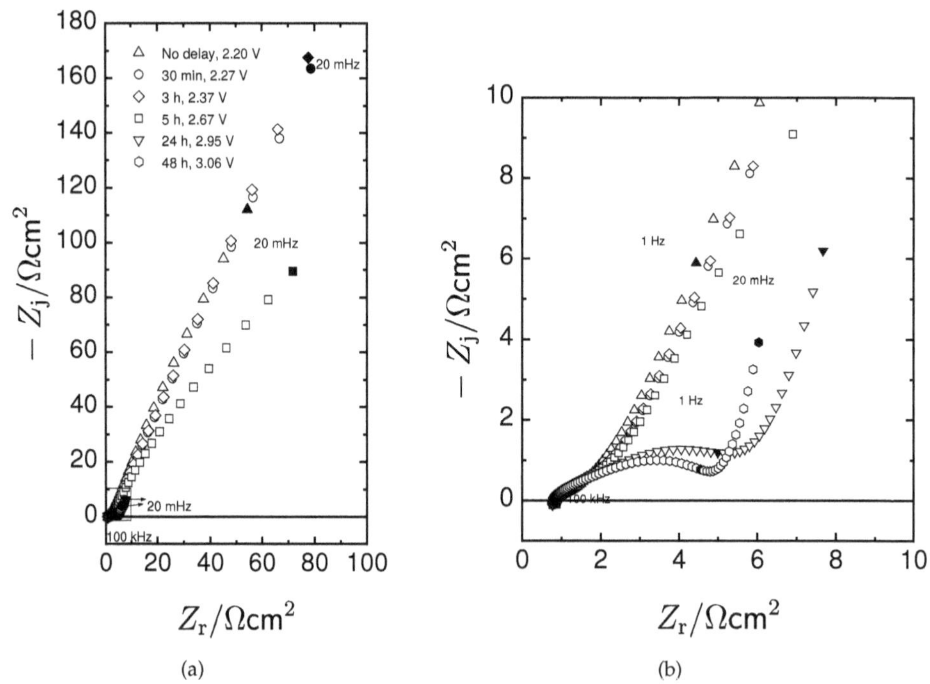

Figure 27.2: Impedance response in Nyquist format for a $LiCoO_2|C$ coin cell, with elapsed time after over-discharge as a parameter: (a) full spectrum showing all the results obtained over a 24-hour period and (b) a zoomed in region emphasizing the results after 24 hours elapsed time. The legend for Figure 27.2(b) is the same as that for Figure 27.2(a). Data taken from Erol et al.[64]

were taken from Erol et al.[64]

The purpose of the present chapter is to examine the information that can be obtained from impedance data collected during an obviously transient process. The first three spectra obtained during the self-charge process were sufficiently similar that they can be used to estimate the standard deviation for the stochastic part of the impedance measurement. The three impedance spectra taken during the self-charge process are presented in Table 27.1 at an elapsed time of 0 minutes and a cell potential of 2.20 V, in Table 27.2 at an elapsed time of 30 minutes and a potential of 2.27 V, and in Table 27.3 at an elapsed time of 180 minutes and a potential of 2.37 V. The experimental protocol was presented by Erol et al.[64,15]

27.1 Initial Inspection

As presented in Chapters 26 and 28–30, an initial inspection involves the use of graphical methods and regression of a general measurement model.

27.1.1 Graphical Analysis

The three spectra given in Tables 27.1, 27.2, and 27.3 are presented in Figure 27.3 in Nyquist format. While the shape of the impedance diagrams are similar, there are clear differences between the spectra evident in the lack of overlap of identified frequencies for the first spectrum at 2.20 V and the second and third spectra. The zoomed presentation in Figure 27.3(b) reveals a high-frequency inductive feature similar to that presented in Figure 8.10.

The zoomed Nyquist plot shown in Figure 27.3(b) can be used to estimate the ohmic resistance. The ohmic resistance may be more easily extracted from the real part of the impedance presented in Figure 27.4. In spite of the inductive feature shown in Figure 27.3(b), the real part of the impedance asymptotically approaches a finite value at high frequency. The ohmic resistance obtained by the graphical analysis is presented in Table 27.4. The ohmic resistance decreased slightly with elapsed time from 0.837 to 0.762 Ωcm^2.

The impedance is presented in Bode format in Figure 27.5. The uncorrected modulus shown in Figure 27.5(a) approaches the ohmic resistance at high frequency as does the real part of the impedance shown in Figure 27.4. The ohmic-resistance-corrected modulus increases at frequencies higher than 50 kHz due to the influence of the high-frequency inductive feature seen in Figure 27.3(b). The phase angle shown in Figure 27.5(b) becomes positive at frequencies higher than 40 kHz; whereas, the ohmic-resistance-corrected phase angle changes sign at a frequency of 40 kHz and approaches +90 degrees at 100 kHz.

The high-frequency inductive loop suggests that the measurement above 40 kHz includes the inductance of the leads. The graphical method is not sufficient to determine the frequency at which the inductance affects the measured impedance. The

Table 27.1: Impedance data obtained for a commercial $LiCoO_2|C$ coin cell after over-discharge to a potentail of 2.2 V. The impedance was monitored during a period of self-charge. The elapsed time was 0 minutes, and the potential was 2.20 V.

f, Hz	Z_r, Ωcm^2	Z_j, Ωcm^2	f, Hz	Z_r, Ωcm^2	Z_j, Ωcm^2
100007	0.837614	0.110547	31.5793	1.79808	−0.701976
79433.6	0.83946	0.073171	24.9911	1.87554	−0.786348
63105.5	0.84498	0.042651	20.0011	1.95619	−0.880737
50123.7	0.851587	0.016543	15.7934	2.05099	−0.9986
39811.2	0.859843	−0.005115	12.5002	2.15486	−1.13505
31634.1	0.869167	−0.023989	9.9987	2.26548	−1.28718
25123.7	0.879709	−0.04006	7.89389	2.39514	−1.47361
19954.4	0.89143	−0.054526	6.31623	2.53368	−1.68882
15852.9	0.904036	−0.067184	4.99748	2.68824	−1.94503
12587.2	0.9174	−0.079104	3.97302	2.85332	−2.24469
9998.2	0.931457	−0.089929	3.15718	3.03535	−2.60534
7942.6	0.946591	−0.100035	2.51085	3.24465	−3.03553
6311.1	0.96292	−0.110224	1.99489	3.47752	−3.5629
5011.18	0.9799	−0.120097	1.58325	3.74847	−4.19985
3980.93	0.99833	−0.130044	1.25729	4.06893	−4.96369
3162.67	1.01721	−0.140151	1.00006	4.44143	−5.88802
2511.16	1.03795	−0.150637	0.7946	4.88095	−6.98216
1995.37	1.05953	−0.161504	0.63106	5.41397	−8.29684
1584.49	1.08293	−0.173048	0.50119	6.05086	−9.8703
1259.04	1.10751	−0.185157	0.39827	6.81341	−11.7398
999.95	1.13366	−0.197837	0.3163	7.72696	−13.9794
794.163	1.16059	−0.210925	0.25121	8.81712	−16.6338
630.984	1.19017	−0.225197	0.19946	10.14	−19.7927
501.303	1.22166	−0.240949	0.15853	11.7153	−23.5432
398.209	1.25344	−0.257707	0.12587	13.6481	−28.0023
316.113	1.28688	−0.276978	0.09999	15.9294	−33.3314
251.185	1.32219	−0.298896	0.07945	18.7096	−39.624
199.495	1.35974	−0.324208	0.0631	22.1173	−47.1868
158.533	1.39999	−0.353001	0.05013	26.2384	−56.1186
125.911	1.44369	−0.386119	0.03981	31.2462	−66.713
100.009	1.49129	−0.423918	0.03163	37.445	−79.3918
79.4083	1.54336	−0.466448	0.02512	45.1069	−94.226
50.0083	1.66173	−0.569143	0.01995	54.4491	−112.076
40.0059	1.7257	−0.629173			

Table 27.2: Impedance data obtained for a commercial $LiCoO_2|C$ coin cell after over-discharge to a potentail of 2.2 V. The impedance was monitored during a period of self-charge. The elapsed time was 30 minutes, and the potential was 2.27 V.

f, Hz	Z_r, Ωcm^2	Z_j, Ωcm^2	f, Hz	Z_r, Ωcm^2	Z_j, Ωcm^2
100007	0.761779	0.104527	31.5793	2.27445	−1.05047
79433.6	0.766251	0.069843	24.9911	2.38441	−1.16897
63105.5	0.771165	0.039067	20.0011	2.49476	−1.30409
50123.7	0.777708	0.012907	15.7934	2.6209	−1.47772
39811.2	0.784682	−0.010078	12.5002	2.75509	−1.68707
31634.1	0.793429	−0.030646	9.9987	2.89684	−1.92861
25123.7	0.803459	−0.049517	7.89389	3.06554	−2.23849
19954.4	0.815098	−0.067815	6.31623	3.24944	−2.59827
15852.9	0.827651	−0.084696	4.99748	3.46704	−3.03982
12587.2	0.842463	−0.101661	3.97302	3.71341	−3.55899
9998.2	0.858756	−0.118652	3.15718	4.00268	−4.18153
7942.6	0.876856	−0.135588	2.51085	4.35472	−4.90793
6311.1	0.896781	−0.153282	1.99489	4.74995	−5.80762
5011.18	0.919237	−0.17189	1.58325	5.22975	−6.86619
3980.93	0.943734	−0.191069	1.25729	5.80381	−8.11359
3162.67	0.97118	−0.211197	1.00006	6.47431	−9.6038
2511.16	1.00107	−0.232568	0.7946	7.27616	−11.3406
1995.37	1.03402	−0.254834	0.63106	8.2403	−13.4036
1584.49	1.07063	−0.278779	0.50119	9.37545	−15.8333
1259.04	1.11061	−0.303601	0.39827	10.7249	−18.69
999.95	1.15363	−0.330177	0.3163	12.3197	−22.0563
794.163	1.20196	−0.357338	0.25121	14.1928	−26.0354
630.984	1.25238	−0.386415	0.19946	16.4128	−30.7413
501.303	1.3085	−0.41518	0.15853	19.0235	−36.2296
398.209	1.36811	−0.446399	0.12587	22.1075	−42.8164
316.113	1.43174	−0.478902	0.09999	25.7392	−50.5517
251.185	1.49873	−0.512904	0.07945	30.0045	−59.6761
199.495	1.56927	−0.549219	0.0631	35.067	−70.5241
158.533	1.64303	−0.58873	0.05013	41.0567	−83.3496
125.911	1.7206	−0.632068	0.03981	48.1317	−98.588
100.009	1.80215	−0.680597	0.03163	56.5474	−116.607
79.4083	1.88752	−0.735458	0.02512	66.7707	−138.014
50.0083	2.07247	−0.869914	0.01995	78.5937	−163.503
40.0059	2.16857	−0.95016			

Table 27.3: Impedance data obtained for a commercial $LiCoO_2|C$ coin cell after over-discharge to a potentail of 2.2 V. The impedance was monitored during a period of self-charge. The elapsed time was 180 minutes, and the potential was 2.37 V.

f, Hz	Z_r, Ωcm^2	Z_j, Ωcm^2	f, Hz	Z_r, Ωcm^2	Z_j, Ωcm^2
100007	0.761598	0.105242	31.5793	2.28122	−1.06531
79433.6	0.766361	0.069719	24.9911	2.39235	−1.18666
63105.5	0.771378	0.039359	20.0011	2.50365	−1.3251
50123.7	0.777605	0.013071	15.7934	2.63068	−1.50368
39811.2	0.784977	−0.00993	12.5002	2.76684	−1.71809
31634.1	0.793711	−0.030574	9.9987	2.91046	−1.96636
25123.7	0.803626	−0.049614	7.89389	3.08206	−2.28376
19954.4	0.815016	−0.067469	6.31623	3.26907	−2.65189
15852.9	0.827914	−0.084798	4.99748	3.49104	−3.10494
12587.2	0.842343	−0.101677	3.97302	3.74352	−3.63813
9998.2	0.858523	−0.118736	3.15718	4.03932	−4.27723
7942.6	0.876457	−0.135877	2.51085	4.39662	−5.03514
6311.1	0.896661	−0.153672	1.99489	4.80631	−5.9396
5011.18	0.918765	−0.172249	1.58325	5.29738	−7.02053
3980.93	0.94351	−0.191702	1.25729	5.88904	−8.29522
3162.67	0.97106	−0.211957	1.00006	6.57504	−9.8118
2511.16	1.00116	−0.233484	0.7946	7.39497	−11.5854
1995.37	1.0344	−0.256113	0.63106	8.379	−13.6828
1584.49	1.07118	−0.280205	0.50119	9.5388	−16.1645
1259.04	1.11079	−0.305247	0.39827	10.9099	−19.0599
999.95	1.15479	−0.331898	0.3163	12.5193	−22.5014
794.163	1.20217	−0.359338	0.25121	14.4318	−26.548
630.984	1.25393	−0.388308	0.19946	16.6716	−31.3032
501.303	1.30981	−0.417679	0.15853	19.2912	−36.9346
398.209	1.36984	−0.449306	0.12587	22.376	−43.6591
316.113	1.43381	−0.482229	0.09999	25.9951	−51.5532
251.185	1.50119	−0.516716	0.07945	30.2484	−60.8922
199.495	1.57201	−0.553599	0.0631	35.261	−71.9804
158.533	1.64612	−0.593368	0.05013	41.1032	−85.0902
125.911	1.72431	−0.63768	0.03981	48.0347	−100.71
100.009	1.80596	−0.68714	0.03163	56.259	−119.254
79.4083	1.89197	−0.743044	0.02512	66.1103	−141.29
50.0083	2.07816	−0.880552	0.01995	77.5912	−167.512
40.0059	2.17481	−0.96265			

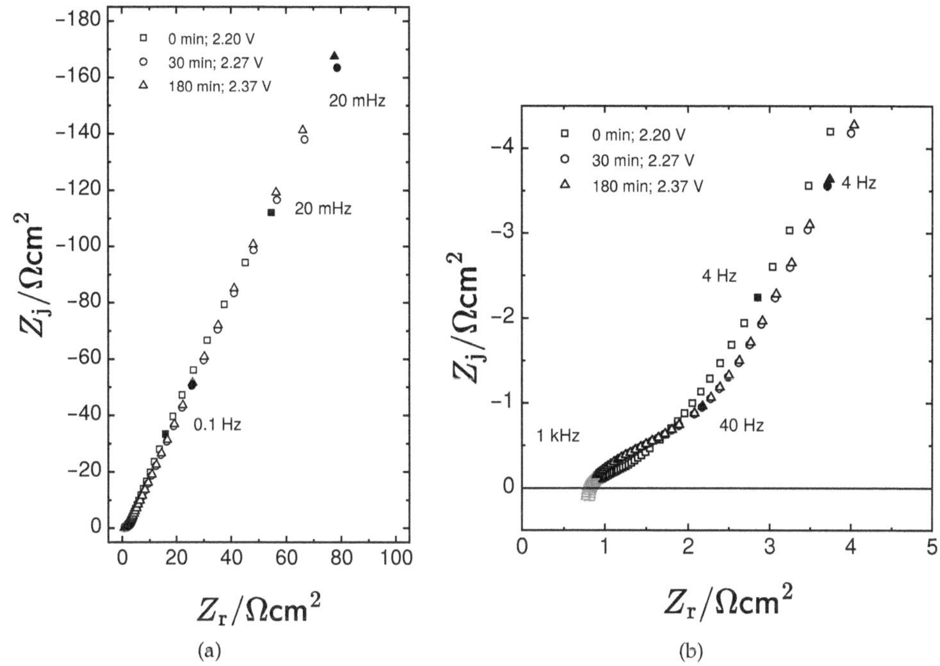

Figure 27.3: Impedance response in Nyquist format for a $LiCoO_2|C$ coin cell with elapsed time after over-discharge as a parameter: (a) full spectrum showing all the results obtained over a 24-hour period and (b) a zoomed-in region emphasizing the results after 24 hour elapsed time. Data taken from Erol et al.[64]

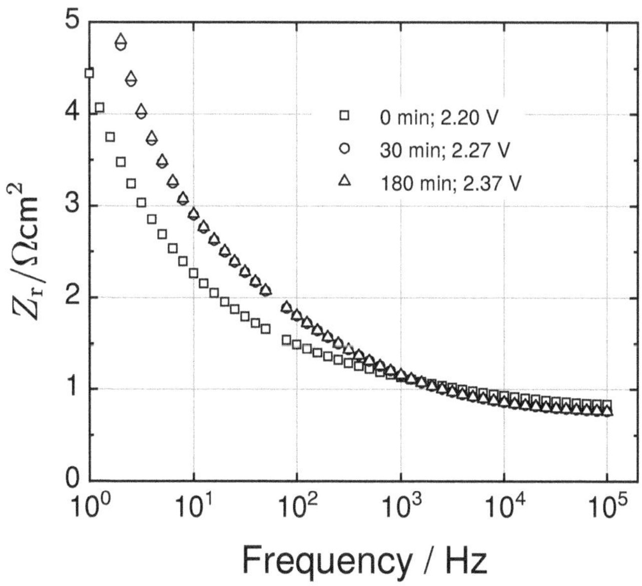

Figure 27.4: Real part of the impedance as a function of frequency for a $LiCoO_2|C$ coin cell with elapsed time after over-discharge as a parameter.

Table 27.4: Ohmic resistance obtained by the initial graphical and measurement model analysis of the data presented in Figure 27.4.

Analysis \ Table	27.1	27.2	27.3
Units	Ωcm^2	Ωcm^2	Ωcm^2
Graphical	0.837	0.762	0.762
Measurement model	0.8724 ± 0.0046	0.7979 ± 0.0046	0.7982 ± 0.0046

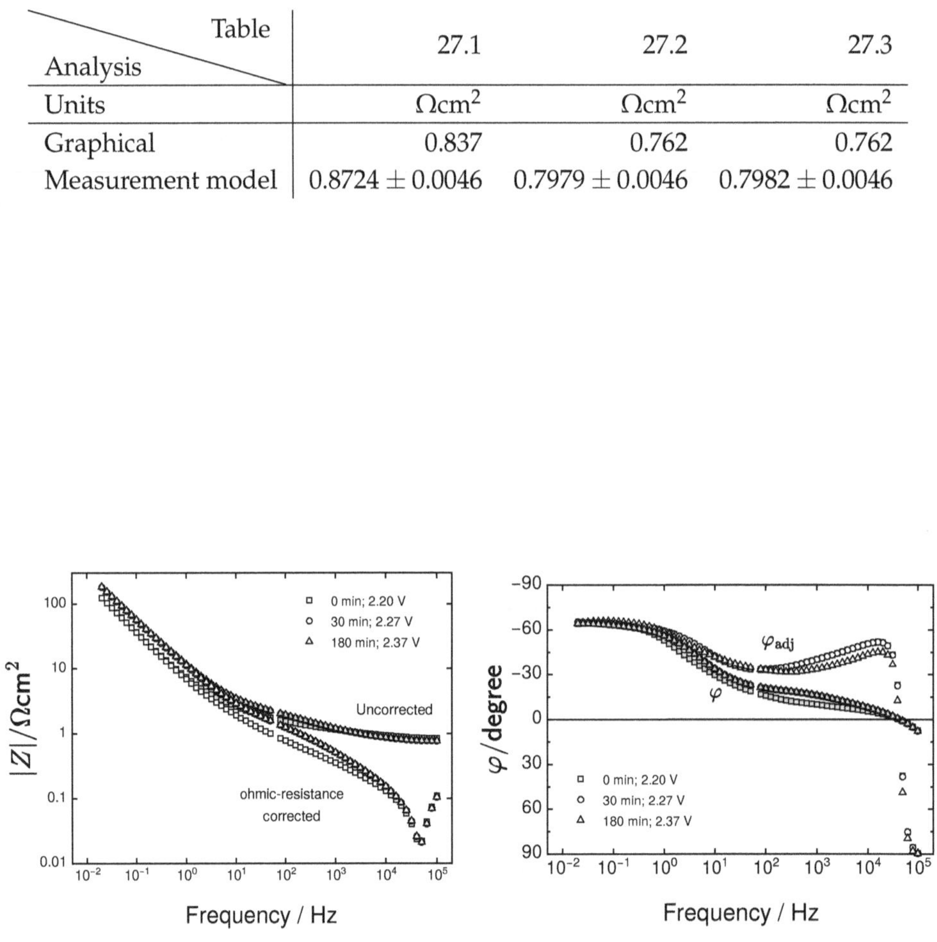

Figure 27.5: Impedance data given in Figure 27.3(a) as a function of frequency: (a) modulus (gray-filled symbols) and ohmic-resistance-corrected modulus (open symbols) and (b) phase angle (gray-filled symbols) and ohmic-resistance-corrected phase angle (open symbols).

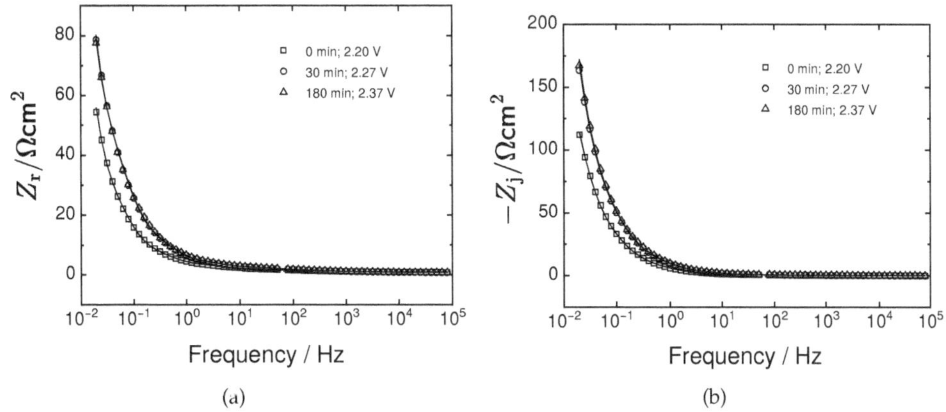

Figure 27.6: Impedance data given in Figure 27.3(a) as a function of frequency: (a) real part of the impedance and (b) imaginary part of the impedance. Lines represent the fit of equation (25.1) to the data with 7 *RC* or Voigt elements.

measurement model analysis described in the subsequent section can be used to assess the suitable frequency range for analysis.

27.1.2 Measurement Model Analysis

The measurement model analysis employed in Chapter 26 was limited to the assessment of consistency with the Kramers–Kronig relations under modulus weighting. As pseudoreplicates are available for the present case, the analysis of the $LiCoO_2|C$ battery data includes, as well, an analysis of the stochastic error structure.

Stochastic Error Analysis

The measurement model, described in Section 26.1.2, was fit to each of the spectra presented in Tables 27.1, 27.2, and 27.3. The results are presented in Figure 27.6, where the lines represent the fit of equation (25.1) to the data with seven statistically significant *RC* or Voigt elements. Modulus weighting was used for the regression. The corresponding residual errors are presented in Figure 27.7 as functions of frequency. Over most of the frequency range, the residual errors were generally less than $Z - \widehat{Z} = 0.02\ \Omega\text{cm}^2$. The residual errors were larger than $0.02\ \Omega\text{cm}^2$ at frequencies larger than 40 kHz. The oscillation in the residual errors can be attributed to the observation that equation (25.1) does not represent a correct model for the battery.

The goal of the present part of the analysis is to determine the standard deviation of the stochastic part of the impedance measurements. The residual error can be expressed as

$$Z(\omega) - \widehat{Z}(\omega) = \varepsilon_{\text{fit}}(\omega) + \varepsilon_{\text{stoch}}(\omega) + \varepsilon_{\text{bias}}(\omega) \tag{27.1}$$

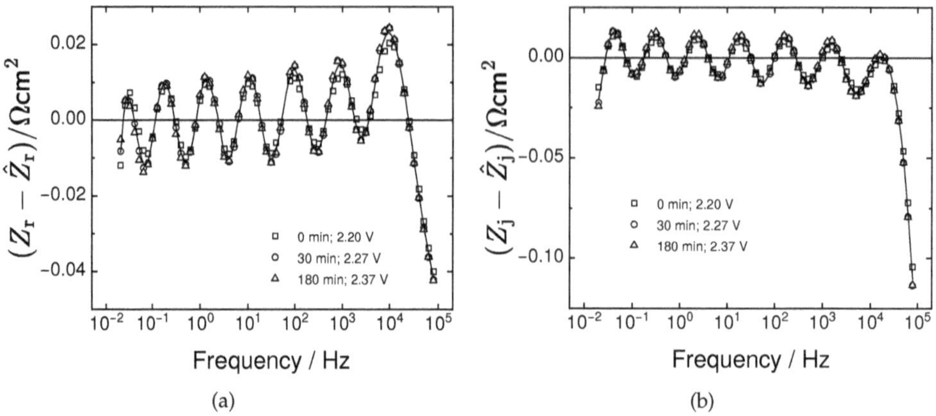

Figure 27.7: Residual errors for the regression of equation (25.1) to the data given in Figure 27.3(a) as a function of frequency: (a) real part of the impedance and (b) imaginary part of the impedance. Lines represent the average residual error at a given frequency.

where $\widehat{Z}$ is the model value, $\varepsilon_{\text{fit}}(\omega)$ is the systematic error that can be attributed to inadequacies of the model, $\varepsilon_{\text{stoch}}(\omega)$ is the stochastic error with expectation $\mathrm{E}\{\varepsilon_{\text{stoch}}(\omega)\} = 0$, and $\varepsilon_{\text{bias}}(\omega)$ represents the systematic experimental bias error that cannot be attributed to model inadequacies. Typically, the impedance is a strong function of frequency and can vary over several orders of magnitude over the experimentally accessible frequency range. The stochastic errors of the impedance measurement are heteroscedastic, which, in this case, means that the standard deviation and variance of the stochastic errors are strong functions of frequency. Selection of an appropriate weighting strategy is, therefore, critical for interpretation of data.

In the present work, the mean value of the residual error is considered to comprise the sum of the bias error and the lack of fit. The sum $\varepsilon_{\text{fit}} + \varepsilon_{\text{bias}}$, represented by lines in Figure 27.7, was estimated from the average residual error calculated on a frequency-by-frequency basis. Thus, the stochastic contribution was obtained from the standard deviation of the residual errors at each frequency. The resulting estimate for the standard deviation is presented as a function of frequency in Figure 27.8(a).

The standard deviations are shown in Figure 27.8(a) to be strong functions of frequency. In addition, the standard deviation for the real part of the impedance at a given frequency is seen to be equal to the standard deviation for the imaginary part of the impedance. The line shown in Figure 27.8(a) represents the fit of

$$\sigma = \alpha|Z_j| + \beta|Z_r| + \gamma|Z|^2 + \delta \tag{27.2}$$

to the estimated standard deviations. The model parameter values are presented in Table 27.5. Note that, while α and β are dimensionless, the units of γ and δ depend on the units of the impedance under analysis.

The standard deviations scaled to the modulus of the impedance are presented in

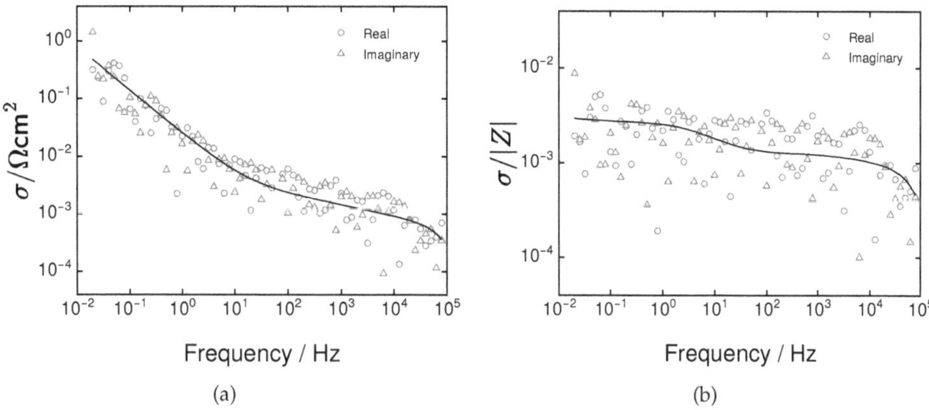

Figure 27.8: Standard deviation of impedance measurements obtained from the residual errors presented in Figure 27.7: (a) real part and (b) imaginary part. The line represents the fit of equation (27.2).

Table 27.5: Regressed constants for the standard deviation of the stochastic part of the error structure given in equation (27.2) for the results presented in Figure 27.8(a).

Parameter	Value	Units
α	$3.008 \times 10^{-3} \pm 3.7 \times 10^{-5}$	Dimensionless
β	—	Dimensionless
γ	$1.35 \times 10^{-6} \pm 3.7 \times 10^{-7}$	$(\Omega cm^2)^{-1}$
δ	$5.82 \times 10^{-4} \pm 1.9 \times 10^{-5}$	Ωcm^2

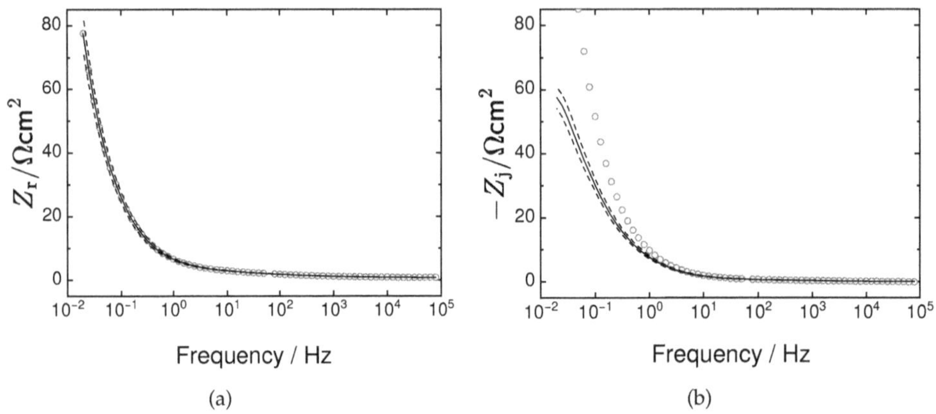

Figure 27.9: Fit of the measurement model equation (25.1) to the real part of the impedance obtained after an elapsed time of 180 minutes (Table 27.3): (a) fit to the real part and (b) prediction of the imaginary part. Dashed lines represent the 95.4% confidence interval for the model.

Figure 27.8(b) as functions of frequency. The standard deviation ranges from 0.05% of the modulus at high frequency to 0.3% of the modulus at low frequency.

Consistency with the Kramers–Kronig Relations

The high-frequency inductive feature in the impedance response appears to be related to the inductance of the leads, as described in Problem 8.11. The question at hand is: what part of the spectrum should be included in the regression of the process model? The measurement model can be employed to identify the useable frequency range. The spectrum selected for this analysis was the impedance obtained after an elapsed time of 180 minutes presented in Table 27.3.

The measurement model given as equation (25.1) was regressed to the real part of the impedance measurement under a weighting based on the standard deviation reported in Figure 27.8(a). The first measured frequency was deleted from the data set regressed, as was measurement at 60 ± 5 Hz and 120 ± 5 Hz. The results are presented in Figure 27.9(a) as a function of frequency. On a linear scale, the fit looks adequate. The parameter values were used to predict the imaginary part of the impedance. The estimated standard deviations of parameter estimated were used in a Monte-Carlo simulation to estimate the 95.4% (2σ) confidence interfacial for the model prediction. The fit to the real part of the measurement yielded a poor prediction of the imaginary part of the impedance at low frequency. This result is expected because the information content of the real part of the impedance is less than the information content of the imaginary part. At high frequency, however, the real part of the impedance approaches a finite value for the ohmic resistance, and the prediction of the imaginary part of the impedance should be valid. On a the linear scale shown in Figure 27.9(b), the prediction appears to be adequate.

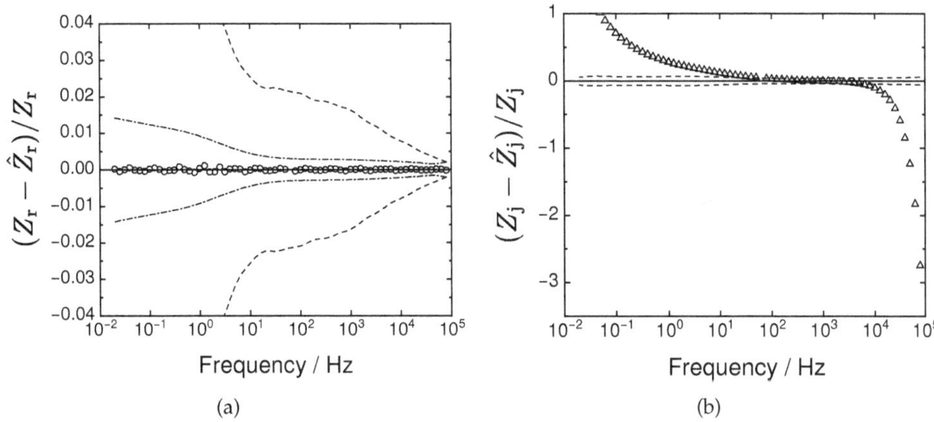

Figure 27.10: Normalized residual errors for the fit of the measurement model equation (25.1) to the real part of the impedance obtained after an elapsed time of 180 minutes (Table 27.3): (a) fit to the real part and (b) prediction of the imaginary part. Dashed lines represent the 95.4% confidence interval for the model. Dash-dot lines represent the scaled stochastic error structure of the measurement (see Figure 27.8(a)).

The residual errors provide a much more sensitive evaluation of the fit to the real part of the measurement and of the prediction of the imaginary part from the regressed parameters. The normalized residual errors are presented in Figure 27.10(a) for the fit of the measurement model to the real part of the impedance obtained after an elapsed time of 180 minutes (Table 27.3). The outer pair of dashed lines represents the confidence interval for the model, and the inner pair of dash-dot lines represent the stochastic error structure of the data. The fit looks quite good. The normalized errors for the prediction of the imaginary part of the measurement given in Figure 27.10(b) show significant error at high frequency. Eleven high-frequency points fell outside the confidence interval for the prediction of the imaginary part of the impedance. These data were removed from the subsequent regression analyses.

The complex regression of the measurement model to the resulting data set is presented in Figure 27.11 with 95.4% confidence intervals included as dashed lines. The number of *RC* Voigt elements included in the model was increased to 11 by deleting the high-frequency data affected by the lead inductance. The logarithmic scale employed in Figure 27.11 for the real and imaginary parts of the impedance serves to emphasize the quality of the fit at high frequency. The Nyquist plots, shown in Figure 27.12, emphasize the quality of the fit at low frequencies.

The normalized residual errors are presented in Figure 27.13. The residual errors fall well within the noise level of the measurement. As the residual errors are also within the 95.4% confidence interval for the model, the data used for the regression satisfy the Kramers–Kronig relations.

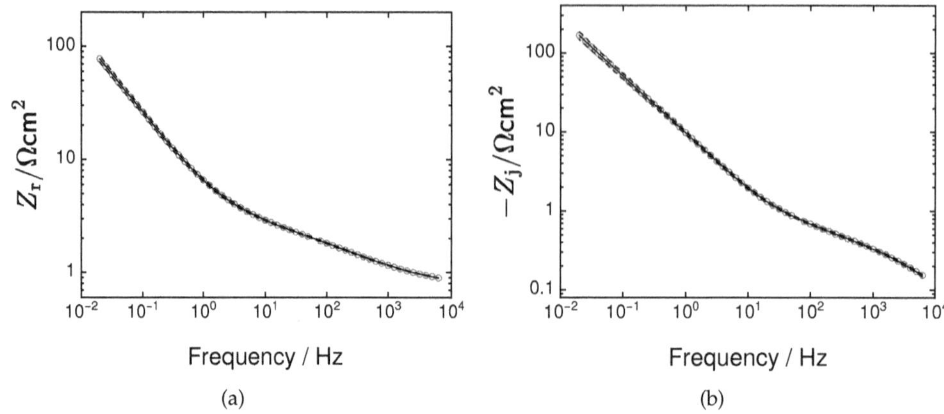

Figure 27.11: Fit of the measurement model equation (25.1) to the Kramers–Kronig-consistent part of the impedance obtained after an elapsed time of 180 minutes (Table 27.3): (a) real part and (b) imaginary part. Dashed lines represent the 95.4% confidence interval for the model.

27.2 Interpretation Model Development

The impedance data presented in Tables 27.1, 27.2, and 27.3 were the result of two-electrode measurements. Thus, the interpretation model needs to account for the contribution of both the anode and the cathode. The corresponding circuit model is presented in Figure 27.14, The resulting model can be expressed as

$$Z = Z_{\mathrm{a}} + R_{\mathrm{e}} + Z_{\mathrm{c}} \tag{27.3}$$

where the ohmic resistance may be expected to account for the resistance of the electrolyte, the separator, and any parasitic contact or lead resistances.

The low-frequency portion of the spectrum shown in Figure 27.1 reveals a straight line suggestive of mass transfer as might be given by the high-frequency limit of equation (11.13) in Problem 11.2. The line, however, has an angle larger than the 45° expected for the Warburg impedance. Erol and Orazem[15] suggested that the large angle could be attributed to anomalous diffusion in the $LiCoO_2$ cathode. As the steep low-frequency behavior is dominant in the Nyquist plots shown in Figure 27.3, the term Z_{c} cannot be neglected in equation (27.3).

Erol and Orazem[15] suggested that the charge-transfer reaction at the carbon electrode is represented by

$$x\mathrm{Li}^+ + x\mathrm{e}^- + 6\mathrm{C} \rightleftarrows \mathrm{Li}_x\mathrm{C}_6 \tag{27.4}$$

Given the slow diffusion of the reacting species through the graphite[114] and large particle sizes in the electrode,[115] a Warburg diffusion impedance was assumed for the anode.

The desired electrochemical reaction at the anode is the intercalation of lithium, but lithium can also react with components of the electrolyte to form a solid-electrolyte in-

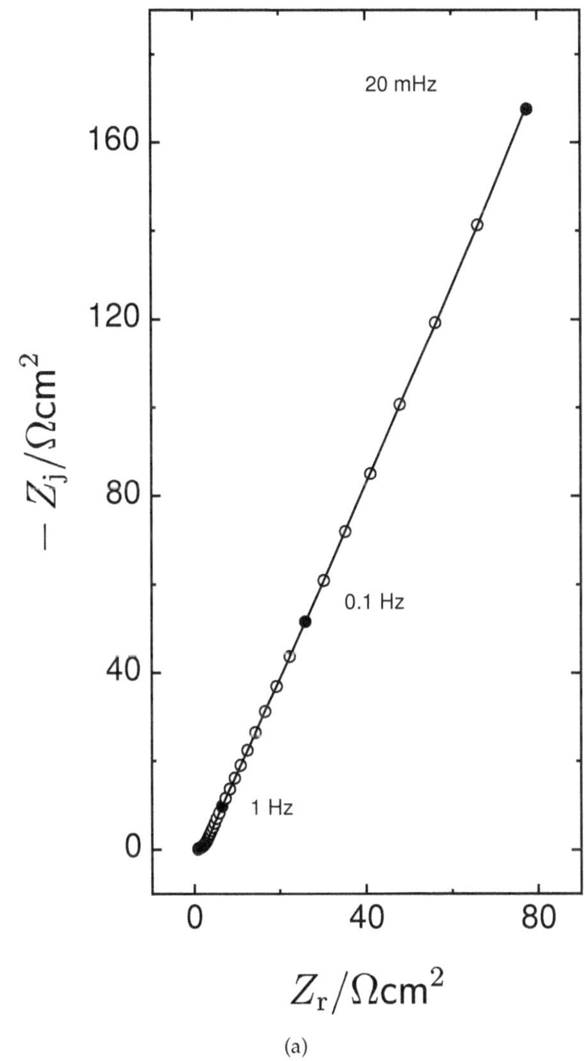

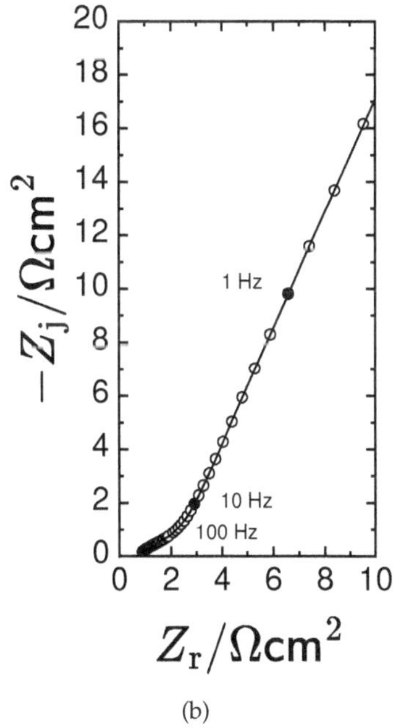

Figure 27.12: Fit of the measurement model equation (25.1) to the Kramers–Kronig-consistent part of the impedance obtained after an elapsed time of 180 minutes (Table 27.3) in Nyquist format: (a) full spectrum and (b) high-frequency region.

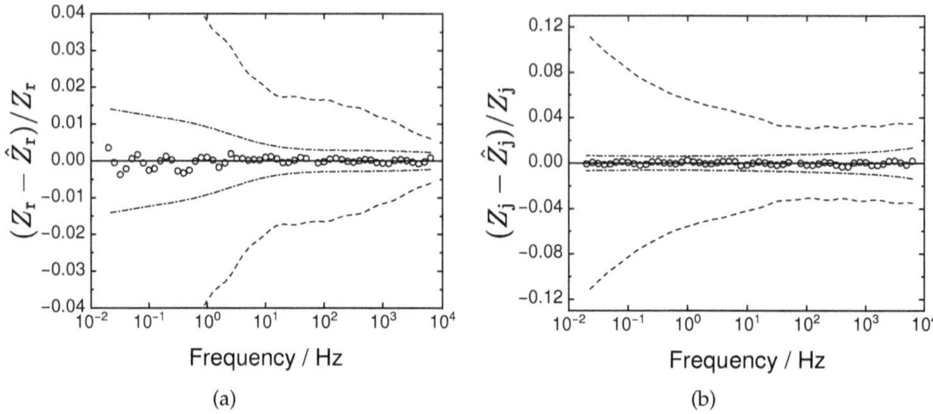

Figure 27.13: Normalized residual errors for the fit of the measurement model equation (25.1) to the Kramers–Kronig -consistent part of the impedance obtained after an elapsed time of 180 minutes (Table 27.3): (a) real part and (b) imaginary part. Dashed lines represent the 95.4% confidence interval for the model. Dash-dot lines represent the scaled stochastic error structure of the measurement (see Figure 27.8(a)).

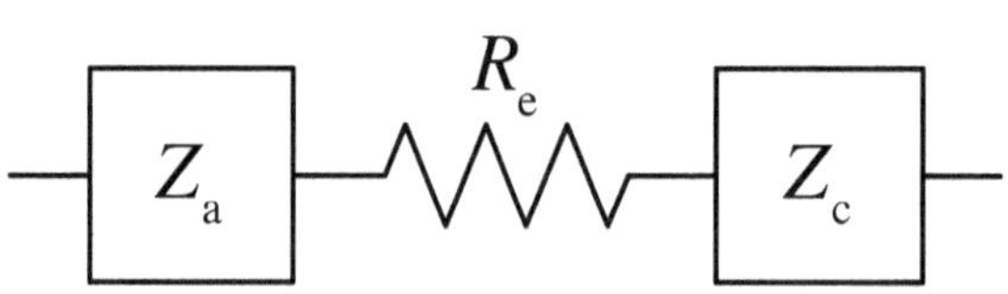

Figure 27.14: Equivalent circuit representation of the model developed for the impedance data presented in Tables 27.1, 27.2, and 27.3.

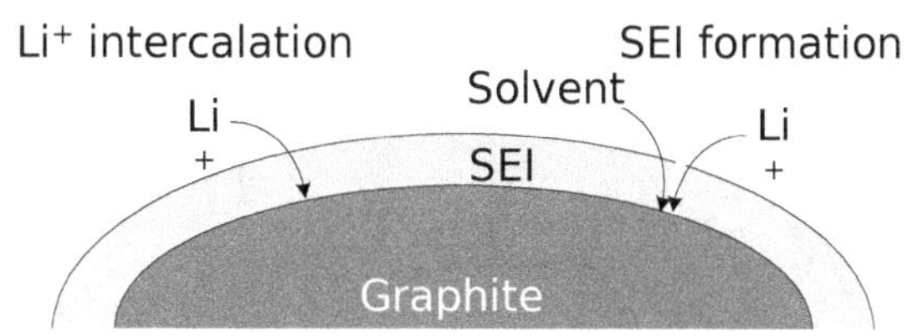

Figure 27.15: Schematic representation of a graphite anode for which the process of SEI formation occurs in parallel with intercalation of lithium ions. Taken from Erol and Orazem[15] as adapted from Pinson and Bazant.[118]

terphase (SEI). The SEI layer is a passivating film that forms at the surface of the active electrode material in the battery, protecting the electrolyte and other battery components (especially graphitic anode) from undesirable reactions. Deshpande et al.[116] suggested that the cracks at the graphite particle surface propagate upon electrochemical cycling. For each cycle, the newly cracked surface is exposed to electrolyte and thus results in SEI formation. As a result of diffusion induced stresses, the cracks at the particle surface propagate with cycling, new surface is exposed to electrolyte, and lithium ions are consumed in the formation of new SEI. Aurbach et al.[117] explained that the lithiation of graphite during cycling changes the particle orientation due to cracking. It allows solution species to penetrate between graphite particles and therefore isolate them electrically from the bulk electrolyte.

As shown in Figure 27.15, adapted from Pinson and Bazant,[118] the processes of lithium intercalation and the SEI formation were envisioned to occur as parallel reactions. Accordingly, the impedances associated with these reactions were considered to be in parallel.

The charge-transfer reaction at the cathode electrode is represented by

$$\mathrm{LiCoO_2} \rightleftarrows \mathrm{Li}_{1-x}\mathrm{CoO_2} + x\mathrm{Li}^+ + x\mathrm{e}^- \tag{27.5}$$

A characteristic feature of the impedance response of the $LiCoO_2|C$ battery is the slope of the straight line, seen at low frequencies in the Nyquist plot, that has an angle with respect to the real axis greater than 45 degrees. This behavior is not seen for all lithium battery chemistries. Greenleaf et al.[119] showed that the $LiFePO_4$ system yields an angle with respect to the real axis of 45 degrees, which is consistent with Fickian diffusion impedance. Osaka et al.[120] suggested that the larger slope can be attributed to diffusion to particles of different sizes. In the present work, an alternative explanation is posed in terms of anomalous diffusion.[121,122]

The constitutive equation for anomalous diffusion is derived from continuous time random walk in the macroscopic limit. The random walkers wait after each jump for a period of time drawn from a broad power-law distribution. The effect is that some ions remain stuck for a long time and the effective diffusion is slower. The impedance

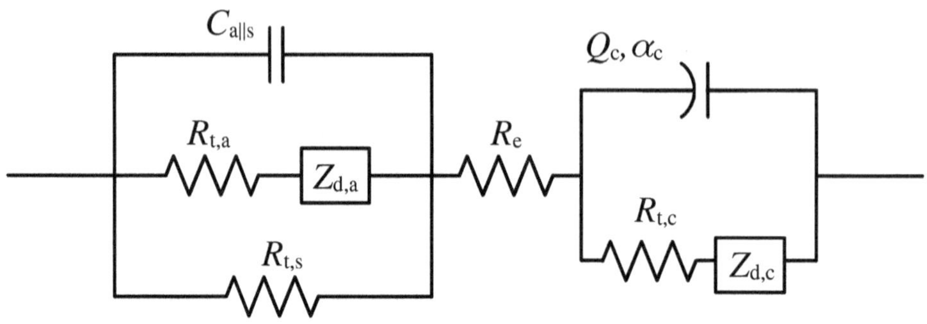

Figure 27.16: Equivalent circuit representation of the model developed in the present work for the complete Li-ion battery system. Taken from Erol and Orazem.[15]

used for this type of anomalous diffusion is associated with systems that show a diffusion hindered by sporadic adsorption of diffusing species to a fixed substrate. The associated impedance response, with a reflecting boundary condition, was derived by Bisquert and Compte.[123] The resulting diffusion impedance is expressed as

$$Z_d(\omega) = \frac{Z_d(0)}{\tau^{\gamma_c/2}} \coth\left[(j\omega\tau)^{\gamma_c/2}\right] \tag{27.6}$$

where $Z_d(0)$ is the zero-frequency asymptote for the real part of the diffusion impedance, and τ is the time constant. In the high-frequency limit, the impedance is proportional to $Z_d(0)/\tau^{\gamma_c/2}$.[1]

The equivalent circuit representation of the model developed by Erol and Orazem[15] is shown in Figure 27.16. The resulting model can be expressed as equation (27.3). The anode impedance Z_a, given as

$$Z_a(\omega) = \frac{(R_{t,a} + Z_{d,a})R_{t,S}}{R_{t,a} + Z_{d,a} + R_{t,S} + j\omega(R_{t,a} + Z_{d,a})R_{t,S}C_a} \tag{27.7}$$

included the parallel contributions of Li^+ intercalation expressed as the sum of the charge-transfer resistance $R_{t,a}$ and diffusion impedance $Z_{d,a}$; SEI formation expressed as a charge-transfer resistance $R_{t,S}$; and a capacitance C_a. The diffusion impedance for the anode, $Z_{d,a}$ was considered to follow a Warburg impedance, i.e.,

$$Z_{d,a} = \frac{R^*_{d,a}}{\sqrt{j\omega}} \tag{27.8}$$

where $R^*_{d,a}$ is the Warburg diffusion resistance. The cathode impedance Z_c was expressed as

$$Z_c(\omega) = \frac{R_{t,c} + Z_{d,c}}{1 + (j\omega)^{\alpha_c}(R_{t,c} + Z_{d,c})Q_c} \tag{27.9}$$

where the capacitive contribution was expressed as a constant-phase element with parameters α_c and Q_c, and the faradaic impedance was expressed as the sum of the charge-transfer resistance $R_{t,c}$ and anomalous-diffusion impedance $Z_{d,c}$. As the measured frequency range did not reveal the sharp increase associated with the reflecting boundary condition, an asymptotic form of the anomalous-diffusion impedance expression was used, i.e.,

$$Z_{d,c}(\omega) = \frac{R^*_{d,c}}{(j\omega)^{1-\gamma_c/2}} \tag{27.10}$$

where $R^*_{d,c}$ is a lumped parameter

$$R^*_{d,c} = \frac{Z_{d,c}(0)}{\tau_c^{\gamma_c/2}} \tag{27.11}$$

containing the diffusion time constant, τ_c

$$\tau_c = \frac{\delta_c^2}{D_c} \tag{27.12}$$

and the zero-frequency asymptote for the real part of the diffusion impedance, $Z_{d,c}(0)$.

27.3 Regression Analysis

The models were regressed to each of the data sets given in Tables 27.1, 27.2, and 27.3 by use of a Levenberg–Marquardt algorithm (Origin® 2020) under error structure weighting. Regression of the model presented by Erol and Orazem,[15] i.e.,

$$\begin{aligned} Z &= R_e + \frac{\left(R_{t,a} + A_a/\sqrt{j\omega}\right) R_{t,S}}{R_{t,a} + A_a/\sqrt{j\omega} + R_{t,S} + j\omega\left(R_{t,a} + A_a/\sqrt{j\omega}\right) R_{t,S} C_a} \\ &+ \frac{R_{t,c} + A_c/(j\omega)^{1-\gamma/2}}{1 + (j\omega)^{\alpha}_{c}\left(R_{t,c} + A_c/(j\omega)^{1-\gamma/2}\right) Q_c} \end{aligned} \tag{27.13}$$

yielded parameters with confidence intervals that included zero. Thus, a simplified approach is needed.

As the impedance shown in Figure 27.2 is dominated by the cathodic impedance, the first model attempted included only the cathodic impedance and the ohmic resistance. Statistically consistent parameters were obtained by regression. In a second effort, the anode impedance, equation (27.7), was represented by

$$Z_a = \frac{R_{t,a,eff}}{1 + j\omega R_{t,a,eff} C_a} \tag{27.14}$$

Thus, the revised model could be expressed as

$$Z = R_e + \frac{R_{t,a,eff}}{1 + j\omega R_{t,a,eff} C_a} + \frac{R_{t,c} + A_c/(j\omega)^{1-\gamma/2}}{1 + (j\omega)^{\alpha}_{c}\left(R_{t,c} + A_c/(j\omega)^{1-\gamma/2}\right) Q_c} \tag{27.15}$$

Table 27.6: Regression results and $\pm\sigma$ confidence intervals of a fit of equation (27.15) to the data sets given in Tables 27.1, 27.2, and 27.3 at a temperature of 20 °C.

Parameter	Table 27.1	Table 27.2	Table 27.3
R_e, Ωcm^2	0.9275 ± 0.00051	0.8191 ± 0.0035	0.8211 ± 0.0040
$R_{t,c}$, Ωcm^2	1.79 ± 0.26	2.032 ± 0.075	1.968 ± 0.089
α_c,	0.710 ± 0.031	0.672 ± 0.011	0.684 ± 0.014
Q_c, $Fcm^{-2}s^{\alpha} - 1$	0.0147 ± 0.0031	0.00758 ± 0.00059	0.00691 ± 0.00071
A_c, $\Omega cm^2 s^{-\gamma/2}$	41.2 ± 5.6	57.4 ± 2.0	56.8 ± 2.4
γ_c,	0.485 ± 0.042	0.517 ± 0.011	0.528 ± 0.013
$R_{t,a,eff}$, Ωcm^2	0.198 ± 0.011	0.1633 ± 0.0075	0.1684 ± 0.0089
C_a, μFcm^{-2}	416 ± 26	406 ± 20	391 ± 22

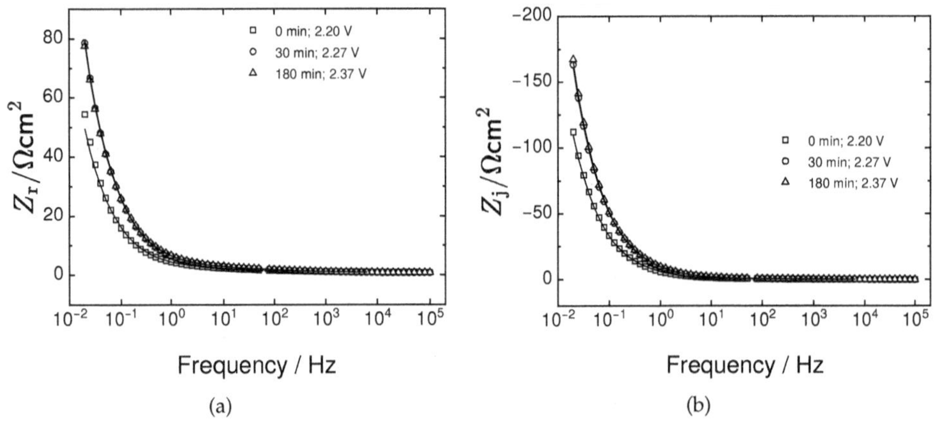

Figure 27.17: Regression results for the fit of equation (27.15) to the Kramers–Kronig-consistent part of the impedance with elapsed time as a parameter: (a) real part and (b) imaginary part. The lines represent the model fit. Truncated data shown in gray symbols.

where

$$R_{t,a,eff} = \frac{R_{t,a} R_{t,S}}{R_{t,a} + R_{t,S}} \tag{27.16}$$

is the effective resistance for intercalation and SEI formation reactions at the anode.

The results are presented in Table 27.6 where regressed values of the parameters are provided with their $\pm\sigma$ confidence intervals. The regression results for the fit of equation (27.15) to the Kramers–Kronig -consistent part of the impedance are presented in Figure 27.17 with elapsed time as a parameter. The corresponding normalized residual errors are presented in Figure 27.18.

The present fitting results may be compared to a summary of results presented by Erol and Orazem,[15] presented in Table 27.7 as a function of potential. The ohmic resistance R_e in Table 27.6 is similar to the values found in Table 27.7. The charge-transfer resistance for the cathodic reaction $R_{t,c}$ is slightly larger than the value determined for 3.2 V, and the values for α_c and Q_c are consistent with the trends seen in Table 27.7

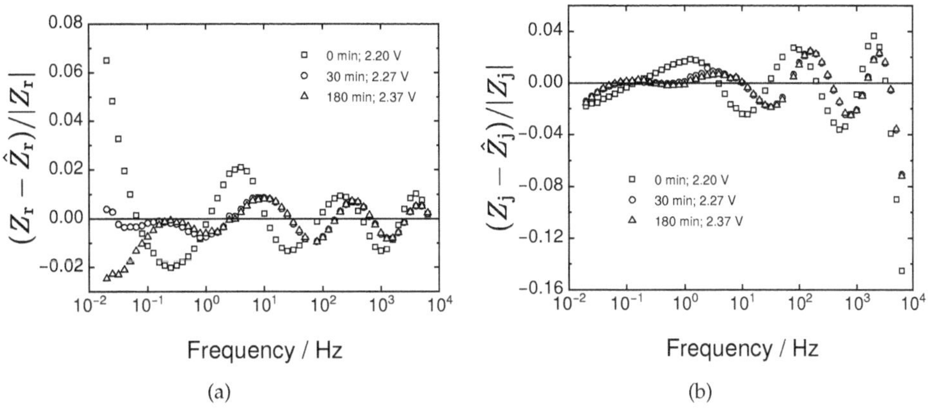

Figure 27.18: Normalized residual errors for the fit of equation (27.15) to the Kramers–Kronig-consistent part of the impedance with elapsed time as a parameter: (a) real part and (b) imaginary part. The lines represent the model fit.

Table 27.7: Regression results and $\pm\sigma$ confidence intervals of a fit of equation (27.13) to the data for cell potential ranging from 4.0 to 3.2 V at a temperature of 20 °C. Taken from Erol and Orazem[15]

Parameter	4 V	3.6 V	3.2 V
R_e, Ω cm^2	0.894 ± 0.0048	0.940 ± 0.0030	0.949 ± 0.0056
$R_{t,c}$, Ω cm^2	0.39 ± 0.019	0.61 ± 0.012	1.21 ± 0.029
α_c	0.91 ± 0.015	0.930 ± 0.0067	0.755 ± 0.0076
Q_c, F cm^{-2} s$^{\alpha_c - 1}$	0.0333 ± 0.00061	0.0527 ± 0.00058	0.0556 ± 0.00092
A_c, Ω cm^2 s$^{-\gamma_{d,c}/2}$	0.1821 ± 0.00098	0.1374 ± 0.00068	0.407 ± 0.0014
γ_c	0.617 ± 0.0043	0.379 ± 0.0045	0.247 ± 0.0039
$A_{W,a}$, Ω cm^2 s$^{-0.5}$	51 ± 1.1	35.9 ± 0.57	37.1 ± 1.3
$R_{t,a}$, Ω cm^2	0.180 ± 0.0069	0.197 ± 0.0044	0.177 ± 0.0097
$R_{t,S}$, Ω cm^2	1.19 ± 0.018	1.15 ± 0.012	0.94 ± 0.025
C_a, µF cm^{-2}	41 ± 2.5	63 ± 2.4	64 ± 5.1

with increasing potential. The cathodic diffusion resistance is much larger in the over-discharged condition than at normal operating potentials. In contrast, the anomalous diffusion exponent γ_c is closer to the ideal value of unity than it is at a potential of 3.2 V. The value of γ_c suggests that lithium ion diffusion was less sticky at lower over-discharged potentials than it was at the nominal 0% state of charge corresponding to a potential of 3.2 V.

27.4 Observations

This work provides an alternative explanation for the slope of the low-frequency impedance response for $LiCoO_2$ cathodes that is steeper than that associated with Fickian diffusion impedance. This effect is clearly associated with the unique properties of the $LiCoO_2$ cathode as the low-frequency impedance response of other Li-ion battery cathodes, e.g., $LiFePO_4$ and $LiMn_2O_4$, shows different behaviors.[119] The group of Osaka[120] have attributed this behavior to the different time constants for diffusion to particles of different sizes. The present work shows that anomalous diffusion may provide an alternative explanation that is specific to the properties of $LiCoO_2$.

The present data and that published in the literature suggest that $LiCoO_2$ possesses a property that causes a diffusion impedance that differs from that of other cathode materials. In situ nuclear magnetic resonance (NMR) might be used in order to detect the anomalous lithium ion diffusion inside the cathode. NMR can be a supportive spectroscopic method to show that the anomalous diffusion occurs in $LiCoO_2$ electrodes. The contribution of anomalous diffusion was diminished at elevated temperatures, suggesting that the free-energy well associated with anomalous or sticky diffusion was shallow.

Chapter 28

Large Double-Layer Capacitance for Iron in the Presence of H_2S

In some particular cases, as for the corrosion of iron in the presence of H_2S, the rapid growth of a conductive and highly porous iron sulphide scale gives rise to a large active area. In this case, the apparent double layer capacitance can become much larger than expected for a typical planar electrode surface. With an increase in elapsed immersion time, pore geometries can appear, which yield different contributions to the impedance.

28.1 Introduction

Corrosion of metallic materials in the oil and gas industry has been a significant concern over the last decades and has major financial and environmental implications. Electrochemical impedance spectroscopy represents an interesting technique to study the evolution with time of corrosion reaction, as it is usually performed at the corrosion potential. The impedance measured at the corrosion potential is composed of an anodic branch in parallel to a cathodic branch, both being in parallel to a double layer capacitance. Analysis of the variation of the impedance around the corrosion potential is a method to determine if one branch can be neglected with respect to the other.

28.2 Experimental Setup

The experimental setup consists in a Devanathan–Stachurski system shown in Figure 28.1. The apparatus is composed of twin cells separated by the iron membrane. The cells are equipped with double jacket in order to maintain the temperature at 24 ± 2 °C. In this experimental system, test specimens consisted in pure iron membranes, annealed under vacuum at 900 °C for 30 min. These membranes were thus exposed on one side to the corrosive environment containing H_2S, where EIS measurements were made (designed as charging side), while the other side was exposed to 0.1 M sodium

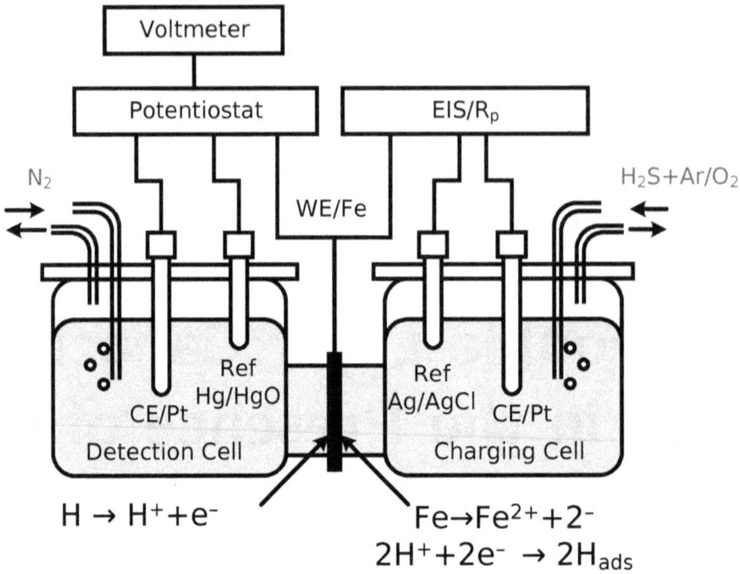

Figure 28.1: Experimental setup for the H_2S experiments using a Devanathan–Stachurski cell. The electrolyte in the detection cell (left) was 0.1 M NaOH, and the electrolyte in the charging cell (right) was 35 g/L NaCl dissolved in distilled water.

hydroxide and maintained at an anodic potential of 250 mV (Hg/HgO|1M KOH) for hydrogen permeation measurement (designated as detection side).

EIS measurements were carried out with the iron membrane as the working electrode, an Ag|AgCl reference electrode, and a platinum mesh as auxiliary electrode. A perturbation of ± 10 mV amplitude around the corrosion potential was applied with a range of frequency of 10 kHz to 1 mHz. Each EIS measurement was preceded by an open circuit potential (OCP) measurement of about 2 h. The test solution at the entry side consisted of 35 g/L NaCl dissolved in distilled water. Continuous bubbling of pure H_2S was maintained during all experiments, resulting in a dissolved H_2S concentration of around 0.1 M and an initial pH of around 3.8–4.

28.3 Initial Inspection

As presented in Chapters 26, 27, 29, and 30, an initial inspection comprises the use of graphical methods and regression of a general measurement model.

28.3.1 Graphical Analysis

The experimental data are given in Table 28.1 as real and imaginary parts of the impedance as functions of frequency. The corresponding Nyquist diagram is presented in Figure 28.2. The characteristic frequency of the high-frequency loop is $f_c = 0.17$ Hz,

Table 28.1: Experimental data for the impedance response of pure iron exposed to a 35 g/L NaCl solution saturated with 1 bar H_2S at 24°C after two hours of immersion.

f, Hz	Z_r, Ω	Z_j, Ω	f, Hz	Z_r, Ω	Z_j, Ω
10000	142.428	−1.18358	2.81	149.732	−23.24
7910	142.428	−0.94454	2.23	151.724	−28.054
6300	142.428	−0.7387	1.76	154.38	−34.03
4960	142.428	−0.57104	1.4	157.866	−40.67
3930	142.428	−0.5229	1.1	162.182	−48.97
3110	142.428	−0.40504	0.875	167.66	−58.1
2460	142.594	−0.35524	0.692	175.96	−68.724
1950	142.428	−0.32038	0.548	187.58	−80.344
1540	142.428	−0.26394	0.434	200.86	−91.632
1220	142.594	−0.24734	0.344	217.46	−101.924
968	142.428	−0.2656	0.272	237.38	−110.224
765	142.428	−0.23406	0.215	260.62	−115.702
606	142.594	−0.1582	0.17	283.86	−117.362
480	142.594	−0.38678	0.135	307.1	−115.37
380	142.594	−0.4482	0.107	327.02	−110.224
301	142.428	−0.4067	0.0847	345.28	−103.584
238	142.594	−0.38844	0.067	360.22	−95.948
189	142.428	−0.5727	0.0531	373.5	−89.308
149	142.594	−0.65238	0.042	381.8	−83.498
118	142.594	−0.85656	0.0332	388.44	−79.348
93.6	142.594	−1.09726	0.0263	395.08	−78.02
74.1	142.594	−1.2865	0.0208	401.72	−78.352
58.6	142.76	−1.6268	0.0165	408.36	−80.51
46.4	142.76	−2.0584	0.0131	415	−85.49
36.8	142.926	−2.4734	0.0103	421.64	−90.968
29.1	142.926	−3.2204	0.00819	429.94	−99.766
23	143.092	−3.901	0.00648	439.9	−111.054
18.3	143.258	−4.731	0.00513	451.52	−123.338
14.4	143.59	−5.8764	0.00406	466.46	−139.772
11.4	143.922	−7.221	0.00322	483.06	−156.87
9.04	144.254	−8.798	0.00255	504.64	−180.94
7.17	144.918	−10.7734	0.00202	527.88	−207.5
5.67	145.748	−13.1638	0.0016	562.74	−242.36
4.49	147.076	−15.9526	0.00126	609.22	−278.88
3.55	148.238	−19.422	0.001	672.3	−318.72

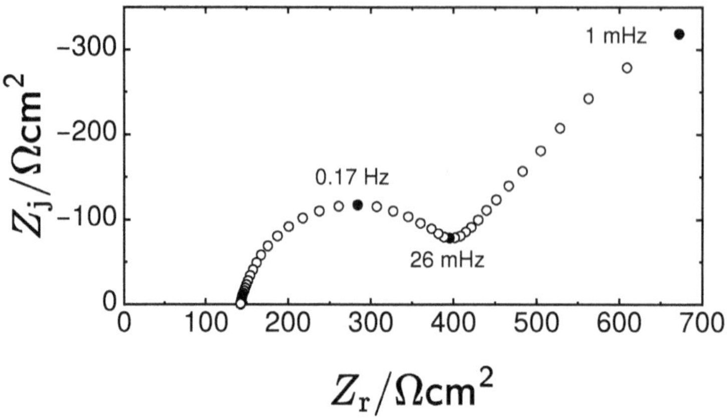

Figure 28.2: Impedance response for pure iron exposed to a 35 g/L NaCl solution saturated with 1 bar H_2S at 24°C after two hours of immersion.

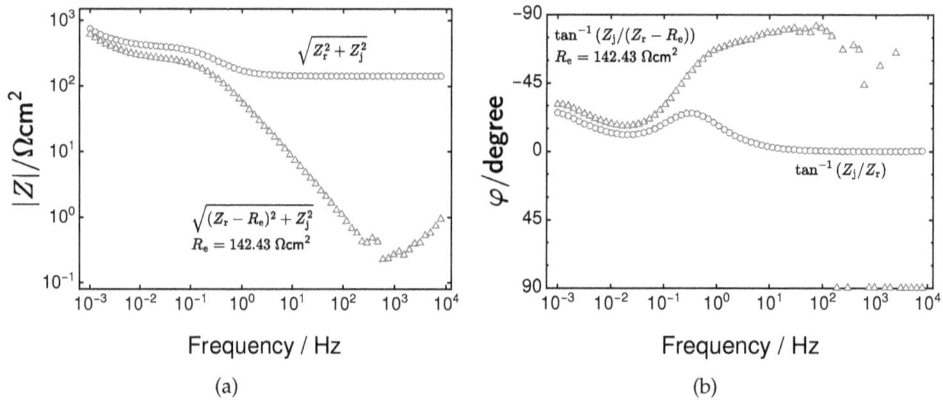

Figure 28.3: Bode and ohmic-resistance-corrected Bode plots for the data presented in Table 28.1: (a) magnitude and (b) phase as functions of frequency.

and the diameter of this loop is about $R = 300\ \Omega cm^2$. The corresponding capacitance value is $C = 1/(2\pi f_c R) = 3\ mF/cm^2$. This huge value will be explained below.

The Bode plots and the ohmic-resistance-corrected Bode plots are given in Figure 28.3. The value of the electrolyte resistance, used to calculate the ohmic-resistance-corrected quantities, was obtained from the high-frequency limit of the real part of the impedance given in Table 28.1. The estimated ohmic resistance was $R_e = 142.43\ \Omega cm^2$. The results shown in Figure 28.3 are very sensitive to the value of the ohmic resistance as Z_r approaches R_e at large frequencies, and noise in the data causes changes in sign in $Z_r - R_e$, resulting in the large shifts seen in Figure 28.3(b). In these corrected Bode plots, the results appear really noisy for frequency larger than 20 Hz.

It is possible to fit this diagram with two capacitive loops or with a capacitive loop

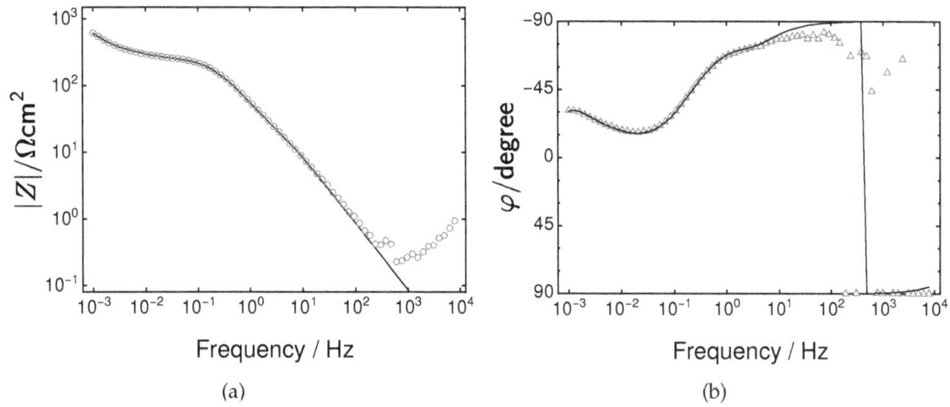

Figure 28.4: Ohmic-resistance-corrected Bode plots for the data presented in Table 28.1: (a) magnitude and (b) phase as functions of frequency. Lines represent the measurement model fit under modulus weighting with six Voigt elements.

Table 28.2: Results of the initial graphical and measurement model analysis of the data presented in Table 28.1.

Parameter	Symbol	Graphical	Measurement model
Ohmic resistance	R_e	142.43 Ωcm^2	142.58$\pm$0.11 Ωcm^2
Capacitance	C_{dl}	3 mF/cm^2	1.84$\pm$0.21 mF/cm^2

in series with a Warburg impedance. Before proposing an equivalent circuit model, it is necessary to consider what electrochemical reactions are involved in this particular case.

28.3.2 Measurement Model Analysis

In the present case, only a single data set is available, which limits the ability to assess the error structure. Nevertheless, the measurement model analysis provides useful information that complements the process model development. The Voigt measurement model was regressed to extract capacitance, ohmic resistance, and polarization resistance values from impedance data. The measurement model is shown to provide a useful means to estimate properties characteristic of the present electrochemical system.

With six Voigt elements, the regression of the measurement model to the data presented in Table 28.1 yielded the fit presented in Figure 28.4. The results shown in Figure 28.4 suggest that the fit is good up to a frequency of about 100 Hz. Then, in the following, the regression will be limited to the points measured below 100 Hz. The values of the electrolyte resistance and of the capacitance are given in Table 28.2 for the measurement model and the graphical analysis. Both the measurement model and the graphical analysis suggest a capacitance that is larger than expected.

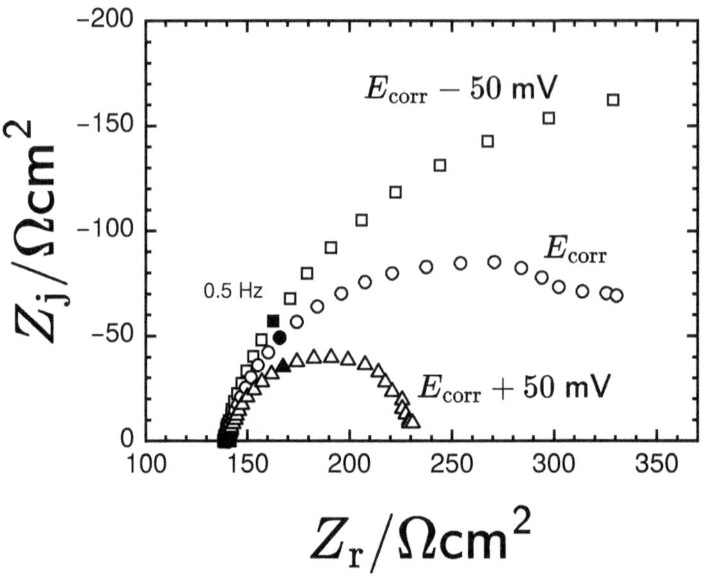

Figure 28.5: Impedance response for pure iron exposed to a 35 g/L NaCl solution saturated with 1 bar H_2S at 24°C. Measurements were collected at $E_{corr} - 50$ mV, E_{corr}, and $E_{corr} + 50$ mV.

28.4 Interpretation Model Development

The general circuit diagram used to analyze impedance data collected at the open-circuit potential is given in Figure 25.11 (page 310). In order to determine the nature of the impedance and to choose the model accordingly, measurements at various potentials around the corrosion potential (E_{corr}) were performed. The results shown in Figure 28.5 show that the diameter of the high-frequency arc decreased as the potential became more anodic, suggesting that the impedance is dominated by anodic reactions.

Ma et al.[124,125] published several papers on pure iron exposed to H_2SO_4/Na_2SO_4 solutions with dissolved H_2S introduced as Na_2S. In these studies, the pH range varied from 0.7 to 3.5 and the dissolved H_2S concentration was 0.4 mmol/L, corresponding to the equilibrium with gaseous H_2S at a partial pressure of 4 mbar. Impedance diagram measured at the corrosion potential showed two capacitive loops and were interpreted with the equivalent circuit shown in Figure 28.6. In this circuit, CPE_{dl} and R_t were, respectively, attributed to the double-layer capacitance and charge-transfer resistance corresponding to the high-frequency capacitive loop. The terms CPE_a and R_a were associated with relaxation of adsorbed species, which was supposed to proceed at a slower rate and to correspond to the low-frequency arc.

In addition to electrochemical measurements, flat coupons were sampled at the end of the test after 620 h exposure and used for surface analysis. The surface of the specimens was covered by a black deposit consisting mainly of mackinawite (FeS) as de-

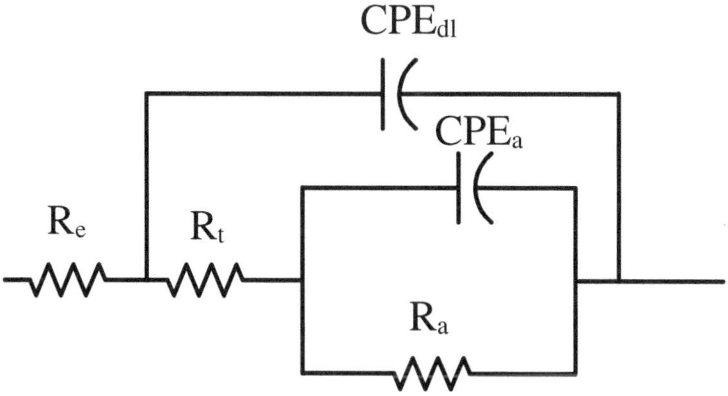

Figure 28.6: Equivalent circuit used by Ma et al.[124,125] to describe iron corrosion in the presence of H_2S.

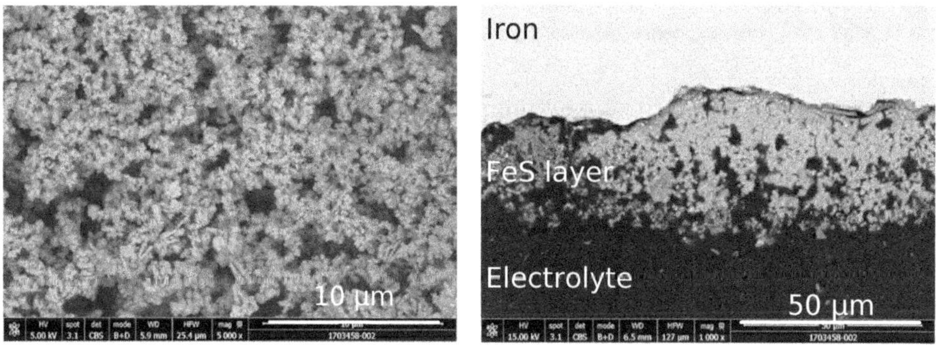

(a) (b)

Figure 28.7: SEM observation of corrosion deposits formed on the iron surface after 620 h exposure to a 35 g/L NaCl solution saturated with 1 bar H_2S at 24°C: (a) top view and (b) side view. Images adapted from Figure 6 of Deffo Allagou et al.[126]

termined by XRD analysis. SEM observations shown in Figure 28.7 revealed that this surface deposit consisted of submicrometric plate-like crystals, resulting in a highly porous scale with a thickness of 20–50 µm.

Typical evolution with time of exposure of Nyquist diagrams is presented in Figure 28.8. For short exposure times (2–137 h), the Nyquist diagrams consisted of a capacitive loop at high and medium frequency and a low-frequency contribution, which corresponds to the start of a second capacitive loop with a very low characteristic frequency. The diameter of the high-frequency capacitive loop increased with time.

At longer immersion time (212 h), some overlap between the high-frequency capacitive loop and the low-frequency contribution occured. It becomes difficult to distinguish the various contributions to the impedance, which seems to be flattened and

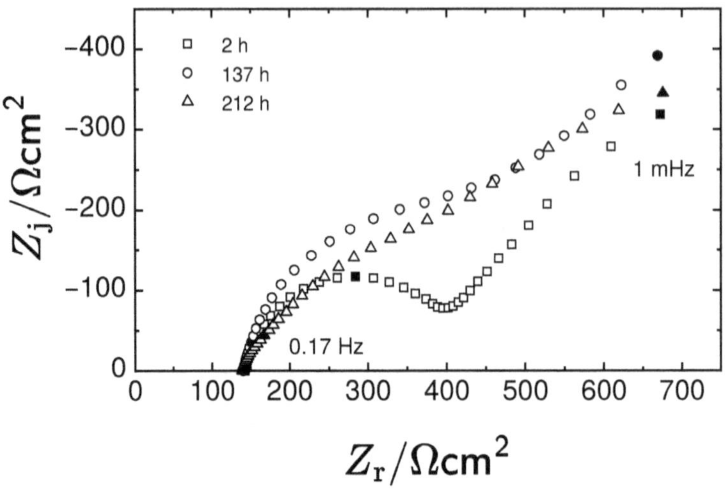

Figure 28.8: Nyquist diagrams for pure iron exposed to a 35 g/L NaCl solution saturated with 1 bar H_2S at 24°C with exposure time as a parameter.

tends towards a single line, with an angle reference to the real axis of 45°.

28.5 Regression Analysis

Impedance results were analyzed with the model of Figure 28.6 proposed in the literature to describe anodic impedance of iron corrosion in the presence of H_2S. In agreement with the graphical analysis and the measurement model results, only the points corresponding to frequencies below 100 Hz were considered. Curve fitting was applied to extract the various physical parameters describing the impedance of the system measurements performed at 2 and 137 h of exposure, as illustrated in Figure 28.8.

The first step in the analysis was the use of the measurement model to extract the stochastic error structure based on measurements performed at 2 and 137 h of exposure. For each measurement, six Voigt elements were extracted, and the corresponding error structure is presented in Figure 28.9. A model for the error structure was obtained as

$$\sigma = 0.00689|Z_j| + 0.000274|Z_r| \tag{28.1}$$

presented as a line in Figure 28.9.

The results of the error-structure-weighted regression of the model shown in Figure 28.6 to measurements obtained at 2 h and 137 h are presented in Figure 28.10. The resulting parameters are shown in Table 28.3. All parameter values are statistically significant. The double-layer capacitance was obtained by use of the Brug formula, i.e.,

$$C = Q^{1/\alpha}\left(\frac{R_e R_t}{R_e + R_t}\right)^{(1-\alpha)/\alpha} \tag{28.2}$$

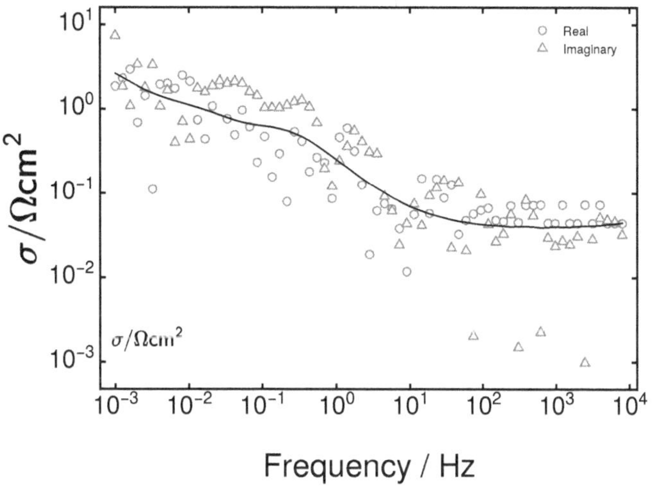

Figure 28.9: Stochastic error structure obtained by use of the measurement model from measurements obtained at 2 h and 137 h. The line represents the model for the error structure.

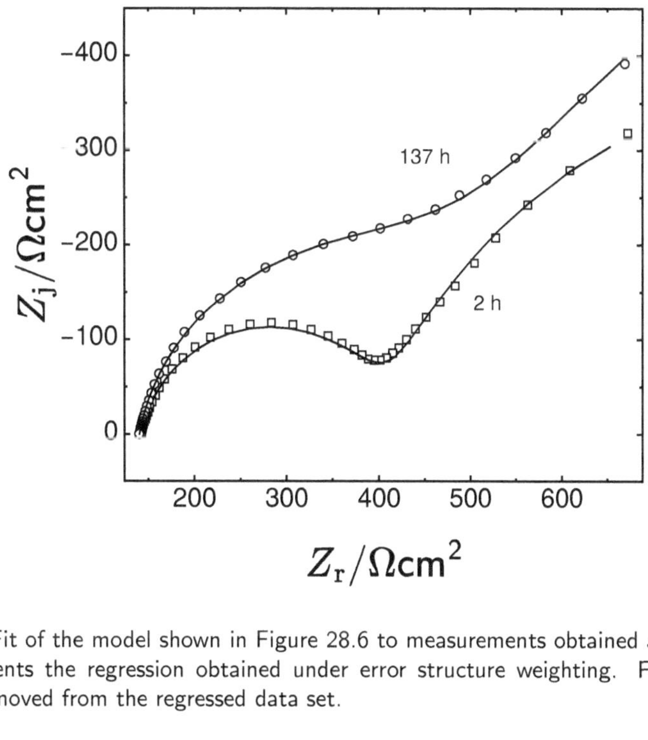

Figure 28.10: Fit of the model shown in Figure 28.6 to measurements obtained at 2 h and 137 h. The line represents the regression obtained under error structure weighting. Frequencies above 100 Hz were removed from the regressed data set.

Table 28.3: Regressed parameter values for the circuit given in Figure 28.6 for measurements obtained at 2 h and 137 h. The regression was weighted in accordance with equation (28.1).

Parameter	Units	2 h	137 h
$Q_{\rm dl}$	$\rm mF/cm^2s^{1-\alpha}$	3.2502 ± 0.0025	24.92 ± 0.10
$\alpha_{\rm dl}$		0.8720 ± 0.0019	0.8816 ± 0.0022
$R_{\rm t}$	$\Omega \rm cm^2$	271.8 ± 1.5	441 ± 12
$R_{\rm a}$	$\Omega \rm cm^2$	1018 ± 68	1900 ± 600
$Q_{\rm a}$	$\rm F/cm^2s^{1-\alpha}$	0.1150 ± 0.0055	0.109 ± 0.023
$\alpha_{\rm a}$		0.793 ± 0.012	0.777 ± 0.046
$R_{\rm e}$	$\Omega \rm cm^2$	142.260 ± 0.018	140.260 ± 0.019
χ^2/ν		18	14
$C_{\rm dl}$	$\rm mF/cm^2$	2.73	28.4

to be 2.73 $\rm mF/cm^2$ after an elapsed time of 2 h and 28.4 $\rm mF/cm^2$ after an elapsed time of 137 h.

The value of the double layer capacitance obtained after fitting is in agreement with the value obtained by graphical analysis of Figure 28.2. It can be immediately noted that $C_{\rm dl}$ values exceed by at least one order of magnitude the usual values for metal corrosion. It is believed that these high values are associated with the rapid formation of a conductive iron sulfide scale. Indeed, this capacitance applies to all conductive surfaces of the electrode. Consequently it increases as the iron sulfide film grows, and its evolution with times gives information on the true surface of the scale. At the beginning of the test, i.e., after less than a few hours immersion, its value is already quite high at 2.73 $\rm mF/cm^2$. Since the typical double-layer capacitance on metal surfaces are on the order of 100 $\rm \mu F/cm^2$, this result implies that iron sulfide scaling has already begun, providing a large effective surface area. Visual observation confirmed that scaling was already present a few minutes after immersion. After 137 h of exposure, the double-layer capacitance reached 28.4 $\rm mF/cm^2$, i.e., 300 times higher than the expected value for a bare electrode. The assumption of the impact of iron sulfide scale on impedance is supported by SEM observations given in Figure 28.7, showing a highly porous structure offering a large effective surface area.

This observation of high capacitance values motivates introduction of a specific component in the impedance model to account for the iron sulfide film. According to the theory developed by de Levie,[50] the impedance of a pore is given by

$$Z_{\rm pore} = \sqrt{R_0 Z_0}\coth\left(\ell\sqrt{\frac{R_0}{Z_0}}\right) \tag{28.3}$$

where ℓ is the pore length, R_0 is the electrolyte resistance for one unit length pore, with units of Ω/cm, and Z_0 is the interfacial impedance (also expressed per surface unit and per unit length of the pore, with units of Ωcm). It is also common to express R_0 and Z_0 as functions of the pore radius r and of the interfacial impedance per surface unit $Z_{\rm eq}$

as

$$R_0 = \frac{\rho}{\pi r^2} \tag{28.4}$$

and

$$Z_0 = \frac{Z_{eq}}{2\pi r} \tag{28.5}$$

respectively, where ρ is the electrolyte resistivity. This resistivity is considered to be uniform inside all pores. Indeed, the test solution contains a relatively high NaCl concentration (0.6 M) that should be sufficient to hinder concentration gradients of reactive species inside the pores. Another simplification made by de Levie is adopted here, i.e., that the axial concentration gradient inside the pores is negligible.

Depending on the value of ℓ and r, the coth function in equation (28.3) admits two limiting situations. If the argument is close to zero (i.e., pores with small length or large radius), then $\coth(X)$ is equal to $1/X$. In such a case. Equation (28.3) simplifies to

$$Z_{pore} = \frac{Z_0}{\ell} = \frac{Z_{eq}}{2\pi r\ell} \tag{28.6}$$

The film behaves as a planar electrode, and its impedance is inversely proportional to the number of pores n and to the surface of pores

$$Z_{film,planar} = \frac{Z_{pore}}{n} = \frac{Z_{eq}}{2n\pi r\ell} \tag{28.7}$$

The other limiting situation corresponds to long and thin pores, for which the argument X in the coth function takes large values. When $X \to \infty$, $\coth(X) \to 1$, meaning that

$$Z_{pore} = \sqrt{R_0 Z_0} \tag{28.8}$$

The film behaves as a semi-infinitely deep porous system, with an impedance proportional to the square root of the interfacial impedance, resulting in a 45° inclination of Nyquist plots. Thus,

$$Z_{film,porous} = \frac{\sqrt{\rho Z_{eq}}}{\sqrt{2}n\pi r^{3/2}} \tag{28.9}$$

These two limiting situations, $Z_{film,planar}$ and $Z_{film,porous}$, correspond to model systems for which all pores have the same geometry. In practice, and as illustrated by SEM observations of the FeS layer in Figure 28.7, there is a distribution of pore size and geometry, leading to the coexistence of both behaviors in parallel. The total impedance of the film is then given by

$$Z_{film} = \frac{Z_{film,planar} Z_{film,porous}}{Z_{film,planar} + Z_{film,porous}} \tag{28.10}$$

As a consequence, as soon as the iron sulfide grows at the metal surface, the film impedance $Z_{film,porous}$ must be added in parallel to the faradaic contribution of Figure 28.6. Models including the parallel contribution of short and deep pores have been applied

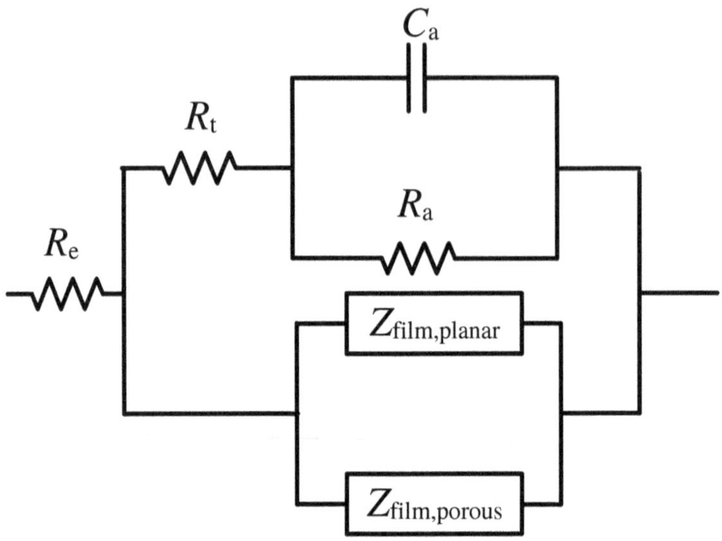

Figure 28.11: Equivalent circuit based on that used by used by Deffo Ayagou et al.[126] to describe iron corrosion at long times in the presence of H_2S.

by Jurczakowski et al.[127] to porous gold electrodes and by Lutz et al.[53] to sputtered iridium oxide electrodes.

For the short immersion times (less than 137 h), the phase tends towards value of about 80°, which can correspond to a CPE on a planar electrode. This case corresponds to a thin film with large pores. This film capacitance is in parallel with the double-layer capacitance of the metal on which the anodic dissolution occurs. The model corresponding to the equivalent circuit of Figure 28.6 is still valid. Introduction of the iron sulfide film contribution, however, provides a physical explanation for the large values of capacitance determined during the analysis of the results at short exposure time.

The model presented in Figure 28.11 was regressed to data collected at elapsed times of 317 h and 437 h. As the exponent for a CPE could not be differentiated from unity, the circuit employed capacitances. The planar contribution to the porous electrode was assumed to be

$$Z_{\text{film,planar}} = \frac{1}{j\omega C_1} \tag{28.11}$$

and the contribution of deep pores was given by

$$Z_{\text{film,porous}} = \sqrt{\frac{1}{j\omega C_2}} \tag{28.12}$$

where the term C_2 is a lumped parameter that includes the contribution of parameters shown in equation (28.9). The film impedance is given by equation (28.10). The impedance associated with faradaic and adsorption reactions, the upper portion of Figure

Table 28.4: Regressed parameter values for the circuit given in Figure 28.11 for measurements obtained at 317 h and 437 h. The regression was weighted in accordance with equation (28.1).

Parameter	Units	317h	437 h
R_a	Ωcm^2	857 ± 52	1700 ± 380
C_a	mF/cm^2	15.0 ± 1.3	8.9 ± 2.9
R_t	Ωcm^2	377 ± 36	810 ± 310
C_1	mF/cm^2	4.310 ± 0.052	11.53 ± 0.16
C_2	mF/cm^2	0.1972 ± 0.0065	0.351 ± 0.020
R_e	Ωcm^2	138.830 ± 0.028	138.830 ± 0.032
χ^2/ν		27	40

28.11, can be expressed as

$$Z_{F,a} = R_t + \frac{R_a}{1 + j\omega R_a C_a} \tag{28.13}$$

and the total impedance is given by

$$Z = R_e + \frac{Z_{F,a} Z_{film}}{Z_{F,a} + Z_{film}} \tag{28.14}$$

The resulting parameter estimates are shown in Table 28.4. All model parameter values were statistically significant. The data and model fit are shown in Figure 28.12.

The ohmic-resistance-corrected modulus for the data collected by Deffo Ayagou et al.[126] at an elapsed time of 437 h is presented in Figure 28.13 as a function of frequency. In the low-frequency range, the slope of −0.5 corresponds to the impedance $Z_{film,porous}$ in Figure 28.11. This means that the two other impedances in parallel with $Z_{film,porous}$, i.e., $Z_{film,planar}$ and Z_f, comprising R_t, C_a, and R_a, can be neglected at low frequencies. For larger frequencies, these two other impedances play a role and the slope decreases to −0.78. Then, the two geometries are both visible in the impedance response, and their impedance are in parallel.

28.6 Observations

The present chapter demonstrates how impedance is affected by the graduate evolution of the electrode surface, in this case, evolving from a rough surface to a porous electrode. An impedance model was proposed, with two distinct contributions: a faradaic anodic impedance consisting of a two-step reaction accounting for charge transfer, adsorption, and desorption and the impedance of the conductive and highly porous iron sulfide (mackinawite) film. The main impact of this scale is to induce extremely large values of double-layer capacitance, due to the huge increase of effective surface, which can reach nearly 600 times that of the geometrical surface of the metallic electrode. The analysis of impedance data also allowed determining the evolution with time of the charge-transfer resistance, from which the corrosion current density could directly be estimated.

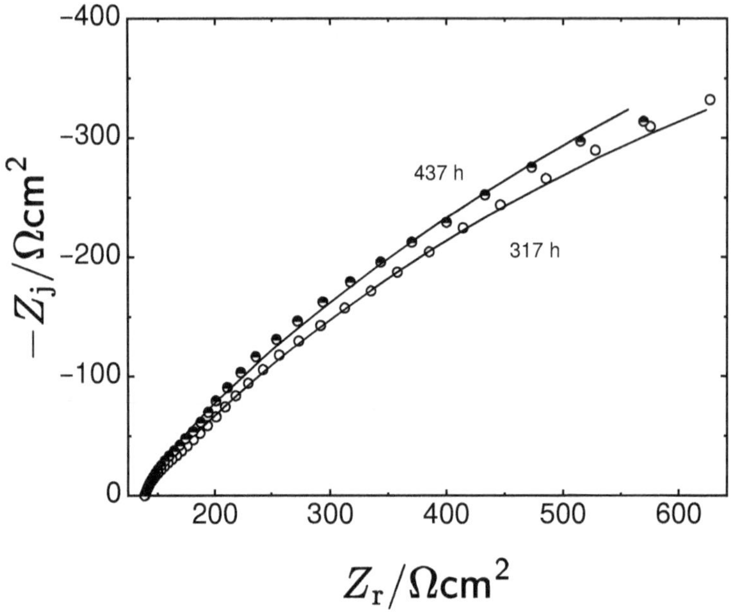

Figure 28.12: Regression of equation (28.14) to data collected by Deffo Ayagou et al.[126] at 317 h and 437 h. Lines represent the fit to the data. Parameter estimates are shown in Table 28.4.

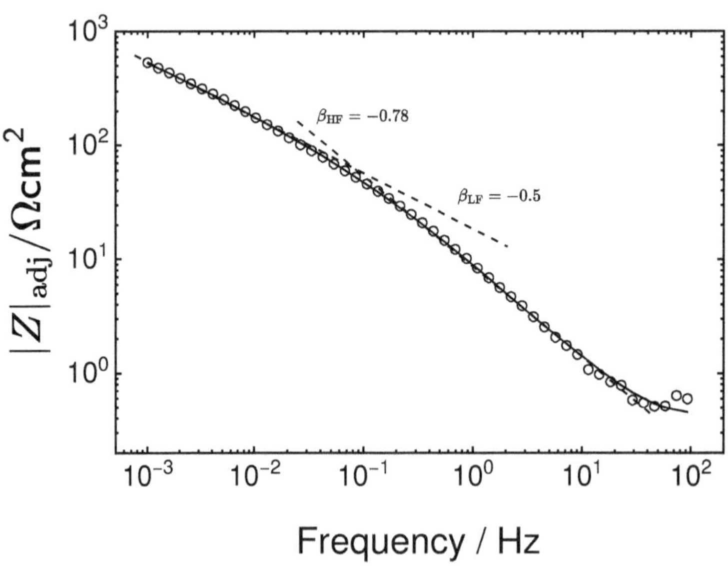

Figure 28.13: Ohmic-resistance-corrected modulus for the data collected by Deffo Ayagou et al.[126] at an elapsed time of 437 h. The solid line represents the fit of equation (28.14) to the data. Dashed lines illustrate the change in slope from -0.78 at high frequency to -0.5 at low frequency.

Chapter 29

Apparent Rotation-Speed Dependence of CPE Parameters

The three data sets reported in this chapter are not replicates; nevertheless, the measurement model may be used to filter lack of replication to identify the stochastic error structure. The data presented here demonstrate the importance of considering both anodic and cathodic reactions when devising a model to fit the data.

Power plants require cooling circuits with water as cooling agent; therefore, there are generally located on seacoasts owing to the ready availability of abundant seawater. Tubes are the basic components of heat exchangers; they provide the heat transfer surface between one fluid flowing inside the tubes and other fluid flowing outside the tubes. In cooling circuits, the water usually circulates, but frequently plant outages cause provisional stagnant conditions that can persist for hours or at worst for days. This condition is of particular concern for corrosion risk especially at the beginning of plant operation. Therefore hydrodynamics is one of the parameters to be studied.

Copper and copper alloys are commonly used in condensers and heat exchangers due to their high thermal conductivity, good resistance to corrosion, and good mechanical workability. Cu–Ni alloys are preferred in marine environments because of their corrosion resistance due to the formation of a thin, adherent, protective surface film, which forms naturally and quickly upon exposure to clean seawater. That surface film is predominantly made up of cuprous oxide, often containing nickel and iron oxides, cuprous hydroxychloride, and cupric oxide. The film becomes more complex when aggressive conditions from the water cause corrosion. The presentation in this chapter is based on work presented by Bautista et al.[128] on corrosion of a 70Cu–30Ni alloy in seawater.

29.1 Initial Inspection

The experimental data were obtained from ring electrodes cut from real new condenser tubes, as shown in Figure 29.1. In Table 29.1, the experimental data are given as real and imaginary parts of the impedance as functions of frequency. From this file, the experimental results are plotted for different rotation speeds. As described in Chapters 26–28 and 30, the initial inspection involves graphical analysis and regression of the measurement model.

29.1.1 Graphical Analysis

At first glance, the impedance data shown in Figure 29.2 can be described by two loops: a depressed high-frequency semicircle followed by a poorly defined low-frequency branch. The characteristic frequencies depend clearly on the rotation speed, but the resistance which corresponds to the loop diameter is more or less constant. Thus, if the data are analyzed with a resistance in parallel to a CPE, one may be tempted to conclude that the CPE parameters are functions of rotation speed. Following Wu et al.,[130] the value of the CPE exponent α may be estimated from the depression of the capacitive loop following

$$\left|\frac{(\Delta Z_{\mathrm{j}})_{\max}}{(\Delta Z_{\mathrm{r}})_{\mathrm{midpoint}}}\right| = \tan(\frac{\alpha\pi}{4}) \tag{29.1}$$

where the term $(\Delta Z_{\mathrm{j}})_{\max}$ represents the height of the depressed capacitive loop and $(\Delta Z_{\mathrm{r}})_{\mathrm{midpoint}}$ represents the half-width of the loop. The values of α may be estimated to be 0.62 for 40 rpm, 0.80 for 160 rpm, and 0.83 for 640 rpm. The CPE, however, is generally associated with a distribution of time constants considered to be a property of the electrode–electrolyte interface. If the resistor corresponds to a charge-transfer resistance, then the CPE must correspond to a double-layer capacitance. As the double-layer capacitance cannot depend on the rotation speed, a deeper analysis is necessary.

The impedance from Table 29.1 is presented in Figure 29.3 in Bode format with rotation speed as a parameter. The ohmic resistance may be extracted from the high-frequency portion of the magnitude as shown in Figure 29.3(a). As discussed in previous chapters, for example, Chapter 26 on page 319, the high-frequency portion of the phase angle shown in Figure 29.3(b) is dominated by the confounding effects of the ohmic resistance.

The ohmic-resistance-corrected magnitude is also presented in Figure 29.3(a), and the ohmic-resistance-corrected phase angle is presented in Figure 29.3(b). The corrected Bode plots show a variation of the modulus with frequency that differs for the three rotation speeds, but all three curves converge in the very-high-frequency range. In the same way, the corrected phases are different for the three rotation speeds but tend toward the same value in the very-high-frequency range. A plateau is apparent at frequencies around 10^3–10^4 Hz with a phase of about 79°. These properties are obscured

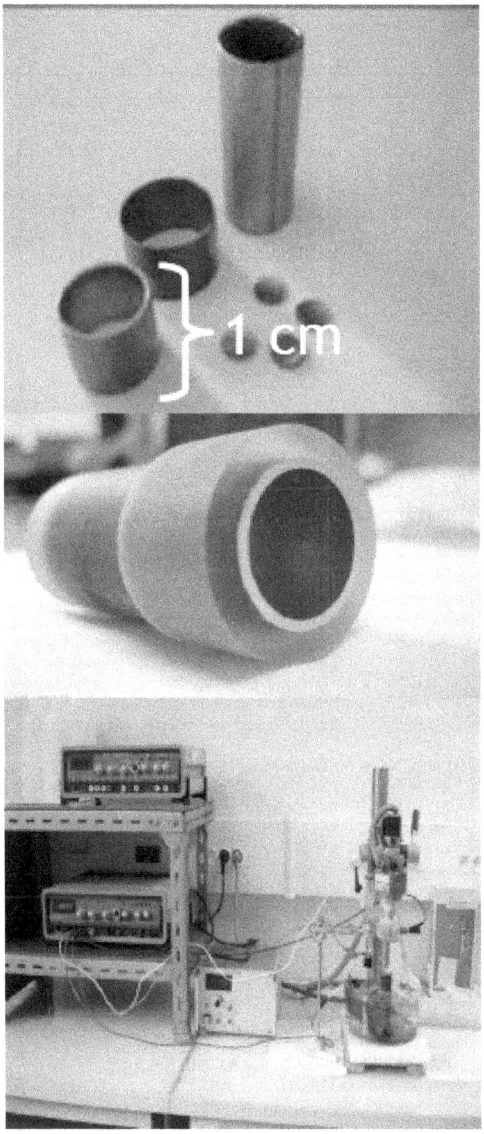

Figure 29.1: 70Cu–30Ni condenser tubes from which circular samples were cut. The electrolyte was filtered natural seawater. Electrodes were used as rotating ring electrodes (RRE). Images adapted with permission from the dissertation of Torres-Bautista.[129]

Table 29.1: Experimental impedance data collected for a 70Cu–30Ni alloy in seawater. Data from Torres-Bautista, et al..[128]

Rotation Speed	40 rpm		160 rpm		640 rpm	
f, Hz	Z_r, Ωcm^2	$-Z_j$, Ωcm^2	Z_r, Ωcm^2	$-Z_j$, Ωcm^2	Z_r, Ωcm^2	$-Z_j$, Ωcm^2
100000	3.70045	0.02803	3.50112	0.02232	3.65923	0.03873
71539	3.70259	0.06125	3.50257	0.05692	3.66175	0.07144
51179	3.71025	0.09836	3.51179	0.09333	3.66951	0.11606
36613	3.70744	0.14066	3.50839	0.13998	3.6698	0.16343
26193	3.72364	0.19557	3.52323	0.19546	3.68445	0.2262
18738	3.74362	0.26497	3.54128	0.26555	3.70375	0.30684
13405	3.76583	0.35425	3.56281	0.35615	3.72635	0.41075
9589.9	3.7992	0.47176	3.59327	0.47683	3.75797	0.55115
6860.6	3.83713	0.62644	3.62916	0.63642	3.79609	0.73877
4908.1	3.88902	0.83331	3.67533	0.85067	3.8444	0.99018
3511.2	3.95935	1.10852	3.73877	1.13781	3.91123	1.32948
2511.9	4.05499	1.47392	3.82617	1.52232	4.0027	1.78548
1797	4.18817	1.95756	3.94877	2.03467	4.12919	2.39803
1285.6	4.37528	2.59407	4.12153	2.71464	4.30544	3.2171
919.68	4.63922	3.42633	4.3649	3.61344	4.55366	4.3101
657.93	5.00607	4.50662	4.70595	4.79452	4.90131	5.76568
470.68	5.51232	5.90352	5.18368	6.34254	5.39359	7.69569
336.72	6.208	7.69831	5.85104	8.35684	6.08995	10.24514
240.89	7.15802	9.99294	6.77623	10.96488	7.06616	13.59261
172.33	8.44346	12.90488	8.04809	14.32302	8.44889	17.97507
123.28	10.18209	16.58506	9.80088	18.63273	10.39161	23.67867
88.198	12.51591	21.20808	12.20066	24.12584	13.10761	31.07977
63.095	15.62379	26.97958	15.47538	31.09529	16.90322	40.59062
45.138	19.7589	34.12751	19.91798	39.86797	22.20621	52.79904
32.292	25.24619	42.95063	25.96399	50.77756	29.62186	68.20749
23.101	32.5144	53.64294	34.09162	64.21982	39.9543	87.53086
16.527	42.0495	66.41396	45.0468	80.50903	54.31709	111.3463
11.823	54.5431	81.45478	59.67634	99.9585	74.18172	140.3784
8.4582	70.76635	98.7363	78.98322	122.6371	102.0828	175.2596
6.0509	91.568	118.0102	104.3429	148.4779	142.0856	214.9229
4.3288	117.9326	138.5936	136.8961	176.7534	197.8412	255.4398
3.0968	150.2239	159.4389	178.0047	207.0756	270.6785	289.0309
2.2154	188.6359	179.2948	230.7242	238.0283	356.5138	307.7519
1.5849	232.9164	196.6966	297.3438	263.5199	443.6586	307.6646
1.1338	281.0769	210.6161	375.5258	275.1308	526.9428	292.4744
0.81113	331.4199	221.1697	457.064	268.8161	603.34	257.632
0.58028	384.8572	230.4041	528.1262	244.9541	658.824	218.832
0.41513	444.3473	235.7779	583.8139	212.4785	701.31	180.6334
0.29698	508.28	231.0346	623.4966	178.6837	731.2733	146.1596
0.21246	568.5073	214.176	654.3426	150.5246	751.3232	116.5261
0.15199	615.6202	190.3431	676.3907	127.1379	766.9596	93.2946
0.10873	651.2677	165.5014	695.3639	109.3578	777.1834	74.66284
0.07779	680.1252	147.5952	710.719	96.65856	785.8261	60.94898
0.05565	702.7262	133.2198	726.2972	87.15353	792.5288	51.19466
0.03981	723.62	123.9951	737.6656	79.8892	799.4934	43.35318
0.02848	747.0455	120.5516	754.175	75.57464	804.2852	39.30149
0.02038	765.0972	118.437	767.7259	74.75014	811.3565	34.88411
0.01458	789.3763	114.5182	785.8746	71.54526	815.576	34.82979
0.01043	814.3247	113.3736	800.4052	73.07883	822.8413	31.71706
0.00746	836.4698	113.2087	817.1377	72.92363	828.1084	35.44186
0.00534	860.0311	110.6867	834.2873	74.5542	837.6144	32.63371
0.00382	881.3032	112.4715	853.6485	75.67455	844.87	40.0222
0.00273	906.1061	112.4715	875.0758	77.91525	857.1017	39.33932
0.00195	926.4858	109.8719	900.1794	82.2463	876.3659	46.18267
0.0014	944.5181	101.2777	934.498	87.87812	902.7111	57.2106
0.0010	961.9102	93.10836	957.2251	101.6366	939.8718	72.83148

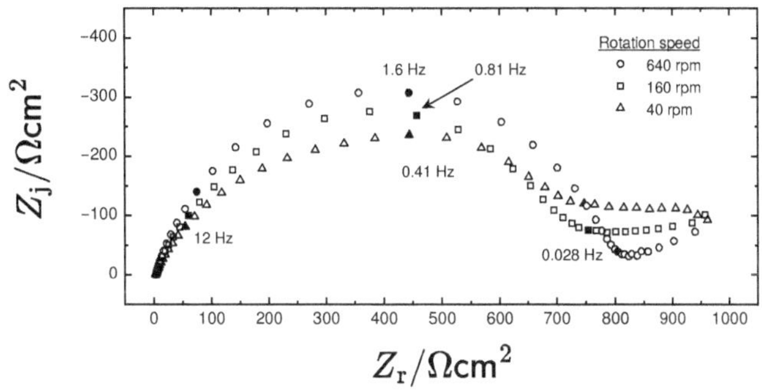

Figure 29.2: Impedance response in Nyquist format for a 70Cu–30Ni alloy in seawater with rotation speed as a parameter.

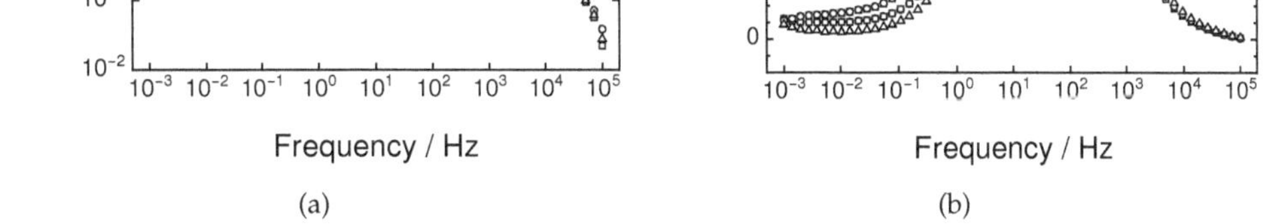

Figure 29.3: Impedance response in Bode format and ohmic-resistance-corrected Bode format with rotation speed as a parameter for a 70Cu–30Ni alloy in seawater: (a) magnitude and (b) phase.

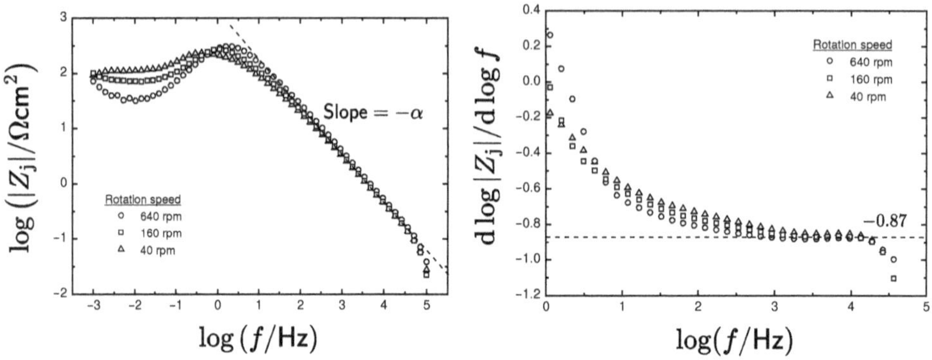

(a) (b)

Figure 29.4: Impedance response for a 70Cu–30Ni alloy in seawater with rotation speed as a parameter: a) the logarithm of the absolute value of the imaginary part of the impedance as a function of the logarithm of frequency and b) the derivative of the logarithm of the absolute value of the imaginary part of the impedance with respect to the logarithm of frequency as a function of the logarithm of frequency.

in the ordinary Bode plots. The ohmic-resistance-corrected results presented in Figure 29.3 suggest that the data at frequencies higher than 10^4 Hz may be unreliable. The measurement model analysis will be used to assess this part of the spectrum.

The results of the corrected Bode plots depend on the determined value for the electrolyte resistance. It is also possible to analyze directly the imaginary part of the impedance. The logarithm of the absolute value of the imaginary part of the impedance is presented in Figure 29.4(a) as a function of the logarithm of frequency. The logarithm of the imaginary parts of the impedance are different at low frequencies for the three rotation speeds but collapse to one curve in the high frequency range. The derivative $\mathrm{d}\log|Z|/\mathrm{d}\log f$, presented in Figure 29.4(b) with rotation speed as a parameter, provides a means of estimating the limiting high-frequency value of α. The derivative $\mathrm{d}\log|Z|/\mathrm{d}\log f$ tends toward a unique value for the three rotation speeds of -0.87, i.e., $\alpha = 0.87$. This value may be used to extract a limiting phase angle $\varphi_{\mathrm{lim}} = \alpha(-90^\circ) = -78.3^\circ$, in good agreement with the high-frequency phase limit of 79° extracted from Figure 29.3(b). The graphical analysis shows that the same CPE exists for the three rotation speeds. It is now necessary to build a physical model to explain the results.

29.1.2 Measurement Model Analysis

The measurement model is described in Sections 26.1.2 (page 324) and 27.1.2 (page 347). As was performed in Section 27.1.2, the measurement model was used to identify the stochastic and bias contributions to the error structure of the impedance measurements.

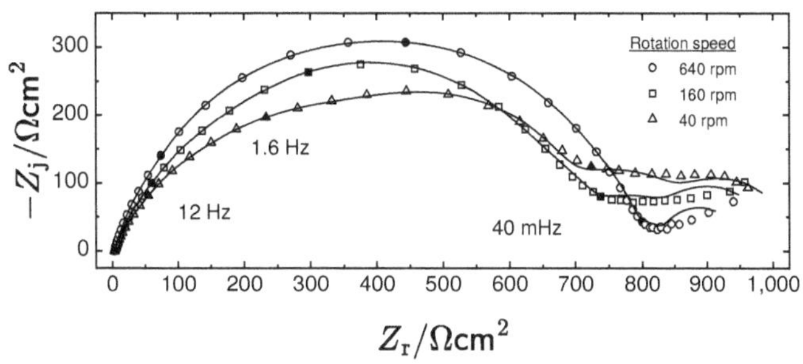

Figure 29.5: Impedance response in Nyquist format for a 70Cu–30Ni alloy in seawater with rotation speed as a parameter. The data are taken from Figure 29.2, and the lines represent the complex fit of the measurement model (equation (25.1)) to the data under modulus weighting.

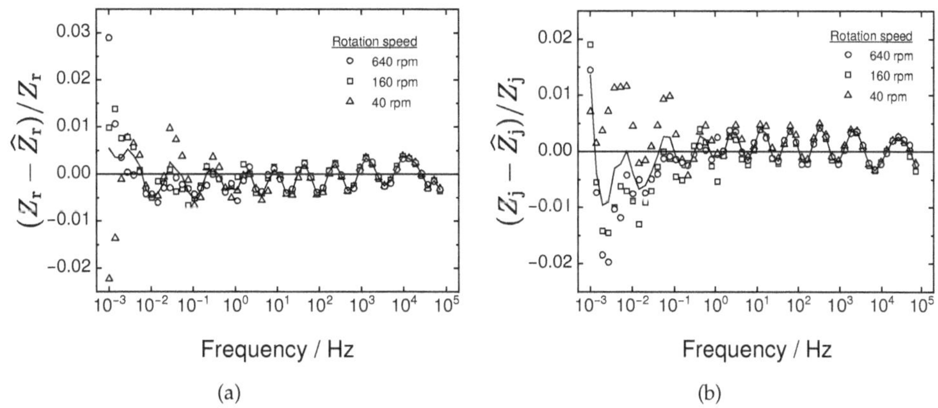

Figure 29.6: Normalized residual errors as functions of frequency for the regression of equation (25.1) to the data given in Table 29.1: (a) real part of the impedance and (b) imaginary part of the impedance. Lines represent the average normalized residual error at a given frequency.

Stochastic Error Analysis

The measurement model, described in Section 26.1.2, was fit to each of the spectra presented in Table 29.1. The results are presented in Figure 29.5. The data are taken from Figure 29.2, and the lines represent the complex fit of the measurement model (equation (25.1)) to the data under modulus weighting. The measurement model provided an adequate fit to the high-frequency capacitive loop, but the low frequency features were not fit as well.

The normalized residual errors corresponding to Figure 29.5 are presented in Figure 29.6 as functions of frequency. Over most of the frequency range, the normalized residual errors were less than 0.02. The residual errors tended to be larger at frequencies less than 0.1 Hz, where the measurement model did not provide a good fit to the data, as shown in Figure 29.5. As mentioned for Figure 27.7, the oscillation in the residual errors can be attributed to the observation that equation (25.1) does not represent a

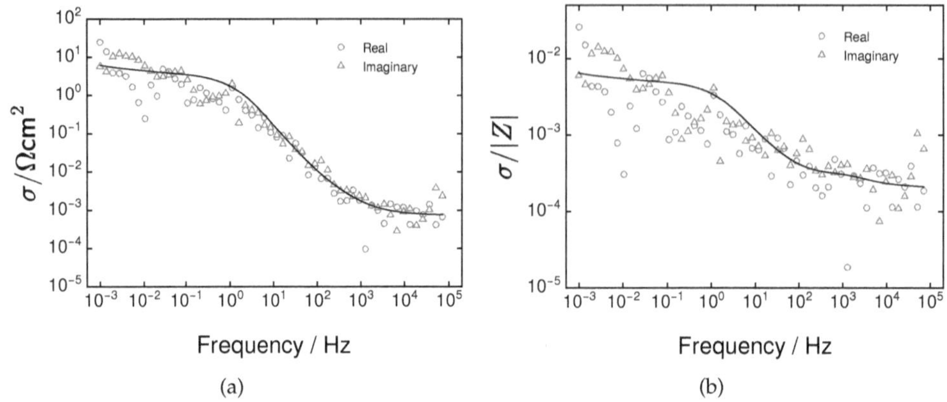

Figure 29.7: Standard deviation of impedance measurements obtained from the residual errors presented in Figure 29.6: (a) real part and (b) imaginary part. The line represents the fit of equation (27.2).

Table 29.2: Regressed constants for the standard deviation of the stochastic part of the error structure given in equation (27.2) for the results presented in Figure 29.7(a).

Parameter	Value	Units
α	$2.46 \times 10^{-4} \pm 1.6 \times 10^{-5}$	
β	—	
γ	$6.713 \times 10^{-6} \pm 5.0 \times 10^{-9}$	$(\Omega cm^2)^{-1}$
δ	$6.65 \times 10^{-4} \pm 2.7 \times 10^{-5}$	Ωcm^2

correct model for the Cu–Ni ring electrode rotating in seawater.

The residual error, given in equation (27.1), can be expressed as the sum of contributions from a lack of fit, bias errors, and stochastic errors. The mean value of the residual error is considered here to be the sum of the bias error and the lack of fit. The sum $\varepsilon_{\text{fit}} + \varepsilon_{\text{bias}}$, represented by lines in Figure 29.6, was estimated from the average residual error calculated on a frequency-by-frequency basis. The stochastic contribution was obtained from the standard deviation of the residual errors at each frequency.

The resulting estimate for the standard deviation is presented as a function of frequency in Figure 27.8(a). The real and imaginary parts of the standard deviation are roughly equal for frequencies higher than 0.1 Hz. The overlap of real and imaginary parts of the standard deviation fails at lower frequencies due to the difficulty of getting a good fit of the measurement model to the data in that frequency range. The standard deviations are shown to be strong functions of frequency, and the scaled standard deviations, given in Figure 29.7(b), are strong functions of frequency. The standard deviations range from 0.6% of the modulus at low frequency to 0.03% of the modulus at high frequency. Constants for the standard deviation of the stochastic part of the error structure given in equation (27.2) are presented in Table 29.2 for the results presented

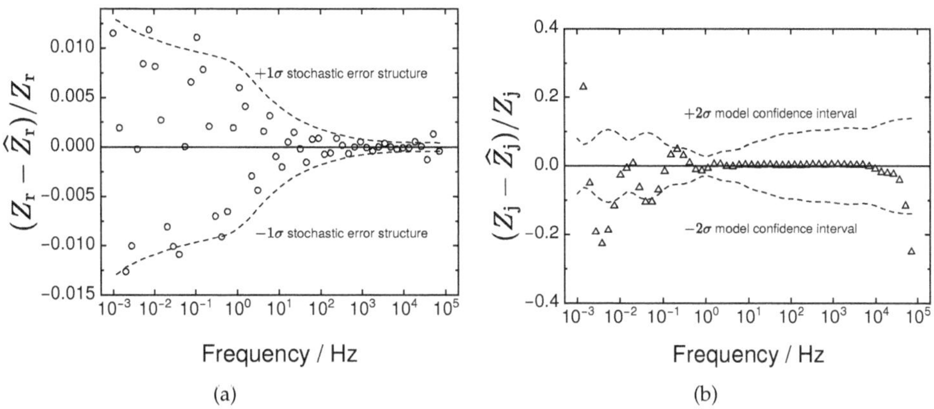

Figure 29.8: Regression results as a function of frequency for the fit of equation (25.1) to the data collected at 160 rpm (given in Table 29.1): (a) normalized residual errors for the fit to the real part of the impedance and (b) normalized errors for the prediction of the imaginary part of the impedance. Dashed lines represent the $\pm 1\sigma$ stochastic error structure in panel(a)and the $\pm 2\sigma$ (95.4%) confidence interval in panel (b).

in Figure 29.7(a). The resulting model is presented as lines in Figure 29.7.

Consistency with the Kramers–Kronig Relations

To assess the potential for artifacts at high frequency, the measurement model was fit to the real part of the impedance measurement. An example result is presented in Figure 29.8(a) for the residual errors as a function of frequency. The regression was weighted by the error structure given in equation (27.2) with the parameters presented in Table 29.2. The first measured frequency was neglected because the first measurement is often influenced by the transient from the steady-state condition to the applied frequency. The parameters obtained by regression of the real part of the impedance were used in equation (25.1) to predict the imaginary part of the impedance. The resulting normalized errors are presented in Figure 29.8(b). One high-frequency point fell outside the confidence interval for the model and could be removed from subsequent regressions. As the fit to the real part of the impedance does not give an adequate prediction for the low-frequency imaginary part of the impedance, no conclusions regarding consistency with the Kramers–Kronig relations at low frequency can be made. Regressions to the imaginary part of the impedance showed that the low-frequency data were consistent with the Kramers–Kronig relations.

Estimation of the Capacitance

The results presented in Figure 29.8(a) show that instrument and wiring artifacts cannot account for the decrease in ohmic-resistance corrected phase angle for frequencies higher than 10 kHz, as revealed in Figure 29.3(b). An alternative explanation could

Table 29.3: Regression parameters obtained by the fit of equation (25.1) to the data collected at 160 rpm (given in Table 29.1) determined to be consistent with the Kramers–Kronig relations.

k	τ_k, s	f_k, Hz	R_k, Ωcm^2	C_k, $\mu F/cm^2$
1	$(1.78 \pm 0.079) \times 10^{-5}$	8,900	0.146 ± 0.0069	122 ± 7.9
2	$(1.09 \pm 0.066) \times 10^{-4}$	1,500	0.508 ± 0.037	215 ± 20
3	$(4.60 \pm 0.29) \times 10^{-4}$	350	2.06 ± 0.15	224 ± 22
4	$(1.75 \pm 0.12) \times 10^{-3}$	91	7.45 ± 0.52	235 ± 23
5	$(6.82 \pm 0.45) \times 10^{-3}$	23	27.4 ± 1.9	248 ± 23
6	$(2.80 \pm 0.16) \times 10^{-2}$	5.7	$106. \pm 5.2$	263 ± 20
7	0.139 ± 0.0061	1.1	457 ± 17	305 ± 17
8	0.617 ± 0.14	0.26	$113. \pm 17$	$(5.46 \pm 1.5) \times 10^3$
9	7.32 ± 1.5	0.022	$102. \pm 10$	$(7.19 \pm 1.6) \times 10^4$
10	90.6 ± 15	0.0018	$170. \pm 14$	$(5.33 \pm 1.0) \times 10^5$
R_e	–	–	3.5074 ± 0.0011	–

be associated with the geometry-induced frequency dispersion, but estimation of the characteristic frequency requires an estimate for the capacitance.

The parameters presented in Table 29.3 were obtained from a complex fit of the measurement model to the data collected at 160 rpm that were determined to be consistent with the Kramers–Kronig relations. Each RC element k yielded values for a time constant τ_k and a resistance R_k. The values were sorted in order of increasing time constant and decreasing frequency $f_k = 1/2\pi\tau_k$. A capacitance was calculated for each element as $C_k = \tau_k/R_k$. The effective capacitance, calculated as

$$C_{eff} = \frac{1}{\sum_k 1/C_k} \tag{29.2}$$

had a value of $C_{eff} = 30.4 \pm 0.97\ \mu F/cm^2$.

The capacitance obtained from the measurement model parameters is insensitive to the contributions from RC elements corresponding to low frequencies. As shown in Table 29.3, the RC element with $\tau_k = 90$ s and $f_k = 1.8$ mHz contributes only 0.0057% to the summation $\sum_k(1/C_k)$; whereas the RC elements $k = 1-7$ in Table 29.3 together contribute 99.4%.

The standard deviations for calculated values were obtained by a linear propagation of errors. For example, the standard deviations provided in Table 29.3 for capacitance were estimated from

$$\sigma_{C_k}^2 = \sigma_{\tau_k}^2\left(\frac{1}{R_k}\right)^2 + \sigma_{R_k}^2\left(\frac{\tau_k}{R_k^2}\right)^2 \tag{29.3}$$

where the standard deviations σ_{τ_k} and σ_{R_k} were obtained from Levenberg–Marquardt regression. A discussion of the methods to estimate propagation of errors in calculations is presented in Section 3.2 of Orazem and Tribollet.[1]

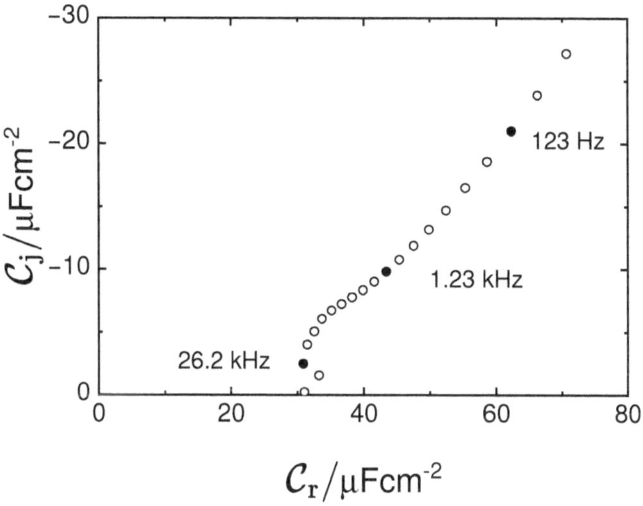

Figure 29.9: The imaginary part of the complex capacitance as a function of the real part of the complex capacitance for the impedance data collected at a rotation speed of 160 rpm and presented in Table 29.1.

Calculation of the complex capacitance provides another approach for obtaining the capacitance. The complex capacitance may be obtained from equation (26.3). As shown in Figure 26.13, the calculation is very sensitive to the estimated value for ohmic resistance. The complex capacitance is presented in Figure 29.9 under the assumption that the ohmic resistance was $R_e = 3.5074\ \Omega\text{cm}^2$, as obtained from the measurement model regression presented in Table 29.3. The high-frequency limit yields a capacitance of $C_r = 30.9\ \mu\text{F/cm}^2$, in good agreement with the value $C_{\text{eff}} = 30.4 \pm 0.97\ \mu\text{F/cm}^2$ obtained from the measurement model analysis.

The characteristic frequency for a disk electrode with capacitance $C_{\text{eff}} = 30.4 \pm 0.97\ \mu\text{F/cm}^2$ and ohmic resistance $R_e = 3.5074\ \Omega\text{cm}^2$ can be obtained from equation (26.2) to be $f_{c,\text{disk}} = 1.2$ kHz. The electrode geometry, however, as shown in Figure 29.1, is that of a ring. The characteristic frequency for geometry-induced frequency dispersion on a ring electrode can be compared to that of a disk electrode, i.e.,[131]

$$\frac{f_{c,\text{ring}}}{f_{c,\text{disk}}} = \frac{r_2\left(1 + (r_1/r_2)^2\right)}{(r_1 - r_2)\,g} \tag{29.4}$$

where the correction factor g, given by

$$\begin{aligned} g &= 1 + 1.0742\left(\frac{r_1}{r_2}\right) - 4.2227\left(\frac{r_1}{r_2}\right)^2 + 8.9244\left(\frac{r_1}{r_2}\right)^3 \\ &\quad -9.4462\left(\frac{r_1}{r_2}\right)^4 + 3.8806\left(\frac{r_1}{r_2}\right)^5 \end{aligned} \tag{29.5}$$

has a value between 1.2 and unity. For the present system, $r_2 = 0.75$ cm and $r_1 =$

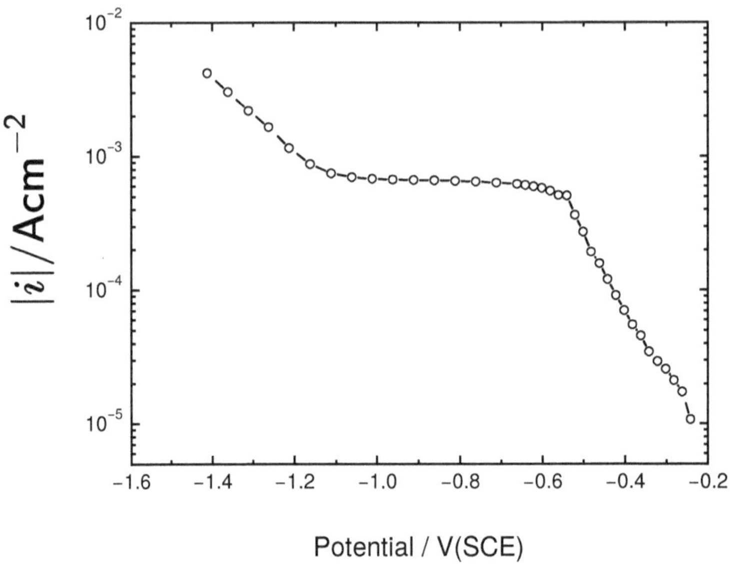

Figure 29.10: Cathodic polarization curve for a 70Cu–30Ni ring electrode at a rotation speed of 160 rpm.

0.65 cm. The exposed area is $A = 0.45$ cm^2. Inclusion of electrode dimensions in equation (29.4) yields $f_{\text{c,ring}}/f_{\text{c,disk}} = 9.3$. Thus, $f_{\text{c,ring}} = 1.1 \times 10^4$ Hz. Thus, all measurements for frequencies above 10 kHz may be assumed to have been influenced by geometry-induced frequency dispersion, in agreement with the discussion of Figure 29.3(b).

29.2 Interpretation Model Development

The cathodic polarization curve is presented in Figure 29.10 for a 70Cu–30Ni ring electrode at a rotation speed of 160 rpm. A current plateau can be observed illustrating the influence of mass-transfer-limited reactions. The measured current is reduced near the open-circuit potential (to the right in Figure 29.10) by the influence of the anodic reaction.

29.2.1 Electrical Circuit Framework

At the corrosion potential, the anodic and cathodic currents have the same magnitude and the net current is equal to zero. By principles of summation of currents, the anodic and cathodic impedances must be in parallel. The corresponding equivalent electrical circuit for the interface is presented in Figure 29.11. As the electrolyte was aerated, the cathodic reaction is the reduction of dissolved oxygen

$$O_2 + 2H_2O + 4e^- \rightarrow 4OH^- \tag{29.6}$$

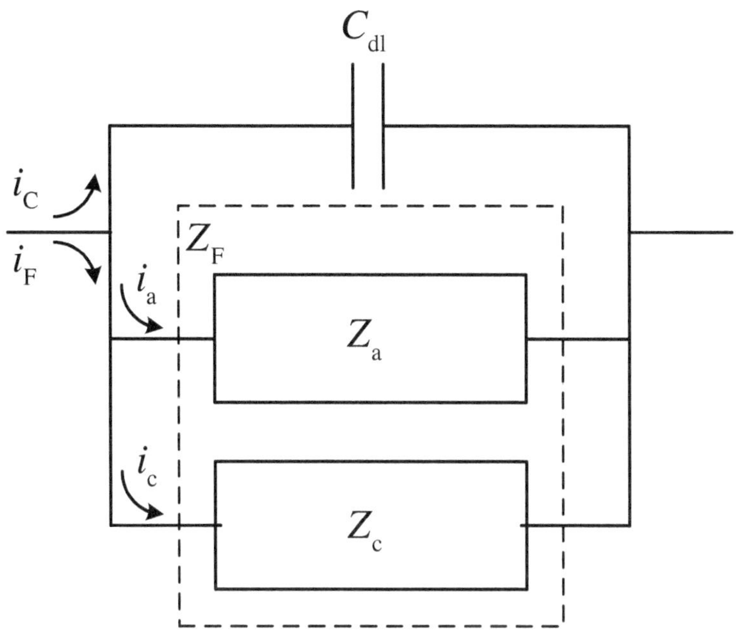

Figure 29.11: Electrical circuit for the electrode–electrolyte interface corresponding to the balance of anodic and cathodic reactions under the open-circuit condition, i.e., at the corrosion potential. Taken from Orazem and Tribollet[1] with permission of John Wiley & Sons.

For a copper electrode in NaCl electrolytes, the anodic reactions involves formation of CuCl

$$Cu + Cl^- \rightarrow CuCl + e^- \tag{29.7}$$

followed by the formation of the soluble cuprous complex

$$CuCl + Cl^- \rightleftarrows CuCl_2^- \tag{29.8}$$

Deslouis et al.[132] showed that the dissolution reaction (29.7) is therefore influenced by diffusion of $CuCl_2^-$ ions to the electrode surface.

The general circuit associated with the corrosion of 70Cu–30Ni at the corrosion potential can be given as Figure 29.12(a) in which a constant-phase element is assumed to be associated with the double layer, kinetic and diffusion impedance terms are associated with surface coverage and diffusion for the anodic reactions (29.7) and (29.8), and kinetic and diffusion impedances are associated with the cathodic reaction (29.6). As the first current plateau, shown in Figure 29.10 and attributed to reduction of dissolved oxygen, is close to the corrosion potential, one may hypothesize that the mass-transfer-limited plateau for the cathodic reaction may be extrapolated to the corrosion potential. Thus, $R_{t,c}$ can be neglected, and the cathodic mass-transfer impedance is a Warburg im-

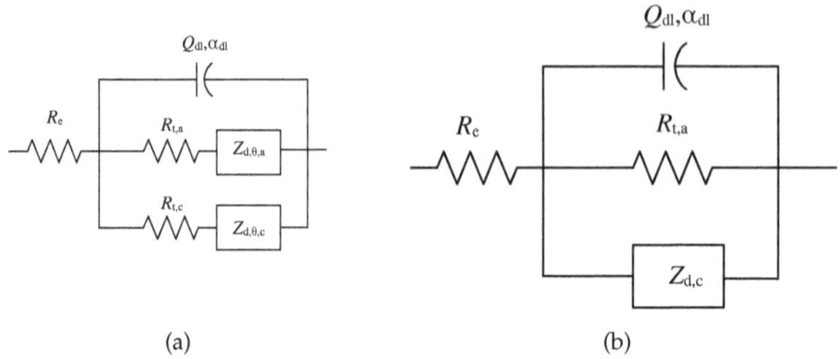

Figure 29.12: Equivalent circuit diagram proposed for corrosion of 70Cu–30Ni at the corrosion potential: (a) diagram accounting for a constant-phase element associated with the double layer, kinetic and diffusion impedance associated with surface coverage and diffusion for the anodic reaction, and kinetic and diffusion impedance associated with the cathodic reaction and (b) simplified diagram obtained under the assumption that the cathodic reaction is diffusion controlled and the anodic reaction is kinetically controlled.

pedance given by

$$Z_{D,c}(\omega) = \frac{1}{k_c (j\omega)^{1/2}} \tag{29.9}$$

where k_c has units of $s^{0.5}/\Omega cm^2$. Under the assumption that the anodic reaction is kinetically controlled, $Z_{D,\theta,a}$ may be neglected in favor of the charge-transfer resistance $R_{t,a}$. The associated electric circuit is given as Figure 29.12(b).

29.2.2 Mathematical Model

The equivalent circuit of Figure 29.12(b) corresponds to the high-frequency loop of the experimental impedance diagrams shown in Figure 29.2. Thus, the high-frequency loop illustrates mainly the anodic charge transfer (with diameter equal to $R_{t,a}$), and its depressed shape is partly due to the parallel cathodic Warburg impedance. The cathodic Warburg impedance can be neglected only in a frequency range higher than 5 kHz (see, for example, Figures 29.3(b) and 29.4). The low-frequency loop is related to contributions to the anodic impedance by mass transfer of chloride ions and by partial blocking by CuCl (i.e., $Z_{D,\theta,a}$).

As the low-frequency loops of the impedance diagrams were not well defined and, indeed, were described only by a few points, the single high-frequency loops were analyzed by regression of the equivalent circuit presented in Figure 29.12(b), in which the double-layer capacitance, in the form of a CPE, is in parallel to the anodic charge transfer resistance and the cathodic diffusion impedance. The model corresponding to Figure 29.12(b) may be expressed as

$$Z = R_e + \frac{1}{(j\omega)^{\alpha} Q + 1/R_{t,a} + k_c (j\omega)^{1/2}} \tag{29.10}$$

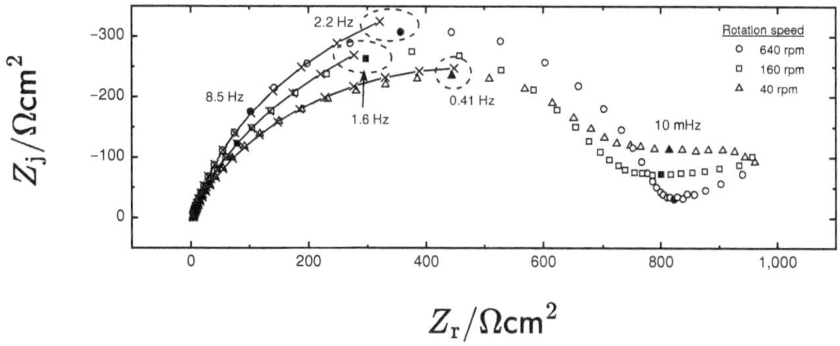

Figure 29.13: Impedance response in Nyquist format for a 70Cu–30Ni alloy in seawater with rotation speed as a parameter. Regression results are included as $- \times -$.

where the CPE is given by $1/(j\omega)^\alpha Q$ and the diffusion term is given by $1/k_c\sqrt{j\omega}$.

29.3 Regression Analysis

Equation (29.10) was regressed to the data presented in Table 29.1 under an error structure weighting based on equation (27.2) with parameters presented in Table 29.2. The data with frequencies greater than 9 kHz were truncated to eliminate the influence of geometry-induced frequency dispersion. Thus, the high-frequency limit for the regressed frequency range was 6.68 kHz. The low-frequency limit, chosen to remove the influence of mass-transfer effects that were not included in equation (29.10), varied with rotation speed. The low-frequency limits were 0.415 Hz at 40 rpm, 1.58 Hz at 160 rpm, and 2.21 Hz at 640 rpm.

The regression results are presented in Figure 29.13 as Nyquist plots and in Figure 29.14 as the absolute value of the imaginary part of the impedance as a function of frequency. As shown in Figure 29.7, the weighting strategy emphasized the high-frequency impedance more than the lower frequency. The regression results are presented in Table 29.4. If the CPE behavior is assumed to be associated with surface distributed time constants for charge-transfer reactions (due to time-constant distribution along the electrode surface), then it is possible to apply the equation derived by Brug et al.[54] to calculate the effective capacitance associated with the CPE, e.g.,

$$C_{\text{eff}} = Q^{1/\alpha} \left(\frac{R_e R_{t,a}}{R_e + R_{t,a}} \right)^{(\alpha-1)/\alpha} \tag{29.11}$$

The capacitance values calculated from the impedance diagrams, taking for R_e and $R_{t,a}$ the values extracted from the regression procedure, are given in Table 29.4. These capacitance values, of the order of several tens of $\mu F/cm^2$, are typical of those for a double layer capacitance. In addition, the values of the CPE exponent α are almost

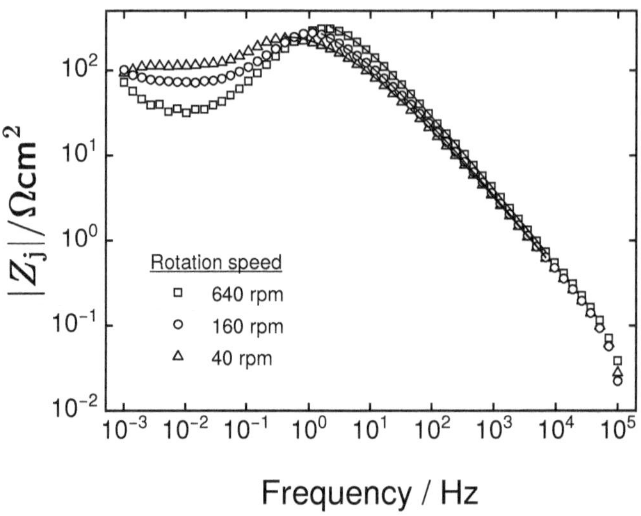

Figure 29.14: Absolute value of the imaginary part of the impedance as a function of frequency with rotation speed as a parameter. Regression results are included as lines.

Table 29.4: Results for the regression of equation (29.10) to the data presented in Table 29.1. High-frequency data were truncated to eliminate the influence of geometry-induced frequency dispersion, and low-frequency data were truncated to eliminate the influence of anodic diffusion.

Rotation Speed, rpm	40	160	640
Potential, V(SCE)	-0.225	-0.244	-0.256
Q, $\mu F/cm^2\ s^{1-\alpha}$	113.8 ± 2.2	99.5 ± 1.5	86.1 ± 1.6
α	0.8848 ± 0.0021	0.8988 ± 0.0016	0.9012 ± 0.0019
R_e, Ωcm^2	3.6814 ± 0.0049	3.4961 ± 0.0036	3.6583 ± 0.0047
$R_{t,a}$, Ωcm^2	1029 ± 28	1470 ± 61	1359 ± 67
k_c, $\mu s^{1/2}/\Omega cm^2$	0.6090 ± 0.0093	0.4710 ± 0.0073	0.2499 ± 0.0083
$C_{eff}/\mu F\ cm^{-2}$	41.3	40.6	35.6

Table 29.5: Comparison of capacitance values obtained from regression of equation (29.10) coupled with the Brug formula (equation (29.11)) and the measurement model applied to data sets truncated such that the maximum frequency was 919 Hz.

Rotation Speed	40 rpm	160 rpm	640 rpm
Brug formula	41.3 μF/cm^2	40.6 μF/cm^2	35.6 μF/cm^2
Measurement model	41.3±2.6 μF/cm^2	40.9±1.6 μF/cm^2	34.6±1.2 μF/cm^2

independent of rotation speed. The average value of 0.895 is close to the value of 0.87 estimated by the graphical analysis.

29.4 Observations

The measurement model and the Cole–Cole representation suggested that the electrode capacitance was on the order of 30 μF/cm^2; whereas, the interpretation model, coupled with the Brug formula, suggested a capacitance on the order of 40 μF/cm^2. The Cole–Cole representation and the measurement model were found in Chapter 26 to give very reliable values for capacitance for a film in which a distribution of film properties gave rise to a distributed-time-constant behavior. In the present case, the value of the capacitance, either 30 or 40 μF/cm^2, suggests that the constant-phase-element behavior can be attributed to a surface distribution of time constants.

A characteristic frequency of 11 kHz was identified above which frequency dispersion is caused by the geometry of the ring electrode. The measurement model was fit to impedance data sets truncated such that the maximum frequency was 919 Hz, about a decade smaller than the characteristic frequency of 11 kHz. The resulting capacitances, presented in Table 29.5, are in excellent agreement with the capacitances obtained from regression of equation (29.10) coupled with the Brug formula (equation (29.11)). The above analysis validates the equivalent circuit proposed in Figure 29.12(b). Hence the high-frequency loop illustrates mainly the anodic charge transfer, and its diameter is equal to $R_{t,a}$.

To test the value of capacitance obtained by the measurement model, a regression analysis was performed for synthetic data obtained by finite-element simulation for a disk electrode in an insulating plane. This system has been shown to yield a frequency dispersion caused by the electrode geometry.[109,133] The simulation parameters were $r_0 = 0.1$ cm, $\rho = 10^5$ Ωcm, $R_t = 10^4$ Ωcm^2, and $C_{dl} = 20$ μF/cm^2. The characteristic frequency for geometry-induced frequency dispersion was 0.8 Hz. Regression of the measurement model to data ranging from 1 mHz to 1 kHz yielded a capacitance of 7.8 ± 0.04 μF/cm^2; whereas, regression of the measurement model to data ranging from 1 mHz to 0.25 kHz yielded a capacitance of 19.5 ± 0.05 μF/cm^2, close to the input value of 20 μF/cm^2. These results demonstrate that the measurement model yields a correct estimate for the capacitance of a system influenced by surface distributions of time constants so long as the frequency range chosen does not include the influence of the ohmic impedance. The Cole–Cole representation, however, does not yield an accurate value for the capacitance of a system influenced by a surface distribution of time constants.

The analysis presented in this chapter shows that a rotation-speed-dependent depressed capacitive loop may be attributed to a constant-phase element in parallel with a diffusion impedance. In the development of interpretation models for measurements performed at the open circuit condition, one must consider the parallel contributions of charging, anodic, and cathodic currents. In some cases, as is shown in Chapter 26, only one faradaic contribution is evident in the impedance response. In the present case, both anodic and cathodic contributions played a role.

Chapter 30

Model Development for Delamination of Coated Silver Mirrors

Thin film deposition of metallic, dielectric, and semiconducting materials is widely used in many applications, including microelectronic devices, optical coatings, solar cells, and thin-film batteries. Silver is the most used metal for industrial optical applications. This metal has the lowest emissivity of all metals in the visible and IR regions. However in aggressive environments, a silver layer is susceptible to deterioration. The deterioration depends on many factors, such as pH, humidity, temperature, type and concentrations of pollutants, and exposure time.

Electrochemical methods and, more specifically, electrochemical impedance spectroscopy (EIS) have merged as a way to study material and interface properties of thin film stacks. The objective of the present work is to investigate the degradation behavior of a model $Si_3N_4/Ag/Si_3N_4$ stack immersed in an alkaline 0.5 M Na_2SO_4 solution. The presentation in this chapter follows work presented by Tribollet and Orazem.[134]

30.1 Experimental

The materials, electrochemical techniques, and regression approaches used are presented in this section. The morphology of the thin film sample was examined by optical observation and scanning electron microscopy (SEM). The data used in this analysis were reported by Amor et al.[135] The analysis makes use of the measurement model program described in Chapter 25.

30.1.1 Electrolyte

Electrolytes were prepared from analytical grade anhydrous sodium sulfate (Na_2SO_4) and sodium hydroxide (NaOH) dissolved in deionized water. The measurements were performed in a 0.5 M Na_2SO_4 electrolyte. The solution pH was adjusted to a value of 10 by the addition of dilute NaOH. The pH did not change during the courses of measurements because the addition of NaOH at about 10^{-4} mol/L made the solution less sensitive to spontaneous pH change associated with the dissolution of CO_2 gas contained in air.

30.1.2 Electrochemical Measurements

Electrochemical impedance spectra were collected using a Gamry G300 potentiostat. A three-electrode cylindrical cell in poly(methyl-methacrylate) was designed for the measurements. A saturated sulfate electrode and a large platinum plate were used for reference and counter electrodes, respectively. To eliminate the experimental artifacts usually observed at high frequency due to the high impedance of the reference electrode was coupled to a platinum wire in series with a 0.47 μF capacitor.[136]

The working electrode was a silver layer embedded between two coatings upon a silicon substrate ($Si_3N_4/Ag/Si_3N_4$). A schematic representation is presented in Figure 30.1. A narrow aluminum tape was thermally fitted upon the silver layer to ensure electrical contact with the working electrode input of the potentiostat.

Repeated impedance measurements at the open circuit potential were carried out during 12 hours of immersion after a 1-hour stabilization period. The applied frequency range was 100 kHz to 10 mHz with 10 points per decade. The amplitude of sinusoidal perturbation was 10 mV rms. The resulting impedance spectra are presented in Tables 30.1–30.4.

30.2 Initial Inspection

As presented in Chapters 26–29, the initial inspection comprises graphical analysis and measurement model analysis.

30.2.1 Graphical Analysis

The first step for a graphical analysis is to evaluate the ohmic resistance. The impedance data are presented in Figure 30.2(a) in a Nyquist format. These data overlap, showing a high degree of reproducibility, especially in the low-frequency region. The high-frequency part of the Nyquist plot, shown in Figure 30.2(b), reveals greater variability. In addition, the shape of the plots suggest frequency dispersion associated with an ohmic impedance, complicating assessment of the ohmic resistance. Values for ohmic resistance, obtained both by use of the measurement model program and by extrapolation of Figure 30.2(b), are presented in Figure 30.3. The complex capacitance,

Table 30.1: Impedance data obtained for a silver electrode coated by a Si_3N_4 layer after an elapsed time of three hours.

f, Hz	Z_r, Ωcm^2	Z_j, Ωcm^2	f, Hz	Z_r, Ωcm^2	Z_j, Ωcm^2
100,000	67.327	−5.13	24.99	826.41	−16,637.00
79,430	67.641	−6.51	20	1,002.30	−20,621.00
63,110	67.824	−8.18	15.63	1,368.90	−26,251.00
50,120	68.033	−10.22	12.5	1,835.10	−32,497.00
39,810	68.283	−12.69	9.999	2,318.30	−40,564.00
31,630	68.578	−15.75	7.937	3,305.70	−50,422.00
25,120	68.936	−19.58	6.331	4,415.00	−62,082.00
19,950	69.232	−24.32	4.997	5,462.30	−79,081.00
15,850	69.536	−30.14	3.967	6,796.90	−97,029.00
12,590	69.935	−37.60	3.165	10,298.00	−123,010.00
9,998	70.241	−46.98	2.513	12,966.00	−151,310.00
7,943	70.497	−58.62	1.993	17,188.00	−188,430.00
6,311	71.247	−73.37	1.589	22,531.00	−233,380.00
5,011	71.529	−91.65	1.259	29,987.00	−290,470.00
3,981	72.426	−114.79	1	39,851.00	−360,510.00
3,163	73.076	−144.36	0.7946	52,920.00	−446,770.00
2,511	74.071	−180.58	0.6316	70,772.00	−552,670.00
1,995	75.475	−225.85	0.5012	94,415.00	−684,260.00
1,584	77.683	−283.36	0.3979	126,930.00	−845,870.00
1,259	79.356	−357.71	0.3163	171,210.00	−1,041,500.00
1,000	82.133	−445.98	0.2512	229,810.00	−1,282,200.00
794.2	83.134	−557.77	0.1995	310,460.00	−1,573,800.00
631	87.033	−705.52	0.1585	419,490.00	−1,924,700.00
501.3	92.074	−881.84	0.1259	575,970.00	−2,402,700.00
398.2	98.7	−1,105.80	0.09999	777,810.00	−2,909,400.00
316.1	107.79	−1,387.70	0.07945	1,044,900.00	−3,501,200.00
251.2	119.65	−1,739.30	0.0631	1,397,900.00	−4,189,000.00
199.5	134.56	−2,180.60	0.05013	1,859,500.00	−4,976,900.00
158.5	154.53	−2,732.30	0.03981	2,457,700.00	−5,864,500.00
125.9	180.54	−3,424.50	0.03163	3,207,500.00	−6,830,800.00
100	214.47	−4,291.80	0.02512	4,126,800.00	−7,898,700.00
79.41	261.33	−5,378.40	0.01995	5,236,400.00	−8,998,800.00
63.1	325.65	−6,733.00	0.01585	6,577,800.00	−10,270,000.00
50.12	402.19	−8,436.50	0.01259	8,127,400.00	−11,372,000.00
38.46	516.16	−10,926.00	0.01	9,869,600.00	−12,527,000.00
31.25	655.56	−13,362.00			

Table 30.2: Impedance data obtained for a silver electrode coated by a Si_3N_4 layer after an elapsed time of six hours.

f, Hz	Z_r, Ωcm^2	Z_j, Ωcm^2	f, Hz	Z_r, Ωcm^2	Z_j, Ωcm^2
100,000.00	68.22	−5.19	24.99	841.11	−16,638.00
79,430.00	68.61	−6.48	20.00	1,045.70	−20,627.00
63,110.00	68.77	−8.14	15.63	1,439.00	−26,210.00
50,120.00	69.03	−10.25	12.50	1,828.30	−32,527.00
39,810.00	69.26	−12.74	10.00	2,347.60	−40,274.00
31,630.00	69.58	−15.71	7.94	3,329.40	−50,463.00
25,120.00	69.87	−19.64	6.33	4,068.60	−62,868.00
19,950.00	70.08	−24.31	5.00	5,639.00	−79,061.00
15,850.00	70.43	−30.20	3.97	7,483.70	−97,590.00
12,590.00	70.79	−37.67	3.17	10,563.00	−122,380.00
9,998.00	71.07	−47.00	2.51	13,498.00	−151,080.00
7,943.00	71.60	−58.62	1.99	17,883.00	−188,000.00
6,311.00	71.94	−73.46	1.59	23,630.00	−232,660.00
5,011.00	72.63	−91.78	1.26	31,565.00	−289,460.00
3,981.00	73.59	−115.00	1.00	42,083.00	−358,640.00
3,163.00	73.91	−144.13	0.79	56,490.00	−443,780.00
2,511.00	74.77	−179.96	0.63	74,799.00	−548,860.00
1,995.00	75.57	−226.30	0.50	99,962.00	−677,860.00
1,584.00	76.17	−284.32	0.40	133,960.00	−836,000.00
1,259.00	78.17	−355.74	0.32	178,760.00	−1,027,900.00
1,000.00	82.69	−446.87	0.25	238,700.00	−1,261,200.00
794.20	82.13	−560.56	0.20	320,080.00	−1,545,900.00
631.00	87.24	−702.76	0.16	436,530.00	−1,939,800.00
501.30	92.86	−882.56	0.13	583,210.00	−2,354,800.00
398.20	99.63	−1,106.70	0.10	778,100.00	−2,844,400.00
316.10	108.85	−1,388.90	0.08	1,035,500.00	−3,420,400.00
251.20	120.73	−1,740.90	0.06	1,374,700.00	−4,089,200.00
199.50	136.24	−2,182.00	0.05	1,815,000.00	−4,860,500.00
158.50	157.12	−2,735.40	0.04	2,372,400.00	−5,732,000.00
125.90	181.83	−3,427.40	0.03	3,085,500.00	−6,699,000.00
100.00	217.20	−4,294.30	0.03	3,955,800.00	−7,769,800.00
79.41	262.42	−5,383.90	0.02	5,022,500.00	−8,888,100.00
63.10	328.29	−6,735.10	0.02	6,290,500.00	−10,084,000.00
50.12	406.22	−8,443.80	0.01	7,777,200.00	−11,319,000.00
38.46	536.18	−10,925.00	0.01	9,521,800.00	−12,461,000.00
31.25	646.46	−13,376.00			

Table 30.3: Impedance data obtained for a silver electrode coated by a Si_3N_4 layer after an elapsed time of nine hours.

f, Hz	Z_r, Ωcm^2	Z_j, Ωcm^2	f, Hz	Z_r, Ωcm^2	Z_j, Ωcm^2
100,000.00	68.62	−5.25	24.99	846.29	−16,638.00
79,430.00	68.95	−6.56	20.00	1,119.30	−20,631.00
63,110.00	69.14	−8.18	15.63	1,345.30	−26,246.00
50,120.00	69.39	−10.20	12.50	1,879.00	−32,666.00
39,810.00	69.56	−12.56	10.00	2,408.70	−40,426.00
31,630.00	69.97	−15.74	7.94	3,164.20	−50,580.00
25,120.00	70.11	−19.57	6.33	4,204.90	−62,482.00
19,950.00	70.44	−24.16	5.00	5,680.20	−79,140.00
15,850.00	70.80	−30.15	3.97	8,230.00	−96,937.00
12,590.00	71.17	−37.63	3.17	11,023.00	−122,120.00
9,998.00	71.56	−46.84	2.51	13,705.00	−150,790.00
7,943.00	71.92	−58.60	1.99	18,119.00	−187,740.00
6,311.00	72.35	−73.27	1.59	24,079.00	−232,220.00
5,011.00	72.82	−91.86	1.26	32,116.00	−288,770.00
3,981.00	73.38	−115.03	1.00	42,948.00	−357,970.00
3,163.00	74.53	−144.12	0.79	57,347.00	−442,600.00
2,511.00	75.84	−180.23	0.63	76,731.00	−546,940.00
1,995.00	75.59	−225.76	0.50	102,050.00	−675,210.00
1,584.00	77.62	−285.45	0.40	136,420.00	−831,670.00
1,259.00	79.25	−358.19	0.32	181,150.00	−1,022,400.00
1,000.00	82.79	−445.76	0.25	243,010.00	−1,253,700.00
794.20	84.64	−557.56	0.20	324,490.00	−1,534,500.00
631.00	90.09	−701.27	0.16	433,250.00	−1,871,100.00
501.30	93.10	−882.38	0.13	587,170.00	−2,334,100.00
398.20	100.17	−1,106.40	0.10	782,970.00	−2,819,300.00
316.10	108.79	−1,388.80	0.08	1,039,100.00	−3,388,800.00
251.20	120.96	−1,740.90	0.06	1,377,600.00	−4,048,300.00
199.50	136.92	−2,182.90	0.05	1,810,400.00	−4,803,300.00
158.50	157.83	−2,734.70	0.04	2,374,800.00	−5,664,900.00
125.90	184.02	−3,426.90	0.03	3,082,500.00	−6,608,700.00
100.00	218.98	−4,294.10	0.03	3,944,900.00	−7,660,500.00
79.41	264.45	−5,381.50	0.02	5,024,500.00	−8,764,100.00
63.10	330.12	−6,737.10	0.02	6,289,500.00	−9,939,200.00
50.12	411.39	−8,438.80	0.01	7,753,500.00	−11,150,000.00
38.46	534.71	−10,930.00	0.01	9,449,400.00	−12,344,000.00
31.25	656.18	−13,368.00			

Table 30.4: Impedance data obtained for a silver electrode coated by a Si_3N_4 layer after an elapsed time of 12 hours.

f, Hz	Z_r, Ωcm^2	Z_j, Ωcm^2	f, Hz	Z_r, Ωcm^2	Z_j, Ωcm^2
100,000	68.26	−5.0655	24.99	818.34	−16,618
79,430	68.617	−6.5487	20	1102.5	−20,624
63,110	68.839	−8.152	15.63	1,519.1	−26,162
50,120	69.138	−10.161	12.5	1,799.9	−32,422
39,810	69.334	−12.574	9.999	2,422.7	−40,184
31,630	69.613	−15.686	7.937	3,447.3	−50,355
25,120	69.892	−19.45	6.331	4,201.7	−62,405
19,950	70.199	−24.255	4.997	5,481.9	−78,613
15,850	70.391	−30.126	3.967	8,176.1	−97,506
12,590	70.719	−37.627	3.165	10,526	−122,030
9,998	71.133	−46.974	2.513	13,806	−150,640
7,943	71.47	−58.573	1.993	18,283	−187,500
6,311	71.877	−73.225	1.589	24,242	−231,970
5,011	72.418	−91.62	1.259	32,554	−288,330
3,981	73.127	−114.81	1	43,372	−357,200
3,163	73.348	−143.86	0.7946	57,791	−441,930
2,511	74.767	−180.65	0.6316	77,485	−545,430
1,995	76.003	−226	0.5012	103,080	−673,170
1,584	76.183	−284.42	0.3979	138,170	−829,560
1,259	78.754	−355.77	0.3163	182,790	−1,018,300
1,000	82.124	−446.49	0.2512	245,050	−1,247,800
794.2	86.84	−564.33	0.1995	326,650	−1,526,600
631	85.706	−699.66	0.1585	435,310	−1,860,700
501.3	92.775	−881.97	0.1259	588,030	−2,318,500
398.2	99.63	−1,105.9	0.09999	782,720	−2,803,100
316.1	108.52	−1,388.3	0.07945	103,6200	−3,364,700
251.2	119.99	−1,739.8	0.0631	137,0800	−4,026,200
199.5	136.28	−2,181	0.05013	1,803,900	−4,772,400
158.5	156.41	−2,732.2	0.03981	2,356,700	−5,630,800
125.9	183.09	−3,425.8	0.03163	3,094,900	−6,584,000
100	218.38	−4,292.3	0.02512	3,934,100	−7,599,900
79.41	266.78	−5,381.6	0.01995	4,986,000	−8,723,200
63.1	328.74	−6,736.6	0.01585	6,240,700	−9,886,300
50.12	411.64	−8,430.6	0.01259	7,696,200	−11,095,000
38.46	526.34	−10,922	0.01	9,386,500	−12,296,000
31.25	687.24	−13,364			

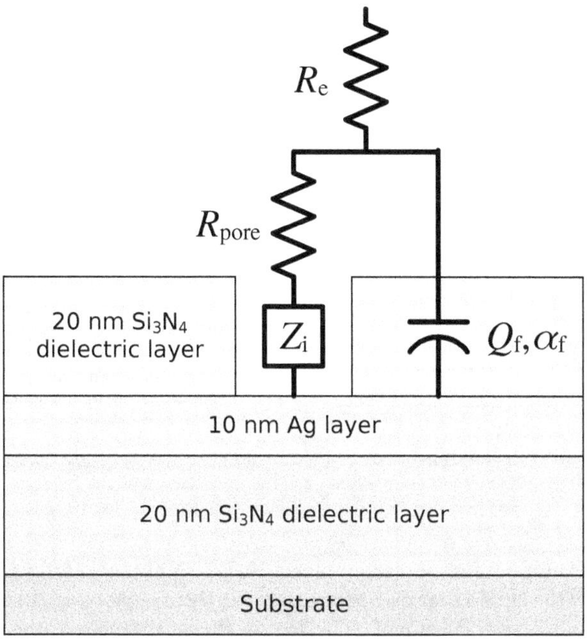

Figure 30.1: Schematic representation of the stack used as a working electrode and equivalent circuit for the interface under investigation. Taken from Tribollet and Orazem.[134]

$$\mathcal{C} = \frac{1}{j\omega(Z - R_e)} \tag{30.1}$$

is given in a complex representation in Figure 30.4. While Figure 30.2(b) shows variability, the high-frequency limit of the complex capacitance plots for the four measurements are in good agreement, yielding a value of 0.31 $\mu F/cm^2$. As shown by Tribollet and Orazem,[134] use of the ohmic resistance obtained by extrapolation of Figure 30.2(b) yields a poorly defined value of capacitance. With a permittivity of 7, the corresponding thickness is 19.99 nm, which is close to the value of 20 nm determined in the coating process. The capacitance shown in Figure 30.4 includes the contribution of the ohmic impedance. Thus, while the graphical method yields a satisfying result, the conclusive result requires regression of a process model.

30.2.2 Measurement Model Analysis

The principle of the measurement model is given in Chapter 25. As measurements were performed from high frequency to low, the highest frequency was removed from the data set analyzed. This step removes the confounding influence of the transient observed when changing from a steady value of potential to a sinusoidally varying potential. Impedance values in the range 50 ± 3 Hz and 100 ± 3 Hz were also removed

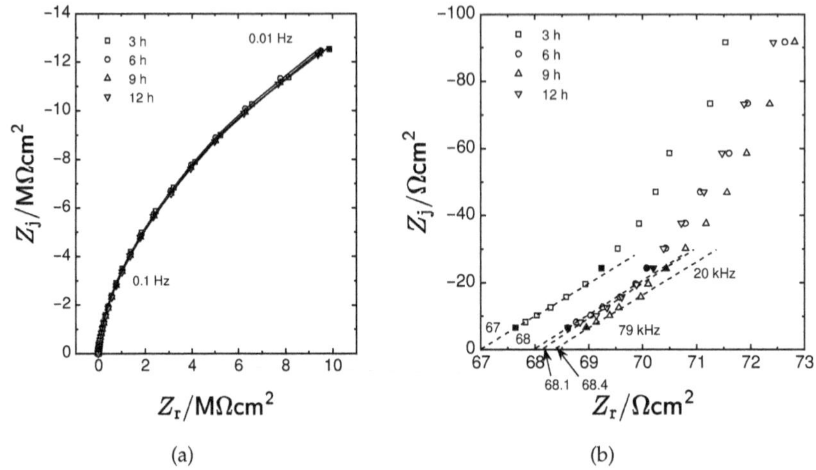

Figure 30.2: Impedance data presented in Tables 30.1–30.4 obtained for a silver mirror coated by a 20 nm Si_3N_4 layer with elapsed time as a parameter: (a) Nyquist plot and (b) zoomed representation, not in a 1 : 1 format, of the Nyquist plot. The lines in panel (a) represent the fit of the measurement model to the respective measurement. Taken from Tribollet and Orazem.[134]

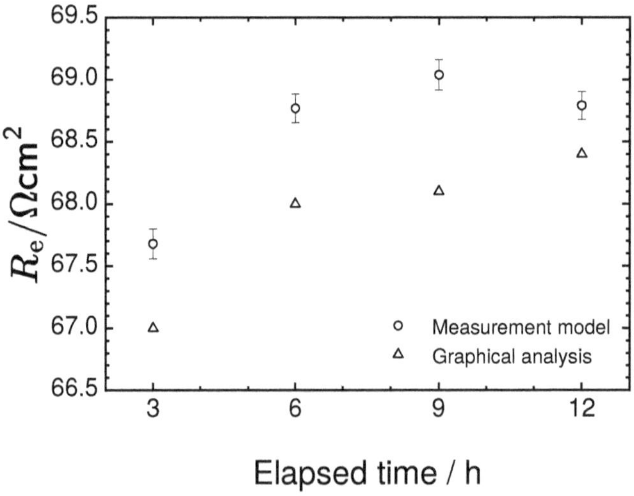

Figure 30.3: The ohmic resistance obtained by the use of the measurement model from the data given in Tables 30.1–30.4 and by extrapolation of Figure 30.2(b). Taken from Tribollet and Orazem.[134]

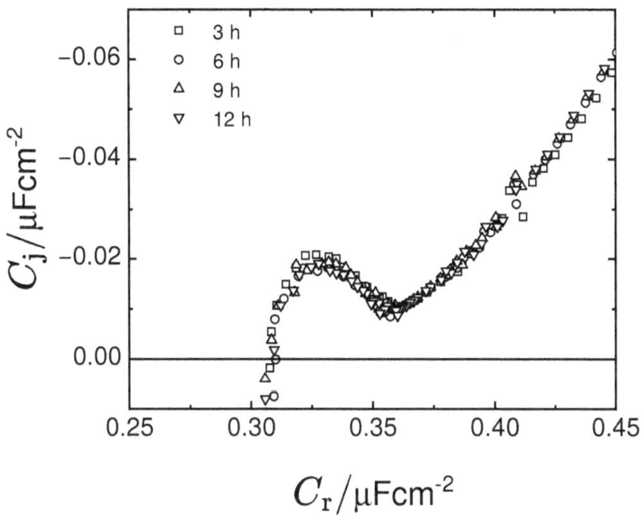

Figure 30.4: The imaginary part of the complex impedance as a function of the real part of the complex capacitance for the impedance data in Tables 30.1–30.4. Taken from Tribollet and Orazem.[134]

as the noise associated with the power line frequency and its first harmonic are inadequately filtered by impedance instrumentation. Initial regressions were performed under modulus weighting as is appropriate for impedance measurements conducted under potentiostatic modulation. Successful regressions were obtained with eight Voigt elements.

The regression of the measurement model to the data presented in Tables 30.1–30.4 yielded the fit presented in Figure 30.2(a). For the present study, four replicated measurements were obtained and the measurement model was used to extract a model for the stochastic error structure. The residual errors for the regression are shown in Figure 30.5 for the real part of the impedance and in Figure 30.6 for the imaginary part. As shown in Tables 30.1–30.4 and in Figure 30.2, the impedance varies from 65 Ωcm^2 to 12 $M\Omega cm^2$, six orders of magnitude. The residual errors vary accordingly from 0.2 Ωcm^2 to 0.8 $M\Omega cm^2$.

The standard deviation of the residual errors provides an estimate for the standard deviation of the stochastic contribution to measurement errors. As shown in Figure 30.7(a), the resulting estimate for the standard deviation is a strong function of frequency. The standard deviation for the real part of the impedance at a given frequency is seen to be equal to the standard deviation for the imaginary part of the impedance. As shown in Figure 30.7(b), the standard deviations are roughly proportional to the magnitude of the impedance and have a value of around 0.1% of the modulus at low frequency and is slightly smaller at high frequency. The lines shown in Figure 30.7

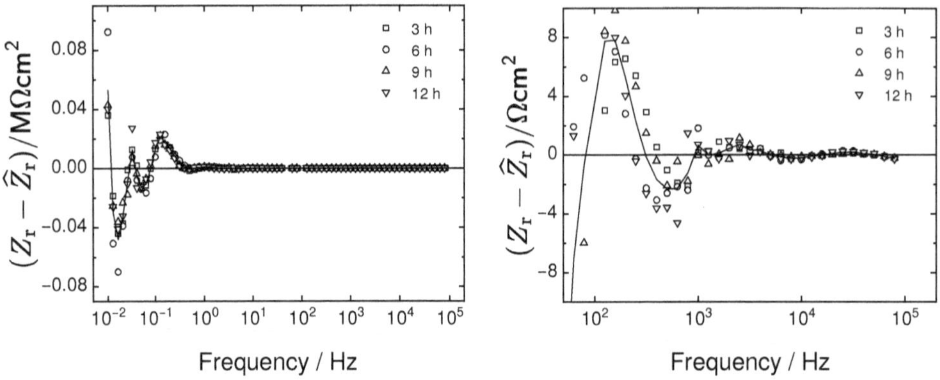

(a) (b)

Figure 30.5: Real residual errors as a function of frequency for complex regression of the measurement model to the data presented in Tables 30.1–30.4: (a) low-frequency range and (b) high-frequency range. The line represents the average residual error. Taken from Tribollet and Orazem.[134]

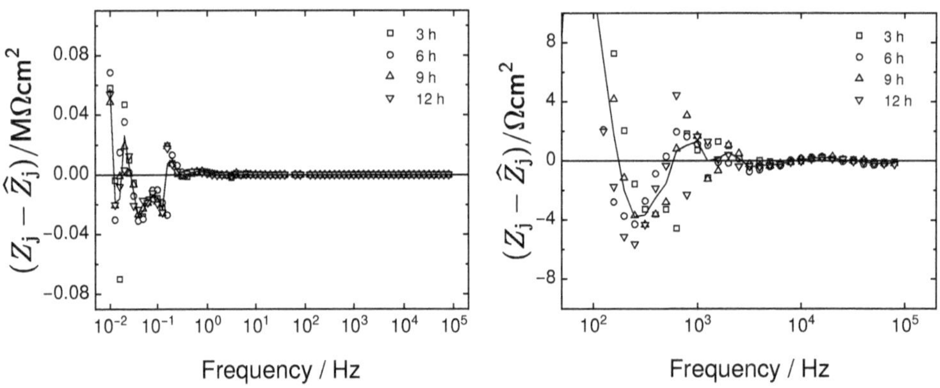

(a) (b)

Figure 30.6: Imaginary residual errors as a function of frequency for complex regression of the measurement model to the data presented in Tables 30.1–30.4: (a) low-frequency range and (b) high-frequency range. The line represents the average residual error. Taken from Tribollet and Orazem.[134]

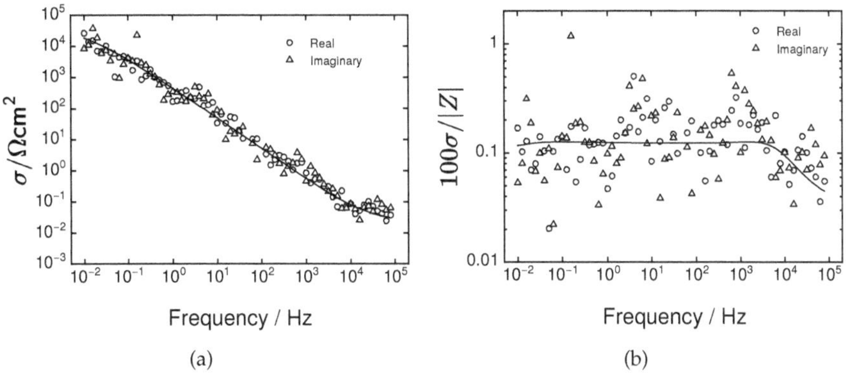

Figure 30.7: Standard deviations for the stochastic error structure of the data presented in Tables 30.1–30.4 as a function of frequency: (a) in units of impedance and (b) scaled by the magnitude of the impedance. The line represents a model for the error structure. Taken from Tribollet and Orazem.[134]

represents the fit of the estimated standard deviations given as

$$\sigma = 0.001216472|Z_j| + 0.0003333912|Z_r| \tag{30.2}$$

Equation (30.2) was used for subsequent regressions of the measurement model and the process models discussed in later section.

The fit of the measurement model under error structure weighting yielded 10 Voigt elements. Normalized residual errors are presented in Figure 30.8. For data showing only the high-frequency limit of a complete semicircle, as shown in Figure 30.2(a), a complex fit of the measurement model provides the most sensitive assessment of the consistency with the Kramers–Kronig relations. The normalized residual errors shown in Figure 30.8 lie well within the $\pm 2\sigma$ confidence interval for the model, and, therefore, the data are consistent with the Kramers–Kronig relations. More than 5% of the normalized residual errors fall outside the $\pm 2\sigma$ noise level of the measurement, a result that is consistent with the regression statistic $\chi^2/\nu = 6.7$.

The ohmic resistance was 68.459 ± 0.071 Ωcm^2, and the capacitance at infinite frequency was 0.300 ± 0.016 μF/cm^2. These values are in good agreement with the values obtained with the graphical analysis. The characteristic frequency

$$f_c = \frac{1}{2\pi R_e C} \tag{30.3}$$

yielded a value of $f_c = 7.75$ kHz. This value suggests that the ohmic impedance influences the impedance at frequencies larger than $f_c = 7.746 \pm 0.068$ kHz.

The impedance data are presented in Bode format in Figure 30.9. A high-frequency plateau, observed in the modulus shown in Figure 30.9(a), corresponds to the ohmic resistance. The ohmic-resistance-corrected modulus appears as a straight line. The phase angle shown in Figure 30.9(b) approaches a high-frequency limit of zero; whereas, the

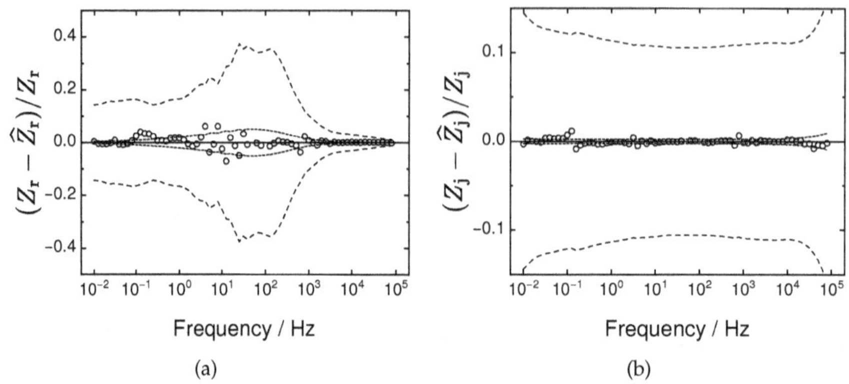

Figure 30.8: Normalized residual errors for the measurement model regression of the data presented in Tables 30.1–30.4 under error-structure weighting as a function of frequency: (a) real part and (b) imaginary part. The upper and lower dashed lines represent the 95.4% ($\pm 2\sigma$) confidence interval for the regression, and the dotted lines represent the $\pm 2\sigma$ noise level of the measurement. Taken from Tribollet and Orazem.[134]

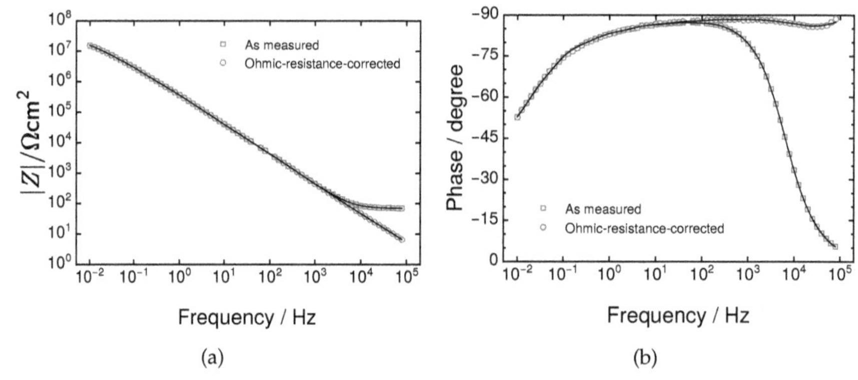

Figure 30.9: Impedance data presented in Table 30.4 obtained for a silver mirror coated by a 20 nm Si_3N_4 layer: (a) modulus and ohmic-resistance-corrected modulus as functions of frequency and (b) phase angle and ohmic-corrected phase angle as functions of frequency. Lines represent the fit of the measurement model under error-structure weighting with 10 Voigt elements. Taken from Tribollet and Orazem.[134]

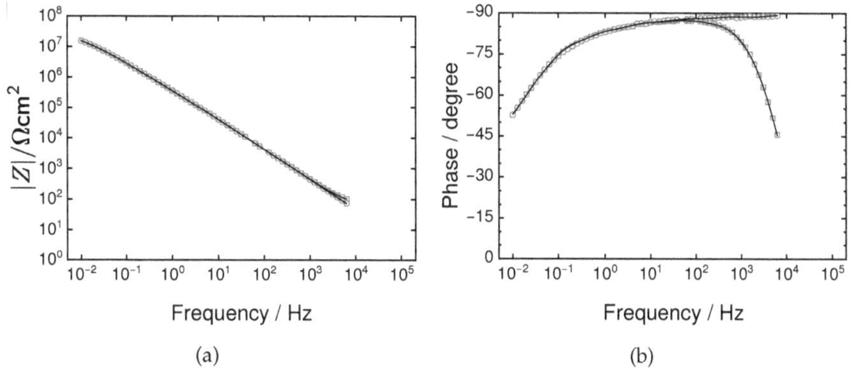

Figure 30.10: Impedance data presented in Table 30.4 obtained for a silver mirror coated by a 20 nm Si_3N_4 layer: (a) modulus and ohmic-resistance-corrected modulus as functions of frequency and (b) phase angle and ohmic-corrected phase angle as functions of frequency. The frequency range was truncated to eliminate frequencies above $f_c = 7.75$ kHz. Lines represent the fit of the measurement model under error-structure weighting with eight Voigt elements. Taken from Tribollet and Orazem.[134]

ohmic-resistance-corrected phase angle approaches a value close to $-90°$. In very high frequency, above 1 kHz, a small decrease follow by an increase is visible on the curve, this behavior is typical of the effect of the ohmic impedance.

To eliminate the influence of the ohmic impedance, the impedance data were truncated by removing frequencies larger than $f_c = 7.75$ kHz. The fit of the measurement model under error structure weighting yielded eight Voigt elements. The corresponding Bode plots are shown in Figure 30.10. The ohmic-resistance-corrected phase angle approaches a value of $-89°$, and the small decrease and increase visible in Figure 30.9(b) has been removed.

The ohmic resistance was 70.798 ± 0.029 Ωcm^2, and the capacitance at infinite frequency was 0.345 ± 0.018 $\mu F/cm^2$. This ohmic resistance is about 3.3% larger than the ohmic resistance obtained with the complete spectrum. It may be considered the low-frequency ohmic resistance that is unaffected by the nonuniform current and potential distributions. The characteristic frequency was $f_c = 6.515 \pm 0.054$ kHz, a value larger than 6.31 kHz, the maximum frequency analyzed.

30.3 Interpretation Model Development

After 12 hours of immersion, optical micrographs of the coated electrode revealed a small region of delamination, as shown in Figure 30.11. As a first approach, the equivalent circuit of a coated electrode, shown in Figure 30.1, was used. The interfacial impedance was represented by a double-layer capacitance in parallel to a charge-transfer resistance. The double-layer capacitance was treated as a constant-phase element to

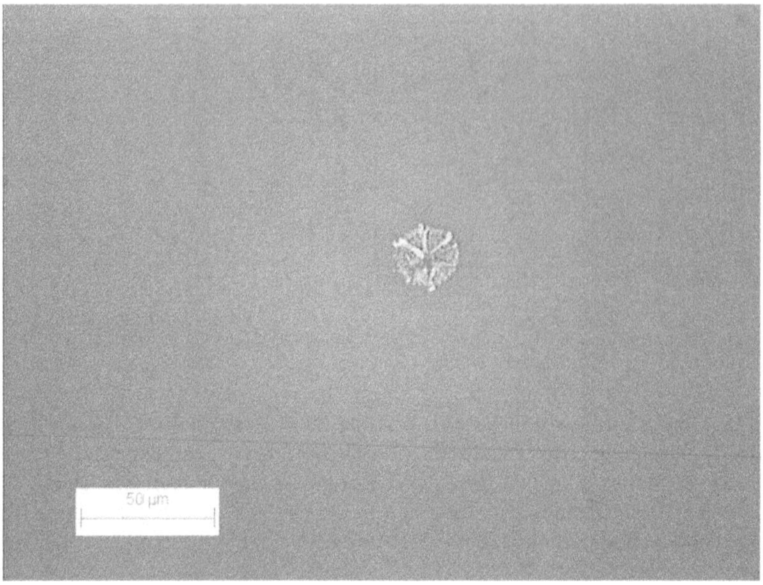

Figure 30.11: Optical characterization of the electrode surface immersed for 12 hours in 0.5 M Na_2SO_4 at pH 10, showing a region of delamination under a pit. Image adapted from Figure 4(c) of Ben Amor et al.[135]

account for frequency dispersion. Thus,

$$Z_{\text{pore}} = R_{\text{pore}} + \frac{R_t}{1 + (j\omega)_p^{\alpha} R_t Q_p} \tag{30.4}$$

The interfacial impedance was in parallel with a CPE used to account for the dielectric response of the Si_3N_4 layer, given as

$$Z_{\text{layer}} = \frac{1}{(j\omega)^{\alpha_f} Q_f} \tag{30.5}$$

The overall impedance was then given by

$$Z = R_e + \frac{Z_{\text{pore}} Z_{\text{layer}}}{Z_{\text{pore}} + Z_{\text{layer}}} \tag{30.6}$$

where R_e represents the ohmic resistance outside the layered structure, and R_{pore} represents the ohmic resistance associated with the pore.

This model was regressed to the impedance data given in Table 30.4 under error-structure weighting. The maximum frequency was 6.31 kHz to avoid the contribution of the ohmic impedance. The results are shown in Table 30.5. The regression provides an excellent fit to the data with $\chi^2/\nu = 28$, close to the value of 6.7 obtained by the measurement model analysis. The parameter R_{pore}, however, has a confidence interval that includes zero and is therefore not statistically significant. This result does not mean

Table 30.5: Results for regression of equations (30.4), (30.5), and (30.6) to the impedance data given in Table 30.4 under error-structure weighting. The maximum frequency was 6.31 kHz to avoid the contribution of the ohmic impedance. The 95.4% confidence interval for the regressed parameters is given as $200\sigma_{x_k}/x_k$ where x_k is the estimated value for parameter k and σ_k is the standard deviation obtained by regression for parameter k.

Parameter, k	Units	Value, x_k	$200\sigma_{x_k}/x_k$, %
R_t	$M\Omega cm^2$	62.7 ± 2.2	7.1%
Q_p	$\mu F/cm^2 s^{1-\alpha}$	0.1346 ± 0.0025	3.8%
α_p	–	0.5131 ± 0.0088	3.4%
Q_f	$\mu F/cm^2 s^{1-\alpha}$	0.4134 ± 0.0077	3.7%
α_f	–	0.9828 ± 0.0019	0.4%
R_e	Ωcm^2	69.76 ± 0.35	1.0%
R_{pore}	$M\Omega cm^2$	0.18 ± 0.26	285%
χ^2/ν	–	28	

Table 30.6: Results for regression of equations (30.4), (30.5), and (30.6) to the impedance data given in Table 30.4 under error-structure weighting. The maximum frequency was 6.31 kHz to avoid the contribution of the ohmic impedance. The parameter R_{pore} was removed from the model used to generate the results in Table 30.5.

Parameter, k	Units	Value, x_k	$200\sigma_{x_k}/x_k$, %
R_t	$M\Omega cm^2$	62.7 ± 2.4	7.8%
Q_p	$\mu F/cm^2 s^{1-\alpha}$	0.1360 ± 0.0022	3.2%
α_p	–	0.5166 ± 0.0084	3.2%
Q_f	$\mu F/cm^2 s^{1-\alpha}$	0.4079 ± 0.0024	1.2%
α_f	–	0.98386 ± 0.00066	0.1%
R_e	Ωcm^2	69.78 ± 0.25	0.7%
χ^2/ν	–	30	

that the model with R_{pore} is incorrect, it means only that the parameter could not be extracted from the impedance data.

The results with R_{pore} removed from the model are presented in Table 30.6. The CPE parameters corresponding to the impedance of the layer has an α_f exponent of 0.984. As this value is close to unity, the Si_3N_4 layer has a behavior close to a pure dielectric. The power-law model may be used to link the CPE parameters to the properties of the film, i.e.,

$$Q_f = \frac{(\varepsilon\varepsilon_0)^{\alpha_f}}{g\delta\rho_{\delta_f}^{1-\alpha_f}} \tag{30.7}$$

where

$$g = 1 + 2.88(1-\alpha_f)^{2.375} \tag{30.8}$$

Under the assumptions that the film thickness is 20 nm and the dielectric constant is $\varepsilon = 7$, equation (30.7) yields a value $\rho_{\delta_f} = 6.4 \times 10^4\ \Omega cm$.

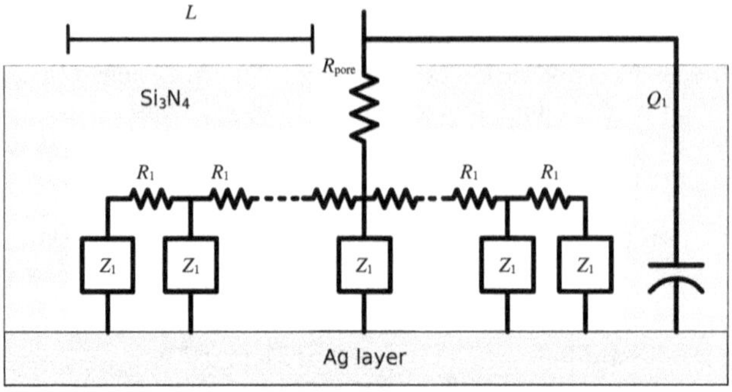

Figure 30.12: Schematic representation of a porous electrode in the $Si_3N_4|Ag|Si_3N_4$ stack. Taken from Tribollet and Orazem.[134]

Examination of the parameter values shows that the 95.4% confidence interval for α_p is 0.4998 to 0.5333. This interval includes $\alpha_p = 0.5$, suggesting that the process may be influenced by mass transfer or by a porous electrode. As the electrochemical reactions observed by EIS on thick silver films were not influenced by mass transfer,[137] the same reactions may be assumed in the present system to be independent of mass transfer. The impedance response was therefore assumed to be associated with a porous electrode.

Following de Levie[50] and Gabrielli et al.,[138] the model may be presented as shown in Figure 30.12. The pore impedance may be given as

$$Z_{pore} = \frac{1}{n_1}\sqrt{R_1 Z_1}\coth\left(L\sqrt{\frac{R_1}{Z_1}}\right) \tag{30.9}$$

where R_1 (Ω/cm) is the resistivity of the electrolyte inside a pore of unit length, n_1 is the number of crossing-pores per unit surface (cm^{-2}), L (cm) is the pore length, and Z_1 (Ωcm) is the interfacial impedance for a unit pore length. The interfacial impedance was assumed to follow a CPE corresponding to a double-layer capacitance linked in parallel with a charge-transfer resistance, i.e.,

$$Z_1 = \frac{R_t}{1 + (j\omega)^{\alpha_p} R_t Q_p} \tag{30.10}$$

As the number of pores n_1 is unknown, equation (30.9) is written in terms of lumped parameters as

$$Z_{pore} = A_1\sqrt{\frac{1}{1 + (j\omega)^{\alpha_p} A_2}}\coth\left(A_3\sqrt{1 + (j\omega)^{\alpha_p} A_2}\right) \tag{30.11}$$

where

$$A_1 = \sqrt{\frac{R_1 R_t}{n_i^2}} \tag{30.12}$$

Table 30.7: Results for regression of equations (30.11) in parallel with the impedance given by equation (30.5), resulting in the overall impedance expressed in equation (30.6), to the impedance data given in Table 30.4 under error-structure weighting. The maximum frequency was 6.31 kHz to avoid the contribution of the ohmic impedance.

Parameter, k	Units	Value, x_k	$200\sigma_{x_k}/x_k$, %
A_1	$M\Omega cm^2$	35.1 ± 1.6	9.2%
A_2	s^{α_p}	24.3 ± 2.0	16.8%
A_3	–	1.311 ± 0.036	5.5%
α_p	–	0.908 ± 0.012	2.6%
Q_f	$\mu F/cm^2 s^{1-\alpha}$	0.4161 ± 0.0018	0.9%
α_t	–	0.98183 ± 0.00054	0.1%
R_e	Ωcm^2	69.50 ± 0.26	0.8%
χ^2/ν	–	34	

Table 30.8: Summary of regression results for the impedance data given in Table 30.4 under error-structure weighting

	χ^2/ν	# parameters
Measurement model	7.4	17
R-CPE	29.6	6
Porous electrode	34.6	7

$$A_2 = R_t Q_p \tag{30.13}$$

and

$$A_3 = \sqrt{\frac{R_1 L^2}{R_t}} \tag{30.14}$$

The impedance given by equation (30.11) was in parallel with the impedance given by equation (30.5), resulting in the overall impedance expressed as equation (30.6).

The results are presented in Table 30.7. The exponent α_p is increased to 0.908. Equation (30.7) yields a value $\rho_{\delta_f} = 1.4 \times 10^5$ Ωcm. The χ^2/ν statistic increased to 34. The number of regressed parameters was 7; whereas, the model shown in Table 30.6 has six parameters.

The fitting of the measurement model, the R-CPE model, and the porous electrode model are presented in Figure 30.13 in Nyquist format. Lines representing the fits of the measurement model, the R-CPE model, and the porous electrode model are indistinguishable. A similar comparison is made in Figure 30.14 in Bode format. Again, lines representing the fits of the measurement model, the R-CPE model, and the porous electrode model are indistinguishable. All three models yield a good fit to the data at the scale shown in Figures 30.13 and 30.14. The differences in the models are shown in Table 30.8. The measurement model yielded the best fit, with $\chi^2/\nu = 6.6$; whereas, the interpretation models had χ^2/ν on the order of 30.

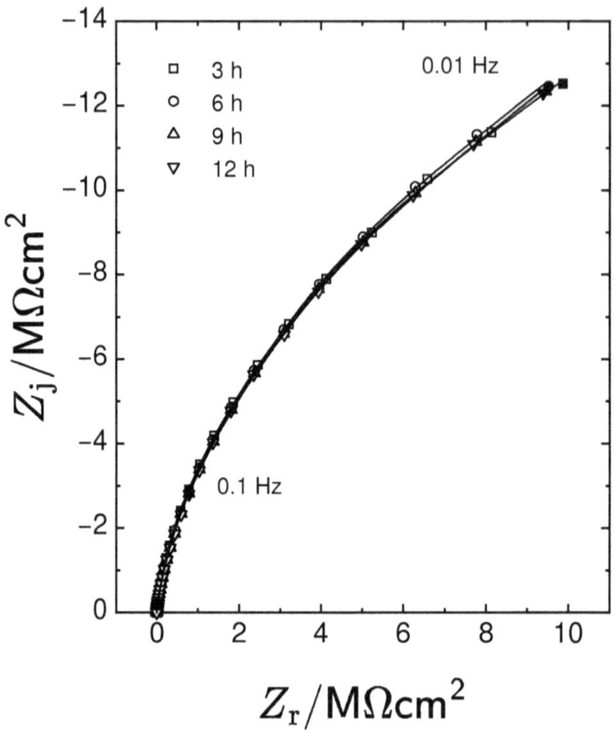

Figure 30.13: Regression results in Nyquist format for the impedance data given in Table 30.4 under error-structure weighting. The maximum frequency was 6.31 kHz to avoid the contribution of the ohmic impedance. Lines representing the fits of the measurement model, the R-CPE model, and the porous electrode model are superposed. Taken from Tribollet and Orazem.[134]

30.4 Observations

The measurement model provided estimates for the ohmic resistance and capacitance that could be used to quantify the frequency range influenced by the ohmic impedance. It allowed quantification of the stochastic error structure and showed that the data were consistent with the Kramers–Kronig relations.

The R-CPE model yielded both an excellent fit and a CPE exponent with a value close to 0.5. But, in this case, the CPE masks the physical meaning of the corresponding impedance. As normal and surface distributions of time constants generally result in CPE exponents between 0.75 and unity, a model yielding $\alpha = 0.5$ was considered incomplete.

The porous-electrode model provided a fit that was of about the same quality as the R-CPE model, and the CPE exponent was on the order of 0.9, providing a coherent model for the dissolution process.

Each of these models provided valuable insight into the nature of the system under study. It is worth remembering that the objective of interpretation modeling is not to

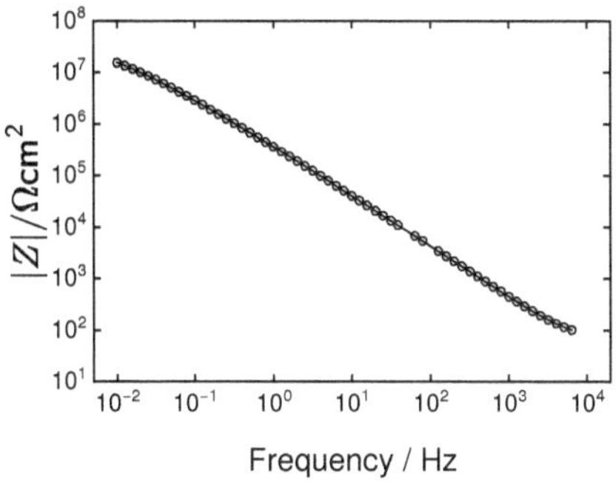

(a)

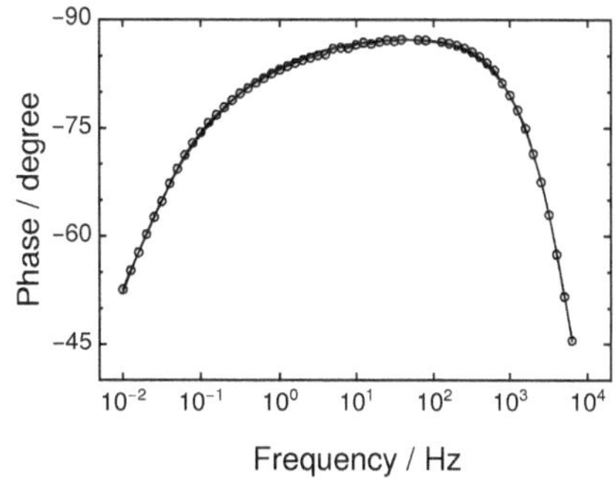

(b)

Figure 30.14: Regression results in Bode format for the impedance data given in Table 30.4 under error-structure weighting: (a) magnitude and (b) phase. The maximum frequency was 6.31 kHz to avoid the contribution of the ohmic impedance. Lines representing the fits of the measurement model, the CPE model, and the porous electrode model are superposed. Taken from Tribollet and Orazem.[134]

provide a good fit with the smallest number of parameters, but, rather, to use the model to learn about the physics and chemistry of the system under study.

Part III

Reference Material

Appendix A

Tables of Reference Material

The following tables provide equations, constants, physical properties, and statistical values that may be useful for the student.

A.1 Complex Variables

A working understanding of complex variables is essential for the analysis of experiments conducted in the frequency domain, such as impedance spectroscopy. The objective of this section is to provide identities useful for working with complex variables. See, for example, Churchill and Brown[139] and Cain.[140] Tables A.1–A.7 were taken from Chapter 1 of Orazem and Tribollet.[1]

Table A.1: Identities for the imaginary number j.

$$j = \sqrt{-1} \tag{A.1}$$

$$j^2 = -1 \tag{A.2}$$

$$j^3 = -j \tag{A.3}$$

$$j^4 = 1 \tag{A.4}$$

$$1/j = -j \tag{A.5}$$

Table A.2: Relationships for complex variables $z = z_\mathrm{r} + \mathrm{j}z_\mathrm{j}$ and $w = w_\mathrm{r} + \mathrm{j}w_\mathrm{j}$.

$$z + w = w + z \tag{A.6}$$

$$zw = wz \tag{A.7}$$

$$a(z + w) = az + aw \tag{A.8}$$

$$a(zw) = (az)w \tag{A.9}$$

$$z + w = (z_\mathrm{r} + w_\mathrm{r}) + \mathrm{j}(z_\mathrm{j} + w_\mathrm{j}) \tag{A.10}$$

$$z - w = (z_\mathrm{r} - w_\mathrm{r}) + \mathrm{j}(z_\mathrm{j} - w_\mathrm{j}) \tag{A.11}$$

$$zw = (z_\mathrm{r}w_\mathrm{r} - z_\mathrm{j}w_\mathrm{j}) + \mathrm{j}(z_\mathrm{r}w_\mathrm{j} + z_\mathrm{j}w_\mathrm{r}) \tag{A.12}$$

$$w\overline{w} = w_\mathrm{r}^2 + w_\mathrm{j}^2 \tag{A.13}$$

$$\begin{aligned} \frac{z}{w} &= \frac{z\overline{w}}{w\overline{w}} \\ &= \frac{(z_\mathrm{r}w_\mathrm{r} + z_\mathrm{j}w_\mathrm{j}) + \mathrm{j}(z_\mathrm{j}w_\mathrm{r} - z_\mathrm{r}w_\mathrm{j})}{w_\mathrm{r}^2 + w_\mathrm{j}^2} \end{aligned} \tag{A.14}$$

Table A.3: Relationships between polar and rectangular coordinates for the complex variable $z = z_\mathrm{r} + \mathrm{j}z_\mathrm{j}$.

$$z_\mathrm{r} = r\cos(\varphi) \tag{A.15}$$

$$z_\mathrm{j} = r\sin(\varphi) \tag{A.16}$$

$$r = |z| = \sqrt{z_\mathrm{r}^2 + z_\mathrm{j}^2} \tag{A.17}$$

$$\varphi = \tan^{-1}\left(\frac{z_\mathrm{j}}{z_\mathrm{r}}\right) \tag{A.18}$$

$$|z| = \sqrt{z\overline{z}} \tag{A.19}$$

$$z = r\left(\cos(\varphi) + \mathrm{j}\sin(\varphi)\right) \tag{A.20}$$

$$z^n = r^n\left(\cos(n\varphi) + \mathrm{j}\sin(n\varphi)\right) \tag{A.21}$$

$$z^{1/n} = r^{1/n}\left[\cos\left(\frac{\varphi}{n} + \frac{2\pi k}{n}\right) + \mathrm{j}\sin\left(\frac{\varphi}{n} + \frac{2\pi k}{n}\right)\right]; \quad k = 0, 1, \ldots, n-1 \tag{A.22}$$

Table A.4: Properties for the complex conjugates of $z = z_r + jz_j$ and $w = w_r + jw_j$.

$$\overline{z+w} = \overline{z} + \overline{w} \tag{A.23}$$

$$\overline{zw} = \overline{z}\overline{w} \tag{A.24}$$

$$\overline{\left[\frac{1}{z}\right]} = \frac{1}{\overline{z}} \tag{A.25}$$

$$\overline{\overline{z}} = z \tag{A.26}$$

Table A.5: Properties for the absolute value of $z = z_r + jz_j$ and $w = w_r + jw_j$.

$$|z| = |\overline{z}| \tag{A.27}$$

$$z\overline{z} = |z|^2 \tag{A.28}$$

$$z_r \leq |z| \tag{A.29}$$

$$z_j \leq |z| \tag{A.30}$$

$$|zw| = |z||w| \tag{A.31}$$

$$\left|\frac{1}{z}\right| = \frac{1}{|z|}; \quad z \neq 0 \tag{A.32}$$

$$|z+w| \leq |z| + |w| \tag{A.33}$$

Table A.6: Trigonometric and hyperbolic relationships for the complex variable $z = z_r + jz_j$.

$$\sin(z) = \left(e^{jz} - e^{-jz}\right)/2j \tag{A.34}$$

$$\cos(z) = \left(e^{jz} + e^{-jz}\right)/2 \tag{A.35}$$

$$\frac{d\sin(z)}{dz} = \cos(z) \tag{A.36}$$

$$\frac{d\cos(z)}{dz} = -\sin(z) \tag{A.37}$$

$$\tan(z) = \frac{\sin(z)}{\cos(z)} \tag{A.38}$$

$$\cot(z) = \frac{\cos(z)}{\sin(z)} \tag{A.39}$$

$$\cos^2 z + \sin^2 z = 1 \tag{A.40}$$

$$\cos(z_1 \pm z_2) = \cos(z_1)\cos(z_2) \mp \sin(z_1)\sin(z_2) \tag{A.41}$$

$$\sin(z_1 \pm z_2) = \sin(z_1)\cos(z_2) \pm \cos(z_1)\sin(z_2) \tag{A.42}$$

$$\sinh(z) = \left(e^{z} - e^{-z}\right)/2 \tag{A.43}$$

$$\cosh(z) = \left(e^{z} + e^{-z}\right)/2 \tag{A.44}$$

$$\tanh(z) = \frac{\sinh(z)}{\cosh(z)} \tag{A.45}$$

$$\coth(z) = \frac{\cosh(z)}{\sinh(z)} \tag{A.46}$$

$$\cos(z) = \cos(x)\cosh(y) - j\sin(x)\sinh(y) \tag{A.47}$$

$$\sin(z) = \sin(x)\cosh(y) + j\cos(x)\sinh(y) \tag{A.48}$$

$$\cosh(z) = \cosh(x)\cos(y) + j\sinh(x)\sin(y) \tag{A.49}$$

$$\sinh(z) = \sinh(x)\cos(y) + j\cosh(x)\sin(y) \tag{A.50}$$

Table A.7: Functional relationships of complex variables commonly encountered in impedance spectroscopy, where x and y are real numbers, and $z = x + jy$.

$$\exp(jx) = \cos(x) + j\sin(x) \tag{A.51}$$

$$\exp(j(x+y)) = \exp(jx)\cdot\exp(jy) \tag{A.52}$$

$$\cos(\omega t + \varphi) = \mathrm{Re}\left\{\exp(j(\omega t + \varphi))\right\} \tag{A.53}$$

$$= \mathrm{Re}\left\{\exp(j\varphi)\cdot\exp(j\omega t)\right\} \tag{A.54}$$

$$\mathrm{Re}\{\ln(z)\} = \ln|z| \tag{A.55}$$

$$\mathrm{Im}\{\ln(z)\} = \arg(z) \tag{A.56}$$

A.2 Statistics

A working understanding of statistics is essential for the analysis of experiments conducted in the frequency domain, such as impedance spectroscopy. Some basic statistical concepts are defined in this section. For a more comprehensive treatment, the reader is directed to standard statistical texts.[141–143] Tables A.8–A.12 were taken from Chapter 3 of Orazem and Tribollet.[1]

Table A.8: Properties of the expectation, where c is a constant and x and y are random variates.

$$E\{c\} = c \tag{A.57}$$

$$E\{x + c\} = E\{x\} + c \tag{A.58}$$

$$E\{cx\} = cE\{x\} \tag{A.59}$$

$$E\{x + y\} = E\{x\} + E\{y\} \tag{A.60}$$

$$E\{(x + y)^2\} = E\{x\}^2 + E\{y\}^2 + 2E\{x\}E\{y\} \tag{A.61}$$

Table A.9: Properties of the variance, where c is a constant and x and y are random variates.

$$\sigma^2(c) = 0 \tag{A.62}$$

$$\sigma^2(x + c) = \sigma_x^2 \tag{A.63}$$

$$\sigma^2(cx) = c^2\sigma_x^2 \tag{A.64}$$

$$\sigma^2(x + y) = \sigma_x^2 + \sigma_y^2 + \sigma_{x,y} \tag{A.65}$$

Table A.10: Student's t-test values. Taken from Chapter 3 in Orazem and Tribollet.[1]

ν \ p	0.6	0.3173	0.10	0.05	0.0455	0.01
3	0.58439	1.1969	2.3534	3.1824	3.3068	5.8409
4	0.56865	1.1417	2.1318	2.7764	2.8693	4.6041
5	0.55943	1.1105	2.0150	2.5706	2.6487	4.0321
10	0.54153	1.0526	1.8125	2.2281	2.2837	3.1693
15	0.53573	1.0345	1.7531	2.1314	2.1812	2.9467
20	0.53286	1.0257	1.7247	2.0860	2.1330	2.8453
25	0.53115	1.0204	1.7081	2.0595	2.1051	2.7874
30	0.53002	1.0170	1.6973	2.0423	2.0868	2.7500
35	0.52921	1.0145	1.6896	2.0301	2.0740	2.7238
40	0.52861	1.0127	1.6839	2.0211	2.0645	2.7045
45	0.52814	1.0113	1.6794	2.0141	2.0571	2.6896
50	0.52776	1.0101	1.6759	2.0086	2.0513	2.6778
55	0.52745	1.0092	1.6730	2.0040	2.0465	2.6682
60	0.52720	1.0084	1.6706	2.0003	2.0425	2.6603
65	0.52698	1.0078	1.6686	1.9971	2.0392	2.6536
70	0.52680	1.0072	1.6669	1.9944	2.0364	2.6479
75	0.52664	1.0067	1.6654	1.9921	2.0339	2.6430
80	0.52650	1.0063	1.6641	1.9901	2.0317	2.6387
85	0.52637	1.0059	1.6630	1.9883	2.0298	2.6349
90	0.52626	1.0056	1.6620	1.9867	2.0282	2.6316
95	0.52616	1.0053	1.6611	1.9853	2.0267	2.6286
100	0.52608	1.0050	1.6602	1.9840	2.0253	2.6259
200	0.52524	1.0025	1.6525	1.9719	2.0126	2.6006
1,000	0.52457	1.0005	1.6464	1.9623	2.0025	2.5808
5,000	0.52443	1.0001	1.6452	1.9604	2.0005	2.5768
10,000	0.52442	1.0001	1.6450	1.9602	2.0003	2.5763
100,000	0.52440	1.0000	1.6449	1.9600	2.0000	2.5759

Table A.11: F-test values for comparison of variance of samples with equal degrees of freedom, i.e., $\nu_1 = \nu_2 = \nu$. Taken from Chapter 3 in Orazem and Tribollet.[1]

ν \ p	0.3	0.2	0.1	0.05	0.025	0.01
3	1.940	2.936	5.391	9.277	15.439	29.457
4	1.753	2.483	4.107	6.388	9.605	15.977
5	1.641	2.228	3.453	5.050	7.146	10.967
10	1.406	1.732	2.323	2.978	3.717	4.849
15	1.318	1.558	1.972	2.403	2.862	3.522
20	1.268	1.466	1.794	2.124	2.464	2.938
25	1.236	1.406	1.683	1.955	2.230	2.604
30	1.213	1.364	1.606	1.841	2.074	2.386
35	1.196	1.332	1.550	1.757	1.961	2.231
40	1.182	1.308	1.506	1.693	1.875	2.114
45	1.170	1.287	1.470	1.642	1.807	2.023
50	1.161	1.271	1.441	1.599	1.752	1.949
55	1.153	1.256	1.416	1.564	1.706	1.888
60	1.146	1.244	1.395	1.534	1.667	1.836
65	1.140	1.233	1.377	1.508	1.633	1.792
70	1.134	1.224	1.361	1.486	1.604	1.754
75	1.129	1.216	1.346	1.466	1.578	1.720
80	1.125	1.208	1.334	1.448	1.555	1.690
85	1.121	1.201	1.322	1.432	1.534	1.663
90	1.117	1.195	1.312	1.417	1.516	1.639
95	1.114	1.189	1.302	1.404	1.499	1.618
100	1.111	1.184	1.293	1.392	1.483	1.598
200	1.077	1.127	1.199	1.263	1.320	1.391
1,000	1.034	1.055	1.084	1.110	1.132	1.159
5,000	1.015	1.024	1.037	1.048	1.057	1.068
10,000	1.011	1.017	1.026	1.033	1.040	1.048
100,000	1.003	1.005	1.008	1.010	1.012	1.015

Table A.12: χ^2-test values for degree of freedom ν and confidence level p. Taken from Chapter 3 in Orazem and Tribollet.[1]

ν \ p	0.5	0.3	0.2	0.1	0.05	0.01
5	4.35	6.06	7.29	9.24	11.07	15.09
10	9.34	11.78	13.44	15.99	18.31	23.21
15	14.34	17.32	19.31	22.31	25.00	30.58
20	19.34	22.77	25.04	28.41	31.41	37.57
25	24.34	28.17	30.68	34.38	37.65	44.31
30	29.34	33.53	36.25	40.26	43.77	50.89
35	34.34	38.86	41.78	46.06	49.80	57.34
40	39.34	44.16	47.27	51.81	55.76	63.69
45	44.34	49.45	52.73	57.51	61.66	69.96
50	49.33	54.72	58.16	63.17	67.50	76.15
55	54.33	59.98	63.58	68.80	73.31	82.29
60	59.33	65.23	68.97	74.40	79.08	88.38
65	64.33	70.46	74.35	79.97	84.82	94.42
70	69.33	75.69	79.71	85.53	90.53	100.43
75	74.33	80.91	85.07	91.06	96.22	106.39
80	79.33	86.12	90.41	96.58	101.88	112.33
85	84.33	91.32	95.73	102.08	107.52	118.24
90	89.33	96.52	101.05	107.57	113.15	124.12
95	94.33	101.72	106.36	113.04	118.75	129.97
100	99.33	106.91	111.67	118.50	124.34	135.81
120	119.33	127.62	132.81	140.23	146.57	158.95
150	149.33	158.58	164.35	172.58	179.58	193.21
200	199.33	209.99	216.61	226.02	233.99	249.45
500	499.33	516.09	526.40	540.93	553.13	576.49
1,000	999.3	1,023	1,037	1,058	1,075	1,107
10,000	9,999	10,074	10,119	10,182	10,234	10,332
100,000	99,999	100,234	100,376	100,574	100,737	101,043

A.3 Physical Properties

The properties presented in this section can be useful for solving some of the problems presented in Part I.

Table A.13: Typical values of diffusion coefficients for ions at infinite dilution in water at $T = 25\ ^\circ C$. See Chapter 5 in Orazem and Tribollet.[1]

Cation	z_i	D_i/cm^2s^{-1}	**Anion**	z_i	D_i/cm^2s^{-1}
H^+	+1	9.312×10^{-5}	OH^-	-1	5.260×10^{-5}
Na^+	+1	1.334×10^{-5}	Cl^-	-1	2.032×10^{-5}
K^+	+1	1.957×10^{-5}	NO_3^-	1	1.902×10^{-5}
Ag^+	+1	1.648×10^{-5}	SO_4^{2-}	-2	1.065×10^{-5}
Mg^{2+}	+2	0.7063×10^{-5}	$Fe(CN)_6^{3-}$	-3	0.896×10^{-5}
Cu^{2+}	+2	0.72×10^{-5}	$Fe(CN)_6^{4-}$	-4	0.739×10^{-5}

Table A.14: Typical ranges of values for dielectric constant, film thickness, and capacitance. Dielectric constants for some other materials are presented in Table A.15 on page 424. See Chapter 5 in Orazem and Tribollet.[1]

System	Dielectric Constant	Thickness	Capacitance
Double layer on bare metal	–	–	$10-50\ \mu F/cm^2$
Fe_2O_3 oxide[60]	12	2.7 nm	$4\ \mu F/cm^2$
		5.9 nm	$1.8\ \mu F/cm^2$
Ni_2O_5 oxide[32]	42	8 nm	$4.6\ \mu F/cm^2$
		29 nm	$1.3\ \mu F/cm^2$
Human skin[144]	21.2	$15\ \mu m$	$1.25\ nF/cm^2$
Epoxy-polyaminoamide paint[145]	4.9	$21\ \mu m$	$0.2\ nF/cm^2$

Table A.15: Values for resistivity and dielectric constant at 298 K for solids. Unless otherwise indicated, values are taken from reference 146. Typical ranges of values for dielectric constant, film thickness, and capacitance are presented in Table A.14 on page 423. See Chapter 12 in Orazem and Tribollet.[1]

Resistivity, Ωcm		Category	Dielectric Constant	Material
1.617×10^{-6}		conductor		Silver, Ag
1.712×10^{-6}		conductor		Copper, Cu
2.255×10^{-6}		conductor		Gold, Au
2.709×10^{-6}		conductor		Aluminum, Al
9.87×10^{-6}		conductor		Iron, Fe
2.11×10^{-5}		conductor		Lead, Pb
1.44×10^{-4}		conductor		Manganese, Mn
5.2×10^{-3}		conductor	20	Magnetite, Fe_3O_4
		semiconductor	12	Hematite, Fe_2O_3-α
0.04	[147]	semiconductor	11.9	Silicon, Si (high purity)
200	[147]	semiconductor	16	Germanium, Ge (high purity)
$1\text{–}5 \times 10^3$		semiconductor	7.6	Cuprite, Cu_2O
6×10^5		semiconductor	18.1	Melaconite, CuO
1×10^8	[148]	insulator	4.1	Quartz, SiO_2
$10^{15}\text{–}10^{17}$	[147]	insulator	2.3	polyethylene

Table A.16: Physical properties for GaAs at 300 K. See Chapter 12 in Orazem and Tribollet.[1]

Energy gap	E_g	1.424	eV
Intrinsic carrier concentration	n_i	2.1×10^6	cm^{-3}
Effective conduction band density of states	N_c	4.7×10^{17}	cm^{-3}
Effective valence band density of states	N_v	9.0×10^{18}	cm^{-3}
Dielectric constant (static)	ϵ	12.9	
Dielectric constant (high frequency)	ϵ	10.89	
Mobility of electrons	μ_n	8500	cm^2/Vs
Mobility of holes	μ_p	400	cm^2/Vs
Diffusion coefficient of electrons	D_n	220	cm^2/s
Diffusion coefficient of holes	D_p	10	cm^2/s

A.4 Selected Electrochemical Notation

Some symbols, units, and relationships for quantities used in electrical circuits are presented in this section. These may facilitate solution of problems in Part I.

Table A.17: Symbols, units, and relationships for quantities used in electrical circuits. See Chapter 4 in Orazem and Tribollet.[1]

Quantity	Symbol	Units	Abbreviation	Relations
Admittance	Y	Siemen	S or mho	$1\ \mathrm{S} = 1/\Omega$
Angular frequency	ω	Radian/Second	s^{-1}	$\omega = 2\pi f$
Capacitance	C	Farad	F	$1\ \mathrm{F} = 1\ \mathrm{s}/\Omega$
Charge	q	Coulomb	C	
Current	I	Amp	A	$1\ \mathrm{A} = 1\ \mathrm{C/s}$
				$1\ \mathrm{A} = 1\ \mathrm{V}/\Omega$
Current density	i	Amp/Area	$\mathrm{A/cm^2}$	$1\ \mathrm{A/cm^2} = 1\ \mathrm{C/s\ cm^2}$
Frequency	f	Hertz	Hz	$f = \omega/2\pi$
Impedance	Z	Ohm	Ω	$1\ \Omega = 1\ \mathrm{V/A}$
				$1\ \Omega = 1\ \mathrm{J/sA^2}$
Inductance	L	Henry	H	$1\ \mathrm{H} = 1\ \mathrm{s}\Omega$
Potential	Φ	Volt	V	$1\ \mathrm{V} = 1\ \mathrm{J/C}$
Resistance	R	Ohm	Ω	

Table A.18: Definitions and notation for potentials used in electrochemical systems. See Chapter 5 in Orazem and Tribollet.[1]

V	Interfacial potential for the working electrode, $V = \Phi_{\mathrm{m}} - \Phi_0$.
$V_{0,\mathrm{k}}$	Interfacial potential at equilibrium for a given reaction k, $V_{0,\mathrm{k}} = (\Phi_{\mathrm{m}} - \Phi_0)_{0,\mathrm{k}}$.
U	Electrode potential with respect to a reference electrode, $U = \Phi_{\mathrm{m}} - \Phi_{\mathrm{ref}}$.
η_{s}	Surface overpotential for a given reaction k, $\eta_{\mathrm{s}} = V - V_{0,\mathrm{k}}$.
η_{c}	Concentration overpotential defined by $\eta_{\mathrm{c}} = \frac{1}{b_{\mathrm{a}}} \ln \left(\frac{c_{\mathrm{i}}(0)}{c_{\mathrm{i}}(\infty)} \right)$.
Φ_{m}	Electrode potential with respect to an unspecified but common reference potential.
Φ_0	Potential of the electrolyte adjacent to the working electrode with respect to an unspecified but common reference potential.
Φ_{ref}	Potential of a reference electrode with respect to an unspecified but common reference potential.
iR_{e}	Ohmic potential drop between the solution adjacent to the working electrode and the location of a reference electrode, i.e., $iR_{\mathrm{e}} = \Phi_0 - \Phi_{\mathrm{ref}}$.

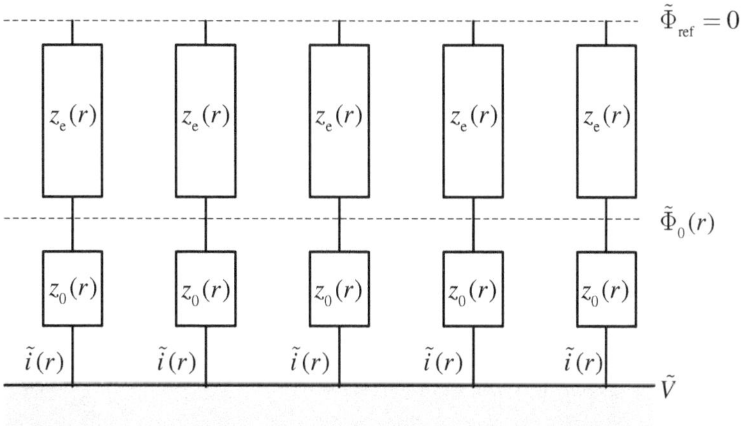

Figure A.1: The location of current and potential terms that make up definitions of global and local impedance.

A.5 Local Electrochemical Impedance Spectroscopy

Local impedance measurements can be considered to be another form of generalized transfer-function analysis. In these experiments, a small probe is placed near the electrode surface. The probe uses either two small electrodes or a vibrating wire to allow measurement of potential at two positions. Under the assumption that the electrolyte conductivity between the two points of potential measurement is uniform, the current density at the probe can be estimated from the measured potential difference ΔV_{probe} by

$$i_{\text{probe}} = \Delta V_{\text{probe}} \frac{\kappa}{d} \tag{A.66}$$

where d is the distance between the potential sensing probes and κ is the conductivity of the electrolyte.

A schematic representation of the electrode–electrolyte interface is given in Figure A.1, where the block used to represent the local ohmic impedance reflects the complex character of the ohmic contribution to the local impedance response. The impedance definitions presented in Table A.19 were proposed by Huang et al.[105] for local impedance variables. They differ in the potential and current used to calculate the impedance. To avoid confusion with local impedance values, the symbol y is used to designate the axial position in cylindrical coordinates.

A.5.1 Global Impedance

The global impedance is defined to be

$$Z = \frac{\widetilde{V}}{\widetilde{I}} \tag{A.67}$$

Table A.19: Definitions and notation for local impedance variables.

Symbol	Meaning	Units
Z	Global impedance (equation (A.67))	Ω or Ωcm^2
Z_r	Real part of global impedance	Ω or Ωcm^2
Z_j	Imaginary part of global impedance	Ω or Ωcm^2
Z_0	Global interfacial impedance (equation (A.74))	Ω or Ωcm^2
$Z_{0,r}$	Real part of global interfacial impedance	Ω or Ωcm^2
$Z_{0,j}$	Imaginary part of global interfacial impedance	Ω or Ωcm^2
Z_e	Global ohmic impedance (equation (A.76))	Ω or Ωcm^2
$Z_{e,r}$	Real part of global ohmic impedance	Ω or Ωcm^2
$Z_{e,j}$	Imaginary part of global ohmic impedance	Ω or Ωcm^2
z	Local impedance (equation (A.69))	Ωcm^2
z_r	Real part of local impedance	Ωcm^2
z_j	Imaginary part of local impedance	Ωcm^2
z_0	Local interfacial impedance (equation (A.71))	Ωcm^2
$z_{0,r}$	Real part of local interfacial impedance	Ωcm^2
$z_{0,j}$	Imaginary part of local interfacial impedance	Ωcm^2
z_e	Local ohmic impedance (equation (A.72))	Ωcm^2
$z_{e,r}$	Real part of local ohmic impedance	Ωcm^2
$z_{e,j}$	Imaginary part of local ohmic impedance	Ωcm^2
$\langle\Phi\rangle$	Spatial average of potential	V
$\overline{\Phi}$	Time average or steady-state value of potential	V
$\langle i\rangle$	Spatial average of current density	A/cm^2
$\overline{i}$	Time average or steady-state value of current density	A/cm^2
y	Axial position variable	cm

where the complex current contribution for a disk electrode is given by

$$\widetilde{I} = \int_0^{r_0} \widetilde{i}(r) 2\pi r \mathrm{d}r \tag{A.68}$$

The use of an uppercase letter signifies that Z is a global value. The global impedance may have real and imaginary values designated as Z_r and Z_j, respectively. The total current could also be represented by $\widetilde{I} = \pi r_0^2 \langle \widetilde{i}(r) \rangle$ where the brackets signify the area-average of the current density.

A.5.2 Local Impedance

The term local impedance traditionally involves the potential of the electrode measured relative to a reference electrode far from the electrode surface.[149,150] Thus, the local impedance is given by

$$z = \frac{\widetilde{V}}{\widetilde{i}(r)} \tag{A.69}$$

The use of a lowercase letter signifies that z is a local value. The local impedance may have real and imaginary values, designated as z_r and z_j, respectively.

The global impedance can be expressed in terms of the local impedance as

$$Z = \left\langle \frac{1}{z} \right\rangle^{-1} \tag{A.70}$$

Equation (A.70) is consistent with the treatment of Brug et al.[54] in which the admittance of the disk electrode was obtained by integration of a local admittance over the area of the disk.

A.5.3 Local Interfacial Impedance

The local interfacial impedance involves the potential of the electrode measured relative to a reference electrode $\Phi_0(r)$ located at the outer limit of the diffuse double layer. Thus, the local interfacial impedance is given by

$$z_0 = \frac{\widetilde{V} - \widetilde{\Phi}_0(r)}{\widetilde{i}(r)} \tag{A.71}$$

The use of a lowercase letter again signifies that z_0 is a local value, and the subscript 0 signifies that z_0 represents a value associated only with the surface. The local interfacial impedance may have real and imaginary values, designated as $z_{0,r}$ and $z_{0,j}$, respectively.

A.5.4 Local Ohmic Impedance

The local ohmic impedance involves the potential of a reference electrode $\Phi_0(r)$ located at the outer limit of the diffuse double layer and the potential of a reference electrode located far from the electrode $\widetilde{\Phi}(\infty) = 0$; see Figure A.1. Thus, the local ohmic impedance is given by

$$z_{\rm e} = \frac{\widetilde{\Phi}_0(r)}{\widetilde{i}(r)} \tag{A.72}$$

The use of a lowercase letter again signifies that $z_{\rm e}$ is a local value, and the subscript *e* signifies that $z_{\rm e}$ represents a value associated only with the ohmic character of the electrolyte. The local ohmic impedance may have real and imaginary values, designated as $z_{\rm e,r}$ and $z_{\rm e,j}$, respectively. The local impedance

$$z = z_0 + z_{\rm e} \tag{A.73}$$

can be represented by the sum of local interfacial and local ohmic impedances.

A.5.5 Global Interfacial Impedance

The global interfacial impedance is defined for a disk electrode to be

$$Z_0 = 2\pi \left(\int_0^{r_0} \frac{1}{z_0(r)} r{\rm d}r \right)^{-1} \tag{A.74}$$

or, in a more general sense,

$$Z_0 = \left\langle \frac{1}{z_0(r)} \right\rangle^{-1} \tag{A.75}$$

The use of an uppercase letter signifies that Z_0 is a global value. The global interfacial impedance may have real and imaginary values, designated as $Z_{0,\rm r}$ and $Z_{0,\rm j}$, respectively.

A.5.6 Global Ohmic Impedance

The global ohmic impedance is defined to be

$$Z_{\rm e} = Z - Z_0 \tag{A.76}$$

The use of an uppercase letter signifies that Z is a global value. The global ohmic impedance may have real and imaginary values, designated as $Z_{\rm e,r}$ and $Z_{\rm e,j}$, respectively.

A.6 Relationships for Impedance

The relationships provided in this section serve to facilitate understanding of impedance spectroscopy.

Table A.20: Some useful relationships for the development of the impedance response associated with faradaic reactions. Taken from Chapter 10 of Orazem and Tribollet.[1]

$$\widetilde{i} = \widetilde{i}_{\mathrm{F}} + \mathrm{j}\omega C_{\mathrm{dl}}\widetilde{V} \tag{A.77}$$

$$\widetilde{U} = \widetilde{i}R_{\mathrm{e}} + \widetilde{V} \tag{A.78}$$

$$Z(\omega) = \frac{\widetilde{U}}{\widetilde{i}} \tag{A.79}$$

Table A.21: Characteristic dimensions and frequencies associated with frequency dispersion caused by nonuniform ohmic resistance or capacitance. Taken from Chapter 13 of Orazem and Tribollet.[1]

Experimental System	ℓ_{c} / cm	f_{c} / Hz
Rectangular electrode ($d_1 \times d_2$)[151]	$\frac{d_1 d_2}{d_1 + d_2}$	$\frac{1}{2\pi}\frac{\kappa(d_1 + d_2)}{C_0 d_1 d_2}$
Disk electrode[133,105]	r_0	$\frac{1}{2\pi}\frac{\kappa}{C_0 r_0}$
Disk electrode (local CPE)[7]	r_0	$\frac{1}{2\pi}\left(\frac{\kappa}{Q r_0}\right)^{1/\alpha}$
Rough disk electrode[47,58]	$f_{\mathrm{r}} r_0$	$\frac{1}{2\pi}\frac{\kappa}{C_0 f_{\mathrm{r}} r_0}$
Ring electrode[131]	$\frac{r_2 - r_1}{1 + (r_1/r_2)^2}$	$\frac{\kappa(1 + (r_1/r_2)^2)}{2\pi C_0 (r_2 - r_1)}$
Roughness[47,58]	$f_{\mathrm{r}}^2 P$	$\frac{\kappa}{2\pi C_0 f_{\mathrm{r}}^2 P}$
Intermittent roughness[58]	$f_{\mathrm{r}} f_{\mathrm{p}} P$	$\frac{\kappa}{2\pi C_0 f_{\mathrm{r}} f_{\mathrm{p}} P}$
Distribution of capacity[57]	P	$\frac{\kappa}{2\pi \langle C_0 \rangle P}$

Table A.22: Approaches used to interpret CPE impedance response associated with normal or surface distributions of time constants. Taken from Chapter 14 of Orazem and Tribollet.[1]

Approach	Formula	Equations
Simple substitution	$C_{\text{eff},Q} = Q$	(14.2)
Characteristic frequency, impedance	$C_{\text{eff,norm}} = Q^{1/\alpha} R_{\|\|}^{(1-\alpha)/\alpha}$	(14.3)
Characteristic frequency, admittance	$C_{\text{eff,surf}} = Q^{1/\alpha} \left(\dfrac{R_e R_{\|\|}}{R_e + R_{\|\|}} \right)^{(1-\alpha)/\alpha}$	(14.4)
Power-law model, normal distribution	$C_{\text{eff,PL}} = gQ \left(\rho_\delta \epsilon \epsilon_0\right)^{1-\alpha}$	(14.5)
	$C_{\text{eff,PL}} = (gQ)^{1/\alpha} \left(\delta \rho_\delta\right)^{(1-\alpha)/\alpha}$	(14.6)

Table A.23: Statistical properties of impedance values obtained by Fourier analysis techniques. Taken from Carson et al.[82] and Chapter 21 of Orazem and Tribollet.[1]

$$\sigma_{Z_r}^2 = \sigma_{Z_j}^2 \tag{A.80}$$

$$\sigma_{Z_r Z_j} = 0 \tag{A.81}$$

$$\sigma_{|Z|\varphi} = 0 \tag{A.82}$$

$$\sigma_{|Z|}^2 = |Z|^2 \sigma_\varphi^2 \tag{A.83}$$

Table A.24: Summary of complex impedance, admittance, and capacitance characteristics for simple blocking and reactive circuits. Taken from Chapter 17 of Orazem and Tribollet.[1]

Circuit	(a)	(b)
Classification	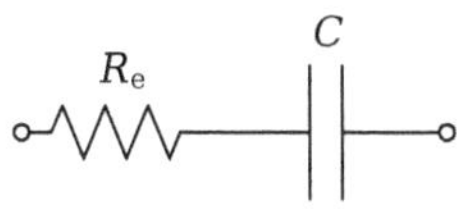 Blocking	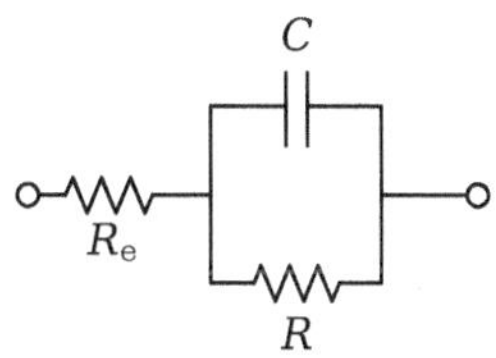 Reactive
Complex impedance		
Z_r	R_e	$R_e + \frac{R}{1+(\omega RC)^2}$
Z_j	$-\frac{1}{\omega C}$	$-\frac{\omega C R^2}{1+(\omega RC)^2}$
Time constant	none	RC
Complex admittance		
Y_r	$\frac{R_e(\omega C)^2}{1+(\omega ReC)^2}$	$\frac{R_e R(\omega C)^2 + 1 + R_e/R}{R\left[(1+R_e/R)^2 + (\omega R_e C)^2\right]}$
Y_j	$\frac{\omega C}{1+(\omega ReC)^2}$	$\frac{\omega C}{(1+R_e/R)^2 + (\omega R_e C)^2}$
Time constant	$R_e C$	$\frac{R_e C}{(1+R_e/R)}$
Complex capacitance‡		
C_r	C	C
C_j	0	$-\frac{1}{R\omega}$
Time constant	None	None
Effective capacitance		
$C_{eff} = -\frac{1}{\omega Z_j}$	C	$C + \frac{1}{\omega^2 R^2 C}$
Time constant	None	RC

‡Complex capacitance is determined for $Z - R_e$.

Table A.25: Compilation of various forms of the Kramers–Kronig relations. Taken from Chapter 22 of Orazem and Tribollet.[1]

$$Z_j(\omega) = \frac{1}{\pi} \int_{-\infty}^{+\infty} \frac{Z_r(x)}{x-\omega} dx \tag{A.84}$$

$$Z_r(\omega) = Z_r(\infty) - \frac{1}{\pi} \int_{-\infty}^{+\infty} \frac{Z_j(x)}{x-\omega} dx \tag{A.85}$$

$$Z_j(\omega) = \frac{2\omega}{\pi} \int_{0}^{\infty} \frac{Z_r(x)}{x^2-\omega^2} dx \tag{A.86}$$

$$Z_r(\omega) = Z_r(\infty) - \frac{2}{\pi} \int_{0}^{\infty} \frac{xZ_j(x)}{x^2-\omega^2} dx \tag{A.87}$$

$$Z_j(\omega) = \frac{2\omega}{\pi} \int_{0}^{\infty} \frac{Z_r(x)-Z_r(\omega)}{x^2-\omega^2} dx \tag{A.88}$$

$$Z_r(\omega) = Z_r(\infty) - \frac{2}{\pi} \int_{0}^{\infty} \frac{xZ_j(x)-\omega Z_j(\omega)}{x^2-\omega^2} dx \tag{A.89}$$

$$\varphi(\omega) = \frac{2\omega}{\pi} \int_{0}^{\infty} \frac{\ln(|Z(x)|)}{x^2-\omega^2} dx \tag{A.90}$$

$$\varphi(\omega) = \frac{2\omega}{\pi} \int_{0}^{\infty} \frac{\ln(|Z(x)|)-\ln(|Z(\omega)|)}{x^2-\omega^2} dx \tag{A.91}$$

$$\varphi(\omega) = \frac{1}{\pi} \int_{0}^{\infty} \ln\left|\frac{x-\omega}{x+\omega}\right| \frac{d\ln|Z(x)|}{dx} dx \tag{A.92}$$

$$\ln(|Z(\omega)|) = -\frac{2}{\pi} \int_{-\infty}^{+\infty} \frac{\varphi(x)}{x-\omega} dx \tag{A.93}$$

$$\operatorname{Re}\{\ln(Z(\omega))\} = \operatorname{Re}\{\ln(Z(0))\} - \frac{2\omega^2}{\pi} \int_{0}^{\infty} \frac{\operatorname{Im}\{\ln(Z(x))\}}{x(x^2-\omega^2)} dx \tag{A.94}$$

$$\operatorname{Im}\{\ln(Z(\omega))\} = +\frac{2\omega}{\pi} \int_{0}^{\infty} \frac{\operatorname{Re}\{\ln(Z(x))\}}{x^2-\omega^2} dx \tag{A.95}$$

Table A.26: Characterization of impedance plot types in terms of sensitivity to discrepancies between model and experimental values. Taken from Chapter 20 of Orazem and Tribollet.[1]

Poor Sensitivity	
Modulus	The Bode magnitude representation is singularly incapable of distinguishing between impedance models unless they provide extremely poor fits to impedance data.
Real	The real impedance representation is similarly insensitive to fit quality.
Modest Sensitivity	
Nyquist	These plots are sensitive only for large impedance values. Impedance data at high-frequency values are typically obscured.
Imaginary	The imaginary impedance representation is modestly sensitive to fit quality.
Log-imaginary	These plots emphasize small values, and values of the slope that differ from ± 1 may suggest the need for new models.
Phase angle	The high-frequency behavior in these plots is counterintuitive due to the role of solution resistance.
Modified phase angle	Ohmic-resistance-corrected phase angle plots may confirm the existence of a CPE.
Excellent Sensitivity	
Residual error plots	Appearance of trending indicates a need to improve the model or to remove data that are inconsistent with the Kramers–Kronig relations.

Table A.27: Results of experiments performed to determine if the simplifications to equation (A.96) expressed as equations (A.97) and (A.98) are justified. Experiments are performed at V_{corr}, $V_{corr} + \Delta V$, and $V_{corr} - \Delta V$. Typical values of ΔV are between 10 and 50 mV. Results are presented in terms of the expected change in the impedance as compared to the value measured at the open-circuit potential.

Observations		
OCP + ΔV	OCP − ΔV	Conclusions
Z decreases	Z decreases	$Z_a \approx Z_c$
Z decreases	Z increases	$Z_a \ll Z_c$
Z increases	Z decreases	$Z_a \gg Z_c$
Z unchanged	Z unchanged	$Z_a \neq f(V)$ and $Z_a \ll Z_c$

A.7 Method to Identify Corrosion Model

At the corrosion or open-circuit potential, the sum of the anodic and cathodic faradaic currents is equal to zero, i.e., $\bar{i}_a + \bar{i}_c = 0$. The faradaic impedance must, therefore, be a parallel combination of Z_a and Z_c. The contribution of the double-layer capacitance is added in parallel. The corresponding expression for the impedance may be given as

$$Z = R_e + \frac{1}{\dfrac{1}{Z_a} + \dfrac{1}{Z_c} + j\omega C_{dl}} \tag{A.96}$$

Expressions for the impedances Z_a and Z_c must be developed separately according to proposed reaction mechanisms.

If $Z_a \gg Z_c$, equation (A.96) can be expressed as

$$Z = R_e + \frac{Z_c}{1 + j\omega C_{dl} Z_c} \tag{A.97}$$

Conversely, if $Z_c \gg Z_a$, equation (A.96) can be expressed as

$$Z = R_e + \frac{Z_a}{1 + j\omega C_{dl} Z_a} \tag{A.98}$$

The present section addresses a method to determine conditions under which the simplifications to equation (A.96) expressed as equations (A.97) and (A.98) are justified.

The recommended approach is to measure the impedance at the open-circuit potential and then at a potential slightly anodic to the open-circuit potential and slightly cathodic to the open-circuit potential, taking care to ensure that a steady-state condition is achieved before making impedance measurements. If both reactions are kinetically controlled, an increase in potential will increase the cathodic impedance Z_c and decrease the anodic impedance Z_a, and a decrease in potential will decrease Z_c and increase the anodic impedance Z_a. The results are presented in Table A.7 in terms of the expected change in the impedance as compared to the value measured at the open-circuit potential. More information is provided in Chapter 9 of Orazem and Tribollet.[1]

It is tempting to assume that, because one is studying corrosion, the impedance response must be due to the corrosion reaction. The analysis presented here shows that the impedance response may provide information on the cathodic reaction rather than the corrosion reaction.

Appendix B

Computer Code

The following code was used to solve some of the problems presented in Part I. The code was written for MATLAB®, but most of the code will run as intended in GNU Octave.

B.1 Fourier Analysis: Nonlinear

The following MATLAB® code may be used to generate the impedance response under potentiostatic modulation. The inputs are

- voltage perturbation in mV,
- steady-state potential in mV,
- capacity in μF/cm^2,
- anodic rate constant in mA/cm^2,
- cathodic rate constant in mA/cm^2,
- anodic transfer coefficient b_a in V^{-1}, and
- cathodic transfer coefficient b_c in V^{-1}

The output is the impedance as a function of frequency, Lissajous curves at every decade in frequency, and current and potential as functions of time at every decade in frequency.

```
function Lissajoubvprint2
%Time to frequency domains - Butler Volmer kinetics
global DV V0 Ka Kc bb bc Cd;
close all;
DV=input('Input: Voltage Perturbation in mV.................. ');
V0=input('Input: Steady-State Voltage in mV.................. ');
Cd=input('Input: Capacity in microF/cm2...................... ');
```

```
Ka=input('Input: anodic rate constant in mA/cm2............. ');
Kc=input('Input: cathodic rate constant in mA/cm2........... ');
bb=input('Input: anode exponential coefficient in 1/V....... ');
bc=input('Input: cathode exponential coefficient in 1/V..... ');

Npoints=600;
DV=DV/1000;V0=V0/1000;
Cd=Cd*1.0e-6;
Ka=Ka/1000;
Kc=Kc/1000;
Rt=1/(bb*Ka*exp(bb*V0)+bc*Kc*exp(-bc*V0));
ff=-3:1:6; ff=10.^ff;
for kk=1:length(ff)
    fname=sprintf('Lissajou-%g.txt',ff(kk));
    fid = fopen(fname, 'wt');
    fprintf(fid,...
     'V0= %g V; DV= %g V; Ka= %g A/cm2; Ka= %g A/cm2; ba= %g V-1;',...
     'ba= %g V-1\n',V0,DV,Ka,Kc,bb,bc);
    fprintf(fid,'Rt= %g W cm2; Cd= %9.2e F/cm2; fchar= %g Hz\n',...
        Rt,Cd,1/(2*pi*Rt*Cd));
    fprintf(fid,...
     'time; potential; current; scaled potential; scaled current \n');
    Tmax=3/ff(kk);
    t=0:Tmax/Npoints:Tmax;
for k=1:Npoints+1
%       V(k)=potential(t(k),ff(kk));
    V(k)=V0+DV*cos(2*pi*ff(kk)*t(k));
    I(k)=faradaicbv2(t(k),ff(kk));
end
Ibar=Ka*exp(bb*V0)-Kc*exp(-bc*V0);Imax=max(abs(I-Ibar));
for k=1:Npoints+1
    fprintf(fid, '%14.6e; %14.6e; %14.6e; %14.6e; %14.6e\n',...
        t(k),V(k),I(k),(V(k)-V0)/DV,(I(k)-Ibar)/Imax);
end
fclose(fid);
KoJ=2*pi*ff(kk)*Rt*Cd;
string=sprintf('Lissajou: f = %g Hz; K/J = %g; Delta V b = %g',...
    ff(kk),KoJ,DV*bb);
figure;set(gcf,'Name',string);
plot((V-V0)/DV,(I-Ibar)/Imax,'k-');title(string);
xlabel('Potential / V');
ylabel('Current Density / A cm^{-2}');
string=sprintf('Time Trace: f = %g Hz; K/J = %g; Delta V b = %g',...
    ff(kk),KoJ,DV*bb);
figure;set(gcf,'Name',string);
plot(t,(I-Ibar)/max(abs(I-Ibar)),'r-');title(string);
xlabel('time / s');
ylabel('Response'); hold on;
plot(t,(V-V0)/DV,'b-');title(string);xlabel('time / s');
ylabel('Response'); legend('Current','Potential');
end
ff=-3:0.1:6; ff=10.^ff;
```

```
IR=zeros(size(ff));
%IR2=IR;
IJ=zeros(size(ff));
%IJ2=IJ;
VR=zeros(size(ff));
VJ=zeros(size(ff));
IC=zeros(size(ff));
%IC2=IC;
VC=zeros(size(ff));
Z=zeros(size(ff));
Zmod=zeros(size(ff));
fname=sprintf('Lissajou-EIS2.txt');
fid = fopen(fname, 'wt');
fprintf(fid,'DV= %g V; Ka= %g A/cm2; Kc= %g A/cm2; b= %g V-1;',...
    'Cd= %9.2e F/cm2\n',DV,Ka,Kc,bb,Cd);
fname=sprintf('Lissajou-complex-IV2.txt');
fid2 = fopen(fname, 'wt');
fprintf(fid2,'DV= %g V; i0= %g A/cm2; b= %g V-1; Cd= %9.2e F/cm2\n',...
    DV,Ka,bb,Cd);
fprintf(fid,'frequency; Zreal; Zimag; Zmodreal; Zmodimag \n');
fprintf(fid2,'frequency; Ireal; Iimag; Vreal; Vimag \n');

for kk=1:length(ff)
    Tmax=1/ff(kk);
    IR(kk)=integral(@(t)currentrealbv2(t,ff(kk)),0,Tmax)/Tmax;
    IJ(kk)=integral(@(t)currentimagbv2(t,ff(kk)),0,Tmax)/Tmax;
    VR(kk)=integral(@(t)potentialreal(t,ff(kk)),0,Tmax)/Tmax;
    VJ(kk)=integral(@(t)potentialimag(t,ff(kk)),0,Tmax)/Tmax;
    IC(kk)=complex(IR(kk),IJ(kk));
    VC(kk)=complex(VR(kk),VJ(kk));
    Z(kk)=VC(kk)/IC(kk);
    Zmod(kk)=Rt/(1+1i*2*pi*ff(kk)*Rt*Cd);
    fprintf(fid, '%14.6e; %14.6e; %14.6e; %14.6e; %14.6e \n',...
        ff(kk),real(Z(kk)),imag(Z(kk)),real(Zmod(kk)),imag(Zmod(kk)));
    fprintf(fid2,'%14.6e; %14.6e; %14.6e; %14.6e; %14.6e \n',...
        ff(kk),real(IC(kk)),imag(IC(kk)),real(VC(kk)),imag(VC(kk)));
end
fclose(fid);
fclose(fid2);
string=sprintf('Rt = %g Ohm cm^2; Delta V b = %g; Z(0)/Rt = %g',...
    Rt,DV*bb,max(real(Z))/Rt);disp(string);
figure;loglog(ff,abs(real(Z)),'bo');title(string);
xlabel('f / Hz');ylabel('Z_r / Ohm cm^{2}');
set(gcf,'Name','Real Impedance');
hold on;loglog(ff,real(Zmod),'k-');
figure;loglog(ff,abs(imag(Z)),'bo');title(string);xlabel('f / Hz');
ylabel('|Z_j| / Ohm cm^{2}');set(gcf,'Name','Imaginary Impedance');
hold on;loglog(ff,abs(imag(Zmod)),'k-');
figure;plot(real(Z),-imag(Z),'bo');axis equal;title(string);
xlabel('Z_r / Ohm cm^{2}');ylabel('-Z_j / Ohm cm^{2}');
set(gcf,'Name','Nyquist Plot');
hold on;plot(real(Zmod),-imag(Zmod),'k-');
```

```
function I=faradaicbv2(t,ffun)
global DV V0 Ka Kc bb bc Cd;
% V=potential(t,ff);
V=V0+DV*cos(2*pi*ffun*t);
IC=-Cd*2*pi*DV*ffun*sin(2*pi*ffun*t);
IF=Ka*exp(bb*V)-Kc*exp(-bc*V);
I=IF+IC;
```

B.1.1 Fourier Integration: Current

```
function I=currentimagbv2(t,ffun)
global DV V0 Ka Kc bb bc Cd;
% V=potential(t,ff);
V=V0+DV*cos(2*pi*ffun*t);
IC=-Cd*2*pi*DV*ffun*sin(2*pi*ffun*t);
IF=Ka*exp(bb*V)-Kc*exp(-bc*V);
I=IF+IC;
Ibar=Ka*exp(bb*V0)-Kc*exp(-bc*V0);
I=-(I-Ibar).*sin(2*pi*ffun.*t);

function I=currentrealbv2(t,ffun)
global DV V0 Ka Kc bb bc Cd;
% V=potential(t,ff);
V=V0+DV*cos(2*pi*ffun*t);
IC=-Cd*2*pi*ffun*DV*sin(2*pi*ffun*t);
IF=Ka*exp(bb*V)-Kc*exp(-bc*V);
I=IF+IC;
Ibar=Ka*exp(bb*V0)-Kc*exp(-bc*V0);
I=(I-Ibar).*cos(2*pi*ffun.*t);
```

B.1.2 Fourier Integration: Potential

```
function V=potentialimag(t,ff)
global DV;
V=DV*cos(2*pi*ff.*t).*sin(2*pi*ff.*t);

function V=potentialreal(t,ff)
global DV;
V=DV*cos(2*pi*ff.*t).*cos(2*pi*ff.*t);
```

B.2 Fourier Analysis: Linear

The following MATLAB® code may be used to generate the impedance response under potentiostatic modulation. The inputs are

- voltage perturbation in mV,
- steady-state potential in mV,
- capacity in μF/cm^2, and

- charge-transfer resistance in Ωcm^2.

The output is the impedance as a function of frequency, Lissajous curves at every decade in frequency, and current and potential as functions of time at every decade in frequency. The code requires the Fourier integration code for potential provided in Section B.1.2.

```
function Lissajoulinprint
%Time to frequency domains - Butler Volmer kinetics
global DV V0 Rt Cd;
close all;
DV=input('Input: Voltage Perturbation in mV................. ');
V0=input('Input: Steady-State Voltage in mV................. ');
Cd=input('Input: Capacity in microF/cm2..................... ');
Rt=input('Input: charge-transfer resistance in Ohm cm2...... ');
Npoints=600;
DV=DV/1000;V0=V0/1000;
Cd=Cd*1.0e-6;
ff=-3:1:6; ff=10.^ff;
for kk=1:length(ff)
    fname=sprintf('Lissajou-%g.txt',ff(kk));
    fid = fopen(fname, 'wt');
    fprintf(fid,...
        'V0= %g V; DV= %g V; \n',V0,DV);
    fprintf(fid,'Rt= %g W cm2; Cd= %9.2e F/cm2; fchar= %g Hz\n',...
        Rt,Cd,1/(2*pi*Rt*Cd));
    fprintf(fid,...
     'time; potential; current; scaled potential; scaled current \n');
    Tmax=3/ff(kk);
    t=0:Tmax/Npoints:Tmax;
for k=1:Npoints+1
%       V(k)=potential(t(k),ff(kk));
    V(k)=V0+DV*cos(2*pi*ff(kk)*t(k));
    I(k)=V(k)/Rt;
end
Ibar=V0/Rt;Imax=max(abs(I-Ibar));
for k=1:Npoints+1
    fprintf(fid, '%14.6e; %14.6e; %14.6e; %14.6e; %14.6e\n',...
        t(k),V(k),I(k),(V(k)-V0)/DV,(I(k)-Ibar)/Imax);
end
fclose(fid);
KoJ=2*pi*ff(kk)*Rt*Cd;
string=sprintf('Lissajou: f = %g Hz; K/J = %g;',ff(kk),KoJ);
figure;set(gcf,'Name',string);
plot((V-V0)/DV,(I-Ibar)/Imax,'k-');title(string);
xlabel('Potential / V');
ylabel('Current Density / A cm^{-2}');
string=sprintf('Time Trace: f = %g Hz; K/J = %g;',ff(kk),KoJ);
figure;set(gcf,'Name',string);
plot(t,(I-Ibar)/max(abs(I-Ibar)),'r-');title(string);
xlabel('time / s');
ylabel('Response'); hold on;
```

```
plot(t,(V-V0)/DV,'b-');title(string);xlabel('time / s');
ylabel('Response'); legend('Current','Potential');
end
ff=-3:0.1:6; ff=10.^ff;
IR=zeros(size(ff));
%IR2=IR;
IJ=zeros(size(ff));
%IJ2=IJ;
VR=zeros(size(ff));
VJ=zeros(size(ff));
IC=zeros(size(ff));
%IC2=IC;
VC=zeros(size(ff));
Z=zeros(size(ff));
Zmod=zeros(size(ff));
fname=sprintf('Lissajou-EIS2.txt');
fid = fopen(fname, 'wt');
fprintf(fid, 'DV= %g V;,Cd= %9.2e F/cm2\n',DV,Cd);
fname=sprintf('Lissajou-complex-IV2.txt');
fid2 = fopen(fname, 'wt');
fprintf(fid2, 'DV= %g V; Cd= %9.2e F/cm2\n',DV,Cd);
fprintf(fid, 'frequency; Zreal; Zimag; Zmodreal; Zmodimag \n');
fprintf(fid2, 'frequency; Ireal; Iimag; Vreal; Vimag \n');

for kk=1:length(ff)
    Tmax=1/ff(kk);
    IR(kk)=integral(@(t)currentreallin(t,ff(kk)),0,Tmax)/Tmax;
    IJ(kk)=integral(@(t)currentimaglin(t,ff(kk)),0,Tmax)/Tmax;
    VR(kk)=integral(@(t)potentialreal(t,ff(kk)),0,Tmax)/Tmax;
    VJ(kk)=integral(@(t)potentialimag(t,ff(kk)),0,Tmax)/Tmax;
    IC(kk)=complex(IR(kk),IJ(kk));
    VC(kk)=complex(VR(kk),VJ(kk));
    Z(kk)=VC(kk)/IC(kk);
    Zmod(kk)=Rt/(1+1i*2*pi*ff(kk)*Rt*Cd);
    fprintf(fid, '%14.6e; %14.6e; %14.6e; %14.6e; %14.6e \n',...
        ff(kk),real(Z(kk)),imag(Z(kk)),real(Zmod(kk)),imag(Zmod(kk)));
    fprintf(fid2,'%14.6e; %14.6e; %14.6e; %14.6e; %14.6e \n',...
        ff(kk),real(IC(kk)),imag(IC(kk)),real(VC(kk)),imag(VC(kk)));
end
fclose(fid);
fclose(fid2);
string=sprintf('Rt = %g Ohm cm^2; Z(0)/Rt = %g',...
    Rt,max(real(Z))/Rt);disp(string);
figure;loglog(ff,abs(real(Z)),'bo');title(string);
xlabel('f / Hz');ylabel('Z_r / Ohm cm^{2}');
set(gcf,'Name','Real Impedance');
hold on;loglog(ff,real(Zmod),'k-');
figure;loglog(ff,abs(imag(Z)),'bo');title(string);xlabel('f / Hz');
ylabel('|Z_j| / Ohm cm^{2}');set(gcf,'Name','Imaginary Impedance');
hold on;loglog(ff,abs(imag(Zmod)),'k-');
figure;plot(real(Z),-imag(Z),'bo');axis equal;title(string);
xlabel('Z_r / Ohm cm^{2}');ylabel('-Z_j / Ohm cm^{2}');
```

```
set(gcf,'Name','Nyquist Plot');
hold on; plot(real(Zmod),-imag(Zmod),'k-');

function I=faradaicbv(t,ffun)
global DV V0 i0 bb bc Cd;
% V=potential(t,ff);
V=V0+DV*cos(2*pi*ffun*t);
IC=-Cd*2*pi*DV*ffun*sin(2*pi*ffun*t);
IF=i0*(exp(bb*V)-exp(-bc*V));
I=IF+IC;

function I=currentimaglin(t,ffun)
global DV V0 Rt Cd;
% V=potential(t,ff);
V=V0+DV*cos(2*pi*ffun*t);
IC=-Cd*2*pi*DV*ffun*sin(2*pi*ffun*t);
IF=V/Rt;
I=IF+IC;
Ibar=V0/Rt;
I=-(I-Ibar).*sin(2*pi*ffun.*t);

function I=currentreallin(t,ffun)
global DV V0 Rt Cd;
% V=potential(t,ff);
V=V0+DV*cos(2*pi*ffun*t);
IC=-Cd*2*pi*ffun*DV*sin(2*pi*ffun*t);
IF=V/Rt;
I=IF+IC;
Ibar=V0/Rt;
I=(I-Ibar).*cos(2*pi*ffun.*t);
```

B.3 Fourier Analysis with Noise

The following MATLAB® code may be used to generate the impedance response under potentiostatic modulation with noise added to the time domain signals. The inputs are

- voltage perturbation in mV,
- steady-state potential in mV,
- capacity in μF/cm^2,
- exchange current density in mA/cm^2,
- anodic transfer coefficient b_a in V^{-1},
- cathodic transfer coefficient b_c in V^{-1},
- noise level in potential signal in percent, and
- noise level in current signal in percent.

The output is the impedance as a function of frequency, Lissajous curves at every decade in frequency, and current and potential as functions of time at every decade in frequency. The Fourier integration is performed within the main program.

```
function Lissajoubvnprint2
global DV V0 i0 bb bc Cd;
close all;
DV=input('Input: Voltage Perturbation in mV.................. ');
V0=input('Input: Steady-State Voltage in mV................. ');
% Re=input('Input: Ohmic Resistance in Ohms cm2:........ ');
Cd=input('Input: Capacity in microF/cm2..................... ');
i0=input('Input: exchange current density in mA/cm2......... ');
bb=input('Input: anode exponential coefficient in 1/V....... ');
bc=input('Input: cathode exponential coefficient in 1/V..... ');
noiseV=input('Input: noise in potential signal in percent....... ');
noiseI=input('Input: noise in current signal in percent......... ');
% ff=input('Input: frequency in Hz...................... ');

Npoints=3000;
Cd=Cd*1.0e-6;
i0=i0/1000;
DV=DV/1000;V0=V0/1000;noiseV=noiseV*DV/100;
Ibar=i0*(exp(bb*V0)-exp(-bc*V0));
Rt=1/(bb*i0*exp(bb*V0)+bc*i0*exp(-bc*V0));
ff=-3:0.1:6; ff=10.^ff;
IR=zeros(size(ff));
IJ=zeros(size(ff));
VR=zeros(size(ff));
VJ=zeros(size(ff));
IC=zeros(size(ff));
VC=zeros(size(ff));
Z=zeros(size(ff));
Zmod=zeros(size(ff));
for kk=1:length(ff);
    noiseIf=noiseI*DV*sqrt(1+(Rt*Cd*2*pi*ff(kk))^2)/(Rt*100);
    if floor(log10(ff(kk)))==log10(ff(kk));
        fname=sprintf('Lissajou-%g.txt',ff(kk));
    fid = fopen(fname, 'wt');
    fprintf(fid, 'V0= %g V; DV= %g V; i0= %g A/cm2; ba= %g V-1;',...
        'ba= %g V-1\n',V0,DV,i0,bb,bc);
    fprintf(fid, 'Rt= %g W cm2; Cd= %9.2e F/cm2; fchar= %g Hz\n',...
        Rt,Cd,1/(2*pi*Rt*Cd));
    fprintf(fid, 'time; potential; current; scaled potential;',...
        'scaled current \n');
    end;
    Tmax=1/ff(kk);
    t=0:Tmax/Npoints:Tmax;
    for k=1:Npoints+1;
    V(k)=V0+DV*cos(2*pi*ff(kk)*t(k))+random('norm',0,noiseV);
    I(k)=faradaicbv(t(k),ff(kk))+random('norm',0,noiseIf);
    VR(kk)=VR(kk)+(V(k)-V0)*cos(2*pi*ff(kk)*t(k));
    VJ(kk)=VJ(kk)-(V(k)-V0)*sin(2*pi*ff(kk)*t(k));
```

```
    IR(kk)=IR(kk)+(I(k)-Ibar)*cos(2*pi*ff(kk)*t(k));
    IJ(kk)=IJ(kk)-(I(k)-Ibar)*sin(2*pi*ff(kk)*t(k));
    end;
if floor(log10(ff(kk)))==log10(ff(kk)); Imax=max(abs(I-Ibar));
for k=1:Npoints+1;
    fprintf(fid, '%14.6e; %14.6e; %14.6e; %14.6e; %14.6e\n',...
        t(k),V(k),I(k),(V(k)-V0)/DV,...
        (I(k)-Ibar)/(DV/Rt+Cd*2*pi*DV*ff(kk)));
end;
fclose(fid);

KoJ=2*pi*ff(kk)*Rt*Cd;
string=sprintf('Lissajou: f = %g Hz; K/J = %g; Delta V b = %g',...
    ff(kk),KoJ,DV*bb);
figure;set(gcf,'Name',string);
plot((V-V0)/DV,(I-Ibar)/(DV/Rt+Cd*2*pi*DV*ff(kk)),'k-');
title(string);
xlabel('Potential / V');ylabel('Current Density / A cm^{-2}');
string=sprintf('Time Trace: f = %g Hz; K/J = %g; Delta V b = %g',...
    ff(kk),KoJ,DV*bb);
figure;set(gcf,'Name',string);
plot(t,(I-Ibar)/(DV/Rt+Cd*2*pi*DV*ff(kk)),'r-');
title(string);
xlabel('time / s');ylabel('Response'); hold on;
plot(t,(V-V0)/DV,'b-');title(string);xlabel('time / s');...
    ylabel('Response'); legend('Current','Potential');
end;
end;
% ff=-3:0.1:6; ff=10.^ff;
fname=sprintf('Lissajou-EIS2.txt');
fid = fopen(fname, 'wt');
fprintf(fid,'DV= %g V; i0= %g A/cm2; b= %g V-1; Cd= %9.2e F/cm2\n',...
    DV,i0,bb,Cd);
fname=sprintf('Lissajou-complex-IV2.txt');
fid2 = fopen(fname, 'wt');
fprintf(fid2,'DV= %g V; i0= %g A/cm2; b= %g V-1; Cd= %9.2e F/cm2\n',...
    DV,i0,bb,Cd);
fprintf(fid, 'frequency; Zreal; Zimag; Zmodreal; Zmodimag \n');
fprintf(fid2, 'frequency; Ireal; Iimag; Vreal; Vimag \n');

  for kk=1:length(ff);
%      Tmax=1/ff(kk);
%      IR(kk)=quadv(@(t)currentrealbv(t,ff(kk)),0,Tmax,1e-10)/Tmax;
%      IJ(kk)=quadv(@(t)currentimagbv(t,ff(kk)),0,Tmax,1e-10)/Tmax;
%      VR(kk)=quadv(@(t)potentialreal(t,ff(kk)),0,Tmax,1e-10)/Tmax;
%      VJ(kk)=quadv(@(t)potentialimag(t,ff(kk)),0,Tmax,1e-10)/Tmax;
%      IR2(kk)=quadv(@(t)currentreal2(t,ff(kk)),0,Tmax,1e-10)/Tmax;
%      IJ2(kk)=quadv(@(t)currentimag2(t,ff(kk)),0,Tmax,1e-10)/Tmax;
    %
IC(kk)=complex(IR(kk),IJ(kk));
VC(kk)=complex(VR(kk),VJ(kk));
Z(kk)=VC(kk)/IC(kk);
```

```
Zmod(kk)=Rt/(1+1i*2*pi*ff(kk)*Rt*Cd);
%       IC2(kk)=complex(IR2(kk),IJ2(kk));
        fprintf(fid, '%14.6e; %14.6e; %14.6e; %14.6e; %14.6e \n',...
            ff(kk),real(Z(kk)),imag(Z(kk)),real(Zmod(kk)),imag(Zmod(kk)));
        fprintf(fid2,'%14.6e; %14.6e; %14.6e; %14.6e; %14.6e \n',...
            ff(kk),real(IC(kk)),imag(IC(kk)),real(VC(kk)),imag(VC(kk)));
end;
fclose(fid);
fclose(fid2);
string=sprintf('Rt = %g Ohm cm^2; Delta V b = %g; Z(0)/Rt = %g',...
    Rt,DV*bb,max(real(Z))/Rt);disp(string);
figure;loglog(ff,abs(real(Z)),'bo');title(string);xlabel('f / Hz');...
    ylabel('Z_r / Ohm cm^{2}');set(gcf,'Name','Real Impedance');
hold on;loglog(ff,real(Zmod),'k-');
figure;loglog(ff,abs(imag(Z)),'bo');title(string);xlabel('f / Hz');...
    ylabel('|Z_j| / Ohm cm^{2}');set(gcf,'Name','Imaginary Impedance');
hold on;loglog(ff,abs(imag(Zmod)),'k-');
figure;plot(real(Z),-imag(Z),'bo');axis equal;title(string);...
    xlabel('Z_r / Ohm cm^{2}');ylabel('-Z_j / Ohm cm^{2}');
set(gcf,'Name','Nyquist Plot');
hold on;plot(real(Zmod),-imag(Zmod),'k-');

function I=currentimagbv(t,ffun)
global DV V0 i0 bb bc Cd;
% V=potential(t,ff);
V=V0+DV*cos(2*pi*ffun*t);
IC=-Cd*2*pi*DV*ffun*sin(2*pi*ffun*t);
IF=i0*(exp(bb*V)-exp(-bc*V));
I=IF+IC;
Ibar=i0*(exp(bb*V0)-exp(-bc*V0));
I=-(I-Ibar).*sin(2*pi*ffun.*t);

function I=currentrealbv(t,ffun)
%global DV V0 Ka Kc bb bc Cd;
global DV V0 i0 bb bc Cd;
% V=potential(t,ff);
V=V0+DV*cos(2*pi*ffun*t);
IC=-Cd*2*pi*ffun*DV*sin(2*pi*ffun*t);
IF=i0*(exp(bb*V)-exp(-bc*V));
I=IF+IC;
Ibar=i0*(exp(bb*V0)-exp(-bc*V0));
I=(I-Ibar).*cos(2*pi*ffun.*t);
```

B.4 Single-Frequency Fourier Analysis with Noise

The following MATLAB® code may be used to generate the single-frequency impedance response under potentiostatic modulation with noise added to the time domain signals. The inputs for the code are as given in Section B.3.

```
function Lissajoubvnsfprint
```

```
global DV V0 i0 bb bc Cd;
close all;
DV=input('Input: Voltage Perturbation in mV.................  ');
V0=input('Input: Steady-State Voltage in mV.................  ');
% Re=input('Input: Ohmic Resistance in Ohms cm2:........  ');
Cd=input('Input: Capacity in microF/cm2.....................  ');
i0=input('Input: exchange current density in mA/cm2.........  ');
bb=input('Input: anode exponential coefficient in 1/V.......  ');
bc=input('Input: cathode exponential coefficient in 1/V.....  ');
noiseV=input('Input: noise in potential signal in percent.......  ');
noiseI=input('Input: noise in current signal in percent.........  ');

Npoints=3000;
Cd=Cd*1.0e-6;
i0=i0/1000;
DV=DV/1000;V0=V0/1000;noiseV=noiseV*DV/100;
Ibar=i0*(exp(bb*V0)-exp(-bc*V0));
Rt=1/(bb*i0*exp(bb*V0)+bc*i0*exp(-bc*V0));
ff=1/(2*pi*Rt*Cd);
noiseIf=noiseI*DV*sqrt(1+(Rt*Cd*2*pi*ff)^2)/(Rt*100);

for ncycle=1:10;
fname=sprintf('Lissajou-%g-%d.txt',ff,ncycle);
fid = fopen(fname, 'wt');
fprintf(fid, 'V0= %g V; DV= %g V; i0= %g A/cm2; ba= %g V-1;',...
        'ba= %g V-1\n',V0,DV,i0,bb,bc);
fprintf(fid, 'Rt= %g W cm2; Cd= %9.2e F/cm2; fchar= %g Hz\n',...
        Rt,Cd,1/(2*pi*Rt*Cd));
fprintf(fid,...
    'time; potential; current; scaled potential; scaled current \n');
Tmax=1/ff;
VR=0; VJ=0; IR=0; IJ=0;
t=0:Tmax/Npoints:Tmax;
for k=1:Npoints+1;
    V(k)=V0+DV*cos(2*pi*ff*t(k))+random('norm',0,noiseV);
    I(k)=faradaicbv(t(k),ff)+random('norm',0,noiseIf);
    VR=VR+(V(k)-V0)*cos(2*pi*ff*t(k));
    VJ=VJ+(V(k)-V0)*sin(2*pi*ff*t(k));
    IR=IR+(I(k)-Ibar)*cos(2*pi*ff*t(k));
    IJ=IJ+(I(k)-Ibar)*sin(2*pi*ff*t(k));
end;
Imax=max(abs(I-Ibar));
for k=1:Npoints+1;
    fprintf(fid, '%14.6e; %14.6e; %14.6e; %14.6e; %14.6e\n',...
        t(k),V(k),I(k),(V(k)-V0)/DV,(I(k)-Ibar)/(DV/Rt+Cd*2*pi*DV*ff));
end;
fclose(fid);
KoJ=2*pi*ff*Rt*Cd;
string=sprintf('Lissajou: f = %g Hz; K/J = %g; Delta V b = %g',...
    ff,KoJ,DV*bb);
figure;set(gcf,'Name',string);
plot((V-V0)/DV,(I-Ibar)/(DV/Rt+Cd*2*pi*DV*ff),'k-');title(string);...
```

```
    xlabel('Potential / V');ylabel('Current Density / A cm^{-2}');
string=sprintf('Time Trace: f = %g Hz; K/J = %g; Delta V b = %g',...
    ff,KoJ,DV*bb);
figure;set(gcf,'Name',string);
plot(t,(I-Ibar)/(DV/Rt+Cd*2*pi*DV*ff),'r-');title(string);...
    xlabel('time / s');ylabel('Response'); hold on;
plot(t,(V-V0)/DV,'b-');title(string);xlabel('time / s');...
    ylabel('Response'); legend('Current','Potential');


fname=sprintf('Lissajou-EIS2-%d.txt',ncycle);
fid = fopen(fname, 'wt');
fprintf(fid,'DV= %g V; i0= %g A/cm2; b= %g V-1; Cd= %9.2e F/cm2\n',...
    DV,i0,bb,Cd);
fprintf(fid, 'frequency; Zreal; Zimag; Zmodreal; Zmodimag \n');

IC=complex(IR,IJ);
VC=complex(VR,VJ);
Z=VC/IC;
Zmod=Rt/(1+1i*2*pi*ff*Rt*Cd);
fprintf(fid, '%14.6e; %14.6e; %14.6e; %14.6e; %14.6e \n',...
        ff,real(Z),imag(Z),real(Zmod),imag(Zmod));
fclose(fid);
end;
```

B.5 Convective Diffusion: Rotating Disk

This code, developed by Morgan Hazelbaker (née Harding),[152] accounts for three terms of the velocity expansion for a rotating disk electrode. It is an extension of the work by Tribollet and Newman[153] that accounted for two terms of the velocity expansion. The program makes use of a MATLAB® implementation of the tridiagonal solver BAND and MATINV developed by Newman.[154]

B.5.1 Zeroth-Order Term

```
% This problem solves for the first term of the convective-diffusion
% equation with a finite Schmidt number
% d^2(theta0)/d(xi)^2+3*xi^2*d(theta0)/d(xi)-j*K*theta0=0

clc; close all;
format longE;
global A B C D G X Y N NJ

h=0.001;                    %Step-size
logK=-3:.05:3;              %Frequency range 0.001 to 1000
K=10.^logK;
K;
x=0:h:10;                   %Range of x
NJ=length(x);
```

```
NJ;
ZZd=zeros(1,length(K));       %Place holder for diffusion impedance

conc_real=zeros(length(x),length(K));
conc_imag=zeros(length(x),length(K));

c_0=1;             %Boundary condition for x=0
c_e=0;             %Boundary condition for x=end (NJ)
tol=0;             %Absolute tolerance

%Define initial guess
conc(1,1:NJ)=0.5;    %Conc(1,-) is the real component of theta0
conc(2,1:NJ)=0.5;    %Conc(2,-) is the imaginary component of theta0
N=2;

for kk=1:length(K)

error=1;
jcount=0;
jcountmax=4;                  %Maximum number of iterations
while error>=tol
X=zeros(N,N);
Y=X;

jcount=jcount+1;
for J=1:NJ
    A=zeros(N,N);
    B=A;
    D=A;
    %Boundary condition at J=1
    if J==1
        G(1)=conc(1,J)-c_0;       %Real equation
        B(1,1)=-1;

        G(2)=conc(2,J)-0;         %Imaginary equation
        B(2,2)=-1;

        error=abs(G(1))+abs(G(2));

    %Region between boundaries
    elseif (J<=(NJ-1))
        G(1)=conc(1,J+1)-2*conc(1,J)+conc(1,J-1)+...
            h^2*K(kk)*conc(2,J)+...
            1.5*h^3*(J-1)^2*(conc(1,J+1)-conc(1,J-1));
        A(1,1)=-1+1.5*h^3*(J-1)^2;
        B(1,1)=2;
        D(1,1)=-1-1.5*h^3*(J-1)^2;
        B(1,2)=-h^2*K(kk);

        G(2)=conc(2,J+1)-2*conc(2,J)+conc(2,J-1)-...
            h^2*K(kk)*conc(1,J)+...
            1.5*h^3*(J-1)^2*(conc(2,J+1)-conc(2,J-1));
```

```
        A(2,2)=-1+1.5*h^3*(J-1)^2;
        B(2,2)=2;
        D(2,2)=-1-1.5*h^3*(J-1)^2;
        B(2,1)=h^2*K(kk);

        error=error+abs(G(1))+abs(G(2));

    %Boundary condition at J=NJ
    else
        G(1)=conc(1,J)-0;      %BC at far away point is always 0
        B(1,1)=-1;

        G(2)=conc(2,J)-0;      %BC at far away point is always 0
        B(2,2)=-1;

        error=error+abs(G(1))+abs(G(2));
    end

band(J)
end
conc=conc+C;          %C comes from band(J) program

if jcount>=jcountmax, break, end
end

%Creating theta from conc(1,-) and conc(2,-) values
%dtheta accurate to h^2
theta0=complex(conc(1,:),conc(2,:));
dtheta0=gamma(4/3)*(-theta0(3)+4*theta0(2)-3*theta0(1))/(2*h);


Zd=-1/dtheta0;              %turning theta into Impedance

% str=sprintf('K = %g: Zd = %.10e + j%.10e', K(kk), real(Zd),...
% imag(Zd));
% disp(str);
% str=sprintf('dtheta0 = %g', dtheta0); disp(str);
str=sprintf('%.10e', K(kk)); disp(str);
str=sprintf('%.10e', real(Zd)); disp(str);
str=sprintf('%.10e', imag(Zd)); disp(str);

figure(1)
plot(x,conc(1,:),'-r')
hold on;
plot(x,conc(2,:),'-g')
axis([0 2 -.4 1]);
legend('theta_r','theta_i');
title('Dimensionless Concentration away from Electrode Surface');
xlabel('length');
ylabel('theta');

for kkk=1:length(x)
```

```
    conc_real(kkk,kk)=(conc(1,kkk));
    conc_imag(kkk,kk)=(conc(2,kkk));
end

ZZd(kk)=Zd;
Ktemp=K(kk);
fname=sprintf('theta0_%i',kk);
save(fname,'theta0','Ktemp','h');
end

fname=sprintf('zero_order_%i.txt',NJ);
fid = fopen(fname, 'wt');
fprintf(fid, '%14.6e; %14.6e \n', h,h^2);
for kk=1:length(K)
fprintf(fid, '%12.6e; %20.12e; %20.12e; \n', K(kk),real(ZZd(kk)),...
    imag(ZZd(kk)));
end
fclose(fid);

figure(2)
plot(real(ZZd),-imag(ZZd),'-ks');hold on; axis equal;
Zfilm=tanh(sqrt(1i*K))./sqrt(1i*K);
plot(real(Zfilm),-imag(Zfilm),'--b');
legend('MatLab Data',...
    'Finite Film Thickness tanh(sqrt(j*K))/sqrt(j*K)');
title('Nyquist plot');
xlabel('Real part of Impedance');
ylabel('Imaginary part of Impedance');

figure(3)
loglog(K, -imag(ZZd),'--k'); hold on;
loglog(K, real(ZZd),'--b');
legend ('-Imaginary','Real');
title('Impedance vs Frequency');
xlabel('Dimensionless Frequency');
ylabel('Impedance');
```

BAND Algorithm

```
% ****************************************************************
% MATLAB version of BAND
% The following is a version of BAND (and MATINV) obtained by
%  translating the FORTRAN version in Newman.
% ****************************************************************

function [y]=band(J) % i.e. 'J' = 'I' from test_band
global A B C D E G X Y NPOINTS N NJ NP1 DETERMINANT
% ****************************************************************
% Case where J negative
% ****************************************************************
if (J-2) < 0 % i.e. image point (I=1) outside boundary (left hand side)
```

```
    NP1=N+1; % I runs from 1:NPOINTS and NPOINTS = NJ, while N = # eqn
    for I=1:N
        D(I,2*N+1)=G(I);
        for L=1:N
            LPN=L+N;
            D(I,LPN)=X(I,L);
        end
    end
    %  ************************************************************
    %  Calling Matinv
    %  ************************************************************
    matinv(N,2*N+1);
    if (DETERMINANT == 0)
        sprintf('DETERMINANT = 0 at J= %6.3f',J)
        return % stod tidligere break
    else
        for K=1:N
            E(K,NP1,1)=D(K,2*N+1);
            for L=1:N
                E(K,L,1) = -D(K,L);
                LPN = L+N;
                X(K,L) = -D(K,LPN);
            end
        end
        return
    end
end
% **************************************************************
% Case where J-2 equal to zero (I=2, i.e. first real point)
% **************************************************************
if((J-2)==0)

    for I=1:N
        for K=1:N
            for L=1:N
                D(I,K)=D(I,K)+A(I,L)*X(L,K);
            end
        end
    end
end
% **************************************************************
% Case where J-2 is greater than zero
% **************************************************************
if (J-NJ) < 0

    for I=1:N
        D(I,NP1) = -G(I);
        for L=1:N
            D(I,NP1)=D(I,NP1)+A(I,L)*E(L,NP1,J-1);
            for K=1:N
                B(I,K)=B(I,K)+A(I,L)*E(L,K,J-1);
            end
```

```
        end
    end
else

    for I=1:N
        for L=1:N
            G(I)=G(I)-Y(I,L)*E(L,NP1,J-2);
            for M=1:N
                A(I,L)=A(I,L)+Y(I,M)*E(M,L,J-2);
            end
        end
    end
    for I=1:N
        D(I,NP1) = -G(I);
        for L=1:N
            D(I,NP1)=D(I,NP1)+A(I,L)*E(L,NP1,J-1);
            for K=1:N
                B(I,K)=B(I,K)+A(I,L)*E(L,K,J-1);
            end
        end
    end
end
% *****************************************************************
%   Calling Matinv
% *****************************************************************

matinv(N,NP1);
if DETERMINANT ~= 0

    for K=1:N
        for M=1:NP1
            E(K,M,J) = -D(K,M);
        end
    end
else
    sprintf('DETERMINANT = 0 at J= %6.3f',J)
    return % stod tidligere break
end

if (J-NJ)<0

    return
else

    for K = 1:N

        C(K,J) = E(K,NP1,J);
    end
    for JJ = 2:NJ

        M = NJ - JJ + 1;
        for K=1:N
```

```
            C(K,M) = E(K,NP1,M);
            for L=1:N
                C(K,M)=C(K,M)+E(K,L,M)*C(L,M+1);
            end
        end
    end
    for L=1:N

        for K=1:N
            C(K,1) = C(K,1) + X(K,L)*C(L,3);
        end
    end
end

return
```

MATINV Algorithm

```
function [DETERMINANT] = matinv(N,M)
global A B C D E G X Y NPOINTS NJ DETERMINANT
DETERMINANT=1.0;
for I=1:N
    ID(I)=0;
end
for NN=1:N
    BMAX = 1.1;
    for I=1:N
        if(ID(I)~=0)
            continue
        else
            BNEXT = 0;
            BTRY = 0;
            for J=1:N
                if(ID(J)==0)
                    if(abs(B(I,J)) > BNEXT)
                        BNEXT=abs(B(I,J));
                        if(BNEXT > BTRY)
                            BNEXT = BTRY;
                            BTRY = abs(B(I,J));
                            JC = J;
                        end
                    end
                end
            end
        end
        if(BNEXT < (BMAX*BTRY))
            BMAX=BNEXT/BTRY;
            IROW=I;
            JCOL=JC;
        end
    end
    if(ID(JC)==0)
```

```
            ID(JCOL)=1;
        else
            DETERMINANT=0;
            return
        end
        if(JCOL==IROW)
            F=1/(B(JCOL,JCOL));
            for J=1:N
                B(JCOL,J)=B(JCOL,J)*F;
            end
            for K=1:M
                D(JCOL,K)=D(JCOL,K)*F;
            end
        else
            for J=1:N
                SAVE=B(IROW,J);
                B(IROW,J)=B(JCOL,J);
                B(JCOL,J)=SAVE;
            end
            for K=1:M
                SAVE=D(IROW,K);
                D(IROW,K)=D(JCOL,K);
                D(JCOL,K)=SAVE;
            end
        end
        for I=1:N
            if(I ~= JCOL)
                F=B(I,JCOL);
                for J=1:N
                    B(I,J)=B(I,J)-F*B(JCOL,J);
                end
                for K=1:M
                    D(I,K)=D(I,K)-F*D(JCOL,K);
                end
            end
        end
end
return
```

B.5.2 First-Order Term

This program requires the output of code presented in Section B.5.1. In addition, this code requires both the BAND and MATINV algorithms presented in Section B.5.1.

```
%This problem solves for the first term of the convective-diffusion
% equation with a finite Schmidt number
%d^2(theta0)/d(xi)^2+3*xi^2*d(theta0)/d(xi)-j*K*theta0=0

clc; close all;
format longE;
global A B C D G X Y N NJ

h=0.001;                        %Step-size
```

```
logK=-3:.05:3;                  %Frequency range 0.001 to 1000
K=10.^logK;
x=0:h:10;                       %Range of x
NJ=length(x);
ZZd=zeros(1,length(K));         %Place holder for diffusion impedance
aa=0.51023;
bb=-0.61592;
cc=(3/aa^4)^(1/3);


conc_real=zeros(length(x),length(K));
conc_imag=zeros(length(x),length(K));

c_0=1;          %Boundary condition for x=0
c_e=0;          %Boundary condition for x=end (NJ)
tol=0;          %Absolute tolerance

%Define initial guess
conc(1,1:NJ)=0.5;  %Conc(1,-) is the real component of theta0
conc(2,1:NJ)=0.5;  %Conc(2,-) is the imaginary component of theta0
N=2;

for kk=1:length(K)
fname=sprintf('theta0_%i',kk);
load(fname);
flag=0;if(abs(Ktemp-K(kk))~=0);flag=3;quit;end

error=1;
jcount=0;
jcountmax=4;                 %Maximum number of iterations
while error>=tol
X=zeros(N,N);
Y=X;

jcount=jcount+1;
for J=1:NJ
    A=zeros(N,N);
    B=A;
    D=A;
    %Boundary condition at J=1
    if J==1
        G(1)=conc(1,J);  %Real equation, BC at electrode is 0
        B(1,1)=-1;

        G(2)=conc(2,J);  %Imaginary equation, BC at electrode is 0
        B(2,2)=-1;

        error=abs(G(1))+abs(G(2));

    %Region between boundaries
    elseif (J<=(NJ-1))
        G(1)=conc(1,J+1)-2*conc(1,J)+conc(1,J-1)+...
```

```
            h^2*K(kk)*conc(2,J)+...
            1.5*h^3*(J-1)^2*(conc(1,J+1)-conc(1,J-1))...
            -cc*(h^4)*((J-1)^3)*(real(theta0(J+1))...
            -real(theta0(J-1)))/2;
        A(1,1)=-1+1.5*h^3*(J-1)^2;
        B(1,1)=2;
        D(1,1)=-1-1.5*h^3*(J-1)^2;
        B(1,2)=-h^2*K(kk);

        G(2)=conc(2,J+1)-2*conc(2,J)+conc(2,J-1)-...
            h^2*K(kk)*conc(1,J)+1.5*h^3*(J-1)^2*(conc(2,J+1)-...
            conc(2,J-1))-cc*(h^4)*((J-1)^3)*(imag(theta0(J+1))-...
            imag(theta0(J-1)))/2;
        A(2,2)=-1+1.5*h^3*(J-1)^2;
        B(2,2)=2;
        D(2,2)=-1-1.5*h^3*(J-1)^2;
        B(2,1)=h^2*K(kk);
        error=error+abs(G(1))+abs(G(2));
    %Boundary condition at J=NJ
    else
        G(1)=conc(1,J)-0;  %BC at far away point is always 0
        B(1,1)=-1;

        G(2)=conc(2,J)-0;  %BC at far away point is always 0
        B(2,2)=-1;

        error=error+abs(G(1))+abs(G(2));
    end

band(J)
end
conc=conc+C;            %C comes from band(J) program

if jcount>=jcountmax, break, end
end

 %Creating theta from conc(1,-) and conc(2,-) values
 %dtheta accurate to h^2
theta1=complex(conc(1,:),conc(2,:));
dtheta1=gamma(4/3)*(-theta1(3)+4*theta1(2)-3*theta1(1))/(2*h);
dtheta0=gamma(4/3)*(-theta0(3)+4*theta0(2)-3*theta0(1))/(2*h);

Zd=dtheta1/(dtheta0^2);              %turning theta into Impedance

%str=sprintf('K = %.10e: Zd = %.10e + j%.10e', K(kk), real(Zd),...
% imag(Zd)); disp(str);
%str=sprintf('dtheta1 = %g', dtheta1); disp(str);

 str=sprintf('%.10e', K(kk)); disp(str);
 str=sprintf('%.10e', real(Zd)); disp(str);
 str=sprintf('%.10e', imag(Zd)); disp(str);
```

```
figure(1)
plot(x,conc(1,:),'-r')
hold on;
plot(x,conc(2,:),'-g')
axis([0 2 -.4 1]);
legend('theta_r','theta_i');
title('Dimensionless Concentration away from Electrode Surface');
xlabel('length');
ylabel('theta');

for kkk=1:length(x)
    conc_real(kkk,kk)=(conc(1,kkk));
    conc_imag(kkk,kk)=(conc(2,kkk));
end


ZZd(kk)=Zd;
Ktemp=K(kk);
fname=sprintf('theta1_%i',kk);
save(fname,'theta1','Ktemp','h');
end

fname=sprintf('first_order_%i.txt',NJ);
fid = fopen(fname, 'wt');
fprintf(fid, '%14.6e; %14.6e \n', h,h^2);
for kk=1:length(K)
fprintf(fid, '%12.6e; %20.12e; %20.12e; \n',...
    K(kk),real(ZZd(kk)),imag(ZZd(kk)));
end
fclose(fid);

figure(2)
plot(real(ZZd),-imag(ZZd),'-ks');hold on; axis equal;
title('Nyquist plot');
xlabel('Real part of Impedance');
ylabel('Imaginary part of Impedance');

figure(3)
semilogx(K, -imag(ZZd),'--k'); hold on;
semilogx(K, real(ZZd),'--b');
legend ('-Imaginary','Real');
title('Impedance vs Frequency');
xlabel('Dimensionless Frequency');
ylabel('Impedance');
```

B.5.3 Second Order Term

This program requires the output of code presented in Sections B.5.1 and RDE-first. In addition, this code requires both the BAND and MATINV algorithms presented in Section B.5.1.

```
%This problem solves for the second term of the convective-diffusion
% equation with a finite Schmidt number
%d^2(theta0)/d(xi)^2+3*xi^2*d(theta0)/d(xi)-j*K*theta0=0

clc; close all;
format longE;
global A B C D G X Y N NJ

h=0.001;                    %Step-size
logK=-3:.05:3;              %Frequency range 0.001 to 1000
K=10.^logK;
K;
x=0:h:10;                   %Range of x
NJ=length(x);
NJ;
ZZd=zeros(1,length(K));     %Place holder for diffusion impedance
aa=0.51023;
bb=-0.61592;
cc=(3/aa^4)^(1/3);

conc_real=zeros(length(x),length(K));
conc_imag=zeros(length(x),length(K));

c_0=1;          %Boundary condition for x=0
c_e=0;          %Boundary condition for x=end (NJ)
tol=0;          %Absolute tolerance

%Define initial guess
conc(1,1:NJ)=0.5;    %Conc(1,-) is the real component of theta0
conc(2,1:NJ)=0.5;    %Conc(2,-) is the imaginary component of theta0
N=2;

for kk=1:length(K)
fname=sprintf('theta0_%i',kk); % load solution for Theta-0
load(fname);
flag=0;if(abs(Ktemp-K(kk))~=0);flag=3;quit;end
fname=sprintf('theta1_%i',kk); % load solution for Theta-1
load(fname);
flag=0;if(abs(Ktemp-K(kk))~=0);flag=3;quit;end

error=1;
jcount=0;
jcountmax=4;              %Maximum number of iterations
while error>=tol
X=zeros(N,N);
Y=X;

jcount=jcount+1;
for J=1:NJ
    A=zeros(N,N);
    B=A;
    D=A;
```

```
    %Boundary condition at J=1
    if J==1
        G(1)=conc(1,J);       %Real equation
        B(1,1)=-1;

        G(2)=conc(2,J);       %Imaginary equation
        B(2,2)=-1;

        error=abs(G(1))+abs(G(2));

    %Region between boundaries
    elseif (J<=(NJ-1))
        G(1)=conc(1,J+1)-2*conc(1,J)+conc(1,J-1)+...
            h^2*K(kk)*conc(2,J)+1.5*h^3*(J-1)^2*(conc(1,J+1)-...
            conc(1,J-1))-...
            bb*(3/aa)^(5/3)*(h^5)*((J-1)^4)*(real(theta0(J+1))-...
            real(theta0(J-1)))/12 ...
            -cc*(h^4)*((J-1)^3)*(real(theta1(J+1))-...
            real(theta1(J-1)))/2;
        A(1,1)=-1+1.5*h^3*(J-1)^2;
        B(1,1)=2;
        D(1,1)=-1-1.5*h^3*(J-1)^2;
        B(1,2)=-h^2*K(kk);

        G(2)=conc(2,J+1)-2*conc(2,J)+conc(2,J-1)-h^2*K(kk)*conc(1,J)+...
            1.5*h^3*(J-1)^2*(conc(2,J+1)-conc(2,J-1))...
            -bb*(3/aa)^(5/3)*(h^5)*((J-1)^4)*(imag(theta0(J+1))-...
            imag(theta0(J-1)))/12 ...
            -cc*(h^4)*((J-1)^3)*(imag(theta1(J+1))-...
            imag(theta1(J-1)))/2;
        A(2,2)=-1+1.5*h^3*(J-1)^2;
        B(2,2)=2;
        D(2,2)=-1-1.5*h^3*(J-1)^2;
        B(2,1)=h^2*K(kk);

        error=error+abs(G(1))+abs(G(2));

    %Boundary condition at J=NJ
    else
        G(1)=conc(1,J)-0;      %BC at far away point is always 0
        B(1,1)=-1;

        G(2)=conc(2,J)-0;      %BC at far away point is always 0
        B(2,2)=-1;

        error=error+abs(G(1))+abs(G(2));
    end

band(J)
end
conc=conc+C;          %C comes from band(J) program
```

```
if jcount>=jcountmax, break, end
end
%Creating theta from conc(1,-) and conc(2,-) values
%dtheta accurate to h^2
theta2=complex(conc(1,:),conc(2,:));
dtheta0=gamma(4/3)*(-theta0(3)+4*theta0(2)-3*theta0(1))/(2*h);
dtheta1=gamma(4/3)*(-theta1(3)+4*theta1(2)-3*theta1(1))/(2*h);
dtheta2=gamma(4/3)*(-theta2(3)+4*theta2(2)-3*theta2(1))/(2*h);

%turning theta into Impedance
Zd=-(1/dtheta0)*((dtheta1/dtheta0)^2-dtheta2/dtheta0);
ZZd(kk)=Zd;

%str=sprintf('K = %.10e: Zd = %.10e + j%.10e', K(kk),...
%   real(Zd),imag(Zd)); disp(str);
%str=sprintf('dtheta2 = %g', dtheta2); disp(str);

str=sprintf('%.10e', K(kk)); disp(str);
str=sprintf('%.10e', real(Zd)); disp(str);
str=sprintf('%.10e', imag(Zd)); disp(str);

figure(1)
plot(x,conc(1,:),'-r')
hold on;
plot(x,conc(2,:),'-g')
axis([0 2 -.4 1]);
legend('theta_r','theta_i');
title('Dimensionless Concentration away from Electrode Surface');
xlabel('length');
ylabel('theta');

for kkk=1:length(x)
    conc_real(kkk,kk)=(conc(1,kkk));
    conc_imag(kkk,kk)=(conc(2,kkk));
end

end

fname=sprintf('second_order_%i.txt',NJ);
fid = fopen(fname, 'wt');
fprintf(fid, '%14.6e; %14.6e \n', h,h^2);
for kk=1:length(K)
fprintf(fid, '%12.6e; %20.12e; %20.12e; \n', ...
    K(kk),real(ZZd(kk)),imag(ZZd(kk)));
end
fclose(fid);

figure(2)
plot(real(ZZd),-imag(ZZd),'-ks');hold on; axis equal;
title('Nyquist plot');
xlabel('Real part of Impedance');
ylabel('Imaginary part of Impedance');
```

```
figure(3)
semilogx(K, -imag(ZZd),'--k'); hold on;
semilogx(K, real(ZZd),'--b');
legend ('-Imaginary','Real');
title('Impedance vs Frequency');
xlabel('Dimensionless Frequency');
ylabel('Impedance');
```

B.6 Confirmation of Kramers–Kronig Relations

The MATLAB® programs presented in this section were used to solve questions posed in Chapter 22.

B.6.1 Equation (A.84)

The following code was used for the solution of Problem 22.1.a.

```
clc;close all;
%Define parameters
Re=10; % ohm cm2
Rt=100; %ohm cm2
C=20e-6; %F/cm2
f=logspace(-1,6,100);
w=2*pi*f;
T=1./f;
eps1=1e-4;
m=zeros(1,length(T));
Z=zeros(1,length(T));
Zr=zeros(1,length(T));
Zj=zeros(1,length(T));
phi=zeros(1,length(T));
ZjK1=zeros(1,length(T));
%Equivalent circuit
for i=1:length(T)
m(i)=complex(0,w(i));
Z(i)=Re+Rt/(1+m(i)*Rt*C);
Zr(i)=real(Z(i));
Zj(i)=imag(Z(i));
end
%Kramers-Kronig, equation 23.1
for i=1:length(T)
a=w(i);
K=@(x)(Re+Rt./(1+(x.*Rt.*C).^2))./(x-a);
int1=integral(K,-1e9,a-eps1);
int2=integral(K,a+eps1,1e9);
ZjK1(i)=(1/pi).*(int1+int2);
end
semilogx(f,-Zj,'o');
hold on;
```

```
semilogx(f,-ZjK1);
xlim([0.1,1e6]);
ax = gca;
ax.XTick = [0.1,1,10,100,1e3,1e4,1e5,1e6];
xlabel('Frequency/Hz');
ylabel('-Z_j / \Omega cm^2');
title('Verification of Equation (A.85)');
```

B.6.2 Equation (A.85)

The following code was used for the solution of Problem 22.1.b.

```
clc;close all;
%Define parameters
Re=10; % ohm cm2
Rt=100; %ohm cm2
C=20e-6; %F/cm2
f=logspace(-1,6,100);
w=2*pi*f;
T=1./f;
eps1=1e-3;
m=zeros(1,length(T));
Z=zeros(1,length(T));
Zr=zeros(1,length(T));
Zj=zeros(1,length(T));
phi=zeros(1,length(T));
ZrK1=zeros(1,length(T));
%Equivalent circuit
for i=1:length(T)
m(i)=complex(0,w(i));
Z(i)=Re+Rt/(1+m(i)*Rt*C);
Zr(i)=real(Z(i));
Zj(i)=imag(Z(i));
end
%Kramers-Kronig, equation 23.2
for i=1:length(T)
a=w(i);
K=@(x)((-x*Rt^2*C)./(1+(x.*Rt.*C).^2))./(x-a);
int1=integral(K,-1e9,a-eps1);
int2=integral(K,a+eps1,1e9);
ZrK1(i)=Re-(1/pi).*(int1+int2);
end
semilogx(f,Zr,'o');
hold on;
semilogx(f,ZrK1);
xlabel('Frequency/Hz');
ylabel('-Z_j / \Omega cm^2');
title('Verification of Equation (A.86)');
```

B.6.3 Equation (A.86)

The following code was used for the solution of Problem 22.1.c.

```
clc;close all;
%Define parameters
Re=10; % ohm cm2
Rt=100; %ohm cm2
C=20e-6; %F/cm2
f=logspace(-1,6,100);
w=2*pi*f;
T=1./f;
eps1=1e-4;
m=zeros(1,length(T));
Z=zeros(1,length(T));
Zr=zeros(1,length(T));
Zj=zeros(1,length(T));
phi=zeros(1,length(T));
ZjK1=zeros(1,length(T));
%Equivalent circuit
for i=1:length(T)
m(i)=complex(0,w(i));
Z(i)=Re+Rt/(1+m(i)*Rt*C);
Zr(i)=real(Z(i));
Zj(i)=imag(Z(i));
end
%Kramers-Kronig, equation 23.3
for i=1:length(T)
a=w(i);
K=@(x)(Re+Rt./(1+(x.*Rt.*C).^2))./(x.^2-a.^2);
int1=integral(K,0,a-eps1);
int2=integral(K,a+eps1,1e9);
ZjK1(i)=(2*w(i)/pi).*(int1+int2);
end
semilogx(f,-Zj,'o');
hold on;
semilogx(f,-ZjK1);
xlabel('Frequency/Hz');
ylabel('-Z_j / \Omega cm^2');
title('Verification of Equation (A.87)');
```

B.6.4 Equation (A.87)

The following code was used for the solution of Problem 22.1.d.

```
clc;close all;
%Define parameters
Re=10; % ohm cm2
Rt=100; %ohm cm2
C=20e-6; %F/cm2
f=logspace(-1,6,100);
w=2*pi*f;
```

```
T=1./f;
eps1=1e-4;
m=zeros(1,length(T));
Z=zeros(1,length(T));
Zr=zeros(1,length(T));
Zj=zeros(1,length(T));
phi=zeros(1,length(T));
ZrK1=zeros(1,length(T));
%Equivalent circuit
for i=1:length(T)
m(i)=complex(0,w(i));
Z(i)=Re+Rt/(1+m(i)*Rt*C);
Zr(i)=real(Z(i));
Zj(i)=imag(Z(i));
end
%Kramers-Kronig, equation 23.4
for i=1:length(T)
a=w(i);
K=@(x)(-x.^2.*(Rt.^2.*C)./(1+(x.*Rt.*C).^2))./(x.^2-a.^2);
int1=integral(K,0,a-eps1);
int2=integral(K,a+eps1,1e9);
ZrK1(i)=Re-(2/pi).*(int1+int2);
end
semilogx(f,Zr,'o');
hold on;
semilogx(f,ZrK1);
xlim([0.1,1e6]);
ax = gca;
ax.XTick = [0.1,1,10,100,1e3,1e4,1e5,1e6];
xlabel('Frequency/Hz');
ylabel('-Z_j / \Omega cm^2');
title('Verification of Equation (A.88)');
```

B.6.5 Equation (A.88)

The following code was used for the solution of Problem 22.1.e.

```
clc;close all;
%Define parameters
Re=10; % ohm cm2
Rt=100; %ohm cm2
C=20e-6; %F/cm2
f=logspace(-1,6,100);
w=2*pi*f;
T=1./f;
eps1=1e-4;
m=zeros(1,length(T));
Z=zeros(1,length(T));
Zr=zeros(1,length(T));
Zj=zeros(1,length(T));
phi=zeros(1,length(T));
ZjK1=zeros(1,length(T));
```

```
%Equivalent circuit
for i=1:length(T)
m(i)=complex(0,w(i));
Z(i)=Re+Rt/(1+m(i)*Rt*C);
Zr(i)=real(Z(i));
Zj(i)=imag(Z(i));

end
%Kramers-Kronig, equation 23.5
for i=1:length(T)
a=w(i);
K=@(x)(Re+Rt./(1+(x.*Rt.*C).^2)-Zr(i))./(x.^2-a.^2);
int1=integral(K,0,a-eps1);
int2=integral(K,a+eps1,1e9);
ZjK1(i)=(2*w(i)/pi).*(int1+int2);
end
semilogx(f,-Zj,'o');
hold on;
semilogx(f,-ZjK1);
xlim([0.1,1e6]);
ax = gca;
ax.XTick = [0.1,1,10,100,1e3,1e4,1e5,1e6];
xlabel('Frequency/Hz');
ylabel('-Z_j / \Omega cm^2');
title('Verification of Equation (A.89)');
```

B.6.6 Equation (A.89)

The following code was used for the solution of Problem 22.1.f.

```
clc;close all;
%Define parameters
Re=10; % ohm cm2
Rt=100; %ohm cm2
C=20e-6; %F/cm2
f=logspace(-1,6,100);
w=2*pi*f;
T=1./f;
eps1=1e-4;
m=zeros(1,length(T));
Z=zeros(1,length(T));
Zr=zeros(1,length(T));
Zj=zeros(1,length(T));
phi=zeros(1,length(T));
ZrK1=zeros(1,length(T));
%Equivalent circuit
for i=1:length(T)
m(i)=complex(0,w(i));
Z(i)=Re+Rt/(1+m(i)*Rt*C);
Zr(i)=real(Z(i));
Zj(i)=imag(Z(i));
```

```
end
%Kramers-Kronig, equation 23.6
for i=1:length(T)
a=w(i);
K=@(x)((-x.^2*Rt^2*C)./(1+(x.*Rt.*C).^2)-w(i)*Zj(i))./(x.^2-a.^2);
int1=integral(K,0,a-eps1);
int2=integral(K,a+eps1,1e9);
ZrK1(i)=Re-(2/pi).*(int1+int2);
end
semilogx(f,Zr,'o');
hold on;
semilogx(f,ZrK1);
xlim([0.1,1e6]);
ax = gca;
ax.XTick = [0.1,1,10,100,1e3,1e4,1e5,1e6];
xlabel('Frequency/Hz');
ylabel('-Z_j / \Omega cm^2');
title('Verification of Equation (A.90)');
```

B.6.7 Equation (A.90)

The following code was used for the solution of Problem 22.1.g.

```
clc;close all;

%Define parameters
Re=10; % ohm cm2
Rt=100; %ohm cm2
Cdl=20e-6; %F/cm2
f=logspace(-1,6,100);
w=2*pi*f;
T=1./f;
eps1=1e-4;
%Equivalent circuit
m=zeros(1,length(T));
Z=zeros(1,length(T));
Zr=zeros(1,length(T));
Zj=zeros(1,length(T));
phi=zeros(1,length(T));
ZjK7=zeros(1,length(T));
for i=1:length(T)
m(i)=complex(0,w(i));
Z(i)=Re+Rt/(1+m(i)*Rt*Cdl);
Zr(i)=real(Z(i));
Zj(i)=imag(Z(i));
phi(i) = atan2(Zj(i),Zr(i));
end

%Kramers-Kronig, equation 22.7
for i=1:length(T)
a=w(i);
K=@(x)(log(abs(Re+Rt./(1+1j*x*Rt*Cdl)))./(x.^2-a.^2));
```

```
int1=integral(K,0,a-eps1);
int2=integral(K,a+eps1,1e10);
ZjK7(i)=(2*a/pi).*(int1+int2); % const before integral
end

semilogx(f,rad2deg(phi),'bo'); % can also use semilogx()
hold on;
semilogx(f, rad2deg((ZjK7)), 'k-') ;

xlim([0.1,1e6]);
ax = gca;
ax.XTick = [0.1,1,10,100,1e3,1e4,1e5,1e6];
xlabel('Frequency/Hz');
ylabel('\Phi / degrees ');
title('Verification of Equation (A.91)');
```

B.6.8 Equation (A.91)

The following code was used for the solution of Problem 22.1.h.

```
clc; close all;

% Define parameters
Re = 10;                  % Ohm cm
Rt = 100;                 % Ohm cm
Cdl = 20e-6;              % F/cm
f = logspace(-1, 6, 100);    % Hz
w = 2 * pi * f;              % rad/s
eps1 = 1e-4;

% Compute impedance
Z = Re + Rt ./ (1 + 1i * w * Rt * Cdl);
Zmag = abs(Z);                       % Magnitude
phi_true = angle(Z);                 % True phase angle
lnZmag = log(Zmag);                  % ln|Z(  )|

% Interpolation function for ln|Z(x)|
lnZ_fun = @(x) interp1(w, lnZmag, x, 'linear', 'extrap');

% Allocate result
phiK_A92 = zeros(size(w));

% Compute  (  ) using Eq. A.92
for i = 1:length(w)
    omega = w(i);
    lnZ_at_omega = lnZmag(i);

    % Define integrand
    K = @(x) (lnZ_fun(x) - lnZ_at_omega) ./ (x.^2 - omega^2);

    % Integrate avoiding singularity at x = omega
    int1 = integral(K, 0, omega - eps1, 'ArrayValued', true);
```

```
    int2 = integral(K, omega + eps1, 1e7, 'ArrayValued', true);

    phiK_A92(i) = (2 * omega / pi) * (int1 + int2);
end

% Plot comparison
figure;
semilogx(f, rad2deg(phi_true), 'bo', 'DisplayName',...
    'True    (from model)');
hold on;
semilogx(f, rad2deg(phiK_A92), 'r-', 'LineWidth',...
    1.5, 'DisplayName', '    from Eq. A.92');
xlabel('Frequency / Hz');
ylabel('\phi (degrees)');
title('Phase Angle from Eq. A.92');
legend('Location', 'best');
grid on;
```

B.6.9 Equation (A.92)

The following code was used for the solution of Problem 22.1.i

```
clc; close all;

% Define parameters
Re = 10;                    % Ohm cm
Rt = 100;                   % Ohm cm
Cdl = 20e-6;                % F/cm
f = logspace(-1, 6, 100);   % Hz
w = 2 * pi * f;             % rad/s
eps1 = 1e-4;

% Compute impedance
Z = Re+Rt./(1+1i*w*Rt*Cdl);
Zmag = abs(Z);
phi_true = angle(Z);

% Compute ln|Z| and numerical derivative d/dw [ln|Z|]
lnZmag = log(Zmag);              % ln|Z|
dlnZdw = gradient(lnZmag, w);
% Derivative w.r.t. angular frequency

% Interpolation function for d/dw [ln|Z(w)|]
dlnZdw_fun = @(x) interp1(w, dlnZdw, x, 'linear', 'extrap');

% Allocate array for   ( ) from A.93
phiK_A93 = zeros(size(w));

% Compute   ( ) using Eq. A.93
for i = 1:length(w)
    omega = w(i);
    kernel = @(x) log(abs((x-omega)./(x+omega)));
```

```
    % log(|(x−   )/(x+   )|)
    integrand = @(x) kernel(x).*dlnZdw_fun(x);
    % Integral over [0,     ), avoiding singularity near x=0
    int_val = integral(integrand, 1e-4, 1e7,...
        'ArrayValued', true); % adjust bounds
    phiK_A93(i) = -(1 / (pi)) * int_val;
end

% Plot comparison
figure;
semilogx(f,rad2deg(phi_true),'bo','DisplayName',...
    'True      (from model)');
hold on;
semilogx(f,rad2deg(phiK_A93),'r-','LineWidth',1.5,...
    'DisplayName', '    from KK Eq. A.93');
xlabel('Frequency / Hz');
ylabel('\phi (degrees)');
title('Phase Angle from Eq. A.93');
legend('Location', 'best');
grid on;
```

B.6.10 Equation (A.93)

The following code was used for the solution of Problem 22.1.j.

```
clc; close all;

% Define parameters
Re = 10; % ohm cm
Rt = 100; % ohm cm
Cdl = 20e-6; % F/cm
f = logspace(-1, 6, 100); % Hz
w = 2 * pi * f;           % rad/s
T = 1 ./ f;
eps1 = 1e-4;

% Equivalent circuit: compute Z and phase phi
Z = zeros(1, length(T));
phi = zeros(1, length(T));
for i = 1:length(T)
    m = complex(0, w(i));
    Z(i) = Re + Rt / (1 + m * Rt * Cdl);
    phi(i) = atan2(imag(Z(i)), real(Z(i)));
end


% Create extended phi function for negative frequencies
% (odd symmetry)
phi_ext = @(x) sign(x) .* interp1(w, phi, abs(x), ...
    'linear', 'extrap');

% Kramers-Kronig: A.94 with negative sign
```

```
lnZK94 = zeros(size(w));
for i = 1:length(T)
    a = w(i);
    % Integrand: phi(x) / (x - a)
    K = @(x) (phi_ext(x) )./ (x - a);
    int1 = integral(K, -1e10, a - eps1);
    int2 = integral(K, a + eps1, 1e10);
    lnZK94(i) = -(1 / pi) * (int1 + int2);
end

% Compare with direct ln|Z|
lnZmag_direct = log(abs(Z));

% Plot results
semilogx(f, lnZmag_direct, 'bo'); % Direct from impedance
hold on;
semilogx(f, lnZK94, 'k-'); % From modified A.94

xlim([0.1, 1e6]);
ax = gca;
ax.XTick = [0.1, 1, 10, 100, 1e3, 1e4, 1e5, 1e6];
xlabel('Frequency / Hz');
ylabel('ln|Z|');
title('Verification of Equation (A.94)');
legend('Direct', 'A.94 (KK)');
grid on;
```

B.6.11 Equation (A.94)

The following code was used for the solution of Problem 22.1.k.

```
clc; close all;

% Parameters
Re = 10;                % Ohm cm
Rt = 100;               % Ohm cm
C = 20e-6;              % F/cm
eps1 = 1e-7;            % To avoid singularity

% Frequency and angular frequency
f = logspace(-1, 6, 100); % Hz
w = 2 * pi * f;

% Impedance and ln(Z)
Z = Re + Rt ./ (1 + 1i * w * Rt * C);
lnZ = log(Z);
Re_lnZ = real(lnZ);     % Reference real part
Im_lnZ = imag(lnZ);     % Imaginary part for input to integral

% Re{ln Z(0)}
Re_lnZ0 = log(Re + Rt);
```

```
% Initialize result
Re_lnZ_KK = zeros(size(w));

% Numerical integration per A.95
for i = 1:length(w)
    wi = w(i);

    % Interpolate Im{ln Z(x)} during integration
    integrand=@(x)interp1(w,Im_lnZ,x,'pchip',...
        'extrap')./(x.*(x.^2-wi^2));

    % Split the integral to avoid singularity at x = wi
    int1 = integral(integrand, 0, wi - eps1);
    int2 = integral(integrand, wi + eps1, 1e7);

    % Apply A.95 with negative sign
    Re_lnZ_KK(i) = Re_lnZ0 - (2 * wi^2 / pi) * (int1 + int2);...
        % fixed sign
end

% Plot
figure;
semilogx(f,Re_lnZ,'b-', 'LineWidth', 2,'DisplayName',...
    'True Re{ln Z}');
hold on;
semilogx(f,Re_lnZ_KK,'r--','LineWidth',2,'DisplayName',...
    'KK Estimate (A.95)');
xlabel('Frequency (Hz)');
ylabel('Re{ln Z(   )}');
title('Verification of Equation A.95');
legend('Location', 'best');
grid on;
```

B.6.12 Equation (A.95)

The following code was used for the solution of Problem 22.1.1.

```
clc; close all;

% Parameters
Re = 10;                % Ohm cm
Rt = 100;               % Ohm cm
C = 20e-6;              % F/cm
eps1 = 1e-3;            % To avoid singularity

% Frequency and angular frequency
f = logspace(-1, 6, 100); % Hz
w = 2 * pi * f;


% Impedance and ln(Z)
Z = Re + Rt ./ (1 + 1i * w * Rt * C);
```

```
lnZ = log(Z);                     % Complex log of impedance
Re_lnZ = real(lnZ);               % Input for KK
Im_lnZ_true = imag(lnZ);          % validation


% A.96: Compute Im{ln Z} from Re{ln Z}

Im_lnZ_KK = zeros(size(w));

for i = 1:length(w)
    wi = w(i);

    % Define integrand using 'pchip' for stable edge behavior
    integrand=@(x)interp1(w,Re_lnZ,x,'pchip','extrap')./(x.^2-wi^2);

    % Numerical Hilbert transform (principal value)
    int1 = integral(integrand, 0, wi - eps1, 'ArrayValued', true);
    int2 = integral(integrand, wi + eps1, 1e9, 'ArrayValued', true);

    % A.96 with corrected positive sign
    Im_lnZ_KK(i) = (2 * wi / pi) * (int1 + int2);
end

% Plot
N = length(f);
idx = round(N * 0.05) : round(N * 0.95);

figure;
semilogx(f(idx), Im_lnZ_true(idx), 'b-', 'LineWidth', 2,...
    'DisplayName', 'True Im{ln Z}');
hold on;
semilogx(f(idx), Im_lnZ_KK(idx), 'r--', 'LineWidth', 2,...
    'DisplayName', 'KK Estimate (A.96)');
xlabel('Frequency (Hz)');
ylabel('Im{ln Z(   )}');
title('Verification of Equation A.96 (Corrected Sign)');
legend('Location', 'best');
grid on;
```

B.6.13 Application to a CPE

The program listings given below were used for the solution of Problem 22.2 and make use of two forms of the Kramers–Kronig relations. The first is equation (A.86), which provides a conversion from real to imaginary parts.

```
clc;close all;
%Define parameters
Re=10; % ohm cm2
Rt=100; %ohm cm2
Q=20e-6; %F/cm2 s^(1-n)
n=0.9;
```

```
f=logspace(-1,6,100);
w=2*pi*f;
T=1./f;
eps1=1e-4;
m=zeros(1,length(T));
Z=zeros(1,length(T));
Zr=zeros(1,length(T));
Zj=zeros(1,length(T));
phi=zeros(1,length(T));
ZjK1=zeros(1,length(T));
%Equivalent circuit
for i=1:length(T)
m(i)=complex(0,w(i));
Z(i)=Re+Rt/(1+(m(i)^n)*Rt*Q);
Zr(i)=real(Z(i));
Zj(i)=imag(Z(i));
end
%Kramers-Kronig, equation 23.1
cn=cos(n*pi/2);
for i=1:length(T)
a=w(i);
K=@(x)(Re+real(Rt./(1+((1i.*x).^n).*Rt.*Q)))./(x.^2-a.^2);
int1=integral(K,0,a-eps1);
int2=integral(K,a+eps1,1e8);
ZjK1(i)=(2*w(i)/pi).*(int1+int2);
end

semilogx(f,-Zj,'o');
hold on;
semilogx(f,-ZjK1);
xlim([0.1,1e6]);
ax = gca;
ax.XTick = [0.1,1,10,100,1e3,1e4,1e5,1e6];
xlabel('Frequency/Hz');
ylabel('-Z_j / \Omega cm^2');
title('Equation (A.87)');
```

The second program listing employs equation (A.87), which provides a conversion from imaginary to real parts.

```
clc;close all;
%Define parameters
Re=10; % ohm cm2
Rt=100; %ohm cm2
Q=20e-6; %F/cm2 s^(1-n)
n=0.9;
f=logspace(-1,6,100);
w=2*pi*f;
T=1./f;
eps1=1e-4;
m=zeros(1,length(T));
Z=zeros(1,length(T));
Zr=zeros(1,length(T));
```

```
Zj=zeros(1,length(T));
phi=zeros(1,length(T));
ZrK1=zeros(1,length(T));
%Equivalent circuit
for i=1:length(T)
m(i)=complex(0,w(i));
Z(i)=Re+Rt/(1+(m(i)^n)*Rt*Q);
Zr(i)=real(Z(i));
Zj(i)=imag(Z(i));
end
%Kramers-Kronig, equation 23.1
cn=cos(n*pi/2);
for i=1:length(T)
a=w(i);
K=@(x)(x.*imag(Rt./(1+((1i.*x).^n).*Rt.*Q)))./(x.^2-a.^2);
int1=integral(K,0,a-eps1);
int2=integral(K,a+eps1,1e8);
ZrK1(i)=Re-(2/pi).*(int1+int2);
end

semilogx(f,Zr,'o');
hold on;
semilogx(f,ZrK1);
xlim([0.1,1e6]);
ax = gca;
ax.XTick = [0.1,1,10,100,1e3,1e4,1e5,1e6];
xlabel('Frequency/Hz');
ylabel('Z_r / \Omega cm^2');
title('Equation (A.88)');
```

List of Symbols

Roman

$\mathcal{A}$ temperature-dependent coefficient used in the development for thermoelectrochemical impedance

A constant A = 0.92486353 used for the velocity expansion for a rotating disk, see Problem 11.5

a constant a = 0.510232618867 used for the velocity expansion for a rotating disk

A lumped parameter in the expression for impedance response of a reaction dependent on potential and surface coverage

A_{i} activation energy for diffusion

A_{IC} Akaike information criterion

a_{IJ} constant used for the velocity expansion for an impinging jet with value determined by experiment

A_{op} open-loop gain for an operational amplifier

A_{PI} Akaike performance index

A_{ν} gain of the operational amplifier

B constant B = 1.20221175 used for the velocity expansion for a rotating disk, see Problem 11.5

b constant b = −0.615922014399 used for the velocity expansion for a rotating disk

B lumped parameter in the expression for impedance response of a reaction dependent on potential and surface coverage

B magnetic field, T

b kinetic parameter, V^{-1}

$\mathbf{C}$ covariance matrix for regression analysis

$\mathcal{C}$ complex capacitance, F or $\mathrm{F/cm^2}$

C capacitance, $\mathrm{F/cm^2}$ or F (1F = 1C/V)

C_0	interfacial capacitance, F/cm^2
C_{dl}	double-layer capacitance, F/cm^2 or F ($1F = 1C/V$)
CE	counterelectrode
C_{eff}	effective capacitance, F or F/cm^2
c_i	volumetric concentration of species i, mol/cm^3
C_Y	oxide capacitance associated with the Young impedance, F/cm^2
d	distance between potential sensing electrodes in the sensor used for local impedance spectroscopy measurements, cm
d_{act}	average diameter of the active portion of a partially blocked electrode, cm
D_i	diffusion coefficient for species i, cm^2/s
E	electronic energy, eV
E_c	conduction-band energy, eV
e_{out}	noise in the output from the current follower, V
E_{out}	potential output from the current follower, V
E_v	valence-band energy, eV
F	statistic used for testing equality of variances
F	Faraday's constant, 96,487 C/equiv
$\widetilde{f}$	complex oscillating variable used in expression for radial velocity for a rotating disk electrode
F	dimensionless radial component of velocity for laminar flow to a disk electrode
f	frequency, $f = \omega/2\pi$, Hz
$F(\omega)$	general frequency-dependent input function
$f(t)$	general time-dependent input function
f_i	activity coefficient for species i, dimensionless
f_{N_k}	characteristic frequency for measurement of impedance at a frequency f_k using N_k cycles, Hz
f_{scan}	characteristic frequency for measurement comprising a range of perturbation frequencies, Hz
f_{set}	characteristic frequency for measurement of a series of spectra, Hz
f^*	normalized frequency
$\widetilde{g}$	complex oscillating variable used in expression for angular velocity for a rotating disk electrode

G	dimensionless angular component of velocity for laminar flow to a disk electrode
g	dimensionless parameter used for the power-law model, see equation (14.26)
g	gravitational acceleration, cm/s^2
$G(\omega)$	general transfer function
$G(W)$	generic transfer function relating input and output quantities for an electrochemical system
$g(t,t')$	function showing a linear relationship between input and output functions
H	hypothesis
$\widetilde{h}$	complex oscillating variable used in expression for axial velocity for a rotating disk electrode
H	dimensionless axial component of velocity for laminar flow to a disk electrode
H_i	generic transfer function relating the time derivative of state quantities for an electrochemical system to the state and input quantities
F_k	generic transfer function relating the output quantities for an electrochemical system to the state and input quantities
I	current, A
i	current density, mA/cm^2
i_0	exchange current density, mA/cm^2
i_0	output current for an operational amplifier, A
i_a	anodic current density, mA/cm^2
i_C	charging current density, mA/cm^2
i_c	cathodic current density, mA/cm^2
ΔI	amplitude of sinusoidal current signal, A
i_F	faradaic current density, mA/cm^2
i_k	kinetically controlled current density, $i_\mathrm{k} = -k_\mathrm{c} n \mathrm{F} c_\mathrm{i}(\infty)\exp(-b_\mathrm{c}\eta_\mathrm{s})$, mA/cm^2
i_lim	mass-transfer-limited current density, $i_\mathrm{lim} = -n\mathrm{F}D_\mathrm{i}c_\mathrm{i}(\infty)/\delta_\mathrm{i}$, mA/cm^2
i_n1	current noise contribution from differential amplifiers, A
i_n2	current noise contribution from differential amplifiers, A
i_n5	current noise contribution from differential amplifiers, A
i_n6	current noise contribution from differential amplifiers, A
i_probe	current density measured by a small local-impedance-spectroscopy sensor located near an electrode, mA/cm^2

iR_e	ohmic potential drop between the solution adjacent to the working electrode and the location of a reference electrode, i.e., $iR_e = \Phi_0 - \Phi_{ref}$, see Table A.18, V
i_{reg}	parasitic current noise arising from noise in the voltage control, A
i_S	power current for an operational amplifier, A
$\mathbf{J}$	identity matrix
j	imaginary number, $j = \sqrt{-1}$
J	dimensionless exchange current density
K	dimensionless frequency associated with the geometry of a disk electrode
K	lumped-parameter rate constant for an electrochemical reaction that includes the equilibrium potential difference, mA/cm^2
K^*	lumped-parameter rate constant for an electrochemical reaction, mA/cm^2
k	rate constant for electrochemical reaction that excludes the exponential dependence on potential
K_i	dimensionless frequency associated, for example, with convective diffusion of species i
k_M	mass-transfer coefficient, cm/s
$K_{x,i}$	dimensionless position-dependent frequency associated, for example, with convective diffusion of species i
ℓ	depth of a pore, cm
L	inductance, H ($1H = 1Vs^2/C$)
L	length of an electrode, cm
ℓ_c	characteristic dimension for geometry-induced frequency dispersion, cm
M	mass, g
M_i	symbol for the chemical formula of species i
N	normal distribution function
N	collection efficiency for a ring–disk electrode
n	electron concentration, cm^{-3}
n	number of electrons transferred in electrochemical reaction following the reaction stoichiometry, dimensionless
N_c	effective density of conduction-band states, cm^{-3}
N_{dat}	number of measured values
N_i	flux of species i, mol/cm^2s
N_i	local mass-transfer flux, mol/cm^2s

N_{k}	number of cycles for an impedance measurement at frequency f_{k}
N_{p}	number of parameters
n_{t}	electron concentration in deep-level states, cm^{-3}
N_{v}	effective density of valence-band states, cm^{-3}
n_{x}	sample size used to calculate the standard error of sample distribution x
P	parameter vector used in regression analysis
P	probability
p	probability of observing the given sample result under the assumption that the null hypothesis is true
p	dimensionless frequency, $p = \omega/\Omega$
p	hole concentration, cm^{-3}
p	pressure, atm
$\boldsymbol{\Delta P}$	parameter vector for nonlinear regression analysis
$p_{\mathrm{i,k}}$	reaction order for anodic reactants i in reaction k
Q	significance level at which the hypothesis that the variances are not equal can be rejected
Q	CPE coefficient, see equation (19.1), $\mathrm{s}^{\alpha}/\Omega\mathrm{cm}^2$
q	charge, $\mathrm{C/cm}^2$
$q_{\mathrm{i,k}}$	reaction order for cathodic reactants i in reaction k
R	universal gas constant, 8.3143 J/mol K
R	resistance, $\Omega\mathrm{cm}^2$ or Ω ($1\Omega = 1\mathrm{Vs/C}$)
R	modulus
r	radial coordinate, cm
r	rate of reaction, $\mathrm{mol/cm}^2\mathrm{s}$
r_0	radius of a disk electrode, cm
Re	Reynolds number, $\mathrm{Re} = \rho v 2R/\mu$, dimensionless
R_{e}	electrolyte or ohmic resistance, Ω or $\Omega\mathrm{cm}^2$
REF	reference electrode
R_{m}	resistance of the current measurement circuit, Ω
R_{s}	resistance of the potential control circuit, Ω
R_{t}	charge-transfer resistance, $\Omega\mathrm{cm}^2$
S	weighted sum of squares
s	sample standard deviation

s^2	sample variance
$\mathrm{Sc_i}$	Schmidt number, $\mathrm{Sc_i} = \nu/D_\mathrm{i}$, dimensionless
SE	standard error
Sh	Sherwood number, $\mathrm{Sh} = k_\mathrm{M}d/D_\mathrm{i}$, dimensionless
$s_\mathrm{i,k}$	stoichiometric coefficient for species i in reaction k
S_m	potential difference across the inputs to the current follower, V
s_m	gain of the current follower
s_m	noise in the potential difference across the inputs to the current follower, V
s_x	standard error of sample distribution x
t	statistic used for testing equality of means
T	temperature, K
T	time corresponding to an integer number of cycles, s
t	time, s
t_cycle	period of a cycle at frequency f, s
U	electrode potential with respect to a reference electrode, $U = \Phi_\mathrm{m} - \Phi_\mathrm{ref}$, see Table A.18, V
u_i	mobility of species i
$\mathbf{V}$	variance–covariance matrix used in regression analysis
V	interfacial potential for the working electrode, see Table A.18, V
V	potential, V
v	velocity, cm/s
V_+	potential at the positive input terminal for an operational amplifier, V
V_-	potential at the negative input terminal for an operational amplifier V
V_0	output potential for an operational amplifier, V
$V_\mathrm{0,k}$	interfacial potential under equilibrium conditions for reaction k, see Table A.18, $V_\mathrm{0,k} = (\Phi_\mathrm{m} - \Phi_0)_\mathrm{0,k}$, V
ν_CE	voltage noise contribution, V
ΔV	amplitude of sinusoidal potential signal, V
ν_e	voltage noise contribution, V
ν_p	voltage noise contribution, V
ν_ref	voltage noise contribution, V
$\nu_{\mathrm{R_m}}$	voltage noise contribution, V
V_S	potential of the power leads for an operational amplifier, V

W	width of an electrode, cm
WE	working electrode
W_i	transfer function associated with electrohydrodynamic impedance of species i
W_k	generic input quantities for an electrochemical system
$X(\omega)$	general frequency-dependent output function
$x(t)$	general time-dependent output function
X_k	generic state quantities for an electrochemical system
Y	admittance, $Y - 1/Z$, $1/\Omega\mathrm{cm}^2$
y	normal coordinate, cm
Y_k	generic output quantities for an electrochemical system
Z	global impedance, see Table A.19, $\Omega\mathrm{cm}^2$
z	local impedance, see Table A.19, $\Omega\mathrm{cm}^2$
Z_0	global interfacial impedance, see Table A.19, $\Omega\mathrm{cm}^2$
z_0	local interfacial impedance, see Table A.19, $\Omega\mathrm{cm}^2$
Z_c	mass-transport transfer function
Z_D	diffusion impedance, $\Omega\mathrm{cm}^2$
Z_e	global ohmic impedance, see Table A.19, $\Omega\mathrm{cm}^2$
z_e	local ohmic impedance, $\Omega\mathrm{cm}^2$
Z_eq	interfacial impedance per unit pore area
Z_F	faradaic impedance, $\Omega\mathrm{cm}^2$
z_i	charge associated with species i
Z_k	tabulated dimensionless values for diffusion impedance where $\mathrm{k} = 1,2,3$, see equation (11.54) for a rotating disk electrode and equation (11.55) for a submerged impinging jet
Z_Y	Young impedance, $\Omega\mathrm{cm}^2$

Greek

α	CPE exponent, see equation (19.1), dimensionless
α	constant $\alpha = 0.88447411$ used for the velocity expansion for a rotating disk, see Problem 11.5
α	probability of incorrectly rejecting the null hypothesis when it is actually true
α	symmetry factor used in electrode kinetics, dimensionless
$\boldsymbol{\alpha}$	tensor corresponding to the second derivative of the objective function with respect to parameter

β	Tafel slope, V/decade of current
$\boldsymbol{\beta}$	vector corresponding to the first derivative of the objective function with respect to parameter
β_{y}	local wall velocity gradient
Γ	maximum surface coverage, mol/cm^2
γ	exponent in power-law model
γ	fractional surface coverage
δ	thickness, cm
ϵ	porosity
ε	dielectric constant
$\varepsilon_{\mathrm{bias}}$	bias error, Ω
$\varepsilon_{\mathrm{drift}}$	bias error associated with nonstationary behavior, Ω
ε_0	permittivity of vacuum, 8.8542×10^{-14} F/cm
$\varepsilon_{\mathrm{fit}}$	error associated with model inadequacy, Ω
$\varepsilon_{\mathrm{inst}}$	bias error associated with instrumental artifacts, Ω
$\varepsilon_{\mathrm{stoch}}$	stochastic error, Ω
ζ	dimensionless position, $\zeta = y\sqrt{\Omega/\nu}$
η_{c}	concentration overpotential, see Table A.18, V
η	dimensionless axial position, $\eta = y\sqrt{a/\nu}$
η_{s}	surface overpotential for reaction k, $\eta_{\mathrm{s}} = V - V_{0,\mathrm{k}}$, see Table A.18, V
Θ_{i}	dimensionless concentration
θ_{i}	dimensionless oscillating part of the concentration, $\theta_{\mathrm{i}}(y) = \widetilde{c}_{\mathrm{i}}/\widetilde{c}_{\mathrm{i}}(0)$
κ	conductivity, see, e.g., equation (13.14), S/cm
λ	Debye length, cm
λ	constant used for analysis of impedance at low frequency
λ	constant used for regression analysis in the method of steepest descent and the Levenberg–Marquardt methods
μ	fluid viscosity, g/cm s
μ_{i}	electrochemical potential of species i, J/mol
μ_x	mean value of sample distribution x
ν	degree of freedom
ν	kinematic viscosity, $\nu = \mu/\rho$, cm^2/s
ν	wave number

ξ	dimensionless position
π	ratio of a circle's circumference to its diameter, $\pi = 3.14159265359$
ρ	fluid density, g/cm^3
ρ	resistivity, Ωcm
ρ_0	resistivity at position $y = 0$, Ωcm, see equation (14.23)
ρ_δ	resistivity at position $y = \delta$, Ωcm, see equation (14.23)
ρ_{sc}	charge density for a semiconductor, C/cm^3
σ_x	standard deviation of sample distribution x
σ_{x_1,x_2}	covariance of sample distributions x_1 and x_2
σ_x^2	variance of sample distribution x
τ	time constant, e.g., $\tau = RC$, s
τ_{RC}	characteristic RC time constant, s
τ_{rz}	shear stress, N/cm^2
τ_{scan}	time required to measure a complete spectrum, s
τ_{set}	time required to measure a series of spectra, s
Φ	potential, V
ϕ	dimensionless stream function for flow to a submerged impinging jet
φ	phase angle for impedance
Φ_0	potential of the electrolyte adjacent to the working electrode with respect to an unspecified but common reference potential, see Table A.18, V
Φ_m	electrode potential with respect to an unspecified but common reference potential, see Table A.18, V
Φ_{ref}	potential of a reference electrode with respect to an unspecified but common reference potential, see Table A.18, V
φ	phase lag
χ^2	sum of squares used in regression analysis
Ω	rotation speed, s^{-1}
ω	angular frequency, $\omega = 2\pi f$, s^{-1}
ω_c	characteristic angular frequency, $\omega_c = 1/RC$, s^{-1}

General Notation

$\mathrm{Im}\{X\}$	imaginary part of X
$\mathrm{Re}\{X\}$	real part of X
$\overline{X}$	steady-state or time-averaged part of $X(t)$

$\overline{z}$	complex conjugate of a complex number z, $\overline{z} = z_r - jz_j$
$\langle X \rangle$	spatially averaged value of X
$\widetilde{X}$	oscillating part of $X(t)$
$E\{X\}$	expectation of $X(t)$
$\widehat{Z}$	model value for Z
X'	first derivative of $X(t)$ with respect to position
θ	pertaining to the angular direction

Subscripts

0	located at the inner limit of the diffuse double layer
adj	pertaining to terms corrected for the ohmic resistance
a	pertaining to anodic reactions
c	pertaining to cathodic reactions
CE	pertaining to the counterelectrode
cell	pertaining to the electrochemical cell
d	associated with the diffuse region of charge
dl	double layer
HF	high frequency
i	pertaining to chemical species i
ihp	located at the inner Helmholtz plane
IJ	impinging jet
j	imaginary
ℓ	pertaining to a porous layer
LF	low frequency
m	located at the electrode surface
mod	model value
ob	observed value
ohp	located at the outer Helmholtz plane
r	pertaining to the radial direction
r	real
ref	pertaining to a reference electrode
WE	pertaining to the working electrode
y	pertaining to the axial direction

Z_j	pertaining to the imaginary part of the impedance
Z_r	pertaining to the real part of the impedance

References

1. M. E. Orazem and B. Tribollet, *Electrochemical Impedance Spectroscopy*, 2nd edition (Hoboken: John Wiley & Sons, 2017).
2. P. Agarwal, M. E. Orazem, and L. H. García-Rubio, "Measurement Models for Electrochemical Impedance Spectroscopy: 1. Demonstration of Applicability," *Journal of the Electrochemical Society*, **139** (1992) 1917–1927.
3. P. Agarwal, O. D. Crisalle, M. E. Orazem, and L. H. García-Rubio, "Measurement Models for Electrochemical Impedance Spectroscopy: 2. Determination of the Stochastic Contribution to the Error Structure," *Journal of the Electrochemical Society*, **142** (1995) 4149–4158.
4. P. Agarwal, M. E. Orazem, and L. H. García-Rubio, "Measurement Models for Electrochemical Impedance Spectroscopy: 3. Evaluation of Consistency with the Kramers-Kronig Relations," *Journal of the Electrochemical Society*, **142** (1995) 4159–4168.
5. S. Chakri, I. Frateur, M. E. Orazem, E. M. M. Sutter, T. T. M. Tran, B. Tribollet, and V. Vivier, "Improved EIS Analysis of the Electrochemical Behaviour of Carbon Steel in Alkaline Solution," *Electrochimica Acta*, **246** (2017) 924–930.
6. A. S. Nguyen, N. Causse, M. Musiani, M. E. Orazem, N. Pébère, B. Tribollet, and V. Vivier, "Determination of Water Uptake in Organic Coatings Deposited on 2024 Aluminium Alloy: Comparison between Impedance Measurements and Gravimetry," *Progress in Organic Coatings*, **112** (2017) 93–100.
7. V. M.-W. Huang, V. Vivier, I. Frateur, M. E. Orazem, and B. Tribollet, "The Global and Local Impedance Response of a Blocking Disk Electrode with Local CPE Behavior," *Journal of the Electrochemical Society*, **154** (2007) C89–C98.
8. M. Sluyters-Rehbach, "Impedances of Electrochemical Systems: Terminology, Nomenclature, and Representation: I. Cells with Metal Electrodes and Liquid Solutions," *Pure and Applied Chemistry*, **66** (1994) 1831–1891.
9. D. D. Macdonald, *Transient Techniques in Electrochemistry* (New York: Plenum Press, 1977).

10. C. Gabrielli, *Identification of Electrochemical Processes by Frequency Response Analysis*, Solartran Instrumentation Group Monograph, The Solartran Electronic Group Ltd., Farnborough, England (1980).

11. J. R. Macdonald, editor, *Impedance Spectroscopy: Emphasizing Solid Materials and Systems* (New York: John Wiley & Sons, 1987).

12. E. Barsoukov and J. R. Macdonald, editors, *Impedance Spectroscopy: Theory, Experiment, and Applications*, 2nd edition (Hoboken: John Wiley & Sons, 2005).

13. M. Itagaki, Y. Hatada, I. Shitanda, and K. Watanabe, "Complex impedance spectra of porous electrode with fractal structure," *Electrochimica Acta*, **55** (2010) 6255–6262.

14. Y. Gourbeyre, B. Tribollet, C. Dagbert, and L.Hyspecka, "A Physical Model for Anticorrosion Behavior of Duplex Coatings: Corrosion, Passivation, and Anodic Films," *Journal of the Electrochemical Society*, **153** (2006) B162–B168.

15. S. Erol and M. E. Orazem, "The Influence of Anomalous Diffusion on the Impedance Response of $LiCoO_2$—C Batteries," *Journal of Power Sources*, **293** (2015) 57–64.

16. W. Watson and M. E. Orazem, "EIS: Measurement Model Program: Version 1.8," ECSarXiv (2023).

17. E. McCafferty, *Introduction to Corrosion Science* (New York: Springer-Verlag, 2010).

18. B. E. Conway, *Electrochemical Data* (Amsterdam: Elsevier, 1952).

19. P. T. Wojcik, P. Agarwal, and M. E. Orazem, "A Method for Maintaining a Constant Potential Variation during Galvanostatic Regulation of Electrochemical Impedance Measurements," *Electrochimica Acta*, **41** (1996) 977–983.

20. P. T. Wojcik and M. E. Orazem, "Variable-Amplitude Galvanostatically Modulated Impedance Spectroscopy as a Tool for Assessing Reactivity at the Corrosion Potential without Distorting the Temporal Evolution of the System," *Corrosion*, **54** (1998) 289–298.

21. *How Cabling and Signal Amplitudes Affect EIS Results*, Technical report, Gamry Instruments, Warminster, PA (2016).

22. *Accuracy Contour Plots: Measurement and Discussion*, Technical report, Gamry Instruments, Warminster, PA (2016).

23. E. Barsoukov and J. R. Macdonald, editors, *Impedance Spectroscopy: Theory, Experiment, and Applications*, 3rd edition (Hoboken: John Wiley & Sons, 2018).

24. S. K. Roy and M. E. Orazem, "Error Analysis for the Impedance Response of PEM Fuel Cells," *Journal of the Electrochemical Society*, **154** (2007) B883–B891.

25. R. Makharia, M. F. Mathias, and D. R. Baker, "Measurement of Catalyst Layer Electrolyte Resistance in PEFCs Using Electrochemical Impedance Spectroscopy," *Journal of the Electrochemical Society*, **152** (2005) A970–A977.

26. R. M. Darling and J. P. Meyers, "Kinetic Model of Platinum Dissolution in PEMFCs," *Journal of the Electrochemical Society*, **150** (2003) A1523–A1527.
27. S. K. Roy, M. E. Orazem, and B. Tribollet, "Interpretation of Low-Frequency Inductive Loops in PEM Fuel Cells," *Journal of the Electrochemical Society*, **154** (2007) B1378–B1388.
28. S. K. Roy, H. Hagelin-Weaver, and M. E. Orazem, "Application of Complementary Analytical Tools to Support Interpretation of Polymer-Electrolyte-Membrane Fuel Cell Impedance Data," *Journal of Power Sources*, **196** (2011) 3736–3742.
29. M. E., J. M. Esteban, K. J. Kennelley, and O. Degerstedt, R. M., "Mathematical Models for Cathodic Protection of an Underground Pipeline with Coating Holidays: 2. Case Studies of Parallel Anode CP Systems," *Corrosion*, **53** (1997) 427–436.
30. J. A. Koutecky and V. G. Levich, "An Application of a Rotating Disc Electrode to the Study of Kinetic and Catalytic Processes in Electrochemistry," *Rossiiskaya Akademiya Nauk. Doklady Akademii Nauk*, **117** (1957) 441–444.
31. J. A. Koutecky and V. G. Levich, "The Use of a Rotating Disc Electrode in the Study of Electrochemical Kinetics and Electrolytic Processes," *Zh Fiz Khim+*, **32** (1958) 1565–1575.
32. B. Hirschorn, M. E. Orazem, B. Tribollet, V. Vivier, I. Frateur, and M. Musiani, "Determination of Effective Capacitance and Film Thickness from CPE Parameters," *Electrochimica Acta*, **55** (2010) 6218–6227.
33. M. T. T. Tran, B. Tribollet, V. Vivier, and M. E. Orazem, "On the Impedance Response of Reactions Influenced by Mass Transfer," *Russ J Electrochem+*, **53** (2017) 932–940.
34. O. E. Barcia, O. R. Mattos, N. Pébère, and B. Tribollet, "Mass-Transport Study for the Electrodissolution of Copper in 1M Hydrochloric Acid Solution by Impedance," *Journal of the Electrochemical Society*, **140** (1993) 2825–2832.
35. I. Epelboin and M. Keddam, "Faradaic Impedances: Diffusion Impedance and Reaction Impedance," *Journal of the Electrochemical Society*, **117** (1970) 1052–1056.
36. L. Péter, J. Arai, and H. Akahoshi, "Impedance of a Reaction Involving Two Adsorbed Intermediates: Aluminum Dissolution in Non-Aqueous Lithium Imide Solutions," *Journal of Electroanalytical Chemistry*, **482** (2000) 125 – 138.
37. R. D. Armstrong, R. E. Firman, and H. R. Thirsk, "AC Impedance of Complex Electrochemical Reactions," *Faraday Discussions*, **56** (1973) 244–263.
38. C. Deslouis, I. Epelboin, M. Keddam, and J. C. Lestrade, "Impédance de Diffusion d'un Disque Tournant en Régime Hydrodynamique Laminaire. Etude Expérimentale et Comparaison avec le Modèle de Nernst," *Journal of Electroanalytical Chemistry*, **28** (1970) 57–63.
39. V. G. Levich, *Physicochemical Hydrodynamics* (Englewood Cliffs: Prentice Hall, 1962).

40. R. E. White, C. M. Mohr, Jr., and J. S. Newman, "The Fluid Motion Due to a Rotating Disk," *Journal of the Electrochemical Society*, **123** (1976) 383–385.
41. D.-T. Chin and C. Tsang, "Mass Transfer to an Impinging Jet Electrode," *Journal of the Electrochemical Society*, **125** (1978) 1461–1470.
42. J. M. Esteban, G. Hickey, and M. E. Orazem, "The Impinging Jet Electrode: Measurement of the Hydrodynamic Constant and Its Use for Evaluating Film Persistency," *Corrosion*, **46** (1990) 896–901.
43. B. A. Boukamp and H. J. M. Bouwmeester, "Interpretation of the Gerischer Impedance in Solid State Ionics," *Solid State Ionics*, **157** (2003) 29–33.
44. H. Gerischer, "Wechselstrompolarisation Von Elektroden Mit Einem Potentialbestimmenden Schritt Beim Gleichgewichtspotential," *Zeitschrift fur Physikalische Chemie*, **198** (1951) 286–313.
45. C. Deslouis, B. Tribollet, M. Duprat, and F. Moran, "Transient Mass Transfer at a Coated Rotating Disk Electrode: Diffusion and Electrohydrodynamical Impedances," *Journal of the Electrochemical Society*, **134** (1987) 2496–2501.
46. M. Gao, M. S. Hazelbaker, R. Kong, and M. E. Orazem, "Mathematical Model for the Electrochemical Impedance Response of a Continuous Glucose Monitor," *Electrochimica Acta*, **275** (2018) 119–132.
47. C. L. Alexander, B. Tribollet, and M. E. Orazem, "Contribution of Surface Distributions to Constant-Phase-Element (CPE) Behavior: 1. Influence of Roughness," *Electrochimica Acta*, **173** (2015) 416–424.
48. C. Deslouis, B. Tribollet, and M. A. Vorotyntsev, "Diffusion-Convection Impedance at Small Electrodes," *Journal of the Electrochemical Society*, **138** (1991) 2651–2657.
49. E. Remita, D. Boughrara, B. Tribollet, V. Vivier, E. Sutter, F. Ropital, and J. Kittel, "Diffusion Impedance in a Thin-Layer Cell: Experimental and Theoretical Study on a Large-Disk Electrode," *The Journal of Physical Chemistry C*, **112** (2008) 4626–4634.
50. R. de Levie, "Electrochemical Responses of Porous and Rough Electrodes," in *Advances in Electrochemistry and Electrochemical Engineering*, P. Delahay, editor, volume 6 (New York: Interscience, 1967) 329–397.
51. L. Young, "Anodic Oxide Films 4: The Interpretation of Impedance Measurements on Oxide Coated Electrodes on Niobium," *Transactions of the Faraday Society*, **51** (1955) 1250–1260.
52. A. Lasia, *Electrochemical Impedance Spectroscopy and Its Applications* (New York: Springer, 2014).
53. H. M. Lutz, Y. Wu, C. C. Eluagu, S. F. Cogan, K. J. Otto, and M. E. Orazem, "Analysis of Electrochemical Impedance Spectroscopy Data for Sputtered Iridium Oxide Electrodes," *Journal of Neural Engineering*, **22** (2025) 036007.

54. G. J. Brug, A. L. G. van den Eeden, M. Sluyters-Rehbach, and J. H. Sluyters, "The Analysis of Electrode Impedances Complicated by the Presence of a Constant Phase Element," *Journal of Electroanalytical Chemistry*, **176** (1984) 275–295.
55. J.-B. Jorcin, M. E. Orazem, N. Pébère, and B. Tribollet, "CPE Analysis by Local Electrochemical Impedance Spectroscopy," *Electrochimica Acta*, **51** (2006) 1473–1479.
56. B. Hirschorn, M. E. Orazem, B. Tribollet, V. Vivier, I. Frateur, and M. Musiani, "Constant-Phase-Element Behavior Caused by Resistivity Distributions in Films: 1. Theory," *Journal of the Electrochemical Society*, **157** (2010) C452–C457.
57. C. L. Alexander, B. Tribollet, and M. E. Orazem, "Contribution of Surface Distributions to Constant-Phase-Element (CPE) Behavior: 2. Capacitance," *Electrochimica Acta*, **188** (2016) 566–573.
58. C. L. Alexander, B. Tribollet, and M. E. Orazem, "Influence of Micrometric-Scale Electrode Heterogeneity on Electrochemical Impedance Spectroscopy," *Electrochimica Acta*, **201** (2016) 374–379.
59. C. L. Alexander, B. Tribollet, and M. E. Orazem, "Contribution of Surface Distributions to Constant-Phase-Element (CPE) Behavior: 3. Adsorbed Intermediate," *Electrochimica Acta*, **251** (2017) 99–108.
60. M. E. Orazem, B. Tribollet, V. Vivier, S. Marcelin, N. Pébère, A. L. Bunge, E. A. White, D. P. Riemer, I. Frateur, and M. Musiani, "Dielectric Properties of Materials Showing Constant-Phase-Element (CPE) Impedance Response," *Journal of the Electrochemical Society*, **160** (2013) C215–C225.
61. C. Cleveland, S. Moghaddam, and M. E. Orazem, "Nanometer-Scale Corrosion of Copper in Deaerated Deionized Water," *Journal of the Electrochemical Society*, **161** (2014) C107–C114.
62. C. Gabrielli and B. Tribollet, "A Transfer Function Approach for a Generalized Electrochemical Impedance Spectroscopy," *Journal of the Electrochemical Society*, **141** (1994) 1147–1157.
63. B. Tribollet, J. S. Newman, and W. H. Smyrl, "Determination of the Diffusion Coefficient from Impedance Data in the Low Frequency Range," *Journal of the Electrochemical Society*, **135** (1988) 134–138.
64. S. Erol, M. E. Orazem, and R. P. Muller, "Influence of overcharge and overdischarge on the impedance response of batteries," *Journal of Power Sources*, **270** (2014) 92–100.
65. R. de L. Kronig, "On the Theory of Dispersion of X-Rays," *Journal of the Optical Society of America and Review of Scientific Instruments*, **12** (1926) 547–557.
66. R. de L. Kronig, "Dispersionstheorie im Röntgengebeit," *Physikalishce Zeitschrift*, **30** (1929) 521–522.
67. H. A. Kramers, "Die Dispersion und Absorption von Röntgenstrahlen," *Physikalishce Zeitschrift*, **30** (1929) 522–523.

68. H. A. Kramers, "La Diffusion de la Lumière par les Atomes," in *Atti Congresso Internazionale Fisici, Como*, volume 2 (1927) 545–557.

69. H. M. Nussenzveig, *Causality and Dispersion Relations* (New York: Academic Press, 1972).

70. M. I. Kozak, V. N. Zhikharev, P. P. Puga, and V. Y. Loya, "The Kramers-Kronig Relations: Validation Via Calculation Technique," *International Journal of Innovative Science Engineering and Technology*, **4** (2017) 152–159.

71. B. Hirschorn and M. E. Orazem, "On the Sensitivity of the Kramers-Kronig Relations to Nonlinear Effects in Impedance Measurements," *Journal of the Electrochemical Society*, **156** (2009) C345–C351.

72. J. M. Goh, C. Eluagu, J. Babauta, and M. E. Orazem, "Comparison of Approaches for Assessing Linearity of Impedance Measurements," *Journal of The Electrochemical Society*, **171** (2024) 036508.

73. J. R. Macdonald, "Power-Law Exponents and Hidden Bulk Relaxation in the Impedance Spectroscopy of Solids," *Journal of Electroanalytical Chemistry*, **378** (1994) 17–29.

74. A. N. Jansen, P. T. Wojcik, P. Agarwal, and M. E. Orazem, "Thermally Stimulated Deep-Level Impedance Spectroscopy: Application to an n-GaAs Schottky Diode," *Journal of the Electrochemical Society*, **143** (1996) 4066–4074.

75. M. E. Orazem, P. Agarwal, A. N. Jansen, P. T. Wojcik, and L. H. García-Rubio, "Development of Physico-Chemical Models for Electrochemical Impedance Spectroscopy," *Electrochimica Acta*, **38** (1993) 1903–1911.

76. K. Deng, S. Huang, X. Wang, Y. Yao, Y. Yang, L. Yu, Y. Pei, J. An, Q. Jiang, X. Liu, S. Yang, and O. Chen, Kevin J., "Deep Level Transient Spectroscopy: Tracing Interface and Bulk Trap-Induced Degradation in AlGaN/GaN-Heterostructure Based Devices," *Information & Functional Materials*, **1** (2024) 282–303.

77. L. H. García-Rubio and N. Ro, "Detailed Copolymer Characterization Using Ultraviolet Spectroscopy," *Canadian Journal of Chemistry*, **63** (1985) 253–263.

78. L. H. García-Rubio, "The Effect of Molecular Size on the Absorbtion Spectra of Macromolecules," *Macromolecules*, **20** (1987) 3070–3075.

79. L. H. García-Rubio, C. Lopez-Menacho, and S. Grossman, "Characterization of Proteins During Aggregation II: Use of Model Molecules for Spectroscopy Analysis," *Chemical Engineering Communications*, **122** (1993) 85–101.

80. P. Agarwal, M. E. Orazem, and L. H. García-Rubio, "Application of the Kramers Kronig Relations to Electrochemical Impedance Spectroscopy," in *Electrochemical Impedance: Analysis and Interpretation*, J. Scully, D. Silverman, and M. Kendig, editors, volume ASTM STP 1188 (Philadelphia: American Society for Testing and Materials, 1993) 115–139.

81. M. Durbha, M. E. Orazem, and L. H. García-Rubio, "Spectroscopy Applications of the Kramers-Kronig Relations: Implications for Error Structure Identification," *Journal of the Electrochemical Society*, **144** (1997) 48–55.

82. S. L. Carson, M. E. Orazem, O. D. Crisalle, and L. H. García-Rubio, "On the Error Structure of Impedance Measurements: Simulation of Frequency Response Analysis (FRA) Instrumentation," *Journal of the Electrochemical Society*, **150** (2003) E477–E490.

83. S. L. Carson, M. E. Orazem, O. D. Crisalle, and L. H. García-Rubio, "On the Error Structure of Impedance Measurements: Simulation of Phase Sensitive Detection (PSD) Instrumentation," *Journal of the Electrochemical Society*, **150** (2003) E491–E500.

84. P. K. Shukla, M. E. Orazem, and O. D. Crisalle, "Validation of the Measurement Model Concept for Error Structure Identification," *Electrochimica Acta*, **49** (2004) 2881–2889.

85. L. Pauwels, W. Simons, A. Hubin, J. Schoukens, and R. Pintelon, "Key Issues for Reproducible Impedance Measurements and Their Wellfounded Error Analysis in a Silver Electrodeposition System," *Electrochimica Acta*, **47** (2002) 2135–2141.

86. O. Blajiev, T. Breugelmans, R. Pintelon, and A. Hubin, "Improvement of the Impedance Measurement Reliability by Some New Experimental and Data Treatment Procedures applied to the Behavior of Copper in Neutral Chloride Solutions containing Small Heterocycle Molecules," *Electrochimica Acta*, **51** (2006) 1403–1412.

87. B. A. Boukamp, "A Linear Kronig-Kramers Transform Test for Immittance Data Validation," *Journal of the Electrochemical Society*, **142** (1995) 1885–1894.

88. C. You, M. A. Zabara, M. E. Orazem, and B. Ulgut, "Application of the Kramers–Kronig Relations to Multi-Sine Electrochemical Impedance Measurements," *Journal of the Electrochemical Society*, **167** (2020) 020515.

89. M. E. Orazem, P. T. Wojcik, M. Durbha, I. Frateur, and L. H. García-Rubio, "Application of Measurement Models for Interpretation of Impedance Spectra for Corrosion," *Ultrafast Phenomena in Semiconductors 2001*, **289-292** (1998) 813–828.

90. M. E. Orazem, P. K. Shukla, and M. A. Membrino, "Extension of the Measurement Model Approach for Deconvolution of Underlying Distributions for Impedance Measurements," *Electrochimica Acta*, **47** (2002) 2027–2034.

91. Y.-M. Chen, A. S. Nguyen, M. E. Orazem, B. Tribollet, N. Pébère, M. Musiani, and V. Vivier, "Identification of Resistivity Distributions in Dielectric Layers by Measurement Model Analysis of Impedance Spectroscopy," *Electrochimica Acta*, **219** (2016) 312–320.

92. B. Hirschorn, M. E. Orazem, B. Tribollet, V. Vivier, I. Frateur, and M. Musiani, "Constant-Phase-Element Behavior Caused by Resistivity Distributions in Films: 2. Applications," *Journal of the Electrochemical Society*, **157** (2010) C458–C463.

93. H. Liao, W. Watson, A. Dizon, B. Tribollet, V. Vivier, and M. E. Orazem, "Physical Properties Obtained from Measurement Model Analysis of Impedance Measurements," *Electrochimica Acta*, **354** (2020) 136747.

94. C. You, A. Titov, B. H. Kim, and M. E. Orazem, "Impedance Measurements on QLED Devices: Analysis of High-Frequency Loop in Terms of Material Properties," *Journal of Solid State Electrochemistry*, **24** (2020) 3083–3090.

95. M. E. Orazem, "Electrochemical Impedance Spectroscopy: The Journey to Physical Understanding," *Journal of Solid State Electrochemistry*, (2020).

96. S. Wang, J. Zhang, O. Gharbi, V. Vivier, M. Gao, and M. E. Orazem, "Electrochemical Impedance Spectroscopy," *Nature Reviews Methods Primers*, **1** (2021) 41.

97. V. Vivier and M. E. Orazem, "Impedance Analysis of Electrochemical Systems," *Chemical Reviews*, **122** (2022) 11131–11168.

98. Y. Sakamoto, M. Ishiguso, and G. Kitigawa, *Akaike Information Criterion Statistics* (Boston: D. Reidel, 1986).

99. C. You, A. Dizon, M. Gao, V. Vivier, and M. E. Orazem, "Experimental Observation of Ohmic Impedance," *Electrochimica Acta*, **413** (2022) 140177.

100. B. Hirschorn, B. Tribollet, and M. E. Orazem, "On Selection of the Perturbation Amplitude Required to Avoid Nonlinear Effects in Impedance Measurements," *Israel Journal of Chemistry*, **48** (2008) 133–142.

101. M. D. Murbach, B. Gerwe, N. Dawson-Elli, and L. kun Tsui, "impedance.py: A Python package for electrochemical impedance analysis," *Journal of Open Source Software*, **5** (2020) 2349.

102. R. Pollard and T. Comte, "Determination of Transport Properties for Solid Electrolytes from the Impedance of Thin Layer Cells," *Journal of the Electrochemical Society*, **136** (1989) 3734–3748.

103. G. Katırcı, M. A. Zabara, and B. Ulgut, "Methods: Unexpected Effects in Galvanostatic EIS of Randles' Cells: Initial Transients and Harmonics Generated," *Journal of the Electrochemical Society*, **169** (2022) 030527.

104. H. W. Bode, *Network Analysis and Feedback Amplifier Design* (New York: D. Van Nostrand, 1945).

105. V. M.-W. Huang, V. Vivier, M. E. Orazem, N. Pébère, and B. Tribollet, "The Apparent CPE Behavior of an Ideally Polarized Blocking Electrode: A Global and Local Impedance Analysis," *Journal of the Electrochemical Society*, **154** (2007) C81–C88.

106. M. E. Orazem, "A Systematic Approach toward Error Structure Identification for Impedance Spectroscopy," *Journal of Electroanalytical Chemistry*, **572** (2004) 317–327.

107. B. Tribollet, V. Vivier, and M. E. Orazem, "EIS Technique in Passivity Studies: Determination of the Dielectric Properties of Passive Films," in *Reference Module in Chemistry, Molecular Sciences and Chemical Engineering* (Elsevier, 2017) 1–14.

108. H. P. R. Frederikse, "Permittivity (Dielectric Constant) of Inorganic Solids," in *CRC Handbook of Chemistry and Physics*, W. M. Haynes, editor, volume 96 (Boca Raton: Taylor and Francis Group, LLC, 2015) 47–55.

109. J. S. Newman, "Frequency Dispersion in Capacity Measurements at a Disk Electrode," *Journal of the Electrochemical Society*, **117** (1970) 198–203.

110. O. Gharbi, A. Dizon, M. E. Orazem, M. T. T. Tran, B. Tribollet, and V. Vivier, "From Frequency Dispersion to Ohmic Impedance: A New Insight on The High-Frequency Impedance Analysis of Electrochemical Systems," *Electrochimica Acta*, **320** (2019) 134609.

111. S. Havriliak and S. Negami, "A Complex Plane Representation of Dielectric and Mechanical Relaxation Processes in Some Polymers," *Polymer*, **8** (1967) 161–210.

112. Y.-M. Chen, N. G. Rudawski, E. Lambers, and M. E. Orazem, "Application of Impedance Spectroscopy and Surface Analysis to Obtain Oxide Film Thickness," *Journal of the Electrochemical Society*, **164** (2017) C563–C573.

113. AA Portable Power Corp., "Rechargeable Li-Ion Button Battery Serial LIR2032," (2006). Accessed: 03-08-2014.

114. J. Thevenin, "Passivating Films on Lithium Electrodes. An Approach by Means of Electrode Impedance Spectroscopy," *Journal of Power Sources*, **14** (1985).

115. J. Song and M. Z. Bazant, "Effects of Nanoparticle Geometry and Size Distribution on Diffusion Impedance of Battery Electrodes," *Journal of the Electrochemical Society*, **160** (2013) A15–A24.

116. R. Deshpande, M. Verbrugge, Y.-T. Cheng, J. Wang, and P. Liu, "Battery Cycle Life Prediction with Coupled Chemical Degradation and Fatigue Mechanics," *Journal of the Electrochemical Society*, **159** (2012) A1730–A1738.

117. D. Aurbach, B. Markovsky, I. Weissman, E. Levi, and Y. Ein-Eli, "On the Correlation between Surface Chemistry and Performance of Graphite Negative Electrodes for Li ion Batteries," *Electrochimica Acta*, **45** (1999) 67–86.

118. M. B. Pinson and M. Z. Bazant, "Theory of SEI Formation in Rechargeable Batteries: Capacity Fade, Accelerated Aging and Lifetime Prediction," *Journal of the Electrochemical Society*, **160** (2013) A243–A250.

119. M. Greenleaf, H. Li, and J. Zheng, "Application of Physical Electric Circuit Modeling to Characterize Li-ion Battery Electrochemical Processes," *Journal of Power Sources*, **270** (2014) 113–120.

120. T. Osaka, T. Momma, D. Mukoyama, and H. Nara, "Proposal of Novel Equivalent Circuit for Electrochemical Impedance Analysis of Commercially Available Lithium Ion Battery," *Journal of Power Sources*, **205** (2012) 483–486.

121. J.-Y. Go and S.-I. Pyun, "A Review of Anomalous Diffusion Phenomena at Fractal Interface for Diffusion-Controlled and Non-Diffusion-Controlled Transfer Processes," *Journal of Solid State Electrochemistry*, **11** (2007) 323–334.

122. J. F. Lutsko and J. P. Boon, "Microscopic Theory of Anomalous Diffusion Based on Particle Interactions," *Physical Review E: Statistical Physics, Plasmas, Fluids, and Related Interdisciplinary Topics*, **88** (2013) 022108.

123. J. Bisquert and A. Compte, "Theory of the Electrochemical Impedance of Anomalous Diffusion," *Journal of Electroanalytical Chemistry*, **499** (2001) 112 – 120.

124. H. Ma, X. Cheng, S. Chen, G. Li, X. Chen, S. Lei, and H. Yang, "Theoretical Interpretation on Impedance Spectra for Anodic Iron Dissolution in Acidic Solutions Containing Hydrogen Sulfide," *Corrosion*, **54** (1998) 634–640.

125. H. Ma, X. Cheng, G. Li, S. Chen, Z. Quan, S. Zhao, and L. Niu, "The Influence of Hydrogen Sulfide on Corrosion of Iron under Different Conditions," *Corrosion Science*, **42** (2000) 1669–1683.

126. M. D. Deffo Ayagou, T. T. Mai Tran, B. Tribollet, J. Kittel, E. Sutter, N. Ferrando, C. Mendibide, and C. Duret-Thual, "Electrochemical Impedance Spectroscopy of Iron Corrosion in H_2S Solutions," *Electrochimica Acta*, **282** (2018) 775–783.

127. R. Jurczakowski, C. Hitz, and A. Lasia, "Impedance of Porous Au Based Electrodes," *Journal of Electroanalytical Chemistry*, **572** (2004) 355–366.

128. B. E. Torres-Bautista, M. L. Carvalho, A. Seyeux, S. Zanna, P. Cristiani, B. Tribollet, P. Marcus, and I. Frateur, "Effect of Protein Adsorption on the Corrosion Behavior of 70Cu–30Ni Alloy in Artificial Seawater," *Bioelectrochemistry*, **97** (2014) 34–42.

129. B. E. Torres-Bautista, *Effect of Biomolecules Adsorption on Oxide Layers Developed on Metallic Materials Used in Cooling Water Systems*, Ph.D. dissertation, Université Pierre et Marie Curie, Paris, France (2014).

130. S.-L. Wu, M. E. Orazem, B. Tribollet, and V. Vivier, "Impedance of a Disk Electrode with Reactions Involving an Adsorbed Intermediate: Local and Global Analysis," *Journal of the Electrochemical Society*, **156** (2009) C28–C38.

131. Y.-M. Chen, C. L. Alexander, C. Cleveland, and M. E. Orazem, "Influence of Geometry-Induced Frequency Dispersion on the Impedance of Ring Electrodes," *Electrochimica Acta*, **235** (2017) 437–441.

132. C. Deslouis, B. Tribollet, G. Mengoli, and M. M. Musiani, "Electrochemical Behaviour of Copper in Neutral Aerated Chloride Solution. I. Steady-State Investigation," *Journal of Applied Electrochemistry*, **18** (1988) 374–383.

133. V. M.-W. Huang, V. Vivier, M. E. Orazem, N. Pébère, and B. Tribollet, "The Apparent CPE Behavior of a Disk Electrode with Faradaic Reactions," *Journal of the Electrochemical Society*, **154** (2007) C99–C107.

134. B. Tribollet and M. E. Orazem, "On Interpretation of Impedance Spectra," *Electrochimica Acta*, **568** (2026) 149009.

135. Y. Ben Amor, E. M. M. Sutter, H. Takenouti, M. E. Orazem, and B. Tribollet, "Interpretation of Electrochemical Impedance for Corrosion of a Coated Silver Film in Terms of a Pore-in-Pore Model," *Journal of the Electrochemical Society*, **161** (2014) C573–C579.

136. A.-T. Tran, F. Huet, K. Ngo, and P. Rousseau, "Artefacts in Electrochemical Impedance Measurement in Electrolytic Solutions Due to the Reference Electrode," *Electrochimica Acta*, **56** (2011) 8034 – 8039.

137. Y. Ben Amor, E. Sutter, H. Takenouti, B. Tribollet, M. Boinet, R. Faure, J. Balencie, and G. Durieu, "Electrochemical Study of the Tarnish Layer of Silver Deposited on Glass," *Electrochimica Acta*, **131** (2014) 89–95.

138. C. Gabrielli, M. Keddam, A. Khalil, R. Rosset, and O. Zidoune, M., "Study of Calcium Carbonate Scales by Electrochemical Impedance Spectroscopy," *Electrochimica Acta*, **42** (1997) 1207–1218.

139. R. Churchill and J. Brown, *Complex Variables and Applications*, 6th edition (New York: McGraw Hill, 1990).

140. G. Cain, *Complex Analysis* (Atlanta: Georgia Institute of Technology, 1999).

141. G. W. Snedecor and W. G. Cochran, *Statistical Methods*, 6th edition (Ames: the Iowa State University Press, 1961).

142. G. E. P. Box, J. S. Hunter, and W. G. Hunter, *Statistics for Experimenters: An Introduction to Design, Data Analysis and Model Building*, 2nd edition (New York: John Wiley & Sons, 2005).

143. D. C. Montgomery, *Design and Analysis of Experiments*, 7th edition (Hoboken: John Wiley & Sons, 2009).

144. E. A. White, M. E. Orazem, and A. L. Bunge, "Characterization of Damaged Skin by Impedance Spectroscopy: Mechanical Damage," *Pharmaceutical Research*, **30** (2013) 2036–2049.

145. A. S. Nguyen, M. Musiani, M. E. Orazem, N. Pébère, B. Tribollet, and V. Vivier, "Impedance Analysis of the Distributed Resistivity of Coatings in Dry and Wet Conditions," *Electrochimica Acta*, **179** (2015) 452–459.

146. W. M. Haynes, editor, *CRC Handbook of Chemistry and Physics*, 97th edition (CRC Press, 2017).

147. J. F. Shackelford, *Introduction to Materials Science for Engineers*, 5th edition (Upper Saddle River, NJ: Prentice–Hall, 2000).

148. L. H. Van Vlack, *Elements of Materials Science and Engineering* (Boston: Addison-Wesley, 1975).

149. R. S. Lillard, P. J. Moran, and H. S. Isaacs, "A Novel Method for Generating Quantitative Local Electrochemical Impedance Spectroscopy," *Journal of the Electrochemical Society*, **139** (1992) 1007–1012.

150. F. Zou, D. Thierry, and H. S. Isaacs, "High-Resolution Probe for Localized Electrochemical Impedance Spectroscopy Measurements," *Journal of the Electrochemical Society*, **144** (1997) 1957–1965.

151. K. Davis, A. Dizon, C. L. Alexander, and M. E. Orazem, "Influence of Geometry-Induced Frequency Dispersion on the Impedance of Rectangular Electrodes," *Electrochimica Acta*, **283** (2018) 1820–1828.

152. M. S. Harding, *Mathematical Models for Impedance Spectroscopy*, Ph.D. dissertation, University of Florida, Gainesville, FL (2017).

153. B. Tribollet and J. S. Newman, "The Modulated Flow at a Rotating Disk Electrode," *Journal of the Electrochemical Society*, **130** (1983) 2016–2026.

154. J. S. Newman, "Numerical Solution of Coupled, Ordinary Differential Equations," *Industrial and Engineering Chemistry Fundamentals*, **7** (1968) 514–517.

Author Index

The references for this book are listed on page 491 in numerical order of appearance. The corresponding authors of the works are presented in this index. The numbers refer to the cited works given on page 491.

Subject Index

Numbers followed by a "p" refer to the page where the corresponding entry can be found in a problem. These problem solutions are presented in Part I. Page numbers without a following letter are either page ranges or are found in Part II. Numbers followed by a "t" refer to the page where the corresponding entry can be found in tables or definitions in Appendix A. Numbers followed by a "c" refer to the page where the corresponding entry can be found as part of computer code. The code is found in Appendix B.

The Electrochemical Society Series

The Electron Microprobe
Edited by T. D. McKinley, K. F. J. Heinrich, and D. B. Wittry

Chemical Physics of Ionic Solutions
Edited by B. E. Conway and R. G. Barradas

High-Temperature Materials and Technology
Edited by Ivor E. Campbell and Edwin M. Sherwood

Alkaline Storage Batteries
S. Uno Falk and Alvin J. Salkind

The Primary Battery (in Two Volumes)
Volume I
Edited by George W. Heise and N. Corey Cahoon

Volume II
Edited by N. Corey Cahoon and George W. Heise

Zinc-Silver Oxide Batteries
Edited by Arthur Fleischer and J. J. Lander

Lead-Acid Batteries
Hans Bode
Translated by R. J. Brodd and Karl V. Kordesch

Thin Films-Interdiffusion and Reactions
Edited by J. M. Poate, M. N. Tu, and J. W. Mayer

Lithium Battery Technology
Edited by H. V. Venkatasetty

Quality and Reliability Methods for Primary Batteries
P. Bro and S. C. Levy

Techniques for Characterization of Electrodes and Electrochemical Processes
Edited by Ravi Varma and J. R. Selman

Electrochemical Oxygen Technology
Kim Kinoshita

Synthetic Diamond: Emerging CVD Science and Technology
Edited by Karl E. Spear and John P. Dismukes

Corrosion of Stainless Steels, Second Edition
A. John Sedriks

Semiconductor Wafer Bonding: Science and Technology
Q.-Y. Tong and U. Göscle

Fundamentals of Electrochemistry, Second Edition
V. S. Bagotsky

Fundamentals of Electrochemical Deposition, Second Edition
Milan Paunovic and Mordechay Schlesinger

Uhlig's Corrosion Handbook, Third Edition
Edited by R. Winston Revie

Fuel Cells: Problems and Solutions
Vladimir S. Bagotsky

Lithium Batteries: Advanced Technologies and Applications
Edited by B. Scrosati, K. M. Abraham, W. A. van Schalkwijk, and J. Hassoun

Modern Electroplating, Fifth Edition
Edited by Mordechay Schlesinger and Milan Paunovic

Electrochemical Power Sources: Batteries, Fuel Cells, and Supercapacitors
By V. S. Bagotsky, A. M. Skundin, and Y. M. Volfkovic

Molecular Modeling of Corrosion Processes: Scientific Development and Engineering Applications
Edited by C. D. Taylor and P. Marcus

Atmospheric Corrosion, Second Edition
Christofer Leygraf, Inger Odnevall Wallinder, Johan Tidblad, and Thomas Graedel

Electrochemical Impedance Spectroscopy, Second Edition
Mark E. Orazem and Bernard Tribollet

Electrochemical Systems, Fourth Edition
John Newman and Nitash P. Balsara

LaQue's Handbook of Marine Corrosion
David A. Shifler *LaQue's Handbook of Marine Corrosion*, Second Edition. Edited by David A. Shifler.

Metallic, Magnetic and Carbon-Based Nanomaterials: Synthesis and Biomedical Applications
Ajit Khosla, Irshad A. Wani and Mohammad N. Lone

Bioelectrocatalysis: From Electron Transfer Processes to Emerging Technological Applications
Edited by Shelley D. Minteer and Matteo Grattieri

Problem Solving in Electrochemical Impedance Spectroscopy: Methods, Data Analysis and Case Studies
Mark E. Orazem and Bernard Tribollet

www.ingramcontent.com/pod-product-compliance
Lightning Source LLC
LaVergne TN
LVHW060446111026
849957LV00029B/391
* 9 7 8 1 1 1 9 7 7 9 2 1 6 *